Advanced Metallizations In Microelectronics

MATERIALS RESEARCH SOCIETY SYMPOSIUM PROCEEDINGS   VOLUME 181

Advanced Metallizations In Microelectronics

Symposium held April 16-20, 1990, San Francisco, California, U.S.A.

EDITORS:

Avishay Katz
AT&T Bell Laboratories, Murray Hill, New Jersey, U.S.A.

Shyam P. Murarka
Rensselaer Polytechnic Institute, Troy, New York, U.S.A.

Ami Appelbaum
Rockwell International Corporation, Newbury Park, California, U.S.A.

MRS

MATERIALS RESEARCH SOCIETY
Pittsburgh, Pennsylvania

CAMBRIDGE UNIVERSITY PRESS
Cambridge, New York, Melbourne, Madrid, Cape Town,
Singapore, São Paulo, Delhi, Mexico City

Cambridge University Press
32 Avenue of the Americas, New York NY 10013-2473, USA

Published in the United States of America by Cambridge University Press, New York

www.cambridge.org
Information on this title: www.cambridge.org/9781107410176

Materials Research Society
506 Keystone Drive, Warrendale, PA 15086
http://www.mrs.org

First published 1990
First paperback edition 2012

Single article reprints from this publication are available through
University Microfilms Inc., 300 North Zeeb Road, Ann Arbor, MI 48106

CODEN: MRSPDH

ISBN 978-1-107-41017-6 Paperback

Contents

*Invited Paper

*Invited Paper

*Invited Paper

*Invited Paper

*Invited Paper

*Invited Paper

Preface

The Symposium on Advanced Metallizations in Microelectronics was held on April 16-20, 1990, in San Francisco, U.S.A., as Symposium B of the 1990 Spring Meeting of the Materials Research Society. This symposium focused on the material and processing issues of metallization in conjunction with microelectronics applications. The purpose of the symposium was to provide a forum for the discussion of the recent advances and processes associated with the metallization of Si, InP and GaAs based electronic, optic and optoelectronic devices and IC's.

More than 150 papers from 19 countries (Australia, Belgium, Canada, China, Czechoslovakia, France, Hungary, India, Israel, Japan, Korea, Poland, Spain, Sweden, The Netherlands, United Kingdom, USA, USSR, West Germany) were presented on fundamental and applied aspects of metallization issues for microelectronics.

The symposium included sessions on overview and concerns of metallization, diffusion barriers, metallization schemes for interconnects, wiring and packaging, silicides, silicides-polysilicon systems, Au-based and non Au-based ohmic contacts for GaAs technology, contacts to InP and related materials, general aspects of variety of deposition techniques, and the special application of W and W-alloys. More than 90 papers, which are included in this proceedings, illustrate both the present state of knowledge in metallization processing and contacts for microelectronic materials and technology and the thrust of current research and development in the field.

The editors wish to thank the sponsors, invited and contributing authors, session chairmen, reviewers, and MRS staff and program officials for their help in organizing this successful symposium and publishing this volume.

Avishay Katz
Shyam P. Murarka
Ami Applebaum

June 1990

Acknowledgments

The editors of this proceedings are grateful to all the following organizations for sponsoring the symposium and raising their kind support:

AT&T Bell Laboratories
A.G. Associates
Kratos Analytical, Inc.
Materials Research Corporation
Nimic, Inc.
Sumitomo Electric, Inc.
Crystacomm, Inc.
North Eastern Analytical Corporation

Recent Materials Research Society Symposium Proceedings

MATERIALS RESEARCH SOCIETY SYMPOSIUM PROCEEDINGS

Earlier Materials Research Society Symposium Proceedings listed in the back.

PART I

Metallization Overview, Concerns and Diffusion Barriers

ELECTRICAL PROPERTIES AND SCHOTTKY BARRIERS OF
METAL-SEMICONDUCTOR INTERFACES

M.O. Aboelfotoh
IBM Research Division, T.J. Watson Research Center,
P.O. Box 218, Yorktown Heights, NY 10598

ABSTRACT

The electrical properties of metal/Si(100) and metal/Ge(100) interfaces formed by the deposition of metal on both n-type and p-type Si(100) and Ge(100) have been studied in the temperature range 77-295 K with the use of current- and capacitance-voltage techniques. Compound formation is found to have very little or no effect on the Schottky-barrier height and its temperature dependence. For silicon, the barrier height and its temperature dependence are found to be affected by the metal. For germanium, on the other hand, the barrier height and its temperature dependence are unaffected by the metal. The temperature dependence of the Si and Ge barrier heights is found to deviate from the predictions of recent models of Schottky-barrier formation based on the suggestion of Fermi-level pinning in the center of the semiconductor indirect band gap.

I. INTRODUCTION

Almost all theoretical models of Schottky-barrier formation at metal-semiconductor interfaces have been based on the original suggestion of Bardeen[1], that the Fermi level at the interface is pinned by states in the semiconductor band gap. The nature of these states is clearly one of the key issues in establishing a microscopic understanding of the barrier height. Recently, new models[2] have been proposed based on the suggestion of Fermi-level pinning in the center of the semiconductor indirect band gap by states intrinsic to the interface. In these models the effect of the metal has been neglected and the barrier height has been related to the semiconductor band-structure properties (i.e., indirect band gaps and spin-orbit splittings). Other models[3] of barrier formation have been proposed based on pinning by states associated with defects in the semiconductor. However, these defect models, which are commonly used to describe pinning on surfaces with submonolayer metal coverages, still provoke much controversy[4]. In an attempt to shed light on the nature of the states responsible for Fermi-level pinning, we have measured the dependence of the barrier height on temperature for metal(silicide)-silicon and metal(germanide)-germanium systems with a wide range in metal electronegativity in the temperature range 77-295 K using current-voltage (I-V) and capacitance-voltage (C-V) techniques. In this paper, we present and discuss the results of these measurements in terms of models of barrier formation based on pinning in midgap.

Mat. Res. Soc. Symp. Proc. Vol. 181. ©1990 Materials Research Society

II. ELECTRICAL PROPERTIES

A. n-Type Si(100)

The forward I-V characteristics plotted in Fig. 1 as a function of temperature show examples of the results obtained for Cu[5] on lightly doped $(10^{14} - 10^{15} cm^{-3})$ n-type and p-type Si(100). The samples display a good ideality factor which, however, increases slowly as the temperature is lowered, reaching a value of 1.19 at 95 K. This temperature dependence of the ideality factor is found to have the form $n = 1 + (T_0 / T)$, where T_0 is a constant, independent of temperature. Figure 2 shows nkT/q plotted against kT/q. It is evident that the data can be fitted to a straight line parallel to the unity slope line. The value of T_0 for these samples is found to be 20 K. Moreover, the dependence of $\ell n (J_0/T^2)$, where J_0 is the saturation current density at zero applied voltage, on 1/T is found to be nonlinear in the temperature range measured; however, if $\ell n(J_0/T^2)$ is plotted against $1/nT = 1/(T + T_0)$, a straight line is obtained with a slope giving a barrier-height value at 0 K of 0.65 eV, as shown in Fig. 3, in very good agreement with the 0-K value reported by Arizumi and Hirose[6] for Cu on n-type Si(111). This shows that the saturation current density J_0 can be described by

$$J_0 = A^*T^2 \exp \left[- q\Phi_{Bn}/k(T + T_0) \right] \tag{1}$$

where A^* is the effective Richardson constant and Φ_{Bn} is the n-type barrier height. Similar results have also been found for Ti-Si[7] and W-Si[8] Schottky barriers formed on lightly doped n-type Si(100) and by other authors for Schottky barriers formed on lightly doped n-type GaAs[9] and n-type InP[10]. The temperature variation of the barrier height calculated using Eq.(1) is shown in Fig. 4 by the solid circles. The solid line shows the temperature variation of the barrier height calculated on the assumption that it is entirely due to the temperature dependence of the indirect band gap in Si with the barrier-height value at 235 K as a reference. It is clear from these data that the barrier height decreases with increasing temperature with a coefficient almost equal to that of the indirect band gap in Si. It is also to be noted that the barrier-height value calculated using Eq.(1) at 295 K is in very good agreement with that reported by Thanailakis[11] for Cu on n-type Si(111) using photoelectric measurements.

In Fig. 5 we show the temperature variation of the barrier height calculated using Eq.(1) (open triangles) for W[8] on n-type Si(100). The barrier-height values obtained from C-V measurements are also shown in Fig. 5 by the solid circles. The solid line again shows the temperature variation of the barrier height calculated on the assumption that it is entirely due to the dependence of the indirect band gap in Si on temperature with the barrier-height value at 215 K as a reference. It is clear that the barrier-height values calculated using Eq.(1) are in very good agreement with those derived from C-V measurements, and that the change in the n-type barrier height with temperature is again almost equal to the change in the Si indirect band gap. This is consistent with the results of Crowell et al.[12] and of Duboz et al.[13] for Au and CoSi$_2$ on

5

n-type Si(111).

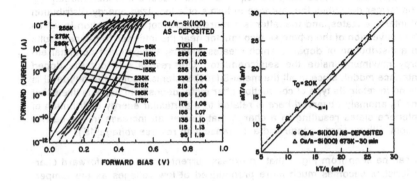

Fig. 1. Forward I-V characteristics of Cu on n-type Si(100) as a function of temperature for samples in the as-deposited state. Junction area is 3.974×10^{-4}cm^2.

Fig. 2. Temperature dependence of the ideality factor for n-type samples in the as-deposited state (open circles) and after annealing at 673 K for 30 min (open triangles).

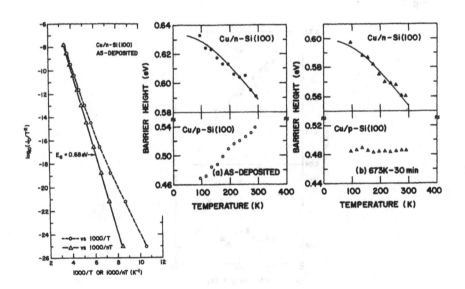

Fig. 3. Temperature dependence of forward current measured at zero applied voltage for n-type samples in the as-deposited state.

Fig. 4. Correlation of the Si indirect band gap change (solid line) and the barrier height variation with temperature for Cu on n-type (solid circles and triangles) and p-type (open circles and triangles) Si(100).

Two models have been proposed to explain the inclusion of the ideality factor in Eq.(1); the interface state model[14,15] and a doped interface model[15]. The former describes the interface in terms of a localized energy distribution of interface states, and the latter is a more macroscopic description in which the conversion of the n-type semiconductor near the interface to p-type results in a distribution of dopants which creates a voltage dependent potential energy maximum inside the semiconductor. Our results rule out the doped interface model, since in all the metal-Si(100) systems studied the n-type Si is found to retain its type of conduction after metal deposition[16]. It is likely that the T_0 anomaly observed here is related to a particular energy distribution of interface states resulting in a charge that causes an increase in the barrier height with forward voltage and a decrease with reverse voltage[15].

It can be seen from Fig. 1 that an excess current region of the forward characteristics becomes much more pronounced at low voltages as the temperature is lowered below 175 K. This excess current exhibits a small temperature dependence, suggesting that it is likely caused by high electric field effects near the edges of the junctions[17]. However, this excess current appears to have little effect on the temperature dependence of the ideality factor at higher forward voltages where thermionic emission dominates, since if the effect is large, then T_0 would be expected to be temperature dependent, i.e., to increase at low temperatures[17]. This is clearly not the case here, as shown in Fig. 2.

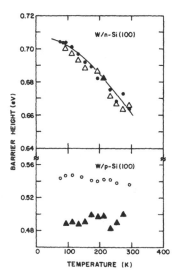

Fig. 5. Correlation of the Si indirect band gap change (solid line) and the barrier height variation with temperature for W on n-type and p-type Si(100). Open and solid triangles: barrier-height values determined from I-V measurements using Eq.(1) (see text). Solid and open circles: barrier-height values determined from C-V measurements.

In Fig. 6 we show examples of the forward I-V characteristics obtained after annealing the Cu/n-type Si(100) samples at 673 K for 30 min. The samples display a good ideality factor which again increases slowly with decreasing temperature, reaching a value of 1.17 at 95 K. Again, this temperature dependence of the ideality factor is found to have the form $n = 1 + (T_0 / T)$, with T_0 having a value of 20 K(Fig. 2). Moreover, the dependence of $\ell n(J_0/T^2)$ on 1/T is again not linear, whereas the dependence of $\ell n(J_0/T^2)$ on 1/nT is linear in the temperature range measured with a slope giving a barrier- height value at 0 K of 0.62 eV, as shown in Fig. 7. The temperature variation of the barrier height calculated using Eq.(1) is shown in Fig. 4 by the solid triangles. The solid line again shows the temperature variation of the barrier height calculated on the assumption that it is entirely due to the temperature dependence of the indirect band gag in Si. It is clear that in these samples the barrier height also decreases with increasing temperature with a coefficient almost equal to the temperature coefficient of the indirect band gap in Si.

X-ray photoemission-spectroscopy measurements[18] on Cu-Si(100) samples with 1000Å Cu showed that an annealing at 473 K for 30 min is sufficient to cause the Cu film to fully react with Si to form a metal-rich Cu-Si compound with stoichiometry close to Cu_3Si. The results in Fig. 4 then indicate that silicide formation has very little or no effect on the barrier height and its temperature dependence.

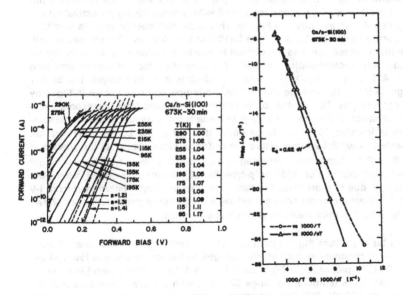

Fig. 6. Forward I-V characteristics of Cu on n-type Si(100) as a function of temperature for samples annealed at 673 K for 30 min.

Fig. 7. Temperature dependence of forward current measured at zero applied voltage for n-type samples annealed at 673 K for 30 min.

It can be seen from Fig. 6 that a region of the forward characteristics with a high value of ideality factor (which increases with decreasing temperature) becomes evident at low voltages as the temperature is lowered below 155 K. A plot of $\ell n(J_o/T^2)$ obtained by extrapolating this low-voltage linear region of the forward characteristics to zero applied voltage, versus $1/T$ (not shown here) is found to be linear in the temperature range 95-135 K with a slope giving an activation energy of 0.26 eV. This low activation energy value is evidently associated with recombination which causes deviations from thermionic emission behavior at low voltages and low temperatures[19]. This excess thermally activated current, however, has essentially no effect the temperature dependence of the ideality factor at higher forward voltages and low temperatures, as shown in Fig. 2.

B. p-Type Si(100)

The forward I-V characteristics plotted in Fig. 8 as a function of temperature show examples of the results obtained for Cu[5] on lightly doped $(10^{14} - 10^{15} cm^{-3})$ p-type Si(100). The samples display a high ideality factor which, however, remains essentially unchanged in the temperature range 98-290 K. In contrast to the n-type samples, the data of nkT/q against kT/q can be fitted to a straight line passing through the origin with a slope equal to 1.23, as shown in Fig. 9. In addition, the dependence of $\ell n(J_o/T^2)$ on $1/T$ is now linear in the temperature range measured with a slope giving an activation energy of 0.43 eV, as shown in Fig. 10. This activation energy value is less than the barrier height for Cu on p-type Si(100) at 0 K (0.49 eV). These results indicate that in these samples the current is due to thermionic emission in combination with recombination. In fact, if the relation for thermionic emission $[J_o = A^*T^2 \exp(-q\Phi_{Bp}/kT)]$ is used to calculate a barrier height, the barrier height is found to increase with increasing temperature, as shown in Fig. 4 by the open circles. This is due to the fact that deviations from thermionic emission behavior due to recombination become more pronounced as the temperature is lowered[19]. It is to be noted that the p-type barrier height is not expected to exhibit a temperature dependence, since almost all the change in the Si indirect band gap is reflected in the change of the n-type barrier height with temperature. For W[8] on p-type Si(100), on the other hand, where the current is due to thermionic emission, the barrier height determined from I-V and C-V measurements does not exhibit a temperature dependence, as shown in Fig. 5 by the solid triangles and the open circles.

It can be seen from Fig. 8 that an excess current region of the forward characteristics becomes evident at low voltages as the temperature is lowered below 195 K. A plot of $\ell n(J_F/T^2)$ at 0.2 V versus $1/T$ (not shown here) is found to be linear in the temperature range 98-155 K with a slope giving an activation energy of 0.14 eV. This low activation energy value is again associated with recombination which causes even more deviations from thermionic emission behavior at low voltages and low temperatures.

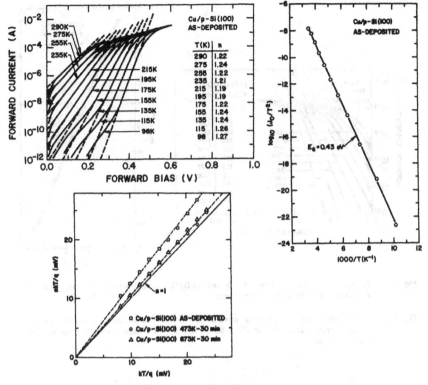

Fig. 8. Forward I-V characteristecs of Cu on p-type Si(100) as a function of temperature for samples in the as-deposited state. Junction area is $1.31 \times 10^{-4} \text{cm}^2$.

Fig. 9. Temperature dependence of the ideality factor for p-type samples in the as-deposited state (open squares) and after annealing at 473 K for 30 min (open cirles) and 673 K for 30 min (open triangles).

Fig. 10. Temperature dependence of forward current measured at zero applied voltage for p-type samples in the as-deposited state.

In Fig. 11 we show examples of the forward I-V characteristics obtained after annealing the Cu/p-type Si(100) samples at 473 K for 30 min. Results of the annealing at 673 K for 30 min are very similar. The samples display a good ideality factor which remains unchanged in the temperature range 95-290 K. Again, for these samples the data of nkT/q against kT/q can be fitted to a straight line passing through the origin with a slope now equal to 1.04 (Fig. 9). Moreover, the dependence of $\ln(J_o/T^2)$ on 1/T is linear in the temperature range measured with slopes giving barrier-height values of 0.48 and 0.49 eV for the 473-K and 673-K annealed samples, respectively, as shown in Fig. 12. It is also clear from the open triangles shown in Fig. 4 that in these samples the barrier height calculated using the relation for thermionic emission does not exhibit a temperature dependence.

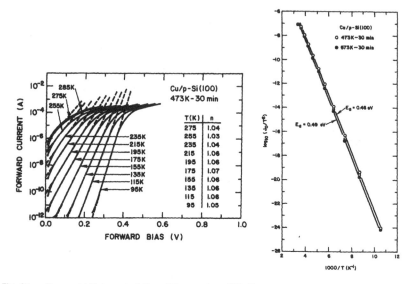

Fig. 11. Forward I-V characteristics of Cu on p-type Si(100) as a function of temperature for samples annealed at 473 K for 30 min.

Fig. 12. Temperature dependence of forward current measured at zero applied voltage for p-type samples annealed at 473 K for 30 min (open circles) and 673 K for 30 min (solid circles).

C. n-Type Ge(100)

The forward I-V characteristics plotted in Fig. 13 as a function of temperature show examples of the results obtained for Ti on n-type Ge(100). The samples display a good ideality factor which remains essentially unchanged with temperature. In addition, the dependence of $\ell n(J_o/T^2)$ on 1/T is linear in the temperature range measured with a slope giving a barrier-height value at 0 K of 0.60 eV, as shown in Fig. 14. The temperature variation of the barrier height calculated using the relation for thermionic emission $[J_o = A^*T^2 \exp(-q\Phi_{Bn}/kT)]$ is shown in Fig. 15 by the open circles. In Fig. 15 we also show the temperature variation of the barrier height obtained from C-V measurements for Cu on n-type Ge(100) (open triangles). The solid lines show the temperature variation of the barrier height calculated on the assumption that it is entirely due to the temperature dependence of the indirect band gap in Ge with the barrier-height value at 215 K as a reference. It is clear from these data that for both Cu and Ti the change in the n-type barrier height with temperature is almost equal to the change in the Ge indirect band gap. It is also to be noted that for Cu, the barrier-height values at 295 K are in very good agreement with those reported by Thanailakis and Northrop[20] for Cu on Ge(111) using I-V and C-V measurements.

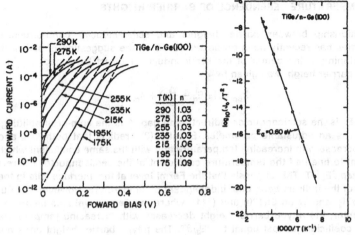

Fig. 13. Forward I-V characteristics of Ti on n-type Ge(100) as a function of temperature for samples annealed at 753 K for 1 h. Results for samples in the as-deposited state are very similar.

Fig. 14. Temperature dependence of forward current measured at zero applied voltage for n-type samples annealed at 753 K for 1 h.

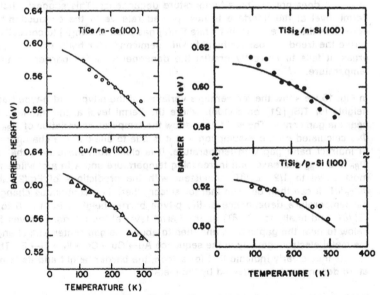

Fig. 15. Correlation of the Ge indirect band gap change (solid line) and the barrier height variation with temperature for Cu on n-type Ge(100) (open circles) and Ti on n-type Ge(100) (open triangles).

Fig. 16. Correlation of one-half the Si indirect band gap change (solid lines) and the barrier height variation with temperature for TiSi₂ on n-type Si(100) (solid circles) and TiSi₂ on p-type Si(100) (open circles).

III. TEMPERATURE DEPENDENCE OF BARRIER HEIGHTS

A relationship between barrier heights and semiconductor band-structure properties has recently been proposed based on the suggestion of of Fermi-level pinning in the center of the semiconductor indirect band gap[2]. The p-type barrier height was given by

$$\Phi_{Bp} = \frac{1}{2}(E_g^i - \frac{\Delta}{3}) + \delta_m \qquad (2)$$

where E_g^i is the semiconductor indirect band gap, Δ is the spin-orbit splitting, and δ_m is an adjustable parameter. Equation(2) predicts that both Φ_{Bn} and Φ_{Bp} decrease with increasing temperature and with the same coefficient which is equal to one-half the temperature coefficient of the semiconductor indirect band gap, $\partial E_g^i/\partial T$. This suggests that the Fermi level at the interface falls in the center of the indirect band gap, independent of temperature. However, for Au, Cu, CoSi$_2$, and W on Si(100) and (111), where the Fermi level falls below the gap center, the n-type barrier height decreases with increasing temperature with a coefficient almost equal to $\partial E_g^i/\partial T$. The p-type barrier height does not exhibit a temperature dependence. This suggests that the Fermi level at the interface is pinned relative to the valence-band edge. Furthermore, for ErSi$_2$[13] on Si(111), where the Fermi level falls above the gap center, the change in the p-type barrier height with temperature is almost equal to the change in the Si indirect band gap. The n-type barrier height(Φ_{Bn} = 0.28 eV at 77 K) does not exhibit a temperature dependence. This suggests that the Fermi level at the interface is now pinned relative to the conduction-band edge. It is therefore clear that while Eq.(2) has been shown[2] to correctly describe the trends in barrier heights with semiconductor band-structure properties, it fails to correctly predict the dependence of the barrier height on temperature.

In Fig. 16 we show the temperature variation of the n-type and p-type barrier heights for TiSi$_2$[21] on Si(100), where the Fermi level at the interface falls near the gap center. The solid lines show the temperature variation of Φ_{Bn} and Φ_{Bp} calculated on the assumption that it is due to one-half the change in the Si indirect band gap with temperature. It is clear from these data that both Φ_{Bn} and Φ_{Bp} decrease with increasing temperature and with a coefficient almost equal to $1/2\ \partial E_g^i/\partial T$, consistent with the predictions of Eq.(2) about $\partial\Phi_{Bp}/\partial T$. It can then be seen that for silicon, there is a continuous change in the temperature dependence of the p-type barrier height (from $\sim$ 0 to 1/2 $\partial E_g^i/\partial T$, and finally to $\partial E_g^i/\partial T$) as the Fermi level at the interface moves from below to near the gap center, and then to above the gap center with changing the metal electronegativity in the sequence Au$\rightarrow$ Cu$\rightarrow$ Co$\rightarrow$ W$\rightarrow$ Ti$\rightarrow$ Er. These results thus clearly indicate that for silicon, the barrier height and its temperature dependence are affected by the metal.

We have also shown that for both Cu and Ti (which represent a wide range in metal electronegativity) on Ge(100), where the Fermi-level position is well below the gap center and is unaffected by the metal(Fig. 15), the n-type barrier height decreases with increasing temperature with a coefficient almost equal to that of the indirect band gap in Ge. This suggests that the Fermi level at the interface is pinned ralative to the valence-band edge. The p-type barrier height is then not expected to show a temperature dependence. These results again clearly deviate from the predictions of Eq.(2) regarding the temperature dependence of the barrier height, $\partial\Phi_{Bp}/\partial T$. Thus, in contrast to silicon, the germanium barrier height and its temperature dependence are unaffected by the metal.

IV. CONCLUSION

We have studied the electrical properties of metal/Si(100) and metal/Ge(100) interfaces formed by the deposition of metal on both n-type and p-type Si(100) and Ge(100) in the temperature range 77-295 K. Compound formation is found to have very little or no effect on the Schottky-barrier height and its temperature dependence. For silicon, the barrier height and its temperature dependence are found to be affected by the metal. For germanium, on the other hand, the barrier height and its temperature dependence are unaffected by the metal. The temperature dependence of the Si and Ge barrier heights is found to deviate from the predictions of recent models of Schottky-barrier formation based on the suggestion of Fermi-level pinning in the center of the semiconductor indirect band gap.

ACKNOWLEDGMENT

It is the author's pleasure to acknowledge B.G. Svensson for helpful discussions and the Central Scientific Services Material Laboratory at Yorktown for the metal depositions.

REFERENCES

1. J. Bardeen, Phys. Rev. **71**, 717 (1947).

2. J. Tersoff, Phys. Rev. Lett. **52**, 465 (1984); Phys. Rev. **B32**, 6968 (1985).

3. W.E. Spicer, I. Lindau, P.R. Skeath, C.Y. Su, and P.W. Chye, Phys. Rev. Lett. **44**, 420 (1980); J. Vac. Sci. Technol. **16**, 1422 (1979); N. Newman, M. van Schilfgaarde, T. Kendelwicz, M.D. Williams and W.E. Spicer, Phys. Rev. **B33**, 1146 (1986).

4. K. Stiles and A. Kahn, Phys. Rev Lett. **60**, 440 (1988); R. Ludeke, G. Jezequel and A. Taleb-Ibrahimi, Phys. Rev. Lett. **61**, 601 (1988).

5. M.O. Aboelfotoh, A. Cros, B.G. Svensson, and K.N. Tu, Phys. Rev. **B** (in press).

6. T. Arizumi and M. Hirose, Japan J. Appl. Phys. **8**, 749 (1969).

7. M.O. Aboelfotoh, J. Appl. Phys. **64**, 4046 (1988).

8. M.O. Aboelfotoh, J. Appl. Phys. **66**, 262 (1989).

9. F.A. Padovani and G.G. Summer, J. Appl. Phys. **36**, 3744 (1965).

10. B. Tuck, G. Eftekhari, and D.M. deCogan, J. Phys. D: Appl. Phys. **15**, 457 (1982).

11. A. Thanailakis, J. Phys. C: Solid St. Phys. **8**, 655 (1975).

12. C.R. Crowell, S.M. Sze, and W.G. Spitzer, Appl. Phys. Lett. **4** , 91 (1964).

13. J.Y. Duboz, P.A. Badoz, F. Arnaud d'Avitaya, and E. Rosencher (unpublished).

14. J.D. Levine, J. Appl. Phys. **42**, 3991 (1971); Solid-State Electron. **17**, 1083 (1974).

15. C.R. Crowell, Solid-State Electron. **20**, 171 (1977).

16. M.O. Aboelfotoh and B.G. Svensson (unpublished).

17. A.N. Saxena, Surf. Sci. **13**, 151 (1969).

18. A. Cros, M.O. Aboelfotoh, and K.N. Tu, J. Appl. Phys. (in press).

19. E.H. Rhoderick, Metal-Semicoductor Contacts (Clarendon, Oxford, 1980).

20. A. Thanailakis and D.C. Northrop, Solid-State Electron. **16** , 1383 (1973).

21. M.O. Aboelfotoh and K.N. Tu, Phys. Rev. **B34**, 2311 (1986); M.O. Aboelfotoh, Phys. Rev. **B39**, 5070 (1989).

PROPERTIES OF OHMIC CONTACTS TO HETEROJUNCTION TRANSISTORS

H Barry Harrison, School of Microelectronic Engineering, Griffith University, Brisbane, Australia 4111

ABSTRACT

The electrical and physical properties of the contact between the active channel of high electron mobility transistors (HEMT's) and the source or drain contacts play an important role in determining the transistor characteristics. This paper considers electrical models that may be applied to the various techniques now available to form these interconnections. Results of electrical (and some physical) studies with regard to these systems are then discussed and compared where possible with predictions made using the electrical models. The comparisons show that the electrical models provide a useful base to identify the important parameters in these interconnections.

1.0 INTRODUCTION

Heterojunction transistors display considerable promise for high frequency analogue and high speed digital devices. The high-electron-mobility transistor (HEMT) variously referred to as a modulation-doped field-effect transistor (MODFET) or selectively doped heterojunction transister (SDHT) is one such heterojunction device that relies on a two dimensional electron gas (2DEG) formed at a selectively doped N^+-n^- heterojunction. The 2DEG formed between N^+ Al_xGa_{1-x}. As to undoped GaAs interface displays high mobility characteristics [1], further the 2DEG layer can be modulated by a Schottky contact and consequently if two additional ohmic contacts (either side of the Schottky) are made to form source and drain interconnections an attractive field effect device is realized [2]. An example of such a structure is depicted schematically in Figure 1 [3].

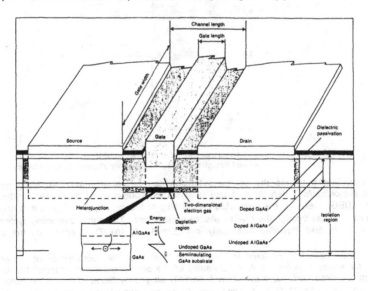

Figure 1 Schematic Section View of a Typical HEMT - FET Structure. [3]

Many parameters effect the electrical performance of the resultant HEMT devices, however the parasitic resistance associated with the source and drain ohmic contacts and with the extension from the modulated 2DEG under the gate to the ohmic contacts, referred to as the source/drain extensions play a significant role in determining the device noise and gain performance in analogue circuits and equivalent effects in digital circuits [4,5].

Attention has been placed on the need to lower ohmic contact resistance, in the main excluding the parasitic source/drain resistance contribution. However the interplay between these various resistances makes it imperative that they be considered in concert [6]. In this paper the various contributions to the resistance losses are explained by way of electrical models. Physical and electrical data obtained from various sources is then used in conjunction with these models to establish the parameters that most significantly effect the parasitic resistances.

2.0 MODELS OF OHMIC STRUCTURES

The schematic of Figure 1 presents one example of a contacting system for a HEMT device. In this example if we consider the source end only (any effects at the source end will be mirrored at the drain end), the implications are that the source ohmic contact has penetrated the heavily doped GaAs ($> 10^{18}/cm^3$) cap, the doped $Al_x Ga_{1-x}As$ ($\sim 10^{18}/cm^3$) large bandgap layer through the heterojunction layer to the 2DEG layer in the undoped (n^-) GaAs. For such a structure ohmic contacts (vertical) are made between both the N^+ GaAs cap as well as the 2DEG layer, it is unlikely that a soft ohmic contact will be made to the $Al_x Ga_{1-x}As$ layer. This is an interesting configuration for FET contacts since most conventional structures are contacted horizontally [7].

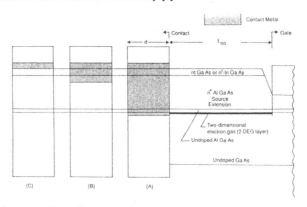

Figure 2   Three different types of contact schemes for HEMTS
A - Alloyed through to 2 DEG   B - Partically alloyed   C- Non-alloyed

For modelling purposes the source extension can be represented electrically by the schematic of figure 3, denoted as L_{SG} [10]. Here σ_1 represents the distributed conductivity of the 2DEG layer, σ_2 the distributed conductivity of the cap layer and g_s the distributed conductance loss through the $Al_x Ga_{1-x}As$. heterostructure. The ohmic contacting systems of Figure 2 are also modelled electrically in Figure 3. In (A), R_{c2} represents the direct

Apart from that depicted in Figure 1 at least two other possibilities of contacting exist, for example in Figure 2, a schematic of the source extension material to (A) the system just described, (B) a system with partially alloyed contacts, where ohmic contact is made only to the top n^+ GaAs cap layer and (C) where a non-alloyed contact is made. In the latter system either a heavily degenerate GaAs layer ($>10^{19}/cm^3$) [8] or a material with a lower barrier height (ideally zero) provides an extremely small specific contact resistance ($<10^{-6}\ \Omega\text{-}cm^2$). $In_x Ga_{1-x}As$ as a top layer is one such example [9].

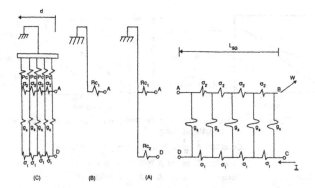

Figure 3   Electrical models, A,B, C,   representing the gate source extension.
(A) Alloyed down to the 2 DEG layer   (B) Alloyed through the top cap
(C) Non-alloyed direct metal contact

contribution of resistance from the 2DEG to the surface, the main contribution presumably being the alloy/2DEG interface. R_{c1} represents the effective contact resistance from the CAP layer to the top of the alloyed contact (with conductive metal overlayed) regarded as an equipotential surface (ground in this case). In (B) it is assumed that the only ohmic contact is made vertically to the top layer and contributes to the resistance R_{c1}. Any fringing under the contact is neglected and as such this should be regarded as an approximation only. Finally the situation in (C) is different to either of the prior cases in that current may enter the 2DEG layer either under the contact or through the extension material.

2.1   ALLOYED TO THE 2-DEG LAYER

The equation that represents the total resistance from point C (Figure 3) to ground for case (A) is non-trivial, however it is possible to make certain assumptions that lead to expressions under appropriate boundary conditions that provide insight into the resultant electrical behaviour.

With reference to Figure 3 consider firstly the extreme case where R_{c1} is large and current is discouraged from exiting from point A. In this case the resistance from terminal C to ground (R_c) is given by the following expression

$$R_c = R_{c2} + \frac{L_t}{\sigma_1 + \sigma_2}\left\{\frac{L_{SG}}{L_t} + \frac{2\sigma_1}{\sigma_2}\tanh\left(\frac{L_{SG}}{2L_t}\right)\right\} \tag{1}$$

where $L_t = \left(g_s\left(\frac{1}{\sigma_1} + \frac{1}{\sigma_2}\right)\right)^{-1/2}$, and has the dimensions of millimetres (mm) such that R_c has the units Ω-mm. (R_{c2} also given in units Ω-mm)

Figure 4 provides a plot of this expression for two values of R_{c2} and two of g_s. For example

Figure 4(a) is a plot for $\frac{1}{\sigma_1} = \frac{1}{\sigma_2} = 1000\Omega/\square$ $\frac{1}{g_s} = 10^{-4}\Omega\text{-cm}^2$ and two values of R_{c2},

(0.1 Ω-mm and 2Ω-mm). For very small values of L_{SG}, R_{c2} dominated, which is to be

expected, in the extreme for example $L_{SG} \le L_t$, $R_c \cong R_{c2} + \frac{L_{SG}}{\sigma_1}$ on the other hand for larger

values of L_{SG}, $R_c = R_{c2} + \frac{2L_t}{\sigma_1 + \sigma_2}\cdot\frac{\sigma_1}{\sigma_2} + \frac{L_{SG}}{\sigma_1 + \sigma_2}$, the first two terms being constant and R_c scaling linearing with L_{SG}. These two extremes are clearly evident from the graph, thus

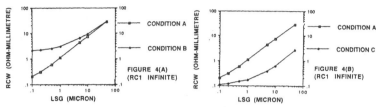

Figure 4   Parasitic source gate resistance as a function of source gate spacing

Condition A $\frac{1}{\sigma_1} - \frac{1}{\sigma_2} = 1000\Omega/\square$, $\frac{1}{gs} = 10^{-4}\Omega\text{-cm}^2$

Condition B as for A with $R_{c1} = 2\Omega\text{-mm}$

Condition C as for A with $\frac{1}{gs} = 10^{-6}\Omega\text{-cm}^2$

for small L_{SG} values it is useful to strive for as small an L_{SG} value as possible along with larger σ_1 values (this of course may not be possible).

Consider now Figure 4B, the value of g_s has been increased (decreasing L_t) by changing its value two orders of magnitude to $\frac{1}{gs} = 10^{-6}\Omega\text{-cm}^2$, this has the effect of producing a much reduced R_c value even for realistic L_{SG} dimensions. Its value will always be below that of the higher tunnelling conductance, value $(10^{-4}\text{x}10^{-4}\Omega\text{-cm}^2)$. Consequently even if L_{SG} is small it is still desirable to strive for a larger value of g_s.

The alternate extreme occurs when R_{c2} takes on a large value (in the limit infinity). In this case the equation relating the resistance from point C to ground becomes, [11]

$$R_c = R_{c1} + \frac{L_{SG}}{\sigma_1 + \sigma_2} + \frac{L_t}{1/\sigma_1 + 1/\sigma_2} \left\{ \frac{2}{\sigma_1 \sigma_2 \sinh\frac{L_{SG}}{L_t}} + \left(\frac{1}{\sigma_1^2} + \frac{1}{\sigma_2^2} \right) \coth\frac{L_{SG}}{L_t} \right\} \quad (2)$$

If we used the prior assumed values for σ_1, σ_2, and $\frac{1}{g_s} = 10^{-4}\Omega - \text{cm}^2$ with an R_{c1} value of

$0.1\Omega\text{-mm}$ then the curve of Figure 5A results. (also show is a curve where

$\frac{1}{\sigma_1} = \frac{1}{\sigma_2} = 100\Omega/\square.$). Those curves have some interesting features, firstly a distinct

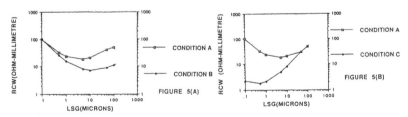

Figure 5.   Parasitic source gate resistance as a function of source gate spacing

Condition A
$= 10^{-6}\Omega\text{-cm}^2$, $R_{c1} = 0.1\Omega\text{-mm}$
$\frac{1}{\sigma_1} - \frac{1}{\sigma_2} = 1000\Omega/\square$, $\frac{1}{gs}$

Condition B as for A but $\frac{1}{\sigma_1} - \frac{1}{\sigma_2} = 100\Omega/\square$

Condition C as for A but $\frac{1}{gs} = 10^{-6}\Omega\text{-cm}^2$

minima occurs for the R_c value, in fact the length at which this occurs is given by

$$L_{opt} = L_t \, Sinh^{-1} \left\{ \left(\frac{1}{\sigma_1} + \frac{1}{\sigma_2} \right)^2 \sigma_1 \sigma_2 + 2 \, \frac{\left(\frac{1}{\sigma_1} + \frac{1}{\sigma_2} \right)}{(1/\sigma_1 \sigma_2)^{1/2}} \right\}$$

(3)

As the conductance of the top and 2DEG layer are increased, the minimum resistance decreases but the value at which it occurs (L_{opt}) increases.

In Figure 5B, a comparison is made for the same conditions as in Figure 5A compared to the case where $\frac{1}{g_s} = 10^{-6} \Omega - cm^2$, the minimum resistance and the optimum length are both dramatically reduced. Other points to note are that at the longer source gate spacings the two resistance values approach one another $\left(\frac{L_{SG}}{\sigma_1 + \sigma_2} \right)$. At the other extreme for very short lengths $R_c \rightarrow R_{c1} + \frac{L_t^2}{L_{SG}} \left(\frac{1}{\sigma_1} + \frac{1}{\sigma_2} \right)$, which is in effect R_{c1} added to the tunnelling conductance term, which reduces to $\frac{1}{g_s \, L_{SG}}$. in the limit. This to be expected since the contact now has a uniform current distribution across its interface.

In both the extreme cases, ($R_{c1} \rightarrow \infty$ or $R_{c2} \rightarrow \infty$) it is obvious that the alternate values of R_{c2} and R_{c1} respectively should be as small as practicable and that $\frac{1}{gs}$ should also be pushed in the same direction. If however true alloying is made to the 2DEG layer even for finite R_{c1} it is still desirable to have the contact length as small as practicable ideally zero so that the only contributing factor to R_c is in fact R_{c2}.

2.2 PARTIALLY ALLOYED CONTACTS (THROUGH THE N+ CAP)

In this case two major assumptions are made (without detailed justification), firstly that no fringing occurs and that no conduction occurs from the contact through the AlGaAs layer under the contact. This is in effect equivalent to the situation in 2.1 where R_{c2} is infinite in value and the curves of Figure 5 for the conditions cited become appropriate. Indeed the conclusion arrived at in 2.1 for this type of contact are applicable, namely that on optimum source gate spacing exists and that to lower this L_{SG} spacing and subsequent parasitic resistance is mainly achieved by increasing the tunnelling conductances (g_s).

2.3 NON-ALLOYED OHMIC CONTACTS

A non-alloyed ohmic contact is depicted schematically in Figure 3(c), this is the most complex system to model completely and accurately, however for the sake of comparison some assumptions are made that will enable an approximate result to be obtained. If we assume that ρ_c could be made to approach zero, ie. a metal contact to a low barrier height heavily doped material such as $In_x Ga_{1-x} As$ where $x \rightarrow 0.5$ then this condition will be approached [13]. Given this condition, by observation the minimum resistance will be obtained when $L_{SG} = 0$, ie. the gate and source contact have zero clearance. (Clearly not physically possible but for the sake of comparison we will assume this to be the case). For such a case σ_2 would be shortened by the metal overlay and σ_1 and g_s would be the only contributers to the parasitic resistance that is

$$R_c = \frac{L_t}{\sigma_1} \, coth \, \frac{d}{L_t}$$

(4)

where $L_t = \left(\dfrac{g_s}{\sigma_1}\right)^{-\frac{1}{2}}$

and d is now the contact pad length (not the gate source extension length). This is the standard expression found for the traditional transmission line model (tlm) for ohmic contacts[14, 15]. The expression is plotted in Figure 6 for the values of σ_1 and g_s previously used ie. Figure 6A compares $1/\sigma_1 = 1000\Omega/\square$ (condition A) and $1/\sigma_1 = 100\Omega/\square$ (condition B) and Figure 6B compares that of condition A and condition C has $1/g_s$ reduced to $10^{-6}\Omega\text{-cm}^{-2}$.

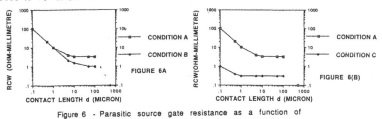

Figure 6 - Parasitic source gate resistance as a function of contact length.

Condition A $\dfrac{1}{\sigma_1}$ $1000\Omega/\square$, $\dfrac{1}{g_s} = 10^{-4}\Omega\text{-cm}^2$

Condition B $\dfrac{1}{\sigma_1}$ $100\Omega/\square$, $\dfrac{1}{g_s} = 10^{-4}\Omega\text{-cm}^2$

Condition C $\dfrac{1}{\sigma_1}$ $1000\Omega/\square$, $\dfrac{1}{g_s} = 10^{-6}\Omega\text{-cm}^2$

In this case (as with normal planar ohmic contacts) when $d > 2.5$ L_T nothing is gained by increasing the contact length (ie. no decrease in resistance value). Furthermore the most dramatic changes in both the minimum contact length and contact resistance are achieved through decreased values of $1/g_s$.

3.0 PRACTICAL CONSIDERATIONS AND EXPERIMENTAL RESULTS

Considerable investigation aimed at providing a better understanding and improving the electrical and physical properties of ohmic contacts to GaAs-FET devices has been carried out in the past [16]. Most attention has been directed to planar type structures that are peculiar to standard GaAs FET's (MESFETS, E-JFETS etc). Perhaps the most popular metal system to date has been based on Au-Ge alloyed to the GaAs. It is believed that two major reactions contribute to this being a successful system, firstly a reaction between the Au and GaAs to form a AuGa compound and a resultant Ga vacancy. This vacancy is believed to be taken up by Ge at the interface. Ge is an amphoteric dopant in GaAs but generally prefers to act as a donor and produces a zone of highly doped degenerate material [17]. This highly degenerate layer makes it possible for an ohmic contact to be formed.

Direct metal contacting to GaAs on the other hand is a somewhat more recent occurance, where techniques for heavily doping the GaAs have become available [7, 8, 18]. However, GaAs has the property of barrier height pinning and therefore must be doped well into degeneracy before a non-alloyed or sintered contact can be made [7]. Techniques based on molecular beam epitaxy (MBE) or high does ion implantation followed by laser beam processing have resulted in (electrically) doping with concentration in excess of $5\times10^{19}/\text{cm}^2$ [8, 19, 20]. These layers have produced specific contact resistance values (ρ_c) of less than $10^{-6}\Omega\text{-cm}^2$. The latter techniques, which showed promise in the early eighties has fallen in favour to the more powerful MBE technology.Furthermore again using MBE lower band gap materials which are better suited to direct ohmic contact formation such as InGaAs on GaAs have been used and shown considerable promise [13].

Unlike its counterparts however the HEMT (outlined schematically in Figure 1) relies on a buried 2DEG conduction layer. As mentioned in section 2 electrical contacts must be made to this layer and to the outside world. Initially various techniques were used to try electrically to assess the contacting techniques, these were based mainly on the transmission line model (tlm) approach [14, 15]. However as outlined by Feuer [21] such experimental techniques could lead to gross errors. Most workers today are aware of the problems involved with the use of the tlm approach, however it is still on occasions the subject of abuse. In the following sections a brief attempt is made to align practical experimental results with the electrical models presented in section 2. It is hoped that this will provide a further understanding of the parameters that play a major role in determining the total source/drain to gate parasitic resistance.

3.1 CONTACTING TO THE 2DEG LAYER

Jones and Eastman [22] picked up on the work of Feuer [21] and performed a systematic study on the effect of providing a metallization (contact) that would, penetrate through the GaAs cap and the Al GaAs layer to contact the 2DEG situated in the lower bandgap material. To establish the conditions that would provide a low resistance for this contact system and to maintain a surface morphology smooth enough to facilitate easy alignment with little or no sideway spreading.

They conducted their study by replacing the highly doped n+ cap by an undoped layer, further the n+ Al GaAs layer that they produced was almost completely devoid of conducting electrons at equilibrium. This is effect provided them with a one layer structure resembling a type of buried vertical contact to the 2DEG layer. The metal system used was that of the alloyed Au-Ge type, however because of the depth of penetration required (compared to conventional contacts) they assumed a time temperature regieme previously felt unacceptable for ohmic contact formation.

They argued that because of this penetration larger amounts of Au-Ge were necessary. In fact they used 100Å-Ni, 1100 Å of 12% Ge-88% Au, 1100Å-Ag and 1100Å of Au. Their best results were obtained at 560° C for a 95 second quartz furnace anneal in a flowing N2 ambient. They claimed that a tlm technique was used to measure a minimum value of R_c of around 0.1Ω-mm, and a transfer length of around 0.2μm. In this case however the use of the term transfer length is a little misleading when compared to a conventional planar contact [12], however it means that the sheet resistance of the 2DEG layer is around 100Ω/□ . Others have since reported similar results [23] and comment on the important role of Ni but provide no conclusive data on the exact role that the Ni plays. Furthermore it is claimed [23] that the higher temperature is not necessary, and that lower temperatures (460°C) for longer times (16 minutes) provides a more controlled contacting technique and that it did not undergo melting, eliminating lateral protrusions in the 2DEG region that had been reportedly to be as large as 0.2μm [24].

Kamada et al. [25] used a similar system and came up with two important observations. Firstly, whilst their use of the tlm as a technique of acquiring accurate data is in question, they none-the-less did observe a distinct orientation dependence of contract resistance. A maximum nearly twice that of the minimum being obtained with a 90° shift in orientation (OTI → 011). Again no suitable explanation was presented for this dependency.

They also observed that during the alloying the Al from the AlGaAs moved to the surface and that the Au in conjunction with the Ni provided almost a continuous path of metal compounds from the 2DEG layer to the surface. The Al GaAs layer being completely destroyed, and presumably the main contribution to the contact resistance is then due to the contacting with the 2DEG layer.

For these types of conditions then, relating back to section 2.1, it would appear that R_{c2} values as low as 0.1Ω-mm can be obtained and that to keep the total R_c as low as possible L_{SG} should be made as short as possible. Further for finite $L_{SG} \frac{1}{gs}$ should also be made as low as practicable, practical devices produced this way should have R_c values approaching 0.1Ω-mm.

3.2 PARTIALLY ALLOYED CONTACTS

In a study on this subject [26] our work shows that at least a trend can be obtained using standard transmission line techniques. If the model of figure 3(B) is modified to that shown below in Figure 7, then current injected through terminal 2 to 3 and voltages across the same terminals will result in a four probe measurement of resistance. This resistance plotted as a function of spacing provided the data of Figure 8.

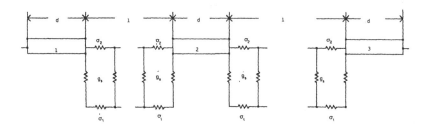

Figure 7 - Equivalent tlm structure for partially alloyed contacts.

Further if the current is injected between 2 and 3 and voltage measured between 1 and 2 a so called end resistance should be recorded. This resistence in all cases was not measurable (too small) leading to the conclusion that the contact is in fact a vertical one. Furthermore the shape of the curve of Figure 8 indicates that for the conditions of contact formation (Au-Ge-Ni) through the top layer but not down to the 2DEG does in fact validate the model applied in section 2.2 ie. for lengths greater than approximately $20\mu m$ the resistance is dominated by the parallel combinations of the cap and the 2DEG layer and below this R_{C1} and the tunnelling conductance begin to dominate. Using a transmission line model however does hide the point of minimum resistance since for zero spacing in this case a finite resistance of $2R_{C1}$ will still be observed. The point of inflection below $20\mu m$ in Figure 8 is an indication of a minimum as predicted in section 2.2. To lower this minimum resistance, requires a decrease in the tunnelling resistance, as detailed by Kachwalla et al.[27].

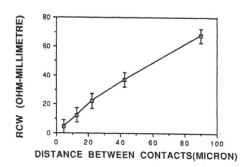

Figure 8 - Results obtained for the resistance between two contacts for the structure of Figure 7.

3.3 NON-ALLOYED CONTACTS

The Technology to produce graded structures such as that depicted in Figure 9A, in which the energy band diagram is shown for a composite structure moving from the metal, through a graded n-InGaAs, to a n AlGaAs hetrostructure which connects to the 2DEG layer. The ability to produce such structures has been available for some time but only recently used to produce HEMTS with credible performance [28]. The main arguments presented against this approach being the difficulty in fabricating the multi-layer structure. However a major advantage is the use of only one final metal layer for gate, source and drain as opposed to the alloying where the gate material is different to the source drain. Furthermore if the mole fraction of InAs is adjusted correctly the specific contact resistance for the metal to In GaAs layer can be made extremely low as shown in Figure 9B [28].

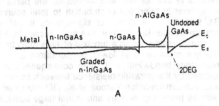

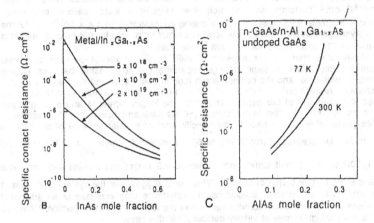

Figure 9 - Some relationships pertaining to multi-layer
structures
(A) - Energy band diagram of graded structure.
(B) - Specific contact resistance as a function of
InAs mole fraction and doping concentration.
(C) Specific contact resistance as a function of
AlAs mole fraction and absolute temperature.

TABLE I
EPITAXIAL PARAMETERS FOR A HEMT WITH NONALLOYED OHMIC
CONTACTS

Material	X-value	Thickness (nm)	Doping (cm-3)
$In_xGa_{1-x}As$	0.5	100	3E19
$In_xGa_{1-x}As$	0 - 0.5	100	3E19 to 1.8E18
GaAs	—	60	1.6E18
$Al_xGa_{1-x}As$	0.2	38	1.4E18
GaAs	—	600	—
GaAs sub.	—	—	—

When forming contacts in this manner and indeed for the partially alloyed case of Section 3.2 it is important to bear in mind the contribution to total source gate resistance bought about by the AlGaAs layer. It is of further significance to consider the effect of temperature on the resistance of this layer. For example shown in Figure 9(C) is the calculated value of specific contact resistance as a function of AlAs mole fraction and absolute temperature [29]. HEMTS are often used at lower temperatures to improve the device performance (lower noise, larger bandwidth etc), it is consequently important that this performance is not compromised by the parasitic resistance between source and gate.

Taking these points into consideration Kuroda et al. [30] produced HEMT structures for analysis purposes as well as producing devices and actual large scale integrated circuits. The parameters of the structures used are summarised in Table 1 [30]. Of importance is the value of InAs mole fraction chosen and the doping of the InGaAs layer ($3 \times 10^{19}/cm^2$), which according to Figure 9(B) would result in specific contact resistance values of less than $10^{-8}\Omega\text{-}cm^2$ certainly an extremely low value if it were realized practically. Furthermore the mole fraction of 0.2 with a doping concentration of $1.4 \times 10^{18}/cm^2$ for the Al GaAs layer again would lead to a low specific contact resistance for this layer of less than $10^{-6}\Omega\text{-}m\text{-}^2$ at 77k and even lower at room temperature.

They presented various aspects of their results, however perhaps the most important for this paper is their findings regarding the relationship between source resistance (source-gate) and the contact length (Loh in their paper and represented as d in this instance) reproduced in Figure 10. The results are strikingly similar to those predicted in section 2.3 of this paper. That is as the length increases above about 1μm. There is no further reduction in resistance (~0.5Ω-mm) and below this the resistance increases presumably scaling back to the specific contact resistance value of the AlGaAs layer. In fact the predicted curve in Figure 6(B), condition C where $\frac{1}{gs} = 10^{-6}\Omega\text{-}cm^2$ and $\frac{1}{\sigma_1} = 1000\Omega/\square$ has a limit above 1μm of around 0.35Ω-mm and moving back toward a final value of $\rho c/L_{SG}$ at 0.1μm of 1Ω-mm. This would suggest that the system of Kuroda et al. has either a larger value of specific contact resistance than expected or an additional parasitic non-scaling resistance contribution. None the less the model of Section 2.3 would appear to have a high degree of validity particularly in this case.

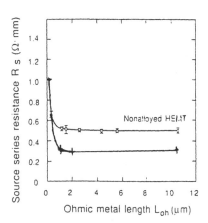

Figure 10 - Results obtained for a graded structure
(O) Parasite resistance as a function of contact length
(+) Results obtained from .2. 3

4.0   CONSLUSION

The area of contacting to multi-layer structures is one where greater understanding of the interplay between the actual contact and contact material needs further clarification. In this paper various electrical models have been presented and their validity tested through the use of published data. No claim is laid to the accuracy of these models, however it does appear under certain practical conditions that they can be used at least as a guide to expected performance of the various contacting systems. Furthermore, they should provide a basis to enable better techniques to acquire the data relevant to the parasitic losses.

ACKNOWLEDGEMENTS

The author would like to acknowledge the fruitful discussion with Drs Kachwalla and King of the Division of Radio Physics, CSIRO, Australia and the financial support provided by the CSIRO through the Griffith University/CSIRO collaborative research grants scheme.

REFERENCES

1.   H. Stormer, R. Dingle, A.C. Gossard, W. Wiegmann and M Sturge, Solid State Comm. Un., Vol 29, 1979, p705.
2.   T. Mimura, S. Hiyamizu, T. Fusii and K. Nanbu, Japan.J.Appl.Phys., vol.19, 1980, pL225.
3.   H. Morkoc and T. Solomon, IEEE Spectrum, Feb. 1984, p28.
4.   R.A. Kiehl, M.D. Fever, R.H. Hendel, J.C.M. Hwang, V.G. Keramidas, C.L. Allyn and R. Dingle, IEEE Electron Device Letters, Vol. EDL-4, 1983, p377.
5.   M. Abe, T. Mimura, N. Yokoyama and H. Ishikana, IEEE Trans. Microwave Theory Tech., Vol.Mtt-30, 1982, p992.
6.   M.D. Feuer, IEEE Trans. on Elec Dev., Vol.Ed-32, 1985, p7.
7.   K. Shewai, IEEE Trans. on Elec.Dev., Vol.Ed-34, 1987, p1642.
8.   P.A. Barnes and A.Y. Cho, Appl.Phys.Lett., Vol 33, Dec. 1978, p1022.
9.   C.K. Peng, G. Ji, N.S. Kumar and H. Morkoc, Appl.Phys.Lett. 53(10), Sept. 1988, p900.
10.   S.J. Lee and C.A. Crowell, Solid-State Elec., Vol 28(7), 1985, p659.
11.   G.K. Reeves and H.B. Harrison, IEEE Electron Devices, Vol 3, 1986, p328.
12.   H.B. Harrison and G.K. Reeves, Electronic Letters, Vol 18(25), Dec. 1982, p1083.
13.   K. Kajiyama, Y. Mizushima and S. Sakata, Appl.Phys.Lett., Vol 23, 1973, p458.
14.   H. Berger, J.Electrochem.Soc., Vol 119, 1972, p507.
15.   G.K. Reeves and H.B. Harrison, IEEE Electron.Dev.Lett., Vol EDL-3(5), 1982, p111.
16.   N. Braslau, J.Vac.Sci.Technol., Vol 19(3), 1981, p803.
17.   Piotrowsha, A. Guivarc'h and G. Pelous, Solid-State Electron., Vol 25(3), 1983, p179.
18.   R. Stall, C.E.C. Wood, K. Board and L.F. Eastman, Electron.Lett., Vol 15(24), Nov. 1979, p800.
19.   P.A. Pianetta, C.A. Stolle and J.L. Hansen, Appl.Phys.Lett., Vol 36(7), 1980, p597.
20.   H.B. Harrison and J.S. Williams, Laser and Electron Bean Processing of Materials, Eds. C.W. White and P.S. Peescy, Academic Press, N.Y., 1980, p
21.   M.D. Feuer, IEEETrans. Electron Devices, Vol Ed-32(1), 1985, p7.
22.   W.L. Jones and L.F. Eastman, IEEE Trans. on Elec.Dev., Vol Ed33(5), 1986, p712.
23.   H. Goronkin, S. Tehrani, T. Remmel, P.L.Fejes and K.J. Johnson, IEEE Elec.Dev., Vol Ed 36(2), 1989, p281.
24.   A. Elis, A.K. Rai and D.W. Langer, Electronics Letters, Vol 23(3), 1987, p113.
25.   M. Kamada, T. Suzuki, F. Nakamura, Y. Mori and M. Arai, Appl.Phys.Lett., 49(19), 1986, p1263
26.   Z.S. Kachwalla, W.D. King and H.B. Harrison, In Press.
27.   Z.S. Kachwalla, W.D. King and J. Wiggins, IREE Con. Digest of Technical Papers, Melbourne 1989, p511.

28. S. Kuroda, N. Harada, T. Katakami and T. Mimura, IEEE Elec.Dev.Lett., Vol Ed 8(9), 1987, p389.
29. A. Kelterson, F. Ponse, T. Henderson, J. Klem and H. Morkoc, J.Appl.Phys., Vol 57, 1985, p2305.
30. S. Kuroda, N. Marada, T. Katakami, T. Mimura and M. Abe, IEEE Transactions of Elec.Dev., Vol 36(10), 1989, p2196.

TAPER ETCHABLE NEON-SPUTTERED MOLYBDENUM FILM

KINYA KATO AND TSUTOMU WADA
Applied Electronics Labs.,Nippon Telegraph and Telephone Corp.,
9-11 Midori-cho, 3-Chome, Musashino-shi,Tokyo 180, Japan

ABSTRACT

Taper etchable molybdenum (Mo) films are successfully obtained by neon (Ne) sputter deposition. Mo films several hundred nanometer thick are deposited using an RF planar magnetron sputtering system. After photolithography, the Ne-sputtered Mo films are etched with a hydrogen peroxide solution. Films deposited in low pressure Ne show an excellent tapered edge with a taper angle of less than 45°, which can eliminate the step-coverage problem. However, higher-pressure Ne-sputtered films and films sputtered in argon (Ar) show steep edge profiles.

Ne-sputtered Mo films show a drastic change in characteristics depending on the gas pressure. The internal stress is strongly compressive for lower pressures and becomes tensile for higher pressures. According to RHEED patterns, lower pressure sputtered film shows a <110> orientation including a halo which indicates the existence of an amorphous-phase, but higher pressure sputtered films show clear <111> orientations. TEM observations confirm the structural change. Higher pressure Ne-sputtered film and Ar-sputtered film show columnar structures, but low-pressure Ne-sputtered film shows the disordered structure.

Taper etchability of the low-pressure Ne-sputtered Mo films is considered to be due to the structural change, especially the included amorphous-phase.

INTRODUCTION

Taper etching is effective at eliminating step coverage problems in multilevel interconnections of micro-electronic devices. Active matrices of flat panel displays, such as liquid crystal displays (LCDs), must be fabricated using low temperature deposition processes such as plasma chemical deposition (PCVD), because of the limits of the substrate glass. Low temperature deposition produces a coarse film structure, especially on steep side walls. These regions are very vulnerable to wet etchant attack.

On active matrix substrates, long data and scan lines are formed, such that each pixel has at least a cross-point. Taper etching of the lower-level metal lines is effective at reducing failures caused at the cross-points [1]. A simple process procedure is also needed to improve process reliability.

Mo is a high-temperature tolerant low-resistive metal which is suitable for use in active matrices. However, conventional Mo films sputtered in ambient Ar exhibit steep edge profiles after a simple wet etching, due to the columnar structure which is common in sputtered metal films [2]. On the other hand, Mo films formed by Ne-sputtering at low pressure show excellent taper etching characteristics.

This paper presents results for taper etchability of Ne-sputtered Mo films and discusses related mechanisms.

Mat. Res. Soc. Symp. Proc. Vol. 181. ©1990 Materials Research Society

EXPERIMENTS

Mo films were deposited by RF magnetron sputtering of a 99.98% pure Mo target. Both Ne and Ar sputtering gases were used for comparison. The gas pressure was controlled to be 0.5 - 6 Pa by the gas flow rate and evacuation speed. Substrates were oxidized 4" diameter Si wafers. The substrate holder was a carrousel that was rotated during deposition. Therefore, Mo was deposited when the substrates passed nearby the target. The substrate holder was water-cooled, and the substrates were not heated. Typical RF power was 2 kW for the 5" x 15" Mo target.

After conventional photolithography using a commercially available positive photoresist, the Mo films were etched with a hydrogen peroxide solution. Edge profiles were observed by a Scanning Electron Microscope (SEM).

The characteristics of the taper etchable Mo films were examined from internal stresses, X-ray lattice parameters, Reflective High Energy Electron Diffraction (RHEED) patterns and Transmission Electron Microscope (TEM) images.

RESULTS AND DISCUSSIONS

Edge Profiles

Mo has extremely poor durability against oxidation. Thus, an oxidizing solution can easily etch Mo film to produce water-solvable oxides or hydro-oxides. We used a hydrogen peroxide solution for the Mo etchant. This etchant dose not attack the other constituents of the active matrices such as amorphous silicon or silicon-nitride insulator. Therefore, the etching is very selective.

Figure 1 shows the edge profiles of Ne-sputtered Mo films against gas pressures. The film thickness is around 200 nm. The film deposited at low pressure shows excellent taper etched profiles with edge angles less than 45°. However, the films deposited at higher pressure shows steep edges with edge angles of almost 90°.

(a) 0.5 Pa   1 μm   (b) 5.4 Pa

Fig. 1   Edge Profiles of Ne-sputtered Mo films.

The Ar-sputtered films have the same profiles as the film deposited in higher pressure Ne, which does not depend on the gas pressure. For Ar, no taper etching was observed.

Figure 2 shows a scanning electron micrograph of the tapered edge sample before resist stripping. The tapered region of the Mo film is formed under the resist. The etchant permeated into the interface between the resist and the Mo film.

Film Structures

The common feature in the taper etchable Mo film characteristics is high compressive stress. The film stresses of the Ne-sputtered Mo films vary according to the gas pressures from highly compressive stress to tensile stress, as shown in Fig. 3. The taper etchable films deposited in low pressure Ne were highly compressive over 1 GPa. On the other hand, films Ne-sputtered at high pressure and Ar-sputtered ones have low compressive or tensile stresses.

1 μm

Fig. 2 Taper etched sample SEM before resist stripping. The taper region is formed under the resist.

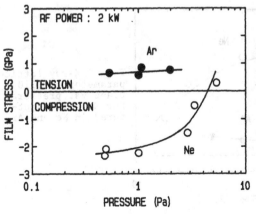

Fig. 3 Film stress vs. pressure for Mo films sputtered in Ne and Ar.

The gas pressure dependence of the film stress for sputtered Mo films has been investigated [3]. If a lighter mass gas, such as Ne, is used for sputtering, the transition pressure P_t from compressive to tensile stress shifts to higher pressure. In this study, P_t for Ne-sputtering was 5 Pa. Conversely, Ar-sputtered films show only tensile stress. These trends suggest that the transition pressure P_t for Ar-sputtering will be lower than 0.4 Pa which was the lowest pressure in this study.

The compressive stress in sputtered metal films have generally been believed to result from the peening effect [4]. The peening effect is due to neutralized sputtering gases reflected from the target surface striking the films. When the compressive stresses result from the peening effect, lattice parameter variations are observed in many metal films [5]. The measured (110) plane lattice parameter variations coincide with the stress change as shown in Fig. 4. The low pressure Ne-sputtered Mo films show larger lattice constants than Ar-sputtered or high pressure Ne-sputtered films. This implies that Mo films that are taper etchable have been subjected to a strong peening effect.

Figure 5 shows the RHEED patterns for the Ne-sputtered Mo films and Ar-sputtered Mo film about 500 nm thick. The Ar-sputtered film shows a <110> orientation which implies that the most densely populated atomic plane (110) of bcc Mo lattice is parallel to the substrate. This is a general rule for vacuum evaporated or sputtered metal films. On the other hand, Ne-sputtered Mo films do not follow this general rule. Film sputtered at low pressure, about 1 Pa, shows a <110> orientation including a halo, which indicates the existence of an amorphous phase. However, high pressure Ne-sputtered films show clear <111> orientations.

Figure 6 shows cross-sectional TEM dark images of Ne-sputtered and Ar-sputtered Mo films. High pressure Ne-sputtered film and Ar-sputtered film have columnar structures, so the columns grow continuously from the substrate to the surface.

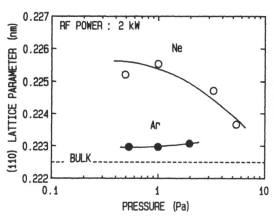

Fig. 4 X-ray lattice parameter of (110) plane vs. pressure for Mo films sputtered in Ne and Ar.

However, low pressure Ne-sputtered Mo film shows the more disordered structure than the above mentioned films. This observation confirms the above mentioned RHEED results. The structure of the taper etchable Mo films sputtered in low pressure Ne is destroyed the structure by the peening effect, so as to include an amorphous phase in the film structure.

These results suggest that the taper etchability of low pressure Ne-sputtered Mo films arises from the amorphous phase in the film structure.

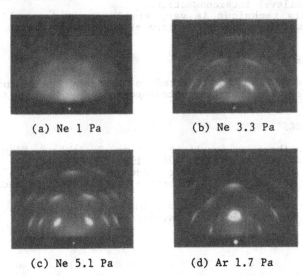

(a) Ne 1 Pa   (b) Ne 3.3 Pa

(c) Ne 5.1 Pa   (d) Ar 1.7 Pa

Fig. 5   RHEED patterns of Ne-sputtered and Ar-sputtered Mo films.   The film thickness is about 500 nm.   RF power is 2 kW.

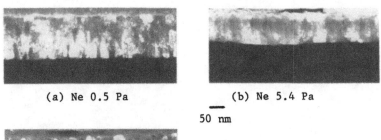

(a) Ne 0.5 Pa   (b) Ne 5.4 Pa

50 nm

Fig. 6   TEM dark images of Ne-sputtered and Ar-sputtered Mo films.   RF power is 2 kW.

(c) Ar 1 Pa

SUMMARY

Ne sputtered Mo films are taper etchable if the Ne pressure is low and a hydrogen peroxide solution is used. Low pressure Ne-sputtering causes structural changes in Mo films so as to include an amorphous-phase by the peening effect. This makes the taper etching of Mo films possible in the same way as amorphous materials.

Taper etched profiles ensure step coverage of the overcoated films and prevent failures caused at the cross-points in the multilevel interconnections.

This technique is very simple so it is applicable to active matrices and other micro-electronic devices.

ACKNOWLEDGMENT

The authors are indebted to Tadamichi Kawada and Shigeto Kohda for their constant encouragement throughout this work.

REFERENCES

1. K. Ishikawa, S. Suzuki, H. Matino, T. Aoki, T. Higuchi and Y.Oana, SID 89 Digest, 226, (1989).
2. J.A. Thornton, J. Vac. Sci. Technol., 11, 666, (1974).
3. D.W. Hoffman and J.A. Thornton, J. Vac. Sci. Technol., 17, 380, (1980).
4. J.A. Thornton, Thin Solid Films, 64, 111, (1979).
5. C.T. Wu, Thin Solid Films, 64, 103, (1979).

APPLICATIONS OF AMORPHOUS TI-P-N DIFFUSION BARRIERS IN SILICON METALLIZATION

E. KOLAWA, L. HALPERIN, P. POKELA, QUAT T. VU, C.W. NIEH
California Institute of Technology, 116-81, Pasadena CA 91125

ABSTRACT

Thin films of amorphous TiP and TiPN$_2$ alloys were deposited by sputtering of a TiP target in an Ar and N$_2$/Ar mixture, respectively. These alloy films were tested as diffusion barriers between Al and Si as well as between Cu and Si and also in metallizations which included TiSi$_2$ as the contacting layer. Rutherford backscattering spectrometry, x-ray diffraction and electrical measurements were used to determine the barrier effectiveness. We find that TiP and TiPN$_2$ films prevent the interdiffusion and reaction between Al and Si up to 500°C and 600°C for 30 minutes annealing, respectively, and between Cu and Si up to 600°C and 700°C, respectively.

1. INTRODUCTION

In silicon integrated circuits, aluminum is commonly used for contact and interconnections, but severe degradation is caused by intermixing and reactions when aluminum and silicon are in direct contact with each other. Thus, diffusion barriers are indispensable in present VLSI contact technologies [1-3]. Many interstitial alloys such as nitrides [4-9], carbides [10-11], borides [12], conductive oxides [13-20], and amorphous alloys [21-27], have been investigated in the past as diffusion barriers in contact metallizations. We have shown most recently, that the idea of using amorphous binary alloys as diffusion barriers, like TaSiN, seems to be very promising [28].

Because of the electromigration problems of Al base interconnection lines, new candidates, like copper [29-30] and tungsten [31] are being evaluated for VLSI applications. Copper reacts with silicon at quite modest temperatures and is also known to be a fast diffusor in silicon. To minimize such deleterious interactions there is a need for very effective diffusion barriers.

We report on properties of amorphous TiP and TiPN$_2$ alloys as diffusion barriers between Al and Si or between Cu and Si. RBS, x-ray diffraction and electrical measurements on planar diodes with shallow p n junctions are used to evaluate the barriers.

2. EXPERIMENTAL PROCEDURES

N-type silicon wafers of <111> orientation either bare or covered with thermally grown SiO2 and carbon tape were used as substrates for the Ti-P-N films. All depositions were performed in an rf sputtering system equipped with a cryopump and a cryogenic baffle. The sputtering chamber was evacuated to a base pressure of about 5×10^{-7} Torr before the deposition. A magnetron-type circular cathode, 7.5 cm in diameter was used as the sputtering source. The substrate holder was placed about 7 cm below the target and was neither cooled nor heated externally.

The Ti-P-N films were deposited by reactive sputtering of a TiP target in an Ar/N$_2$ gas mixture. The flow ratios of Ar to N$_2$ and the total gas pressure were adjusted by flow controllers and monitored with a capacitive manometer in a feedback loop. All Ti-P-N films were sputtered with 10 mTorr total pressure and 300 W forward sputtering power. The ratio of nitrogen to argon gas flow was in the 0 to 30% range.

The atomic composition of the Ti-P-N films was measured by

34

backscattering spectrometry of films on carbon substrates. An x-ray Read Camera was used to determine the phases and crystallographic structure of the films. The film resistivities at room temperature were determined for films on oxidized silicon substrates from sheet resistivities obtained from four point probe data. The film thickness was measured with a profilometer. To test the diffusion barrier properties of TiPN films, samples were prepared on <111> <Si> wafers with the film sequences <Si>/TiPN/Al, <Si>/TiSi$_2$/TiPN/Al, <Si>/TiPN/Cu and <Si>/TiSi$_2$/TiPN/Cu. These same metallizations were deposited also on <100> <Si> wafers with shallow As$^+$ implanted n$^+$p diodes. Prior to loading into the deposition chamber, the silicon substrates were etched with dilute HF. To sputter the titanium silicide films, a composite TiSi$_2$ target and pure argon were used. The TiSi$_2$ film thickness was about 30 nm.

The thickness of the Ti-P or Ti-P-N films were about 100 nm and 80 nm respectively. Aluminum and copper (250 and 500nm) overlayers were sputter-deposited in argon at a total pressure of 5 mTorr, a forward sputtering power of 300W, and a substrate bias of -50V. All films were deposited without breaking the vacuum in the sputtering system. The samples were then annealed in a vacuum furnace at a pressure of about 5 x 10^{-7} Torr in the range of 450°C-700°C for 30 minutes. The effectiveness of these barriers was evaluated from electrical measurements, backscattering spectrometry, and x-ray diffraction analysis. Shallow n$^+$p junction diodes with a 300nm junction depth, an arsenic surface concentration of about 10^{21} atoms cm^{-3}, and 230 x 230 microm^2 contact areas were used for electrical measurements. The contact metallizations were delineated by the lift-off technique.

3. RESULTS AND DISCUSSIONS

The composition of the Ti-P and Ti-P-N films was measured by backscattering spectrometry using the samples deposited on carbon. The atomic ratio of Ti/P in samples deposited in pure argon is about 1 (+8%). All samples deposited in argon/nitrogen gas mixtures (from 5 to 30% N$_2$/Ar flow ratio) exhibit the TiPN$_2$ composition regardless of the value of the flow ratios used for the depositions. The TiPN$_2$ films contain about 5 at.% of oxygen. The amount of oxygen in TiP films tends to be slightly higher (about 7 at.%). All deposited films were x-ray amorphous as revealed by Read Camera diffraction measurements. To determine the compositional stability and the crystallization temperature of the TiPN$_2$ alloys, films deposited on silicon dioxide were annealed for 30 minutes in vacuum at temperatures ranging from 600°C to 1000°C and examined by RBS and Read Camera. After annealing at 700°C, the samples remained unchanged, possessing still the same composition and an amorphous structure. Backscattering spectra of samples annealed at 800°C show a complete loss of phosphorous for all samples and partial loss of nitrogen (Figure 1). The structure of these annealed films is still amorphous, even after annealing at 900°C for 30 minutes.

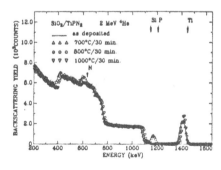

Fig. 1. 2.0 MeV ^{4}He^{++} backscattering of the Si/SiO$_2$/TiPN$_2$ sample before and after annealing in vacuum for 30 min. in the 600°C-1000°C temperature range.

The resistivity of the TiP <100nm> films is about 280 microohmcm as compared to the 60 microohmcm value reported in the literature for bulk samples [32]. The significantly higher resistivity of the TiP films may be partly attributed to the presence of oxygen impurities. The resistivity of the as-deposited $TiPN_2$ <80-100nm> samples varies between 1800 and 2200 microohmcm, which is still acceptable for diffusion barrier applications. Assuming a current flow of 10E6A/cm^2 perpendicular to the plane of a 100nm film and a maximum acceptable voltage drop of 25mV (=kt/q), a maximum acceptable resistivity on the order of 2500 microohmcm is sufficient to qualify a diffusion barrier material electrically. After the 900°C annealing, the resistivity of the remaining amorphous TiN is about 450 microohmcm.

We tested both TiP and $TiPN_2$ films as diffusion barriers between Al and Si or Cu and Si, using RBS, electrical measurements and x-ray diffraction. As a reference, Si/Al and Si/Cu samples were also tested using the same experimental methods as samples containing diffusion barriers.

The Si/TiP <100nm>/Al<250nm> metallization is stable only up to 500°C as monitored by RBS. After annealing at 550°C the layers react. X-ray diffraction reveals that $TiAl_3$ forms. The addition of nitrogen to the titanium phosphide raises the stability of the barrier to 600°C, as indicated by both RBS and measurements of the dc current-voltage characteristics of shallow junctions with the <Si>/$TiPN_2$ (80nm)/Al(500nm) contact scheme. All diodes annealed at 650°C are shorted and have degraded surface morphology. The RBS analysis of the large area <Si>/$TiPN_2$(80nm)/Al(250nm) sample annealed at 650°C indicates a laterally non-uniform reaction between the layers with Ti present on the sample surface.

On the other hand, diodes with the Si/Al metallization, i.e. without diffusion barrier, were all shorted already after annealing at 500°C for 30 minutes. A contacting layer is usually added between a semiconductor and a diffusion barrier to optimize the electrical characteristic of the contact. In the present study a thin layer of about 30nm of $TiSi_2$ was used for this purpose. Figure 2 shows RBS spectra of the <Si>/$TiSi_2$(30nm)/$TiPN_2$(80nm)/Al(250nm) sample before and after annealing at 600°C and 650°C. As can be seen there is no visible difference between the spectra of the as-deposited and 600°C annealed samples. It means that within the resolution limit of backscattering spectrometry the Al and Si layers do not interact during annealing. The backscattering spectrum of the sample annealed at 650°C is typical of a structural collapse of the metallization. X-ray data reveals the presence of the $TiAl_3$ phase in the sample.

Figure 3 shows the reverse current of the shallow n$^+$p junctions with the <Si>/$TiSi_2$(30nm)/$TiPN_2$(80nm)Al(500nm) metallization. After 30 minutes annealing at 600°C, the reverse current of the diodes at 600°C remains unchanged compared to that of an un-annealed sample. Annealing at 650°C destroys the rectification of the junction. For comparison we find that the <Si>/$TiSi_2$/Al contact is electrically stable only up to 500°C for 30 minutes. These combined results show that a 80nm thick reactively sputtered $TiPN_2$ layer elevates the thermal stability of the contact significantly beyond the eutectic temperature of the Al-Si system (577°C).

The failure temperature of the <Si>/Cu system is significantly lower than the failure temperature of the <Si>/Al system. The interdiffusion between the Si substrate and Cu layers in the <Si>/Cu (250nm) sample is already observed by RBS after annealing the sample at 200°C for 30 minutes. After 30 minutes annealing at 300°C, copper oxide (CuO) and copper silicide (Cu_3Si) are present in the sample as determined by x-ray. In fact, RBS of the as-deposited sample shows signs of slight indiffusion of Cu into the Si substrate. The n$^+$p shallow junctions with <Si>/Cu (250nm) metallization are already shorted after annealing at 300°C for 30 minutes.

The presence of the 80nm thick $TiPN_2$ layer in the <Si>/$TiPN_2$/Cu metallization increases the stability of the system up to 600°C. After annealing at 650°C the metallization fails both structurally (RBS) and

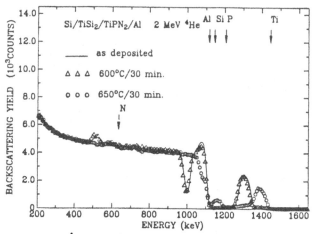

Fig. 2. 2.0 MeV ^{4}He++ Rutherford backscattering spectra of the
Si/TiSi$_2$(30 nm)/TiPN$_2$(80 nm)/Al(250 nm) sample before
and after vacuum annealing at 600°C and 650°C for 30 min.
(Beam incident at 7° from sample normal; backscattered
^{4}He detected at a scattering angle of 170°).

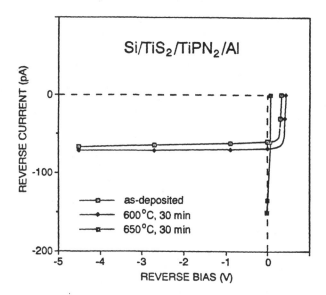

Fig. 3. The reverse current of a shallow n$^+$p junction with a
Si/TiSi$_2$(30 nm)/TiPN$_2$(80 nm)/Al(500 nm) metallization
as a function of temperature for 30 min. annealing
in vacuum.

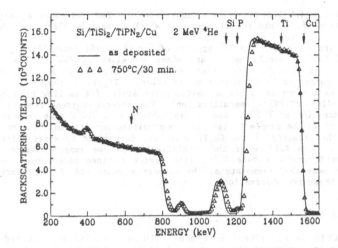

Fig. 4. 2.0 MeV $^4He^{++}$ Rutherford backscattering spectra of the Si/TiSi$_2$(30 nm)/TiPN$_2$(80 nm)/Cu(250 nm) sample before and after vacuum annealing at 750°C for 30 min.

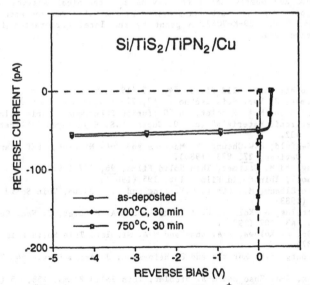

Fig. 5. The reverse current of the shallow n$^+$p junction with the Si/TiSi$_2$(30 nm)/TiPN$_2$(80 nm)/Cu(500nm) metallization as a function of temperature for 30 min. vacuum annealing.

electrically (shallow junction measurements). RBS indicates a very deep
penetration of Cu into the Si substrate which causes the electrical shorting
of the shallow junctions.
Figure 4 presents the RBS spectra of the <Si>/$TiSi_2$(30nm)/$TiPN_2$
(80nm)/Cu(250nm) sample before and after annealing at 700°C. There is no
change in the sample after heat treatment. Electrical measurements on
shallow junctions confirm this observation. Figure 5 shows the reverse
current as a function of the annealing temperature of a shallow junction with
the <Si>/$TiSi_2$/$TiPN_2$/Cu metallization. The reverse current remains stable
after annealing at 700°C. The diodes subjected to 750°C heat treatment are
all shorted. The sample is laterally non-uniform and its color is changed to
grey. The presence of the Cu_3Ti phase was determined by x-ray diffraction
analysis. The failure of the metallization can be expected since, as is
shown in Figure 1 a bare $TiPN_2$ film begins to lose phosphorous and some
nitrogen near that temperature. We suspect a cause and effect relationship
between these two observations.

4. CONCLUSIONS

   A thin (about 100nm) amorphous film of reactively sputtered $TiPN_2$
successfully suppresses the reaction of an Al overlayer with a Si substrate
for 30 minutes annealing in vacuum up to 600°C, which exceeds the Al-Si
eutectic temperature (577°C). In this application, reactively sputtered $TiPN_2$
films compare favorably with the popular TiN diffusion barrier, except for
the higher but still acceptable resistivity.

Acknowledgements

   We thank Rob Gorris and Bart Stevens for technical assistance. The
financial support for this work was provided by the U.S. Army Research Office
under Contract DAAL03-89-K-0049. A grant by the Intel Corporation is also
gratefully acknowledged.

REFERENCES:

1. M-A. Nicolet, Thin Solid Films, 54, 415 (1978).
2. M. W ittmer, J. Vac. Sci. Technol., A2, 273 (1984).
3. H. Kattelus and M-A. Nicolet, in "Diffusion Phenomena in Thin Films and
   Microelectronic Materials" (ed. D. Gupta, P.S. Ho), Noyes Publications,
   1988, p. 432.
4. H. von Seefeld, N.W.Cheung, M. Maenpaa and M-A. Nicolet, IEEE Trans. on
   Electron Devices, 27, 873 (1980).
5. C.Y. Ting and M. Wittmer, Thin Solid Films, 96, 327 (1982).
6. S. Kanomori, Thin Solid Films, 136, 195 (1985).
7. L. Krusin-Elbaum, M.Wittmer, C.Y. Ting and J.J. Cuomo, Thin Solid Films,
   104, 81 (1983).
8. H.P. Kattelus, E. Kolawa, K. Affolter and M-A. Nicolet, J. Vac. Sci.
   Technol., A3, 507 (1985).
9. F.C.T. So, E. Kolawa, X-A. Zhao and M-A. Nicolet, Thin Solid Films, 153,
   507 (1987).
10. M. Eizenberg, S.P. Murarka and P. Heinemann, J. Appl. Phys., 54, 3195
    (1983).
11. H-Y. Yang, X-A. Zhao and M-A. Nicolet, Thin Solid Films, 158, 45 (1988).
12. J.R. Shappirio, J. Finnegan, R. Lux and D. Fox, Thin Solid Films, 119,
    23 (1984).
13. M.L. Green, M.E. Gross, L.E. Papa, K.Y. Schnoes and D. Brasen, J. of
    Electrochem. Soc., 132, 2077 (1985).

14. E. Kolawa, F.C.T. So, E. T-S. Pan and M-A. Nicolet, Appl. Phys. Lett., 50, 854, (1987).
15. L. Krusin-Elbaum, M. Wittmer and D.S. Yee, Appl. Phys. Lett, 50, 1879, (1987).
16. F.C.T. So, E. Kolawa, C.W. Nieh, X-A. Zhao and M-A. Nicolet, Appl. Phys., A45, 265 (1988).
17. E. Kolawa, C. Garland, L. Tran, C.W. Nieh, J.M. Molarius, W. Flick, M-A. Nicolet and J. Wei, Appl. Phys. Lett., 53, 2644 (1988).
18. C.W. Nieh, E. Kolawa, F.C.T. So and M-A. Nicolet, Mat. Lett., 6, 177 (1988).
19. J.M.E. Harper, S.E. Hornstrom, O. Thomas, A. Charai and L. Krusin-Elbaum, J. Vac. Sci. Technol., A7, 875 (1989).
20. A. Charai, S.E. Hornstrom, O. Thomas, P.M. Fryer and J.M.E. Harper, J. Vac. Sci. Technol., A7, 784 (1989).
21. M-A. Nicolet, I. Suni and M. Finetti, Solid State Technol., 26, 129 (1983).
22. M-F. Zhu, F.C.T. So and M-A. Nicolet, Thin Solid Films, 130, 245 (1985).
23. J.D. Wiley, J.H. Perpezko, J.E. Nordman and K-J. Guo, IEEE Trans. Id. Electr., 29, 154 (1982).
24. F.C.T. So, X-A. Zhao, E. Kolawa, J.L. Tandon, ·M.F. Zhu and M-A Nicolet, Mater. Res. Soc. Symp. Proc., 54, 139 (1986).
25. S.E. Hornstrom, T. Lin, O. Thomas, P.M. Fryer and J.M.E. Harper, J. Vac. Sci. Technol, A6, 1650 (1988).
26. L.S. Hung, F.W. Saris, S.Q. Wang and J.W. Mayer, J. Appl. Phys., 59, 2416 (1986).
27. S.Q. Wang and J.W. Mayer, J. Appl. Phys., 65, 1957 (1989).
28. E.kolawa, J.M. Molarius, C.W. Nieh, and M-A. Nicolet, J.Vac.Sci.Technol.A (in print).
29. P.L. Pai and C.H. Ting, in Proc.IEEE VLSI Multilevel Interconnection Conf. (Santa Clara, CA) 1989, p.258.
30. P.L. Pai and C.H. Ting, IEEE Electron Devices Lett. 10,423 (1989).
31. Proc. First and Second Workshops on Tungsten and Other Refractory Metals for VLSI Applications, R.S. Blewer, ed. (MRS, Pittsburgh PA, 1986).
32. G.V. Samsonov and I.M. Vinitskii, Handbook of Refractory Compounds, translated from Russian by K. Shaw (IFI/Plenum, new york, 1980).

TANTALUM AND TANTALUM NITRIDE AS DIFFUSION BARRIERS BETWEEN COPPER AND SILICON

KAREN HOLLOWAY AND PETER FRYER
T.J. Watson Research Center, Yorktown Heights, NY 10598

ABSTRACT

We have investigated the effectiveness of thin tantalum layers as diffusion barriers to copper. Fifty nm Ta films were sputtered onto unpatterned single crystal Si wafers and overlaid with 100 nm Cu. Material reactions in these films were followed as a function of annealing temperature by in-situ resistance measurements, and characterized by by Rutherford Backscattering and cross-section TEM. The effect of the incorporation of nitrogen was explored by reactively depositing Ta(5 at.% N) and Ta_2N. Pure Ta prevents Cu - silicon interaction to at least 600 °C. At higher temperatures, reaction of the Si substrate with Ta forms a planar $TaSi_2$ layer. Cu rapidly penetrates to the Si substrate, forming Cu silicide precipitates at the $TaSi_2$ - Si interface. A study performed on the Si/Ta(N)/Cu film had very similar results. Ta_2N is an even more effective barrier to copper penetration, preventing Cu reaction with the substrate for temperatures up to at least 700 °C.

INTRODUCTION

Current semiconductor technology demands the use of low-resistivity metals such as Al or Cu for VLSI conduction lines and contact structures. Penetration of these metals to device areas must be prevented, however, and this may be accomplished through the use of thin conducting layers which act as diffusion barriers between Si and these metals. Metallurgical differences in the stability of a barrier with aluminum, which is highly reactive, and copper, which is more noble, may be expected. Al often induces failure by reacting with the barrier layer, consuming it in the formation of aluminides [1-3]. Copper is quite mobile at elevated temperatures [4], and may penetrate through a barrier layer without reacting with it. Tantalum is a likely candidate for a barrier to copper reaction with Si as it is a refractory metal which forms no known compounds with copper, so a Cu-Ta contact should be stable to high temperatures. An investigation by Hu, et al. has identified Ta as a good barrier to Cu on oxidized Si substrates to 750 °C [5]. Also, a radiotracer study has shown that the diffusion of Cu in Ta is quite slow at temperatures in the 400 - 700°C range [6]. Barrier properties might be futher improved by the addition of impurities which may segregate to the grain boundaries, obstructing a fast pathway to copper diffusion in Ta. The nitrides of tantalum also have high melting points and belong to the class of relatively dense interstitial compounds [7], thus may provide even greater stability and lower reactivity than elemental Ta. The efficacy of Ta_2N between Al, and Pd and Co silicides have been investigated [8]; however, its resistance to Cu penetration has yet to be shown. We have investigated the structure and effectiveness of thin tantalum, Ta(N), and Ta nitride barrier layers sputtered onto unpatterned single crystal Si wafers and overlaid with Cu. In the present work, the Si substrate provides a sink for copper diffusion; its use also directly addresses the technological concerns. The formation of a copper silicide phase at a Cu-Si interface begins at 200°C [9,10], and in some instances a Cu_3Si skin forms at room temperature [11]. The temperature at which a barrier layer prevents Cu-Si interaction for a reasonable annealing time is one measure of diffusion barrier effectiveness.

Mat. Res. Soc. Symp. Proc. Vol. 181. ©1990 Materials Research Society

EXPERIMENT

Fifty nm thick tantalum films were sputtered onto 5-inch <100> Si substrates, then 1000 Å Cu was deposited without breaking vacuum in a multi-target MRC 643 magnetron sputtering system. The nitrogen-alloyed tantalum and Ta_2N films were obtained by reactively sputtering Ta with 1/4%, and 5% N_2, respectively, in the Ar gas. The base pressure for each run was about 1 x 10^{-7} torr, and total sputter gas pressure was about 13 microns during deposition of the Ta layers, and 15 microns for the Cu. The Si substrates were given a buffered HF dip before loading into the vacuum system. RBS spectra of similar reactively sputtered films showed that a 1/4%N_2/Ar mixture gives a Ta(N) film with 5 at.% nitrogen content; and 5%N_2/Ar gives a film with 33 at.% nitrogen. Electron diffraction patterns taken from through-foil TEM specimens allowed the identification of fine-grained β-Ta in the former case, and fine-grained hexagonal Ta_2N in the latter. The Ta film deposited in pure Ar was also found to be β-Ta. Henceforth the films deposited in pure Ar, with 1/4 % N_2, and 5% N_2 will be denoted as Ta, Ta(N), and Ta_2N, respectively.

Reactions in these films were followed by measuring their sheet resistances with a four-point probe as the temperature was ramped at a 3 °C/min. rate from 25 °C to 750 °C in flowing He purified over a Ti getter. Since the top 100 nm Cu film can be expected to carry nearly all the sensor current, this technique monitors the condition and quantity of Cu over the Ta barrier until a reaction involving copper occurs. The microstructure of samples at various stages in the anneal were studied with Rutherford Backscattering Spectrometry (RBS) and cross-section Transmission Electron Microscopy (TEM), including electron diffraction for phase identification. TEM sample preparation followed the Bravman-Sinclair process [12].

RESULTS

The resistivity-temperature traces for the Ta, Ta(N), and Ta_2N are shown in Figure 1(a-c). In all three cases the sheet resistivity initially rises with temperature in a nearly linear manner consistent with that of the metallic, conductive Cu top layer. The traces for the Cu/Ta/Si and Cu/Ta(N)/Si films are similar (Figure 1 a,b). At about 660 °C, both undergo an abrupt rise, indicating that a more resistive structure has formed. The cooling curves indicate that this change is irreversible. At high temperatures, the resistivity rises with decreasing temperature since the reacted structure allows contact with the semiconducting silicon substrate. As the samples cool further, the curve 'bends over' to the metallic behavior of the reacted film. Both samples, when removed from the furnace, showed a silvery, somewhat rough appearance. The Cu/Ta_2N/Si film, however, shows no such change, and the cooling curve in Figure 1c nearly overlays the heating curve, suggesting that the the Cu layer is nearly unaffected by the anneal. This sample, when removed from the furnace, still appeared as metallic copper, although it was somewhat hazy.

In order to identify the reaction associated with the abrupt resistivity rise, a Cu/Ta/Si film was ramped to 630 °, just below the transition, and allowed to cool. The film structure of the unannealed sample, that annealed to 630 °C, and that annealed to 750 °C was studied by RBS and cross-section TEM, including electron diffraction. Rutherford backscattering spectra taken from these three samples are overlaid in Figure 2. The unannealed film and that rampled to 630 °C exhibit two discrete layers, Ta underneath Cu, without evidence of intermixing. Exposure to temperatures up to 750 °C, on the other hand, had induced the motion of all the elements involved. Both Ta and Si have moved to the surface; the heights of the respective peaks are consistent with the formation of $TaSi_2$. The Cu signal is rather complex. This element is apparently present at maximum concentration at two depths - just below the surface and at about 120 nm; and it is present at significant concentrations at greater depths.

The Cu/Ta/Si films from which RBS spectra were taken were also observed by cross-section TEM in a JEOL 200CX microscope. Before heat treatment, the thin film structure is uniform and planar (Figure 3a). There is no evidence of reaction or intermixing at any interface. The β-Ta grain size is very fine (5-10 nm). After annealing to 630 °C, cross-section TEM shows little apparent microstructural change. Some grain growth in the Cu had occurred, but there is no evidence of intermixing or Cu penetration of the Ta barrier layer. The microstructure the sample annealed to a temperature above the transition, 750 °C, shown in Figure 3b, is quite complex. The 50 nm Ta layer had reacted with the Si substrate to form a 120 nm planar layer of h-TaSi$_2$. In most areas, no copper remains on top of the Ta silicide. Instead, it has penetrated to the TaSi$_2$-Si interface, forming large (1-2 micron deep) copper-containing precipitates in the form of inverted square or rectangular pyramids bounded by Si < 111 >

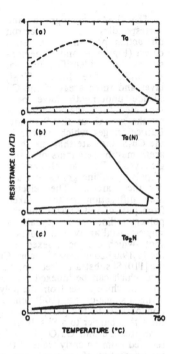

Figure 1 Sheet resistance versus temperature of the Cu/Ta, Cu/Ta(N), and Cu/Ta$_x$N films on < 100 > Si.

planes. Many of the precipitates are truncated on the bottom by a Si < 100 > plane. Electron diffraction performed in a JEOL 4000FX with 300 kV accelerating potential, which allowed penetration of these precipitates, identified them as single-crystal η''-Cu$_3$Si. This phase has been reported to form when copper contaminated single-crystal Si is annealed to high temperatures [13], and it has also been found to be the first silicide phase to form by thermal reaction at the Cu-Si interface.[9-11,13-15]. The TEM images of a large number of the Cu$_3$Si precipitate structures show a 200-400 nm amorphous layer of light contrast between the Cu-containing material and the Si substrate. Such a layer is visible in Figure 3b; a very thick amorphous area is present between Cu$_3$Si and Si in Figure 4a as well. This is amorphous silicon oxide, which has been recently reported associated with Cu-Si interaction and subsequent O$_2$ exposure [16]. The rapid oxidation of silicon, believed to be catalyzed by Cu$_3$Si has been observed even at room temperature. This reaction is characterized by the formation of thick layers of SiO$_2$ underneath a Cu silicide film and the presence of a very thin silicide layer at the SiO$_2$ - Si interface. SiO$_2$ growth, which proceeds through a local incorporation of a large amount of oxygen only under the Cu$_3$Si precipitates, causes the area around the precipitate to swell and protrude above the surface of the sample. Possibly, the oxygen is incorporated from room air after Cu silicide formation.

The TEM micrograph and corresponding electron diffraction pattern of one such defect (Figure 4a and b) demonstrates of identification of η''-Cu_3Si. The 150nm $TaSi_2$ layer and three areas of the Cu silicide underneath have been separated from the Si substrate by the growth of the Si oxide. Lattice fringes which appear in the Cu_3Si indicate that the three areas may be remnants of a single crystal. There are also groups of polycrystalline grains just under these areas. The selected area diffraction pattern taken from part of this crystal and the polycrystalline area reveals a single-crystal array of diffraction spots which can be indexed for

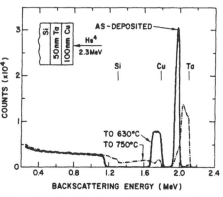

Figure 2   Overlaid Rutherford Backscattering spectra of the Cu/Ta film as deposited onto < 100 > Si, that ramp-annealed to 630 °C, and to 750 °C.

the [1,19,0] zone axis of the η''-Cu_3Si with the [001] direction roughly parallel to the [110] Si substrate direction. Superimposed onto this pattern are rather intense arcs which can be indexed for very highly textured Cu metal. Indeed the Cu pattern, which arises from the polycrystalline grains underneath the Cu_3Si, nearly forms a single-crystal (100) zone axis of fcc Cu, oriented with the [011] Cu direction parallel to the [001] of η''-Cu_3Si. In similar cases of copper penetration through a barrier to a Si substrate, the copper silicide had apparently oxidized to form free Cu and Cu_2O [17]. The Cu metal observed in Figure 4 might have resulted from an early stage of the oxidation of the Cu_3Si; and it retains some of the orientation of the crystalline reactant.

DISCUSSION

While the report of a maximum failure temperature of a Si/barrier/Cu structure is useful for the choice of metallization schemes, an understanding of the failure mechanism of these systems is essential to the design of a maximally effective barrier layer. In the present case, it appears that a complex reaction in which Cu penetration to the Si sink is concomitant with the formation of $TaSi_2$. It occurs rather suddenly; the resistivity rise observed is quite sharp. Also, if a Cu/Ta/Si sample is ramped just to the transition temperature and allowed to cool, some areas of the sample surface appears to be completely reacted, while neighboring areas are shiny and copper-colored, with a sheet resistance that is close to that of an unannealed sample, indicating that little Cu had reacted or diffused into the Ta. RBS studies of Cu/Ta films deposited onto SiO_2 [5] and on < 100 > Si [18] have shown that the outdiffusion of Ta to the Cu surface precedes any motion of Cu or Si, including Cu diffusion into Ta. While further work is necessary to determine the details of this reaction we have shown that pure Ta may be applied successfully as a copper diffusion barriers to rather high temperatures (T > 600 °C) relative to the latter states of VLSI processing. Certainly, the Cu-Si interaction, which occurs at 200 °C if no barrier is present, is prevented until relatively adverse conditions occur. Preliminary indications are that the addition of a small amount of nitrogen through reactive sputtering has little or no effect in terms of barrier effectiveness; however, if enough is incorporated to form the

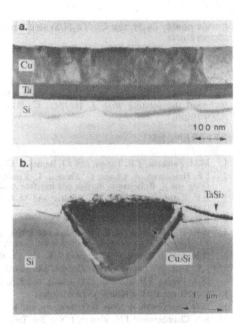

Figure 3 Cross-section TEM micrographs of (a) the unannealed Cu/Ta/Si film, and (b) that annealed to 750 °C. The central feature is a large η''-Cu$_3$Si precipitate, bounded by Si {111} planes. Note that there is a 100-150 nm thickness of silicon oxide under the Cu silicide.

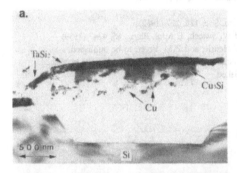

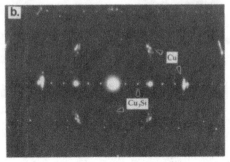

Figure 4 TEM micrograph (a) and corresponding electron diffraction pattern (b) demonstrating identification of η''-Cu$_3$Si and the initial stage of its oxidation.

nitride phase, Ta_2N, the $Cu/Ta_2N/Si$ structure is thermally stable to much higher temperatures.

ACKNOWLEDGEMENTS

The authors would like to acknowledge G. Coleman for the RBS spectra and T. Gallo for assistance with the resistivity furnace. Also, we benifitted from technical discussions with J. Harper, C. Ransom, A. Charai, L. Stolt, and O. Aboelfotoh.

REFERENCES

1. M.M. Farahani, T.E. Turner, and J.J. Barnes, J. Electrochem. Soc. **136**, 1484 (1989).

2. S.E. Hornstrom, A. Charai, O. Thomas, L. Krusin-Elbaum, P.M. Fryer, J.M.E. Harper, S. Gong and A. Robertsson, Surface and Interface Anal. **14**, 7 (1989).

3. H.P. Kattelus, E. Kolawa, K. Affolter, and M-A. Nicolet, J. Vac. Sci. Tecnhol. A3, 2246 (1985).

4. E.R. Weber, Appl. Phys. A, 1 (1983).

5. C-K. Hu, S. Chang, M.B. Small, and J.E. Lewis, in *Proceedings of the Third International VLSI Multilevel Interconnection Conference*, June 9, 1986, Santa Clara, CA.

6. H.M. Spit, D. Gupta, K.N. Tu, and C-K. Hu, private communication.

7. L.E. Toth, in *Transition Metal Carbides and Nitrides*, Academic Press, New York, 1971.

8. M.A. Farroq, S.P. Murarka, C.C. Chang, and F.A. Baiocchi, J. Appl. Phys. **65**, 3017 (1989).

9. L. Stolt and F.M. d'Heurle, to be published.

10. M.O. Aboelfotoh, A. Cros, B.G. Svensson, and K.N. Tu, to be published.

11. S.A. Chambers and J.H. Weaver, J. Vac. Sci. Technol. A3, 1929 (1985).

12. J. Bravman and R. Sinclair, J. Electron Microsc. Tech. 1, 53 (1984).

13. J.K. Solberg, Acta Cryst. A34, 684 (1978).

14. W.J. Ward and K.M. Carroll, J. Electrochem. Soc. **129**, 227 (1982).

15. M. Mundschau, E. Bauer, W. Telieps, and W. Swiech, J. Appl. Phys. **65**, 4747 (1989).

16. J.M.E. Harper, A. Charai, L. Stolt, F.M. d'Heurle, and P.M. Fryer, to be published.

17. K. Holloway, C. Ransom, E. Colgan, J. Gambino, P. Fryer, and R. Schad, to be published.

18. K. Holloway, C. Cabral, P. Fryer, unpublished.

CHEMICAL STABILITY OF VB$_2$ AND ZrB$_2$ WITH ALUMINUM

L.E. Halperin, E. Kolawa, Z. Fu, and M-A. Nicolet
California Institute of Technology, Pasadena, CA 91125

ABSTRACT

The chemical stability of boride thin films with aluminum is investigated. Only two diborides, VB$_2$ and HfB$_2$ have a positive heat of reaction which makes them potential candidates for thermodynamically stable diffusion barriers between Al and Si. Thin films of VB$_2$, and ZrB$_2$ for comparison, prepared by rf sputtering of composite targets were chosen for this study. Multilayer samples of these borides and aluminum were investigated by differential scanning calorimetry to determine if, according to calculations, a reaction between Al and the borides takes place, and to measure the heat of reaction. We find that an exothermic chemical reaction occurs between ZrB$_2$ and Al and that an exothermic crystallization reaction takes place in the VB$_2$ and Al sample. The reaction products were determined using X-ray diffraction.

1. INTRODUCTION

The potential application of diboride films as diffusion barriers in metallizations between Al and Si has previously been investigated [1-6]. The chemical stability of Al with VB$_2$, which has a positive heat of reaction, and Al with ZrB$_2$, which has a negative heat of reaction (Figure 1), are investigated using differential scanning calorimetry as we would like to experimentally confirm the thermodynamic predictions for these structures and check if thermodynamic stability with Al improves the diffusion barrier performance. Promising results with amorphous diffusion barriers and barriers constructed of ternary compounds [7,8] give further incentive for understanding the behavior of the boride films as diffusion barriers in relation to their predicted thermodynamic behavior based on calculated heats of formation. In this paper, we report the heat of enthalpy from experiments with VB$_2$/Al and ZrB$_2$/Al multilayer thin-film samples. The reaction products have been studied by X-ray diffraction. Applications of the films as diffusion barriers between Al and Si are also reported.

2. EXPERIMENTAL PROCEDURES

Thin films of VB$_2$ and ZrB$_2$ were deposited by rf sputtering on pieces of carbon tape covered by approximately 1 micron of Al to determine the film composition and thickness with Rutherford Backscattering Spectrometry (RBS). The aluminum was added to shift the carbon signal to lower energies in the backscattering spectra and allow the boron signal to be detected. The amount of boron present was determined by integrating the area of the elemental peaks, a procedure that is simple to use in spite of the presence of the two distinct ^{10}B and ^{11}B signals. Multilayer samples of Al and the diboride films were also prepared by rf

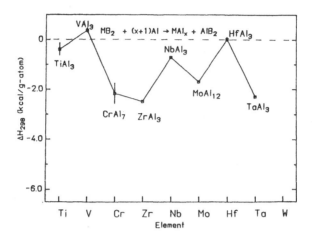

FIGURE 1 - Heat Of Reaction With Al For Various Diborides [1]

sputtering for the calorimetry experiments. A magnetron-type circular cathode, 7.5 cm in diameter, with a cathode-to-substrate spacing of 7 cm was used as the sputtering source. VB_2 and ZrB_2 films were sputtered in pure argon at 10 mTorr using a forward sputtering power of 300 W. The Al films were sputtered in pure argon at 5 mTorr using a forward sputtering power of 300 W and a dc target bias of -50 V. Sputter rates were determined by RBS to calculate film thickness.

The films were deposited on pieces of NaCl totalling 3 cm^2 and floated from the salt in deionized water. The samples were placed into the calorimeter to increase the mass of the sample being evaluated. The total mass of the VB_2 sample was 5 mg and of the ZrB_2 sample was 6 mg. The structure of each multilayer sample was approximately 660 nm of Al at the top and bottom of the sample for rigidity with six layers of boride, approximately 50 nm each, and five layers of Al, approximately 520 nm each, within each sample. The quoted thicknesses of the layers is based on sputter time and sputtering conditions and has been extrapolated from the deposition of single layers of Al and of the borides.

The calorimetry was performed with a Perkin-Elmer DSC-4 Differential Scanning Calorimeter. The calorimeter consists of two sample holders mounted independently in solid aluminum blocks which contain heater and sensor elements. The sample under investigation and a reference sample are placed into Al cups before being placed into the calorimeter. The cups were sealed with an Al lid but not made airtight. The calorimeter is programmed to perform a controlled increase in temperature from 50° C to 600° C at a rate of 20° C/minute. This temperature increase is performed while keeping the sample at the same temperature as the reference. The power required to maintain the sample at the same temperature as the reference is recorded. An increase or decrease in the power provided to maintain the sample at the programmed temperature indicates an endothermic or exothermic reaction. The heat of reaction is then recorded for the sample in millicalories per second. The heat of enthalpy is obtained by integrating the heat of reaction over time and is reported as normalized by the sample mass.

Concurrently with the calorimetric evaluation of the diboride and aluminum system, single layer samples were prepared on both Si/SiO$_2$ and Si substrates. The SiO$_2$ was grown reactively in an oxidation furnace on a <111> Si wafer. The diboride films were again prepared by an rf sputtering process as previously described but the aluminum was added later by an e-beam evaporation. Diboride films on Si/SiO$_2$ substrates, both with and without evaporated aluminum were vacuum annealed at 500°, 550°, 600°, and 650° C for 30 minutes. Annealing of the samples without aluminum was performed to measure the crystallization temperature. The samples with evaporated Al were annealed to investigate a chemical or thermodynamic reaction between the diboride and the Al films. Samples deposited on the Si substrate were also annealed at the same temperatures, except for 650° C, to evaluate the diboride films as a diffusion barrier between Al and Si.

Reaction products formed during the calorimetry were analyzed using a powder method x-ray diffractometer. The samples subjected to annealing were analyzed by RBS and x-ray diffraction using a Read camera. All available compounds of vanadium or zirconium with aluminum or boron as well as aluminum and boron were examined as possible by-products of a chemical reaction.

3. EXPERIMENTAL RESULTS AND DISCUSSION

The composition of the diboride films as determined varied during sample preparation as the exact composition of a sputtered film depends on sputtering conditions. For VB$_2$, the composition varied from a B:V ratio of 2:1 to 2.2:1 and for ZrB$_2$ it varied from a B:Zr ratio of 2.1:1 to 2.3:1. The variance of the ratios is within the ± 10% experimental accuracy of the RBS method. The as-deposited VB$_2$ film on silicon was x-ray amorphous using the Read camera while an as-deposited ZrB$_2$ film on silicon was crystalline.

Calorimetry showed an exothermic heat of reaction for both the VB$_2$ and ZrB$_2$ multilayered structures. This heat of reaction could be generated by either a chemical reaction or a phase transition such as the crystallization of an amorphous material. The vanadium diboride sample experienced a continuous exothermic reaction during calorimetry although an abrupt transition can be seen at about 475° C in the graph of heat flow versus temperature (Figure 2a). The total enthalpy of the reaction peak is -64 cal/g. The x-ray diffractometer shows peaks for Al and VB$_2$ meaning that VB$_2$ crystallized during calorimetry (Figure 3a). No peaks that would correspond to new compounds having formed are visible in the spectrum. Read camera diffraction patterns of a single film show that the film has crystallized after annealing for 30 minutes at 600° C.

The zirconium diboride sample also experienced a continuous exothermic reaction during calorimetry. It differs from the vanadium diboride sample as the curve displays no abrupt transition in the heat of reaction (Figure 2b). The total enthalpy of the reaction peak is -80 cal/g. The x-ray

50

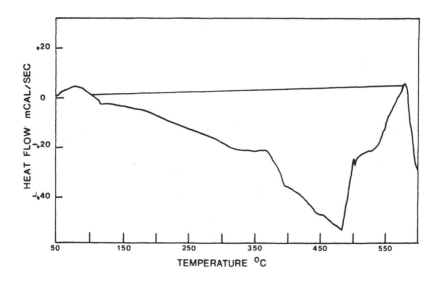

FIGURE 2a - Calorimetry For VB$_2$/Al Multilayer Structure

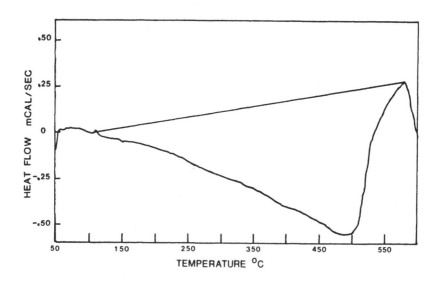

FIGURE 2b - Calorimetry For ZrB$_2$/Al Multilayer Structure

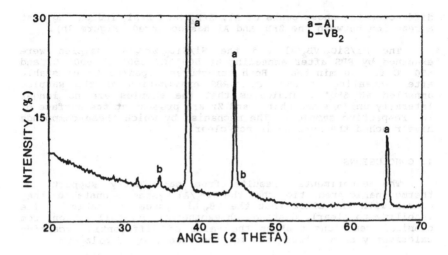

FIGURE 3a - X-Ray Diffractometer Peaks For VB$_2$/Al Calorimetry

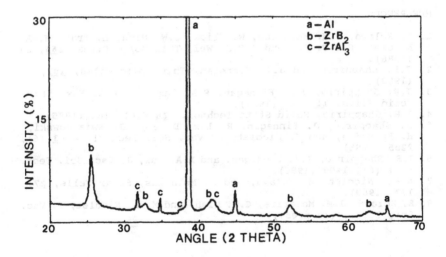

FIGURE 3b - X-Ray Diffractometer Peaks For ZrB$_2$/Al Calorimetry

diffractometer shows peaks of Al, ZrB_2, and $ZrAl_3$ indicating that a reaction between the ZrB_2 and Al has occurred (Figure 3b).

The $Si/SiO_2/VB_2/Al$ and the $Si/SiO_2/ZrB_2/Al$ samples were examined by RBS after annealing at 500° C, 550° C, 600° C, and 650° C for 30 minutes. Both structures appeared to be stable after annealing at 600° C. RBS examination of the samples annealed at 650° C indicates that the samples are no longer laterally uniform and that V and Zr are present at the surface of the respective samples. The mechanism by which these compounds have reached the surface is not clear.

4. CONCLUSIONS

The experimental results from calorimetry support the thermodynamic prediction that the ZrB_2/Al system is unstable; they concur with the claim that the VB_2/Al system is stable. The inability to clearly distinguish exothermic crystallization from chemical reactions weakens the value of differential scanning calorimetry as a litmus test for the stability of bilayers.

ACKNOWLEDGEMENTS

We thank Tom Workman for his help in sample preparation as well as Rob Gorris and Bart Stevens for their technical assistance. The financial support for this work was provided by the Army Research Office. The first author thanks the Amoco Foundation for the support of his graduate studies.

REFERENCES

1. E. Kolawa, J.M. Molarius, W. Flick, C.W. Nieh, L. Tran, M.-A. Nicolet, F.C.T. So, and J.C.S. Wei, Thin Solid Films, **166**, 29 (1988).
2. J.R. Shappirio and J.J. Finnegan, Thin Solid Films, **107**, 81 (1983).
3. J.R. Shappirio, J.J. Finnegan, R.A. Lux, and D.C. Fox, Thin Solid Films, **119**, 23 (1984).
4. J.R. Shappirio, Solid State Technol., **28** (10), 161 (1985).
5. J. Shappirio, J. Finnegan, R. Lux, D. Fox, J. Kwiatkowski, H. Kattelus, and M. Nicolet, J. Vac. Sci. Technol. A, **3** (6), 2255 (1985).
6. J.R. Shappirio, J.J. Finnegan, and R.A. Lux, J. Vac. Sci. Tech. B, **4** (6), 1409 (1986).
7. M.-A. Nicolet, E. Kolawa, and J. Molarius, Solar Cells, **27**, 177 (1989).
8. E. Kolawa, J.M. Molarius, C.W. Nieh, and M.-A. Nicolet, J. Vac. Sci. Tech., (in press).

PART II

Metallization Schemes for Interconnects, Wiring and Packaging

LOW TEMPERATURE PROCESSING FOR MULTILEVEL INTERCONNECTION AND PACKAGING

T.-M. LU, J. F. McDONALD, S. DABRAL, G.-R. YANG, L. YOU, AND P. BAI

Center for Integrated Electronics, Rensselaer Polytechnic Institute, Troy, NY 12180

ABSTRACT

The future high density multilevel interconnection and packaging requires that the combination of the insulator and conductor layers has a low RC value. Thermal stress and diffusion during processing are issues of great concern in the high density multilevel structures. The problem can be alleviated by a proper choice of materials and processes that do not require high temperature. In this paper we propose to use parylene and its derivatives (dielectric constant 2.3-2.6) as the possible interlayer dielectrics and Cu (bulk resistivity ~1.7 $\mu\Omega$-cm) as the conductor. Parylene can be vapor-deposited and cured at room temperature. The metallization of Cu has been achieved at room temperature using the newly developed partially ionized beam deposition technique. This technique has been shown to grow high quality metal films with low resistivity at low substrate temperatures. The interaction between Cu and parylene, including adhesion and diffusion, is also discussed.

I. INTRODUCTION

Thermal stress and diffusion are two of the most outstanding problems in fabricating thin film layered structures for multilevel interconnection and packaging. For many conventional materials it is necessary to process at high temperatures. For example, polyimides require high temperature curing and polymerization. Also, sputtered metal films very often require high temperature annealing to obtain good grain structure and conductivity. Because of the different thermal expansion coefficients in different layers of materials, stress is introduced. Stress can cause wafer bowing and cracking. It can also cause diffusion and degradation of the interfaces between the layers of the materials.

One possible solution is to look for a thermal matching between materials. However, in the future high density interconnection and packaging, the number of different materials and layers are likely to increase. It is very difficult, if not impossible, to obtain a good thermal matching between ALL the layers. The best way to solve the problem is perhaps to avoid high temperature excursions. In this paper, we describe a possible combination of materials and processing techniques which do not require the usual high temperature excursions. The main dielectric material we proposed to use is parylene (poly-p-xylylene) which has a dielectric constant ranging from 2.3 to 2.6 and can be cured at room temperature[1]. The conducting material is copper which has a resistivity of 1.7 $\mu\Omega$-cm. Copper is deposited using a low temperature processing technique known as the partially ionized beam (PIB) deposition technique[2,3]. Both parylene and Cu can be processed practically at room temperature and they represent a combination of materials which give one of the lowest possible RC values.

II. PARYLENE FOR DIELECTRIC LAYERS

Polymerization of parylene was introduced about four decades ago[4]. It has been used to encapsulate printed circuit boards for about two decades[5]. Vapor-deposited parylene films are pinhole free. They are highly resistant to corrosion and chemically inert to acids and solvents. Despite of the very attractive features, parylene has remained largely unexplored for its potential use as the interlayer dielectric material. High temperature thermal instability of these materials is perhaps the main concern.

They are four types of parylene, known as the normal (N-type), chlorinated (C-type), dichlorinated (D-type), and fluorinated (F-type) parylene. The N-type parylene has the structure

$$[CH_2\text{-}\langle\bigcirc\rangle\text{-}CH_2]n.$$

The C-type and D-type parylene contain one and two chlorine atoms on each aromatic ring in the polymer, respectively. In the F-type parylene, the CH_2 complex is replaced by CF_2. The dielectric constant ranges from 2.3 of the F-type parylene to 2.6 of the N-type parylene. Oxidative degradation occurs in the N-type parylene when it is heated in air at about 270°C. However, in nitrogen ambient, the degradation occurs at above 450°C. Figure 1a shows a plot of thermogravimetry measurements of the N-type parylene weight loss as a function of temperature both in air and in nitrogen ambient[6]. A much better thermostability has been found in the F-type parylene. Figure 1b shows the weight loss of the F-type parylene as a function of temperature[6]. The catastrophic weight loss occurs at temperatures as high as 500° in air and 530°C in nitrogen ambient. As far as the temperature is concerned, the F-type parylene is compatible with the conventional IC processing temperature. For low temperature processing, all types of parylene are viable candidates.

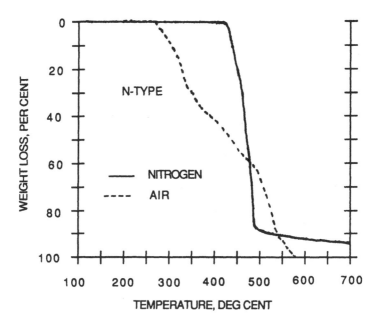

Fig. 1a. Thermogravimetry of N-type parylene at 10°C/min.

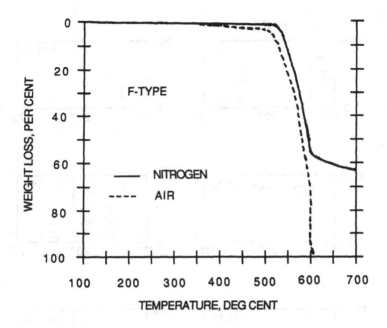

Fig. 1b. Thermogravimetry of F-type parylene at 10°C/min.

Parylene can be deposited from the vapor phase and is carried out in a vacuum chamber. This leads to an inherently cleaner polymerization process as compared to the more conventional processes that use solvents. The deposition of parylene involves two steps. Parylene is first sublimed at above 100°C. The vapor contains di-para-xylylene (dimers). The vapor is then pyrolyzed to form monomers at ~680°C. Polymerization and deposition on the substrate occur at room temperature or below. The parylene deposited at room temperature is polycrystalline having a monoclinic crystal structure. It is known as the alpha-phase. The crystallinity is reduced if deposited at below room temperatures. Polymerization can occur at a temperature as low as -196°C[7]. No deposition takes place on surfaces above 70°C.

A non-reversible structural change occurs when the parylene film is annealed to ~250°C [8,9]. The new phase has a hexagonal crystal structure and is known as the beta-phase. Figure 2 shows a sequence of the measured X-ray diffraction intensity as the deposited parylene samples were annealed in vacuum for 30 minutes at temperatures ranging from room temperature to 400°C. All the measurements were performed at room temperature. The beta-phase parylene is stable at room temperature.

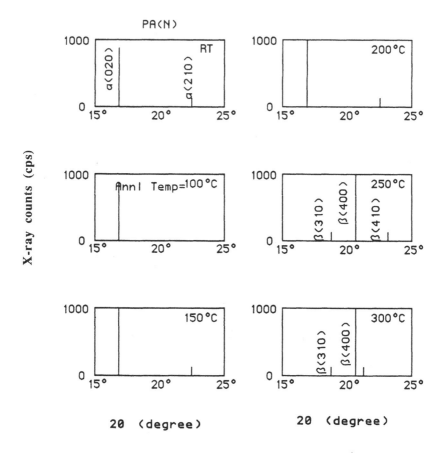

Fig. 2. X-ray intensity scans of N-type parylene annealed at 100, 150, 200, 250, and 300 °C.

Deposition of parylene from the vapor phase is conformal over any practical substrate structure. Reactive ion etching (RIE) techniques using oxygen plasma can be used to remove parylene[1,10]. Various masking techniques can be used to define fine structure patterns. The etch rate of parylene in oxygen plasma is comparable to that of polyimide. Figure 3 shows a comparison of the etch rates of N-type and C-type parylene as a function of power with that of polyimide at a fixed pressure of 50 mT[1]. The etched surface appears to be rough for both parylene and polyimide compared to the as-deposited surface.

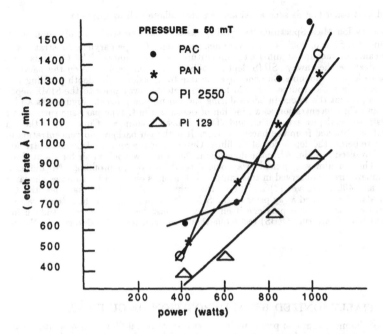

Fig. 3. The measured etch rate is plotted as a function of power with the oxygen pressure maintained at 50 mT.

III. CU METALLIZATION ON PARYLENE

The resistivity of bulk copper is ~1.7 $\mu\Omega$-cm. The only other metal that has a lower (by 5%) resistivity is Ag. Both Cu and Ag are attractive candidates for parylene metallization. Here we shall focus on Cu.

We have found that Cu adheres extremely well to the N-type parylene surface. Cu was deposited using both the conventional thermal evaporation and the partially ionized beam deposition technique. Pure parylene contains neither oxygen nor nitrogen. Oxygen concentration in the parylene films as a result of possible contamination in the deposition process has been studied using the electron probe technique. Within the detection limit (0.1%), we did not observe any oxygen impurity in the films. However, we do expect some oxygen contamination at the surface of the parylene films. The parylene films were left in air for many days before the metallization took place. Therefore chemical bonding through Cu-O-C complex is a possible explanation for the observed good adhesion. Another possible mechanism that leads to the good adhesion is the interaction of Cu d sub-orbitals with the the pi electron system of the benzene in a way similar to the Cr-polyimide interaction[11].

We found that other metals such as Al and Ag also adhere well to parylene.

Secondary Ion Mass Spectrometry (SIMS) has been used to study the diffusion of Cu in parylene. Cu films of 2000 Å thick were deposited onto C-type parylene substrates and were annealed in vacuum at different temperatures for 30 minutes. The Cu films were then chemically etched away. SIMS depth profiling was then performed on the parylene substrate to look for possible Cu concentration due to diffusion. The depth profiling was carried out using a Cs+ source. The Cu films were etched away prior to the SIMS depth profiling to prevent the knock-in effect during the ion milling process. Figure 4 is a plot of the C and Cu concentration as a function of depth in the C-type parylene surfaces for bare substrate, as-deposited, 350°C, and 400°C annealed samples. Curve (1) is the Cu concentration obtained from the bare substrate. It is the Cu background concentration in the substrate before the deposition of Cu films. Curve (2) is the measured Cu concentration of the as-deposited sample. After the deposition, the sample was left in air for about three months. The Cu film was etched away before the SIMS depth profiling. No diffusion (room temperature) is observed in this sample. Diffusion is observed for samples annealed at 350°C and 400°C (curves (3) and (4)). Note that for the annealed samples, the Cu films had also been etched away prior to the SIMS depth profiling. At these temperature, the structure of the C-type parylene itself becomes unstable. More recent study using Rutherford Back Scattering (RBS) technique shows no evidence of Cu diffusion in N-type parylene below 300°C.

IV. PARTIALLY IONIZED BEAM DEPOSITION OF CU FILMS

Two of the most common problems in the growth of metal films at low temperatures are high resistivity and columnar structures. Conventional deposition techniques at low substrate temperatures always produce thin film resistivity considerably higher than the bulk value because the deposited metal films are normally less dense and contain numerous structural defects. Post annealing reduces the defect density and thus improve the conductivity. This process, however, is not compatible with our proposed low temperature processing scheme.

Recently we have employed a newly developed technique, known as the partially ionized beam (PIB) deposition, to grow thin Cu films with a bulk-like resistivity at room temperature[12]. In this technique, the depositing beam contains a small amount of self-ions (0.1-2%). A bias potential of 0.5-3 kV is applied to the substrate during deposition to bombard the growing surface. The technique has been shown to possess a self-cleaning capability[3,13]. The impurity levels both at the interface and in the thin film are dramatically reduced. The impact of the self-ions on surface in the PIB deposition technique also enhances the surface mobility of adatoms during growth. Denser and more oriented thin films with a better grain structure are obtained at low substrate temperatures[14].

Figure 5 is a plot of the resistivity of the as-deposited Cu films on the SiO_2 surface as a function of the percentage self-ions in the vapor during deposition. The substrate bias was fixed at 2 kV. The deposition was performed at room temperature and the films were 2500 Å thick. When no ions are employed (zero ion percentage case), the resistivity of the Cu film is close to 4 $\mu\Omega$-cm. The resistivity reduces when self-ions are employed during deposition. It reaches a minimum value of ~1.8 $\mu\Omega$-cm when the ion percentage in the beam increases to close to 1%. If foreign ions such as Ar+ are used during deposition, the resistivity would increase monotonically as a function of the ion percentage due to inert gas incorporation[15]. The reduction of resistivity is unique to the PIB technique. The reduction of resistivity is believed to be due to the reduction of impurities residing in the Cu grain boundaries and improvement in the grain orientation. Further increase of the ion percentage to larger than 1% would create extensive ion-induced defects in the films and the resistivity would then increase.

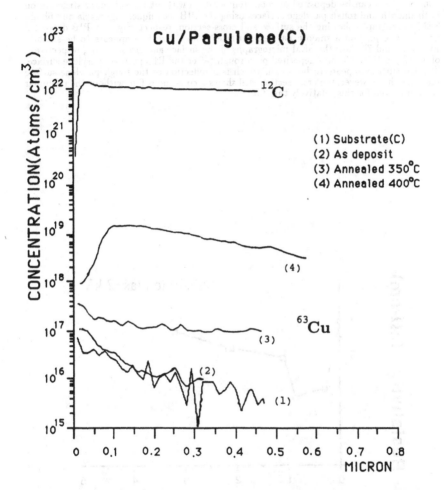

Fig. 4. The concentration of Cu in the parylene substrates measured using SIMS is plotted as a function of depth for different annealing temperature. The Cu films have been etched away before the SIMS depth profiling.

The PIB technique can also help to eliminate thin film columnar structures at low substrate temperatures[1,16]. The columnar growth becomes more pronounced when the substrate surface is rough, such as the etched parylene or polyimide surfaces. It has been shown that the PIB deposition of Al on rough polyimide surface at a low substrate temperature of ~120°C has no columnar structure whereas the conventional evaporation at the same deposition conditions results in columnar growth. In the Cu/parylene case, continuous Cu films can be deposited at room temperature without the columnar structure on both smooth and rough parylene surfaces using the PIB technique. Figures 6a and 6b are SEM micrographs showing the surface and cross-section, respectively, of a PIB deposited Cu film (~4 μm) on a smooth parylene surface. The Cu surface appears to be smooth. Figure 7a and 7b show the SEM micrographs of the surface and cross-section, respectively, of a 1/2 μm thick Cu film deposited on a rough (after the RIE process) parylene surface. The Cu surface appears to be rough, which is a reflection of the rough parylene surface underneath. However, the cross-section still shows a continuous film without the columnar structure even for this relatively thin film.

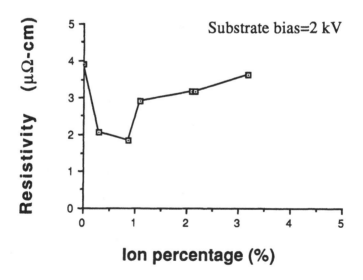

Fig. 5. The resistivity of the PIB deposited Cu films is plotted as a function of ion percentage at a fixed bias potential of 2 kV.

—— $1\,\mu m$

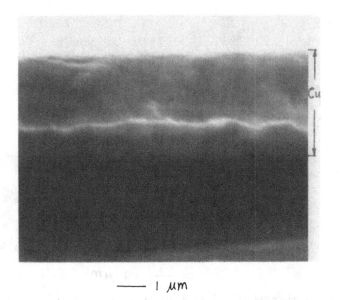

—— $1\,\mu m$

Fig. 6. The SEM micrographs showing the top view (a) and cross-section (b) of the PIB deposited Cu film on the smooth (unetched) parylene surface. The Cu film is 4 um thick.

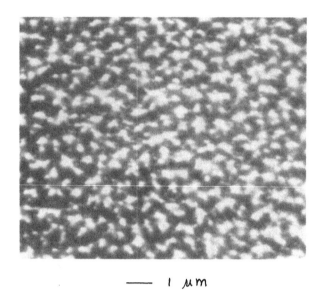

— I μm

— I μm

Fig. 7. The SEM micrographs showing the top view (a) and cross-section (b) of the PIB deposited Cu film on the rough (etched in an oxygen plasma) parylene surface. The Cu film is 1/2 μm thick.

V. SUMMARY

We have studied a low temperature metallization processing scheme using parylene and its derivatives as the dielectric layers and Cu as the conductor. The combination possesses one of the lowest RC values among the combinations of the interconnect and packaging materials. The adhesion of Cu to parylene appears to be excellent. No room temperature or near room temperature diffusion of Cu in parylene has been observed. The partially ionized beam deposition technique has been used to deposit bulk-like conductivity Cu films at room temperature. The Cu films are continuous without the usual columnar structure.

REFERENCES

[1] For a review, see N. Majid, S. Dabral, and J. F. Mcdonald, J. Electronic Materials, 18, 301 (1989).
[2] S.-N. Mei and T.-M. Lu, J. Vac. Sci. Technol. A6, 9 (1988).
[3] G.-R. Yang, P. Bai, T.-M. Lu, and W. M. Lau, J. Appl. Phys. 66, 4519 (1989).
[4] C. J. Brown and A. C. Farthing, J. Chem. Soc. 3270 (1953).
[5] W. F. Gorham and J. T. C. Yeh, J. Organic Chem. 34, 2366 (1965); S. W. Chow, L.A. Pilato and W. L. Wheelwright, J. Organic Chem. 35, 20 (1966).
[6] B. L. Joesten, J. Appl. Polym. Sci. 18, 439 (1974).
[7] S. Kubo and B. Wunderlich, J. Polym. Sci. 10, 1949 (1972).
[8] W. D. Niegisch, J. Appl. Phys. 37, 4041 (1966).
[9] S. Kubo and B. Wunderlich, J. Appl. Phys. 42, 4565 (1971).
[10] J. T. C. Yeh and k. R. Grebe, J. Vac. Sci. Technol. A1, 604 (1983).
[11] R. Haight, R. C. White, B. D. Silverman, and P. S. Ho, J. Vac. Sci. Technol. A6, 2188 (1988).
[12] P. Bai, G.-R. Yang, and T.-M. Lu, Appl. Phys. Lett. 56, 198 (1990).
[13] A. S. Yapsir, T.-M. Lu, and W. A. Lanford, Appl. Phys. Lett. 52, 1962 (1988).
[14] A. S. Yapsir, L. You, T.-M. Lu, and M. Madden, J. Mater. Res. 4, 343 (1989).
[15] R. A. Roy , J. J. Cuomo, and D. S. Yee, J. Vac. Sci. Technol. A6, 1621 (1988).
[16] R. Selvaraj, S.-N. Yang, J. F. McDonald, and T.-M. Lu, Proc. IEEE VLSI Multilevel Interconnection Conference (Electron Devices Society, New York, 1987), p. 440.

ANNEALING OF AMORPHOUS Ni-Nb/Cu OVERLAYER FILMS - THERMAL GROOVE KINETICS.

S.N. Farrens ,Univ. of Calif.-Davis, Mech. Engr. Dept., Davis, CA 95616; and J. H. Perepezko, Univ. of Wisc.-Madison, Dept. of Mat'ls. Sci. and Engr., 1509 University Ave., Madison, WI 53706.

1. ABSTRACT

Recent studies have shown that sputter deposited amorphous Ni_xNb_{1-x} ($60 \leq x \leq 75$) alloy films are effective diffusion barriers for operation at temperatures up to 600 °C[1]. The thermal stability of the amorphous films has been established on Si and GaAs substrates and may be modified by interactions with polycrystalline metal overlayer films. In the current work a detailed microstructural examination by cross-sectional TEM was performed on the annealing behavior of 50 nm polycrystalline Cu films on amorphous $Ni_{60}Nb_{40}$. Specifically, the development of grooves at the intersection of grain boundaries in the Cu film with both the free surface and the amorphous $Ni_{60}Nb_{40}$ base was monitored to evaluate the low temperature interfacial diffusion processes. Based upon measurements of the groove dimensions and dihedral angle during vacuum annealing at 450 °C, the surface diffusion coefficient of Cu was determined as 7.4×10^{-20} m^2/sec and the surface energy for Cu derived from groove kinetics was in good agreement with other reported determinations. In addition, the grain boundary grooves that were present at the Cu/amorphous $Ni_{60}Nb_{40}$ interface in the as-deposited condition were observed to be eliminated during annealing. Again, by following the kinetics of groove healing the interfacial diffusivity was determined to be of the order of 10^{-28} m^2/sec at 500 °C and the interfacial energy for the amorphous film/Cu surface was estimated to be ≥ 340 mJ/cm^2. The benefits of utilizing amorphous Ni-Nb alloys as underlays to retard thermal grooving and electromigration failure in copper are discussed.

2. INTRODUCTION

Two of the principal mechanisms responsible for failures in integrated circuits are substrate interdiffusion and electromigration. Some of the recent advances in the development of reliable diffusion barrier materials have centered on amorphous alloy thin films. These alloys have been found to be resistant to diffusion process up to temperatures of 600 °C for various combinations of substrate and overlayer materials. Recently, these alloys have also been found to decrease thermal groove formation in Cu at the amorphous alloy/copper interface [1]. The result indicates that in addition to serving as diffusion barriers the amorphous Ni-Nb alloy may reduce the time to failure due to electromigration effects.

3. METHODS

Transmission electron microscopy was used to study groove formation in Cu at the vacuum and the alloy interface. The amorphous $Ni_{60}Nb_{40}$ alloys were DC magnetron sputtered under high vacuum conditions in a chamber equipped with an in-situ evaporator. This allowed the Cu to be evaporated without breaking vacuum and ensured a clean interface. The samples were deposited onto Si<111> substrates and annealed in an ion pumped vacuum furnace. Typical annealing pressure ranged from 1.3×10^{-6} to 1.3×10^{-5} Pa. The thin film samples were prepared for examination in the TEM using a standard thin film cross-sectioning technique. The cross-sectional method allows for

A. Geometrical Definitions   B. Energy Balance

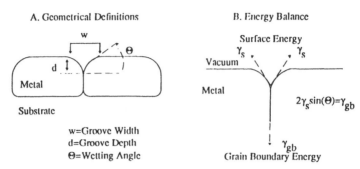

w=Groove Width
d=Groove Depth
Θ=Wetting Angle

Figure 1. A. Groove Geometry B. Energy Balance

simultaneous examination of both the copper/vacuum interface and the copper/alloy interface.

4. BACKGROUND

Thermal groove formation or decay is the physical response of a material to achieve interfacial equilibrium. The development of the groove geometry is determined by the surface and grain boundary energy balance at the triple point where two vacuum-solid interfaces intersect a grain boundary termination at the surface and the kinetics of surface and volume diffusion. Figure 1, is a schematic representation of a groove generated at a grain boundary-surface intersection.

A thermal groove is described by the groove width (w), the groove depth (d), and the equilibrium dihedral angle (ϕ). The equilibrium dihedral angle ($\phi=180-2\Theta$) is determined by the wetting condition (also shown in the figure) and is necessarily different at the surface interface and the alloy interface.

The groove depth and width relationships, as determined by Mullins[2], are given in equations (1) and (2). These geometric factors are expressed in terms of the surface diffusion coefficient, the groove curvature, and physical constants. Finite grain size considerations [3] were not included in initial analysis.

$$W = 4.6 \, (Bt)^{1/4} \tag{1}$$

$$d = 0.973 \, M \, (Bt)^{1/4} \tag{2}$$

Where:

W=Groove Width, d= Groove Depth, M=slope at the groove root=tan(θ)
γ_s=surface energy, γ_{gb}=Grain Boundary Energy, B=$D_s \nu \gamma_s \Omega^2/kT$
D_s=Surface Diffusion Coefficient ν=Atomic Density, Ω=Atomic Volume
k=Boltzmann's Constant, T=Absolute Temperature

Experimentally, the groove depth or width is measured as a function of annealing time for several temperatures to determine if $t^{1/4}$ kinetics is obeyed. If the coefficient B(T) is Arrhenius the activation energy can be evaluated for the rate limiting atomic transport mechanism.

5. RESULTS

These experiments were quite unique because the copper/vacuum interface served as an internal standard which allowed comparison with bulk single crystal data. Figure 2 shows the as-deposited thin film layer of Cu on the amorphous $Ni_{60}Nb_{40}$ alloy. In the micrograph it is possible to observe thermal grooves at both the copper/vacuum and copper/alloy interface that formed during the deposition process. This micrograph is intentionally underfocused to cause dark fringes to appear and accent the copper/alloy interfacial grooves. The Cu layer is approximately 60 nm thick and the initial groove depths ranged from 5.0-7.5 nm.

6. ANALYSIS

6.1. -Copper/Vacuum Interface-

The initial groove geometry which developed during the deposition of the Cu during evaporation was in a non-equilibrium configuration. The initial population (not shown in the figure) of grooves was characterized by a bimodal distribution of groove angles centered around 107° and 145°. Using the expression for the grain boundary energy of Cu γ_{gb}=776-0.123 x T mJ/m^2 (671<T<889) where T in is Kelvin[4], and a value of the Cu surface energy of 1670 mJ/m^2[5] interfacial equilibrium predicts an equilibrium dihedral angle of about 156° at 450 °C and 157° at 550 C°. Therefore, the initial as-deposited groove configuration does not reflect equilibrium.

Figure 2. As-Deposited Microstructure of Cu overlayers on $Ni_{60}Nb_{40}$ Thin films.

Figure 3 is a histogram plot of the experimentally observed dihedral groove angles. These plots graphically illustrate the dynamic changes in the geometry of the groove populations. For all three annealing temperatures the number of grooves with non-equilibrium dihedral angles decreases and ultimately the population is centered between 150-160° . It is therefore, concluded that the Cu thin film was establishing interfacial equilibrium at the vacuum surface consistent with theoretical predictions.

The kinetics of the thermal groove growth is shown in Figure 4 as a plot of the change in the groove depth from an initial depth d_o versus time$^{1/4}$. The kinetic measurements at 450 °C were made on grooves which had dihedral angles of 144-146°. As predicted by theory a linear relationship appears to exist between the measured groove depth and time$^{1/4}$. Using the relationship given by equation 1 it was possible to determine a surface diffusion coefficient for Cu of 7.4×10^{-20} m^2/sec at 450 °C.

6.2. -COPPER/ALLOY INTERFACE-

The consistent agreement between classical theory and experimental data at the vacuum surface provided confirmation of the validity of the experimental approach and analysis. The thermal groove profile at the copper/amorphous alloy interface that was shown in Figure 2 in the as-deposited structure was found to smooth when the samples were annealed. The changes in the groove depth during annealing at 500 and 550 °C is plotted in Figure 4 and reveals an approximate fit to $t^{1/4}$ kinetics.

The decay of non-migrating symmetric grooves in this work was used to estimate an order of magnitude value of ~10^{-18} m^2/sec for the interfacial diffusivity between 500-550 °C. A more complete analysis of surface smoothing in thin films in terms of the analysis for bulk samples [6,7] is in progress. In addition, using the expression given above for the grain boundary energy and noting that the final configuration was approaching complete wetting of the amorphous alloy a limit for the minimum value of the interface energy was determined to be 340 mJ/m^2.

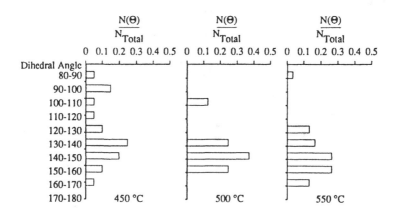

Figure 3. Dihedral Groove Angle in Annealed Sample Population

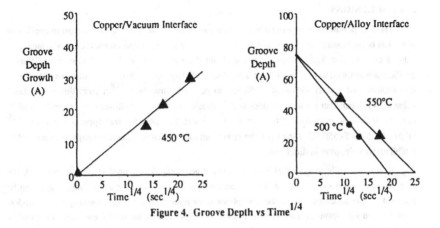

Figure 4. Groove Depth vs Time$^{1/4}$

7. DISCUSSION

These measurements of the copper/vacuum surface diffusion coefficient are the first quantitative measurements on thin films of copper or in multilayer structures. Because of the limited temperature range available for measurements with the amorphous material (anneals must be performed below crystallization temperature for the alloy, 650 °C/1hr for this material) it is not reasonable to extract a reliable measure of an activation energy for the surface diffusion kinetics from this data. The measured groove depths at all the temperatures when compared with groove depths predicted by the extrapolation of high temperature data are approximately a factor of 10-100 times smaller than expected. This is most likely a result of a change in the diffusion mechanism at the lower temperatures which can suppress diffusion[7,8]. However, the equilibrium groove angle as predicted by the wetting condition was approached during annealing.

Electromigration or metal migration in metal conductors occurs whenever the current density exceeds a critical value and metal ions begin to migrate[9]. This will typically occur at a narrowing or constriction point in the conducting line such as at a grain boundary leading to a flux divergence. Unfortunately, as the atoms migrate away from the defect the constriction increases causing a further enhancement of the current density and the process exhibits positive feedback. In these experiments there is evidence to support the hypothesis that the amorphous Ni-Nb alloy will cause groove healing in Cu and prevent necking of the metal conductor at grain boundary triple points. This will the help maintain a constant current density along the conductor and prevent the divergence of the flux which leads to electromigration.

In previous experiments on the diffusive behavior of Cu and amorphous alloys it has been shown that the interdiffusion of Cu and amorphous Ni-Nb alloys is negligible for anneals at 600 °C for up to 1 hour[10]. Thus it is possible to envision that a diffusion barrier and a protective coating of $Ni_{60}Nb_{40}$ could serve to pacify two major pathways for failure in metallizations; namely metal/substrate interdiffusion and electromigration. Further experiments are underway to explore typical metallizations systems such as Au or Al.

8. CONCLUSIONS

These experiments have extended the temperature over which the surface diffusion coefficient of Cu has been measured to as low as 450 °C. The copper surface diffusion coefficient was measured using a cross-section TEM technique and applying thermal groove analysis. The surface diffusion coefficient was found to be 7.4×10^{-20} m^2/sec at 450 °C. High temperature extrapolations of bulk studies estimate the surface diffusion coefficient to be approximately 10^{-18} m^2/sec; this extrapolated value does not consider possible changes in the diffusion mechanism at lower temperatures. In addition, the interfacial diffusion coefficients at the copper/alloy interface was approximately 1×10^{-18} m^2/sec for $500 < T < 550$ °C. Moreover, the groove smoothing that was observed suggests a method to inhibit electromigration in thin films.

It is worthwhile to note that with the sample geometry design used in the present work the assessment of interfacial equilibria between an amorphous film and a crystalline metal layer can be obtained during annealing. Since the amorphous layer represents a configurationally frozen, undercooled liquid this approach can be used to evaluate liquid-crystal interfacial energies in the highly undercooled state. Such measurements have not been possible previously, but are of critical importance in understanding crystal nucleation behavior. For the case of crystalline copper/amorphous $Ni_{60}Nb_{40}$ a lower bound to γ is 340 mJ/m2. This approach can be extended to a variety of crystal/metal interfaces.

The support of DOE under grant (DE-FG-84ER45096) is gratefully acknowledged.

9. REFERENCES

1. Sharon N. Farrens, Ph.D. thesis, University of Wisconsin-Madison, 1989.

2. W.W. Mullins, J. Appl. Phys. 28, 333 (1957).

3. S.A. Hackne and G.C Ojard, Scripta Met. 22, 1731 (1988).

4. G. Barreau, et al., Mem. Sci. Rev. Metall. 68, 357 (1971).

5. W.W. Mullins and P. G. Shewmon, Acta Metall. 7, 163 (1959).

6. R.T. King and W.W. Mullins, Acta Metall. 10, 601 (1962).

7. N.A. Gjostein, Trans. Metall. Soc. AIME, 236, 1267 (1966).

8. K.T. Miller, F.F. Lange, and D.B. Marshall, J. Mater. Res. 5, 151 (1990).

9. T.E. Dillinger, VLSI Engineering, (Prentice-Hall Publishers, Englewood Cliffs, New Jersey, 1988),p. 388.

10. R.E. Thomas, et al., Thin Solid Films, 150, 245 (1987).

CHEMICAL VAPOR DEPOSITION OF COPPER FROM AN ORGANOMETALLIC SOURCE

DAVID B. BEACH,* WILLIAM F. KANE,* FRANCOISE K. LEGOUES* AND CHRISTOPHER J. KNORS**
*IBM Research Division, Thomas J. Watson Research Center, P.O. Box 218, Yorktown Heights, NY 10598.
**IBM General Technology Division, East Fishkill, Hopewell Junction, NY, 12533.

ABSTRACT

High purity copper has been deposited from trialkyl phosphine complexes of cyclopentadienyl and methylcyclopentadienyl copper(I) by thermal chemical vapor deposition (CVD). Films as thick as 4.4 μm have been deposited at growth rates of up to 2000 Å/min with resistivites typically 2.0 $\mu\Omega$ cm, just slightly higher than bulk copper. Depositions were carried out at substrate temperatures between 150 and 220 °C on a variety of substrates including Si, SiO$_2$, polyimide, and Cr/Cu. At low substrate temperatures, copper film growth appears to show some selectivity for transition metal surfaces. An activation energy of 18 kcal/mole has been measured for film growth on Cu seeded substrates. CVD copper films have been characterized by Auger spectroscopy which showed that carbon and oxygen levels are below the limits of detection. Transmission electron microscopy revealed that the copper grain size was ~0.6μm and the grain boundaries are free of precipitates. Films show good conformality.

INTRODUCTION

The deposition of device quality copper by CVD has recently been reported using bis(1,1,1,5,5,5-hexafluoroacetylacetonato)copper(II), Cu(hfa)$_2$, reduced by hydrogen at temperatures between 250-350 °C.[1] We wish to report the preparation of high-purity films from organometallic sources, trialkylphosphine cyclolpentadienyl and methylcyclopentadienyl copper(I), at temperatures 100 °C lower than those used to deposit copper from Cu(hfa)$_2$. The structure of the triethylphosphine complex is shown below:[2]

The triethylphosphine derivative has been used for the CVD of GaCuS$_2$,[3] and by us to deposit copper using a laser.[4] In a preliminary account of this work, summarized in the Abstract, we described deposition of high purity copper from the trialkylphosphine cyclopentadienyl copper(I) complexes, a simple reactor, growth rate as a function of temperature studies, and transmission electron microscopic and Auger spectroscopic analysis of the copper films.[5] In this Paper, we report extension of this work to include the synthesis of new methylcyclopentadienyl derivatives, a more advanced reactor capable of metallizing 125 mm wafers, and studies of selectivity and conformality.

EXPERIMENTAL

The five trialkylphosphine cyclopentadienylcopper(I) complexes prepared and used for deposition in this study were the tributyl-, triethyl-, and trimethylphosphine complexes of cyclopentadienylcopper(I), and the trimethyl-

and triethylphosphine complexes of methylcyclopentadienylcopper(I). The syntheses of these compounds are outlined below:

$$MR + [CuXPR'_3]_4 \rightarrow RCuPR'_3 + MX$$

where: $M = Na, Tl, R = C_5H_5, CH_3C_5H_4,$

$X = Cl, I, Br,$ and $R' = CH_3, C_2H_5, C_4H_9.$

The general procedure is the same as that given by Cotton and Marks,[6] with the exception that diethyl ether or tetrahydrofuran was used as the reaction solvent when the reactants were sodium salts or trimethylphosphine complexes, owing to the limited solubilities of these reactants in hydrocarbon solvents. All of these compounds are white crystalline solids, easily purified by vacuum sublimation, with room temperature vapor pressures of $\sim$ 1mtorr

The reactor used in these studies is shown schematically in Figure 1. The main feature of this design is the use of a gas-conduction wafer chuck necessitated by the low deposition temperatures. The chuck is simply a resistively heated aluminum block with a number of gas passages drilled through the block. Helium gas at 5 torr pressure is allowed to slowly flow from the center of the chuck to the edges where it is exhausted through four pumping ports. Gas leakage was minimal and pressures in the deposition chamber of less than 1 mtorr were routinely obtained. Experiments without the use of the heat conduction gas indicated that the wafer heating was nonuniform, and that the wafer temperature was up to 35 °C cooler that the chuck temperature at 180 °C.

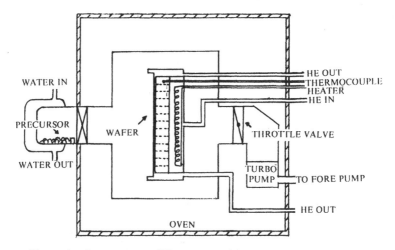

Figure 1. Low-pressure CVD reactor with gas conduction chuck.

In a typical experiment, the reactor was loaded with $\sim$5 grams of precursor, the oven was heated to 80 °C, and the precursor reservoir was warmed to 60 °C with water from a recirculating constant temperature bath to initiate transport of the precursor. Pressure was measured using an MKS Model 315 capacitance manometer which was also used to control the throttle valve on the pumping line. Typically, pressures between 5 and 30 mtorr were used. At a substrate temperature 185 °C, a 2 μm film could be deposited in 20 minutes.

RESULTS AND DISCUSSION

Methylcyclopentadienyl derivatives

Kaesz and coworkers[7] have recently shown that methyl substitution on the cyclopentadienyl ring of platinum cyclopentadienyl complexes increases their volatility. In an effort to determine if this a general trend in the chemistry of the late transition metals, we have prepared methylcyclopentadienyl copper(I) complexes of trimethyl- and triethylphosphine. These compounds are easily prepared by the reaction of sodium methylcyclopentadienide in tetrahydofuran (generated in situ) with a solution of trialkylphosphine-copper halide complex. The work-up and purification is the same as for the cyclopentadienyl compounds.

Unfortunately, the physical properties of the methylcyclopentadienyl complexes are not significantly different from those of the cyclopentadienyl complexes. Both compounds sublime at the same temperature (60-70 °C) as their nonmethylated counterparts, and, more importantly, require the same reservoir temperature to achieve the same transport rate. The chemical properties of these compounds are somewhat different, although they are still excellent sources of high purity copper. The methylcyclopentadienyl complexes were significantly less stable, and decompose 15-20 °C lower than the nonmethylated complexes. This would suggest that substitution of an electron releasing substituent on the cyclopentadienyl ring has the effect of strengthening the copper-carbon bonds at the expense of the copper-phosphorus bond. We have previously suggested[5] that the decomposition temperature of these compounds is governed by the strength of the bond to the 2 electron donor ligand. The methylcyclopentadienyl complexes are also much more reactive with air, implying that the apparent short-term stability of the nonmethylated complexes is related cyrstal packing.

Reactor Design

The improved reactor described above allowed the metallization of large wafers and demonstrated the overall feasibility of the process. Film properties using this reactor were identical to those obtained using the simple reactor, (no detectable carbon, oxygen, or phosphorus impurities, room temperature resistivity below 2 $\mu\Omega$ cm, ~0.6 μm grain size) as was the growth rate as a function of temperature. Unfortunately, this reactor also demonstrated major problems with reproducibility, probably stemming from persistence of the organophosphine ligand in the stainless steel reactor between depositions. After the first one or two depositions, the deposition rate dropped drastically, and could only be restored by complete cleaning of the reactor and replenishing of the source reservoir with fresh precursor. Engineering changes (load-lock, wall material) or chemical changes to the precursor will be required to overcome this problem.

Selectivity

As we have stated previously, these compounds appear to preferentially decompose on transition metal surfaces. In an effort to exploit this selectivity, we have conducted the following experiment: A Si wafer with a 5000 Å oxide layer was patterned to expose 1 μm trenches. The wafers were then exposed to neat tungsten hexafluoride at 400 °C to selectively deposit 1000 Å of tungsten. The wafers were then cooled to 175 °C and triethylphosphine cyclopentadienyl copper was introduced into the reactor and a 3000 Å of copper was grown. The results are shown in Figure 2. For the most part, the copper only deposited on the tungsten seed. The most unusual aspect of this experiment was the extensive growth of copper on the sidewalls where there was no seeding of tungsten. We believe that this may be indicative of high surface mobility of some reactive species that diffuses across the surface of the oxide until finding a reactive site, although we have no proof of such a mechanism.

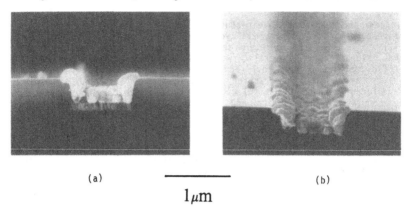

(a)   (b)

1 μm

Figure 2.   Selective CVD copper deposited on selective CVD tungsten on a patterned Si/SiO2 wafer: (a) cross section, (b) 60° tilt.

Conformality

The utility of any metal CVD process is the ability to conformally fill small dimension features. As a test of conformality, we have deposited copper on to a patterned Si/SiO2 wafer seeded with Cr/Cu (500 Å each, sputtered) at 180 °C into a 2 by 4 μm trench (Figure 3). Although the film tore when the wafer was cleaved and the overhang partially obscures the image, the excellent conformality can still be seen.

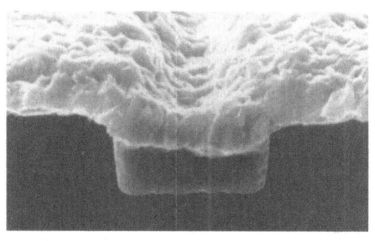

1 μm

Figure 3.   Conformal deposition of CVD copper on patterned Si/SiO2.

CONCLUSION

Deposition of high purity copper films from organometallic sources has been demonstrated. These films are conformal and show selective deposition on transition metal surfaces. A reactor for the deposition of copper films on large wafers has been developed although serious reproducibility problems have yet to be corrected.

REFERENCES

1. A. E. Kaloyeros, A. Feng, J. Garhart, K. C. Brooks, and F. Luehrs, J. Electron. Mater. 19, 271 (1990).

2. F. A. Cotton and J. Tatakas, J. Amer. Chem. Soc. 92, 2353 (1970).

3. K. Hara, T. Kojima and H. Kukimoto, Jpn. J. Appl. Phys. 26, L1107 (1987).

4. C. G. Dupuy, D. B. Beach, J. E. Hurst and J. M. Jasinski, Chem. of Mater. 1, 16 (1989).

5. D. B. Beach, F. K. LeGoues and C. K. Hu, Chem. of Mater. 2, 0000 (1990).

6. F. A. Cotton and T. J. Marks, J. Amer. Chem. Soc. 92, 5114 (1970).

7. X. Ziling, M. J. Strouse, D. K. Shuh, C. B. Knobler, H. D. Kaesz, R. F. Hicks and W. R. Williams, J. Amer. Chem. Soc. 111, 8779 (1989).

BLANKET AND SELECTIVE COPPER CVD FROM Cu(FOD)2 FOR MULTILEVEL METALLIZATION

Alain E. Kaloyeros,* Arjun N. Saxena,+ Kenneth Brooks,** Sumanta Ghosh,+ and Eric Einsenbraun*
*Physics Department, The University at Albany-SUNY, Albany, NY 12222
+Electrical, Computer, and Systems Engineering Department, Rensselaer Polytechnic Institute, Troy, NY 12084
**Chemistry Department, University of Illinois at Urbana-Champaign, Urbana, IL 61801

ABSTRACT

As the focus of integration technology inevitably shifts from the present very large scale integration (VLSI) to ultra large scale integration (ULSI) schemes, thus leading to continuous decrease in circuit dimensions, the limitations of present multilevel metallization technologies become increasingly important. Because of the appreciably higher speeds and more complex multi-functional layering involved in the newest ULSI circuits, the electrical resistance and capacitance of presently used interconnects and their electromigration and stress resistance stand as major limiting factors to signal processing throughput. In this paper, some recent results achieved by the present investigators in their studies of blanket and selective low-temperature metal-organic chemical vapor deposition (LTMOCVD) of copper for potential use in multilevel metallizations in ULSIC's are presented. The films were produced at 300-400°C in atmospheres of pure H_2 or Ar from the β-diketonate precursor bis(6,6,7,7,8,8,8-heptafluoro-2,2-dimethyl-3,5-octanediono)copper(II), Cu(fod)2. The films were analyzed by x-ray diffraction (XRD), Rutherford Backscattering (RBS), Auger electron spectroscopy (AES), scanning electron microscopy (SEM), energy-dispersive x-ray spectroscopy (EDXS), and four-point resistivity probe. The results of these studies showed that films deposited on metallic substrates were uniform, continuous, adherent, highly pure, and exhibited very low resistivity, as low as 1.8 $\mu\Omega$cm for films deposited in pure H_2 atmosphere. Preliminary investigations of selective LTMOCVD of copper showed that selectivity is indeed possible, but is a function of a wide range of parameters that include reactor geometry, substrate type and temperature, working pressure, type of carrier gas, and precursor chemistry.

INTRODUCTION

Aluminum has been widely used as interconnection and via filling metal in microelectronic circuits.[1] However, aluminum use poses severe problems as the focus of integration technology shifts from the present very large scale integration (VLSI) to ultra large scale integration (ULSI) schemes, thus leading to continuous decrease in circuit dimensions. Aluminum limitations include its low electromigration and stress resistance, relatively high electrical resistivity, and reliability problems (e.g., the Kirkendall effect).[2,3] However, aluminum use in ULSI for several interconnect layers will continue, although modifications in its deposition and cladding techniques will be necessary.[4] Metals other than aluminum are thus preferred for the other interconnect layers, e.g., tungsten for first layer metal and via filling, and copper for bus lines and ground planes.[5] Copper, because of its superior electromigration resistance and higher electrical conductivity, has been proposed as better alternative to aluminum for certain interconnect layers in emerging metallization schemes for microelectronic applications.[6,7,8,9,10,11,12,13,14,15]

This paper is the second in a series of reports by the present investigators on the growth of high quality copper films for multilevel metallization by low-temperature metal-organic chemical vapor deposition (LTMOCVD).[16] Results are presented for copper films produced at 350°C from the β-diketonate precursor bis(6,6,7,7,8,8,8-heptafluoro-2,2-dimethyl-3,5-octanediono)copper(II), Cu(fod)$_2$. The films were analyzed using Auger electron spectroscopy (AES), x-ray diffraction (XRD), Rutherford backscattering (RBS), energy-dispersive x-ray spectroscopy (EDAX), scanning electron microscopy (SEM), and four-point resistivity probe. The results of these analyses indicate that the copper films produced are very smooth and uniform, highly pure and have very low resistivity--close to that of OFHC copper. Preliminary investigations of selective LTMOCVD of copper showed that selectivity is indeed possible, but is a function of a wide range of parameters that include reactor geometry, substrate type and temperature, working pressure, type of carrier gas, and precursor chemistry.

EXPERIMENTAL PROCEDURE

A stainless-steel cold-wall type vertical CVD reactor coupled to a Pyrex glass sublimator was used for LTMOCVD of copper from the precursor bis(6,6,7,7,8,8,8-heptafluoro-2,2-dimethyl-3,5-octanediono)copper(II), Cu(fod)$_2$. The vaccum pumping system was based on a high speed turbomolecular pump capable of 10^{-9} torr pressure and 850 l/s pumping speed. An oil free sorption pump was used for the initial roughing down of the sublimator after introduction of the precursor. In a typical deposition run, the reactor was first pumped down to a base pressure of 5×10^{-7} or less. The source compound was then introduced into the sublimator which was heated to temperatures in the range 40-75°C. An electronic mass flow controller was employed to control the flow of the mixture carrier gas-precursor into the reactor. Deposition was carried out, using argon and hygrogen as carrier gases, on a variety of substrates which included aluminum, stainless steel, Pyrex, silicon, silicon dioxide, and various semiconductor test structures. The samples were heated to temperatures in the range 300-400°C and LTMOCVD was carried out at pressures ranging from 1 to 10 torr and gas flows in the range 30-55 sccm.

METHODS OF CHARACTERIZATION

X-ray diffraction was carried out using a Rigaku RAD-B diffractometer with a copper target. The primary voltage and current were set, respectively, at 45 kV and 20 mA. Measurements were performed with a scanning speed of 4°/min and with a sampling step of 0.02°. Auger spectra were recorded using a Physical Electronics PE595 instrument comprising a two-stage, retarding field/cylindrical mirror electron-energy analyzer, a coaxially contained electron gun with a beam normally incident on the sample surface and an electron multiplier. These components are contained in a demountable vacuum system evacuated to below 10^{-10} torr. The analyzer transmission and resolution were 10% and ~0.6% respectively for all data. A beam energy of 3 keV was used with a beam current of 50nA, adjusted to provide convenient measurement conditions, but kept low enough to avoid specimen damage. Scanning electron microscopy (SEM) measurements, employing a 20 keV primary electron beam, were performed using an ISI SMS.2.2 scanning electron microscope, while EDXS was performed using a LINK/Nucleus EX-2030 x-ray system. The primary voltage and current were set, respectively, at 16kV and 0.05 mA.

RESULTS AND DISCUSSION

Film Morphology

The films deposited on metallic substrates were of uniform thickness, continuous, adherent, and mirror-bright. The film thicknesses, determined by Rutherford backscattering, ranged from 0.5 to 1 μm.

XRD scans (figure 1) of blanket copper films on Pyrex showed that, in the 10°-75°C (2θ) region, the films were polycrystalline.

AES Results

The AES results were standardized using a bulk copper standard sample. All samples were sputter cleaned with a xenon ion beam before AES data collection to remove surface contaminants. The choice of a standard of composition and chemical environment similar to that of the films allowed high accuracy to be achieved in AES quantitative analysis. Our results are based on the expectation that any chamical or structural changes induced during the sputter cleaning process are basically the same in the standard and the films.

Figure 2 shows the atomic concentration profiles, after extensive sputtering of surface contaminants, of a film deposited in pure argon (fig.2A), and a film produced in pure hydrogen (fig. 2B). The results indicate the presence of high concentrations of carbon, oxygen and fluorine in the films produced in pure argon, even after extensive sputtering of surface contaminants. The corresponding atomic composition is 76% copper, 13% carbon, 8% oxygen, and 3% fluorine. The corresponding differentiated AES spectrum (figure 3A) gave signals for copper, carbon, oxygen, and fluorine.

The AES atomic concentration profile, on the other hand, of the LTMOCVD produced copper films using pure H2 as carrier gas show high oxygen and carbon contents near the films' surface, which

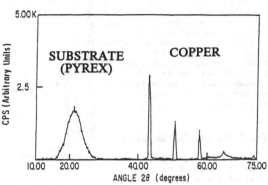

Figure 1   Typical XRD scan of an LTMOCVD-produced copper film.

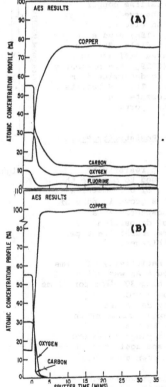

Figure 2   AES atomic concentrations profiles of Cu films produced in: (A) pure Ar, and (B) pure H2.

fall off to well below the detection limits of AES in the films, thus indicating that the C and O signals result only from surface contaminants. No fluorine was detected in the films. Quantitative AES atomic composition analysis gave ~100 at% copper. The corresponding differentiated AES spectrum gave only a copper signal (fig. 3B).

SEM of Blanket Cu Deposition

The SEM micrograph, taken at a magnification of 1200X for a copper film deposited on TiN in pure H$_2$, is shown in Figure 4. A granular texture, indicating a polycrystalline copper phase, is observed. Transmission electron microscopy (TEM) studies using both bright field (BF) and selected-area diffraction patterns (SADP), are presently under way to determine film grain size. The TEM results will be presented in a subsequent report.

Four-Point Resistivity Probe

A four-point resistivity probe was used to measure the resistivity of 0.5 μm-thick copper films grown in pure H$_2$ and Ar atmospheres as a function of deposition temperature, substrate type, and film thickness.

The resistivity of films grown in pure Ar was quite high, reaching 30 μΩcm for films grown below 250°C. The resistivity decreased to ~10μΩcm for Cu films grown in pure Ar in the temperature range 300-500°C, but increased again to 20μΩcm for films grown above 500°C.

Films grown in pure hydrogen exhibited a resistivity behavior similar to that of films grown in pure argon, but the resistivity was much lower: a resistivity

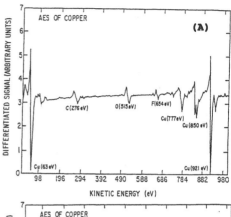

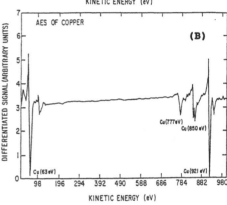

Figure 3   AES differentiated spectra of Cu films produced in: (A) pure Ar, and (B) pure H$_2$.

Figure 4   Typical SEM micrograph of LTMOCVD-produced Cu film.

of ~3 μΩcm was measured for temperatures below 250°C, which dropped off to ~1.8 μΩcm for films produced in the range 300-450°C, then started rising again for temperatures above 450°C.

In both cases (deposition in pure Ar and pure H_2), the higher resistivity at low deposition temperatures was attributed to structural defects (grain boundaries, impurities, vacancies...) due to the low deposition temperature. The increase in resistivity on the high end side was attributed to possible interdiffusion with the underlying substrate. Also, the higher resistivity values for the films grown in pure argon is due to the incorporation of carbon, oxygen, and fluorine impurities in the films.

Selective Deposition of Copper

Studies of selective copper deposition by LTMOCVD of $Cu(fod)_2$ were performed on various test structures as a function of several deposition parameters.

Our initial investigations seem to indicate that selectivity is highly dependent on a wide range of of parameters that include reactor geometry, substrate type and temperature, working pressure, type of carrier gas, and precursor chemistry. Nevertheless, our preliminary findings seem to indicate that selectivity is indeed possible in a narrow window of deposition parameters in a given reactor.

As an illustration of the possibility of selective LTMOCVD of copper, we present in figures 5A (1400X magnification) and 5B (2100X magnification) SEM micrographs of LTMOCVD-produced copper selectively filling holes in the test structure shown infigure 6. Cross-sectional SEM and TEM studies are presently underway to determine the uniformity and conformality of selective copper deposition.

(A)

(B)

Figure 5 SEM micrographs at (A) 1400X, (B) 2100X magnification of copper selectively filling holes in test structure of figure 6.

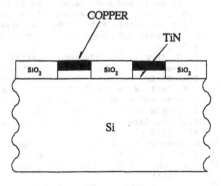

Figure 6 Test structure used in LTMOCVD of copper.

CONCLUSIONS

Preliminary results on blanket and selective low-temperature chemical vapor deposition (LTMOCVD) of copper for potential use in multilevel metallization in ULSIC were presented and summarized. The films were produced at 300-400°C in atmospheres of pure H_2 or Ar from the β-diketonate precursor bis(6,6,7,7,8,8,8-heptafluoro-2,2-dimethyl-3,5-octanediono)copper(II), $Cu(fod)_2$. The films were analyzed by x-ray diffraction (XRD), Rutherford Backscattering (RBS), Auger electron spectroscopy (AES), scanning electron microscopy (SEM), energy-dispersive x-ray spectroscopy (EDXS), and four-point resistivity probe. The results of these studies showed that films deposited on metallic substrates were uniform, continuous, adherent to metallic surfaces, highly pure, and exhibited very low resistivity, as low as 1.8 μΩcm for films deposited in pure H_2 atmosphere. Preliminary investigations of selective LTMOCVD of copper showed that selectivity is indeed possible, but is a function of a wide range of parameters that include reactor geometry, substrate type and temperature, working pressure, type of carrier gas, and precursor chemistry.

ACKNOWLEDGEMENTS

The work was partially supported by the New York State Sematech Center for Excellence in Semiconductor Research (SCOE) at RPI under grant# 88-MC-508 and the State University of New York at Albany under grant#450300. The authors are greatly indebted to Professor James Corbett of SUNY-Albany and Dr. Fred Luehrs of Jordan Valley for their invaluable advice and help, and to Dr. Dipankar Pramanik of VLSI Technology Inc. (San Jose, California) for providing TiN/SiO_2/Si samples. Thanks are due to Dr. Fabio Pintchovsky of Motorola (Austin, Texas) and to Professor Krishna Saraswat of Stanford University (Stanford, California) for providing preprints of some relevant papers.

REFERENCES

[1] See, for instance, **VLSI Technology: Fundamentals and Applications**, edited by Y. Tarui (Springer-Verlag, Berlin, 1986), p.308.

[2] A.N. Saxena, Keynote Address in the **Proceedings of the First International IEEE Conference on VLSI Multilevel Interconnections**, IEEE Catalog#84CH1999-2, June 21-22, 1984, p.1.

[3] A.N. Saxena, in **Proceedings of the Seminar on VLSI State-of-the Art Multilevel Interconnection**, (Santa Clara, CA, 1989) p.2.

[4] D. Pramanik and A.N. Saxena, Solid State Technology 33, 73 (1990).

[5] A.N. Saxena, Keynote Adress of **1st International Workshop on Tungsten and Other Refractory Metals for VLSI Applications** (MRS, Pittsburgh, 1986).

[6] E. K. Yung, L.T. Romankiw, and R.C. Alkire, J. Electrochem. Soc. 136, 206 (1989).

[7] W. C. Shumay, Jr. Advanced Materials Processes 135, 43 (1989).

[8] N. Awaya and Y. Arita, In the **Proceedings of the 1989 Symposium on VLSI Technology**, Digest of Technical Papers, 1989, to be published.

[9] P.M. Jeffries and G.S. Girolami, Chemistry of Materials 1, 8 (1989).

[10] C. Oehr and H. Suhr, Appl. Phys. A45, 151 (1988).

[11] D. Temple and A. Reisman, J. Electrochemical Soc. 136, 3525 (1989).

[12] R.L. Van Hemert, L.B. Spendlove, and R.E. Sievers, J. Electrochem. Soc. 112, 1123 (1965).

[13] C. Trundle and C. J. Brierley, Appl. Surf. Sci. 36, 102 (1989).

[14] S. Poston and A. Reisman, J. Electronic Materials 18, 79 (1989).

[15] C.R. Jones, F.A. Houle, C.A. Kovac, and T.H. Baum, Appl. Phys. Lett. 46, 97 (1985).

[16] A.E. Kaloyeros, A. Feng, J. Garhart, S. Ghosh, A. Saxena, and F. Luehrs, J. Electron. Mater. 19, 271 (1990).

A MICROSTRUCTURAL INVESTIGATION INTO THE EFFECT OF THE AMBIENT ATMOSPHERE ON CHROMIUM/POLYIMIDE INTERFACES

S.R. PEDDADA, I.M. ROBERTSON and H.K. BIRNBAUM
Dept. of Materials Science and Engineering and Materials Research Laboratory,
University of Illinois, Urbana, Illinois 61801.

ABSTRACT

This paper describes the effect of the ambient atmosphere (air, vacuum and deuterium) during thermal cycling on chromium/polyimide (Cr/PI) interfaces. Cross-section TEM has been employed to reveal the interface microstructure and morphology. The extent of interdiffusion across the metal/polyimide interface was determined by Energy Dispersive Spectroscopy in the STEM. The nature of the mechanical failure was identified by SEM and the locus of failure determined by Auger Electron Spectroscopy. It was observed that in the Cr/Cu/Cr/PI/Si system, tensile cracks form in the metal layers that originate at the Cr/PI interface after thermal cycling in any of the ambient atmospheres. Also, after annealing, the Cr/PI interface remains sharp, with no significant diffusion of Cr across the interface, irrespective of the ambient atmosphere.

INTRODUCTION

Polyimides are finding wide applications in the micro-electronic packaging industry as an insulator due to their low dielectric constant and ability to deform plastically. The adhesion at the metal/PI interface is very critical for the overall integrity of the structure. During processing, the metal/PI thin film laminates may be subjected to thermal cycling in various ambient atmospheres, some of which may be detrimental to interface adhesion.

The aim of this investigation is therefore, to understand the effect of the gaseous atmosphere during annealing at 593 K on microstructure and microchemistry of the Cr/PI interface in a Cr/Cu/Cr/PI/Si thin film multilayer structure.

EXPERIMENTAL PROCEDURE

Polyimides used in this study are of the PMDA-ODA type which were prepared on Si substrates. The thickness of the PI layers was approximately 10 micrometers. Metal films were deposited onto PI by physical vapor deposition at a substrate temperature of 450 K. The PI was bombarded with low energy ($\approx$1 keV) Ar^+ ions to sputter clean the surface prior to metallization. In the Cr/Cu/Cr/PI/Si multilayer structure, the layer thicknesses are (approximately): Si, 420 micrometers; PI, 10 micrometers; Cr, 20 nm; Cu, 250 nm; and Cr, 20 nm. These samples were annealed at 593 K for 48 hours in three different environments:
1. HT1 - deuterium gas(D_2), p_{D2} =101.33 kPa
2. HT2 - vacuum, p = 1X10^{-5} Pa
3. HT3 - air
These heat treated samples were characterized by using several analytical

techniques; SEM, TEM, Scanning Auger Electron Spectroscopy (SAES). By using XTEM it was possible to identify the interface structure and morphology. Energy Dispersive Spectroscopy (EDS) was performed in a STEM with a probe size of 1nm to determine the extent of diffusion of the metal into the PI. Mechanical failures, such as tensile cracks and decohesion, were identified by a SEM while the location of the failures was determined by SAES.

RESULTS AND DISCUSSIONS

Figure 1 shows the XTEM image of the thin film structure Cr/Cu/Cr/PI/Si in the as deposited condition. This micrograph shows that the Cr/PI interface is sharp. Chemical analysis by EDS reveals that there is no appreciable diffusion of Cr into PI in the as deposited condition. This suggests a very strong interaction between Cr and PI as concluded by several other investigators[1,2]. The SEM images indicate the absence of any mechanical failure in the as deposited condition.

Figure 2, a SEM image of a sample annealed in a D_2 atmosphere, indicates the formation of tensile cracks in the mutlilayer structure. Surveys by SAES inside these cracks indicate that the cracks reach the Cr/PI interface. The cracks form due to thermal stresses generated during thermal cycling. Similar cracks have been observed in samples annealed in air and in vacuum. Thus, annealing at 593 K leads to the formation of tensile cracks irrespective of the ambient atmosphere in the Cr/Cu/Cr/PI/Si multilayer structure.

Figures 3, 4 and 5 show the XTEM images of samples annealed in D_2 gas, vacuum and air, respectively. These images reveal that the Cr/PI interface is sharp in all three cases. Figures 6 and 7 show EDS profiles obtained from a region about 10 nm from the interface inside the PI and from the Cr layer, respectively. Comparison of the peak height of Cr in these spectra reveals that there is no significant diffusion of Cr into the PI (within the resolution of the EDS). Also, it can be seen that a 20 nm thick Cr film is effective in inhibiting extensive diffusion of Cu into PI. Previous studies[3,4] have shown that extensive diffusion of Cu into PI occurs in the absence of such a diffusion barrier. Although these profiles were taken from a sample annealed in the D_2 atmosphere, similar profiles were obtained from samples heat treated in vacuum and air.

CONCLUSIONS

1. There is no significant diffusion of Cr into PI in the as deposited condition.
2. Tensile cracks form (without decohesion occurring) on annealing at 593 K irrespective of the gaseous environment.
3. The Cr/PI interface remains sharp with no extensive diffusion of Cr into PI on annealing, irrespective of the ambient gas.
4. A 20 nm thick Cr film is an effective barrier between Cu and PI.

ACKNOWLEDGEMENTS

This work was supported by the D.O.E. through contract DOE AC02-76-ER01198. We would like to thank Dr Seshan of IBM for supplying the samples

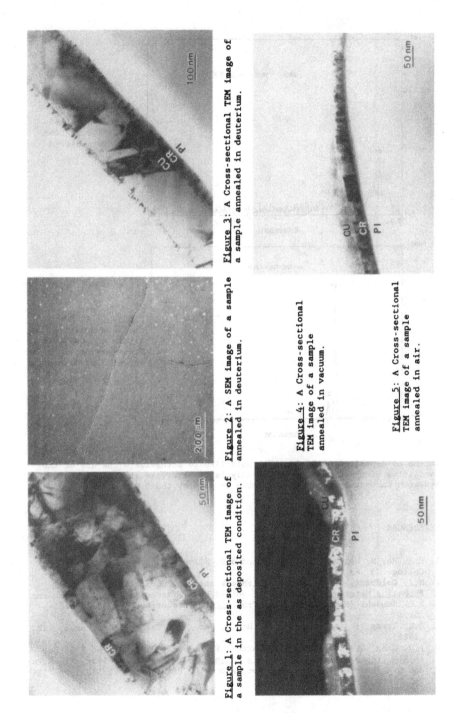

Figure 1: A Cross-sectional TEM image of a sample in the as deposited condition.

Figure 2: A SEM image of a sample annealed in deuterium.

Figure 3: A Cross-sectional TEM image of a sample annealed in deuterium.

Figure 4: A Cross-sectional TEM image of a sample annealed in vacuum.

Figure 5: A Cross-sectional TEM image of a sample annealed in air.

88

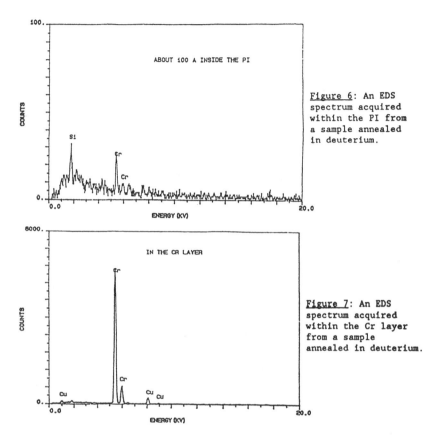

Figure 6: An EDS spectrum acquired within the PI from a sample annealed in deuterium.

Figure 7: An EDS spectrum acquired within the Cr layer from a sample annealed in deuterium.

in the as deposited condition. We would like to acknowledge the use of the facilities of the Center for Microanalysis of Materials at the Materials Research Laboratory which is supported in part by the D.O.E.

REFERENCES

1. P.S. Ho, R. Haight, R.C. White and B.D. Silverman, J. de Physique, Colloq. C5, 49 (1988).
2. M.J. Goldberg, J.C. Clabes, A Viehbeck and C.A. Kovac, in Electronic Packaging Materials Science III, edited by R. Jaccodine, K.A. Jackson and R.C. Sundahl (Mater. Res. Soc. Proc. 108, Pittsburgh, PA 1988) pp 225-232.
3. R.M. Tromp, F. Legoues and P.S. Ho, J. Vac. Sci. Technol A, 3, no.3, 782 (1985).
4. P.O. Hahn, G.W. Rubloff, J.W. Bartha, F. Legoues, R. Tromp and P.S. Ho, in Electronic Packaging Materials Science, edited by E.A. Giess, K. -N.Tu and D.R. Uhlman, (Mater. Res. Soc. Proc. 40, Pittsburgh, PA 1985) pp 251-263.

PART III

Silicides

IN-SITU TRANSMISSION ELECTRON MICROSCOPY OF THE FORMATION OF METAL-SEMICONDUCTOR CONTACTS

J.M. GIBSON, D. LORETTO* and D. CHERNS**
AT&T Bell Laboratories, 600 Mountain Avenue, Murray Hill, NJ 07974
* now at Max Planck Institut fur Metallforschung, Seestrasse 92, D-7000, Stuttgart, FRG.
** permanent address: University of Bristol, H. H. Wills Physics Laboratory, Royal Fort, Tyndall Ave, Bristol BS8 1TL, U.K.

ABSTRACT

We have studied the formation of metal silicides in-situ in an ultra-high vacuum transmission electron microscope. Metals were deposited on in-situ cleaned, reconstructed silicon surfaces and annealed. For the metals Ni and Co, we find that the phase sequence in ultra-thin films is different from that seen in $\approx$1000 Å thick films, and attribute this to the high surface-to-volume ratio. In general reactions occur at room temperature, to form an epitaxial phase if possible. We report preliminary new results on the formation of Pd_2Si.

INTRODUCTION

Metal silicides provide model systems for the study of solid-phase thin film reactions, yet have practical relevance in contacts and interconnects for integrated circuits. Although the growth of such silicides for practical purposes does not presently involve the use of ultra-high vacuums and atomically clean surfaces, these conditions do provide the best controlled experiments for understanding thin film reactions. Furthermore the silicide/silicon interface in real systems is often clean on the atomic scale because the growth of silicides almost invariably leads to the consumption of some Si from the substrate, moving the interface from the original contaminated surface. Results of studies on ultra-thin films under clean conditions may therefore be partly relevant to the interfacial regions of thicker reacted silicides.

From the initial work of Tung,[1] a large body of work has confirmed that very thin silicide growth on clean surfaces can give rise to unique film microstructure. In particular, epitaxial silicide growth is very dependent on the conditions of deposition and annealing in very thin films. For example, the "template" phenomenon, originally observed with $NiSi_2$ formation, has been found to be more widespread. A very thin low-temperature predeposited layer ("template") influences the microstructure of subsequently deposited silicide films. For example, $NiSi_2$ of two different orientations (A and B) can be grown dependent on the thickness of template used on Si(111).[1]

The "template" phenomenon occurs with room temperature deposits of order 10 Å of highly reactive material. Since the structural behavior of these deposits is clearly pivotal to their efficacy, in-situ microstructural characterization is required. Transmission electron microscopy is a very useful tool for such characterization, since it allows the study of crystallography and defects. For in-situ studies, it is imperative that relatively flat surfaces of extended area be used, mandating the plan-view specimen geometry. This is to ensure that results are typical of other thin film experiments. Recently, it has been shown that monolayers can be effectively studied by TEM,[2,3] even when buried below the surface, in the plan-view mode. In order to understand the results of Tung[1] and others, it is essential to have an ultra-high vacuum environment around the specimen and in-situ heating and evaporation capabilities.

In this paper we briefly describe experiments carried out with such a machine. It is a JEOL 200CX TEM which has been modified to achieve of order 10^{-9} torr pressure around

Mat. Res. Soc. Symp. Proc. Vol. 181. ©1990 Materials Research Society

the specimen.[4] The methods used to prepare and clean thin Si specimens and for metal evaporation are described elsewhere in detail.[3,4] The results of Ni, Co and Pd deposition are described in this paper. In general we find that the phase formation sequence in ultra-thin films, such as "templates", is entirely different from that seen before in thicker films, and that epitaxy is predominant because it generally lowers the interface and surface free energies. This fact can be used efficaciously, as in the template phenomenon, to improve the crystalline quality of thicker films and their interfaces.

NICKEL/SILICON REACTIONS

$NiSi_2$ and $CoSi_2$ have the cubic CaF_2 structure. The type-A orientation is aligned with the substrate, whereas type-B has 180° relative rotation about the <111> growth direction. A detailed account of the formation of $NiSi_2$ has been previously reported.[5] Briefly, we found that the epitaxial disilicide with type-B orientation appears to form immediately on room temperature deposition of < 10 Å Ni. This is consistent with more recent findings of Tung,[6] that type-B $NiSi_2$ can be grown by co-deposition on such thin Ni templates at room temperature. Our in-situ diffraction experiments showed that coverages of Ni in excess of approximately 10 Å at room temperature left disordered Ni on the surface of the type-B silicide layer. On mild annealing to about 300°C, this layer would convert partially to a metastable phase: $\theta–Ni_2Si$. This phase is hexagonal and has a very close lattice match to Si (111). It is stable in the bulk phase diagram only above 800°C. Under the right conditions, i.e. a template of thickness 16-20 Å Ni, the entire film converts to $\theta–Ni_2Si$. On subsequent higher temperature annealing (> 400°C) the $\theta–Ni_2Si$ phase converts to type-A $NiSi_2$. This fact was established by dynamic studies using dark-field imaging. At higher initial Ni coverages (> 25Å) the $\theta–Ni_2Si$ phase is not observed. Instead, the normal bulk-stable $\delta–Ni_2Si$ phase forms. This phase is orthorhombic and not epitaxial. After higher temperature annealing, a mixed A and B-type disilicide is formed. Additional insight into these phenomena is provided by in-situ studies of Ni on Si (100). Here, we found that for sufficiently thin templates the disilicide phase forms directly after mild annealing. However, for room-temperature deposited layers greater than 15 Å thickness, the stable $\delta–Ni_2Si$ phase forms almost explosively, as found in thicker multilayer Ni/Si films.[7] The $\theta–Ni_2Si$ phase is never observed.

These results can be simply explained. The nucleation of phases in very thin clean films is dramatically affected by epitaxy. As a result, for optimum conditions, templates can stabilize well-behaved microstructure even in thicker films. In these experiments, the major effect of templates is to control the nucleating phases. We have reason to believe that under other circumstances, an increases in the density of nucleation sites is another desired effect of templates.[8]

COBALT DISILICIDE FORMATION

Early experiments with our instrument showed that on (111) Si, uniform layers of $CoSi_2$ could be grown at low temperature (400°C) by room temperature deposition and annealing.[9] After higher temperature annealing we observed that pinhole nucleation occurred and led to the eventual unwetting of the substrate by the disilicide film. This is significant in the growth of epitaxial heterostructures for metal-base or permeable base transistor applications.[10] Recent studies have centered on Si (100) substrates, since it has been predicted that the band-structure properties are more suited for ballistic transport for films grown on this orientation.[11] We have found that the $CoSi_x$ phase sequence on Si (100) is also strongly thickness dependent, and that this impacts the final microstructure of thicker films. It was known that, in the absence of a predeposited template, $CoSi_2$ films grown by deposition of over 20 Å Co and annealing to 500°C are not epitaxial and contain grains of several orientations.[12] In contrast, Tung and Yalisove[13] showed that it is possible to grow very high quality films of one [(100) or two (100) and (110)] orientations using templates. Our in-situ experiments have shown that for a 5 Å template, annealing leads directly to the formation of epitaxial $CoSi_2$. In contrast, a 20 Å Co layer reacts to form the non-epitaxial Co_2Si phase, almost explosively, at 250°C. On higher temperature annealing, nucleation of several different epitaxial variants of $CoSi_2$ occurs. Figure 1 shows selected-area electron diffraction patterns of an as-deposited ≈ 20 Å Co, revealing polycrystalline Co (a), followed by ≈ 300°C annealing to the Co_2Si phase (b).

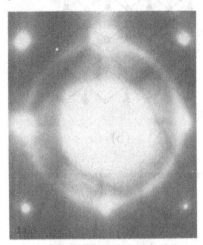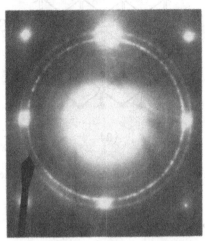

Fig. 1   Diffraction patterns from a 20 Å Co layer on Si (100) deposited in-situ: (a) at room temperature (b) annealed to 300°C;

This is another example of the dominance of epitaxy. In this case, as for Ni on Si (100), non-epitaxial phases can be *suppressed* for very low thickness films, with a dramatic improvement in the epitaxial quality of thicker silicide films grown on templates. Clearly, the "template" phenomenon can be used advantageously to stabilize metastable epitaxial phases and avoid the deleterious effects of non-epitaxial phases.

INTERFACE RECONSTRUCTION OF $CoSi_2$/Si (100)

We recently observed that the interface between Si and $CoSi_2$ (100) (and that for $NiSi_2$), exhibits a periodic reconstruction[14] which is expected to influence the electronic properties (e.g. Schottky barrier height). This reconstruction is seen in both plan-view diffraction patterns and high-resolution cross-sectional micrographs. We have found this reconstruction, which has a 2×1 unit cell, to be ubiquitous at this interface. Only the domain size depends on the preparation conditions. It is easier to observe this reconstruction in plan-view diffraction, because it is only visible in thicker regions of cross-section specimens, where steps are likely to average over 2×1 and 1×2 domains. Dark-field images in plan view have been successfully used to image domains.[14] Domain boundaries represent monatomic steps, and can therefore be used to identify these over large areas at this buried interface in plan-view. By quantitative analysis of both the diffraction data and the high resolution images, we have identified a likely structure for the interface reconstruction, which is an array of dimers similar to the Si (100) surface[14] (Fig. 2). The analysis relied heavily on the attractive feature of transmission diffraction from monolayers: that it is relatively kinematical so that conventional structure determination methods used in X-ray and neutron diffraction are valid.

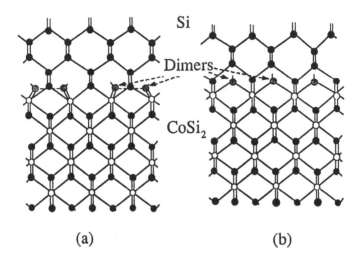

Fig. 2   A model of the CoSi$_2$/Si (2×1) interface reconstruction, seen in two orthogonal <110> projections (a) and (b).

PALLADIUM SILICIDE FORMATION

We have carried out preliminary experiments on the interaction of Pd and Si (111) at room temperature, using our in-situ microscope. In agreement with previous studies, we find that deposits of ≈ 10 Å lead to the growth of epitaxial Pd$_2$Si. However, we observe that this reaction is not instantaneous at room temperature. Fig. 3(a) shows that a disordered ring, possibly corresponding to fine grain polycrystalline Pd, appears initially on deposition. Within 100 seconds we observe the appearance of some diffraction spots characteristic of epitaxial hexagonal Pd$_2$Si (Fig. 3(b)). Within five minutes the layer has converted completely to the epitaxial silicide. Clearly the kinetics of Pd$_2$Si formation could be measured at room temperature. Dark-field images taken during the reaction allow the role of misfit dislocations in Pd$_2$Si formation to be analyzed. These experiments are, however, preliminary. For example, at this stage we are unsure of the effect of the high-energy electron beam.

DISCUSSION

The results presented here on metal-silicide reactions in very thin films show that the nucleation of phases can be a strong function of film thickness and orientation.[15] This we believe to be a simple consequence of the thickness dependence of the surface-to-volume ratio. Epitaxial phases would be expected to have relatively low interfacial (and surface) free energies, thus favoring them on equilibrium free energy grounds over non-epitaxial phases at low thickness. The use of very thin predeposited layers, "templates", has been shown to greatly influence silicide formation through this effect, because of the *suppression* of non-epitaxial phases. We show that this effect can be used by the crystal grower to his advantage. It is also likely that similar effects might occur in interfacial zones of thicker films, and thus affect phase nucleation and interfacial properties.

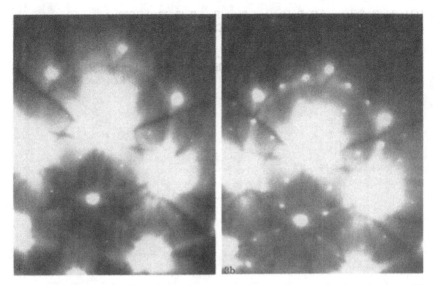

Fig. 3   Diffraction patterns observed at room temperature after the deposition of approx. 10
Å Pd on Si (111): (a) after 10s (b) after 100s.

We show that in-situ TEM is very well-suited for such in-situ studies. It has both monolayer sensitivity, and penetration power to study film microstructure and interfacial structure. It is not well appreciated that interfacial structure can be studied in plan-view geometry, but the example of the well-ordered interfacial reconstruction at the Si/CoSi$_2$ (100) interface demonstrates that well.

The author is grateful for the technical assistance of M. L. McDonald, F. C. Unterwald and D. Bahnck. The assistance of J. L. Batstone was also invaluable.

REFERENCES

[1]   for a review see R. T. Tung and J. M. Gibson, J. Vac. Sci. Technol. A **3**, 987 (1985).

[2]   K. Takayanagi, Y. Tanishiro, M. Takahashi and S. Takahashi, J. Vac. Sci. Technol. A **3**, 1502 (1985).

[3]   J. M. Gibson, in "Surface and Interface Characterization by Electron Optical Methods", ed. A. Howie and U. Valdre (Plenum, 1988).

[4]   M. L. McDonald, J. M. Gibson and F. C. Unterwald, Rev. Sci. Inst. **60**, 700 (1989).

[5]   J. M. Gibson and J. L. Batstone, Surf. Sci. **208**, 317 (1988).

[6]   R. T. Tung and F. Schrey, Appl. Phys. Lett. **55**, 256 (1989).

[7]   L. A. Clevenger, C. V. Thompson, R. C. Cammarata and K. N. Tu, Appl. Phys. Lett. **52**, 795 (1988).

[8]   R. T. Tung, J. M. Gibson and A. F. J. Levi, Appl. Phys. Lett. **48**, 1264 (1986).

[9]   J. M. Gibson, J. L. Batstone and R. T. Tung, Appl. Phys. Lett. **51**, 45 (1987).

[10]   R. T. Tung, A. F. J. Levi and J. M. Gibson, Appl. Phys. Lett. **48**, 635 (1986).

[11] M. D. Stiles and D. R. Hamann, Phys. Rev. B **40**, 1349 (1989).

[12] C. W. T. Bulle Lieuwma, A. H. vanOmmen and L. J. vanIjzendoom, Appl. Phys. Lett. **54**, 249 (1989).

[13] S. M. Yalisove, D. J. Eaglesham and R. T. Tung, Appl. Phys. Lett. **55**, 2075 (1989).

[14] D. Loretto, J. M. Gibson and S. M. Yalisove, Phys. Rev. Lett. **63**, 298 (1989).

[15] R. M. Nemanich, C. M. Doland and F. A. Ponce, Mat. Res. Soc. Proc. **94**, 139 (1987).

FORMATION OF COBALT SILICIDES IN ARSENIC IMPLANTED COBALT ON SILICON SYSTEM.

A.R. SITARAM * AND S.P. MURARKA
Materials Engineering Department/CIE, Rensselaer Polytechnic Institute, Troy, NY 12180.
*now with APRDL, Motorola Inc., 3501, Ed Bluestein Boulevard, Austin, TX 78721.

ABSTRACT

The importance of self aligned cobalt disilicide technology for gate and interconnection, and contact metallization cannot be overemphasized. Simultaneously, the concept of forming shallow junctions by using the metal or silicide layer as a dopant source is gaining prominence. In this work, we will present and discuss the results of the effect of arsenic, implanted into cobalt films on silicon, on the Co-Si reaction. Arsenic redistribution during the reaction, both during furnace annealing and RTA, and the effect of ion implantation and dose and energy will also be included.

INTRODUCTION

Of all the self aligned silicides (or salicides) available, $CoSi_2$ and $TiSi_2$ seem to be the most viable options for use in metallization [1,2]. While $TiSi_2$ has been studied extensively [3] in the past, certain problems, such as compound formation with impurities such as C,N, and O, and dopants, have recently been identified which raise questions about the maintenance of junction integrity during post metallization processing. Although the information regarding cobalt disilicide is not as comprehensive, it appears to suffer less from dopant and impurity interactions than $TiSi_2$ [2,4-6], which make it highly attractive for a self aligned metallization scheme.

The focus of current research in silicides appears to be moving towards utilizing the metal/silicide layer as a dopant source to form the junction [5,7]. This is advantageous, since it results in the formation of a junction conformal with the silicon/silicide interface which reduces the probability of junction shorts, results in low contact resistance due to high dopant concentration at the interface, and restricts implant damage to the silicide layer. Since arsenic is a popular choice for shallow junction formation, this study concentrated on the effect and behavior of arsenic on the reaction of cobalt with (100) silicon, when it was implanted into the cobalt layer.

EXPERIMENTAL

Silicon wafers, 100mm, (100), qnd n-type, with a resistivity of 5-50 Ω-cm were subjected to an RCA clean [8], a dilute HF dip, and an *in situ* backsputter etch. Approximately 675 Å of cobalt was then sputtered onto the wafers. The metal deposition was followed by implants of arsenic into the films. Boron and phosphorus were also implanted into the films for the sake of completeness of the study. TRIM [9], which has been found to simulate implants into cobalt with

Table I: Implant Conditions for Arsenic, Boron, and Phosphorus (from TRIM).

Specie	Matrix	Energy (keV)	Range Å	Straggle Å	Dose cm^{-2}	Vac/Ion	Co thickness (Å)
As	Si	30	265	94	3E15, 1E16 3E16	452	~1400
		100	713	242	3E15, 1E16 1E16	1314	
	Co	30	97	46	3E15, 3E16	471	~675
		100	261	117	3E15, 3E16	1410	
P	Co	25	147	78	3E15, 3E16	365	~675
		100	566	309	3E15, 3E16	1166	
B	Co	30	505	253	3E15, 3E16	278	~675

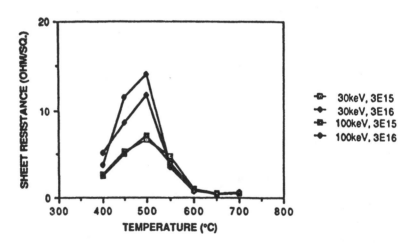

Figure 1. The variation of sheet resistance with temperature following 20 minute furnace anneals, for various arsenic implant conditions.

reasonable accuracy [10], was used to select implant energies. The range of energies available was equipment limited, and so boron was implanted at 30 keV, phosphorus at 25 and 100 keV, and arsenic at 30 and 100 keV. In the case of all three dopants, doses of 3E15 and 3E16 cm^{-2} were used. The details are presented in Table I. Samples from these wafers were subsequently annealed in a furnace, or rapid thermally annealed at various temperatures and for different times. The annealing ambient in all cases was argon containing 3% hydrogen. The reactions were monitored by sheet resistance measurements, RBS and x-ray diffraction. As a consequence of the differences in the resistivities of the three silicides of cobalt occurring during the reaction of Co with Si, it is possible to follow the interaction in a semi-quantitative fashion, with the maximum in the sheet resistance indicating the completion of the CoSi phase. This has been dealt with in detail elsewhere [11,12]. RBS provided the variation in the composition of the film and substrate with depth and, using simulation, provided phase identification. It was also used to study the redistribution of arsenic in a limited fashion. X-ray diffraction was used to confirm the phases identified by RBS.

RESULTS

Figures 1 and 2 show the variation of sheet resistance of the samples following furnace anneals of 20 and 80 minutes, respectively. From these, it is evident that a 20 minute anneal results in the specimens being mostly CoSi following a 500° C anneal. On increasing the anneal time to 80 minutes, the maxima indicating CoSi occurs after a 450° C anneal, and $CoSi_2$ formation begins around 500° C. The effect of increasing the implant dose from 3E15 to 3E16 cm^{-2} results in a slight retardation of the reaction.The variation of the implant depth does not appear to induce much of a change in the reaction rate, though the high dose, shallow implant consistently results in a higher sheet resistance than the remaining samples.

RBS was used to confirm the sequence of phase formation. It also revealed that arsenic accumulated at the top surface of the forming silicide layer at temperatures as low as 450° C after a 20 minute furnace anneal (Figure 3). As can be observed from Figure 4, the arsenic in the silicide layer was barely detectable after a 650° C anneal, implying considerable evaporative loss to the ambient. Similar results were obtained from RBS studies of samples annealed for 80 minutes.

All the samples containing CoSi had an uncharacteristically high sheet resistance. Since RBS revealed that the samples were CoSi, it was possible that implant damage resulted in a high sheet resistance. This, however, poses no problem, since implant damage in silicides has been shown to anneal out at high temperatures [13]. In all cases, complete disilicide formation took place by 600-650° C, depending on the duration of the anneal.

In order to compare the effect of different dopants on the Co-Si reaction, cobalt films implanted with boron and phosphorus were also studied. Table I lists the implant energies and the doses for each specie, while Figures 5 and 6 present the sheet resistance variation for each of the cases following a 60 minute anneal. Figure 5 shows the effect of a 3E15 cm^{-2} dose at all the energies, while Figure 6 does the same for a 3E16 cm^{-2} dose of the three species. It can be seen that the reaction proceeded fastest in the arsenic implanted films in all cases. In the instance of the lower dose, the reaction is not retarded to a great extent by boron or phosphorus. On increasing the dose, it can be seen that boron and phosphorus retard the reaction considerably, whereas it is virtually unimpeded by the presence of arsenic.

Samples with arsenic implanted into cobalt films were rapid thermally annealed in order to compare the reaction with the furnace anneal reactions. The variation of the sheet resistance with temperature following a 30 second anneal is shown in Figure 7. RBS studies were also performed on these samples, but are not presented due to space constraints. From these studies it was evident that the Co-Si reaction proceeded with little retardation due to arsenic, and was similar to earlier RTA studies performed by the same authors involving Co on lightly doped silicon, without the effects of high concentrations of dopants [14]. The chief difference between RTA and furnace anneals lay in the fact that the RTA reaction took seconds whereas the furnace anneal ran into tens of minutes. The dependence of the reaction on temperature did not differ with the method of annealing. As with the furnace anneals, RBS revealed that considerable arsenic accumulation occurred at the top surface of the film following 450° C anneals, and this was lost during anneals at higher temperatures. Also, the reaction proceeded faster with films implanted at 100 keV than with those implanted at 30 keV.

DISCUSSION AND CONCLUSIONS

As is evident from Table I, arsenic, by virtue of being the heaviest atom amongst the three dopants, inflicts the maximum damage on the cobalt film, and has the least penetration depth into it. It is also the slowest diffuser in cobalt disilicide [15]. Simultaneously, boron and phosphorus cause lesser damage, penetrate to greater depths and diffuse more rapidly in $CoSi_2$. The boron implants, along with the 100 keV phosphorus implants, also tail into the metal/silicide interface. Hence, one could expect a greater influence on the reaction due to these two dopants, whereas arsenic would tend to have a relatively smaller effect. This is in contrast to arsenic present in similar concentrations in silicon, which results in considerable retardation of the Co-Si interaction [12]. Its proximity to the top surface, low diffusivity in the film and high vapor pressure would result in high evaporative losses. That this is so is shown by the R_S and RBS data. The high resistivity observed in all the implanted films can be ascribed to implant damage in the metal/silicide films which anneal out at higher temperatures. RTA does not appear to be very different from furnace annealing, except that the reaction rates are considerably faster, and appear to show a greater dependency on the damage caused in the metal film.

Hence, in order to effectively use arsenic as a dopant and the silicide film as a dopant source, it is essential that a capping layer of nitride or LTO be used to restrict dopant losses. The dopant by itself, when implanted into the metal film, has little effect on the reaction. The reaction is highly reproducible, and is an excellent choice for self aligned silicides.

ACKNOWLEDGEMENTS

The authors would like to thank Dr. C.J. Chou of Phillips Laboratories for the implants into the cobalt films, Dr. X.S. Guo of SUNY-Albany for RBS work,

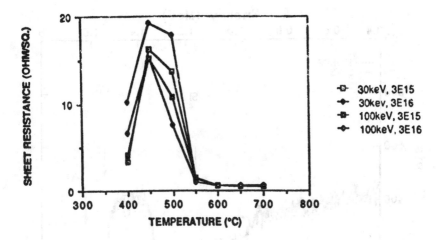

Figure 2. R$_S$ variation with temperature, for the different arsenic implants into cobalt, following 80 minute furnace anneals.

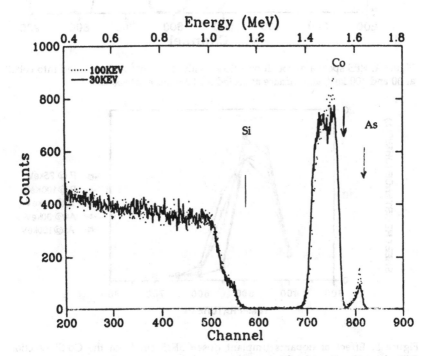

Figure 3. RBS spectra of cobalt on silicon, with 3E16 cm^{-2} As implanted into cobalt at 30 and 100 keV, and furnace annealed for 20 minutes at 450°C.

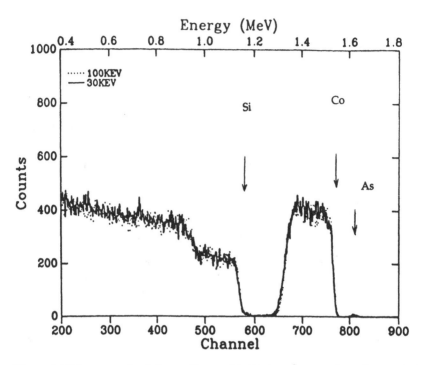

Figure 4. RBS spectra of cobalt on silicon, with 3E16 cm⁻² As implanted into cobalt at 30 and 100 keV, and furnace annealed for 20 minutes at 650°C.

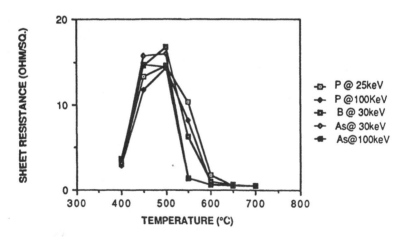

Figure 5. Effect of dopants (implant dose= 3E15 cm⁻²) on the Co-Si reaction following 60 minute furnace anneals.

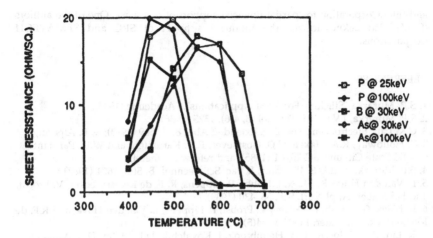

Figure 6. Effect of dopants (implant dose= 3E16 cm^{-2}) on the Co-Si reaction, following 60 minute furnace anneals.

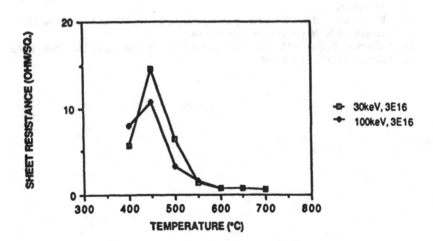

Figure 7. RTA of cobalt on silicon, with As implants into silicon (dose = 3E16 cm^{-2}) for 30 second anneals.

and Intel Corporation for partial financial support of this work. One of the authors (SPM) also acknowledges the financial support of SRC and SEMATECH Corporations.

REFERENCES

1. S.P. Murarka "Silicides For VLSI Applications", Academic (1983).
2. S.P. Murarka, J.Vac. Sci. Technol. B, 4(6), 1325 (1986).
3. C.K. Lau, Electrochem. Soc. Dig. Extended Abs., 83-1, 569 (1983); M.E. Alperin, T.C. Holloway, R.A. Haken, C.D. Gosmeyer, R.V. Karnaugh, and W.D. Parmantie, J. Solid State Circuits, SC-20, 61 (1985), and references therein.
4. S.P. Murarka and D.S. Williams, J. Vac. Sci. Technol. B, 5(6), 1674 (1987).
5. L. Van den Hove, K. Maex, L.Hobbs, P. Lippens, R. F. de Keersmaecker, V. Probst, and H.Schaber, Appl. Surf. Sci., 38, 430 (1989).
6. K. Maex, G. Ghosh, L. Delaey, V. Probst, P. Lippens, L. Van den Hove, and R.F. de Keersmaecker, J. Mater. Res. Soc., 4(5), 1209 (1989).
7. R. Liu, F.A. Baiocchi, L.A. Heimbrook, J. Kovalchik, D.L. Malm, D.S. Williams, and W.T. Lynch, in "ULSI Science and Technology 1987", eds. S. Broydo and C.M. Osburn, The Electrochemical Society, Pennington, NJ (1987); F.C. Shone, K.C. Saraswat, and J.D. Plummer, IEDM Tech. Dig., 407 (1985).
8. W. Kern, Solid State Technology, 15, 34 (1972).
9. J.P. Biersack and L.P. Hoggmark, Nucl. Instr. Methods, 174, 257 (1980).
10. M. Delfino, A.E. Morgan, P. Maillot, and E.K. Broadbent, J. Appl. Phys., 64(2), 607 (1988).
11. A. Appelbaum, R.V. Knoell, and S.P. Murarka, J. Appl. Phys., 57, 1880 (1985).
12. A.R. Sitaram, Ph.D. thesis, Rensselaer Polytechnic Institute, Troy, NY (1990).
13. Y. Shorimachi, H. Ishiwara, H. Yamamoto, and S. Furukawa, Jap. J. Appl. Phys., 21(5), 752 (1982).
14. A.R. Sitaram and S.P. Murarka, paper presented at the 174th Meeting of The Electrochemical Society, Chicago, October 1988.
15. O. Thomas, P. Gas, F.M. d'Huerle, F.K. LeGoues, A. Michel, and G. Scilla, J. Vac. Sci. Technol. A, 6(3), 1736 (1988).

INTERACTION OF Cu AND CoSi$_2$

C. L. Shepard and W. A. Lanford, Physics Dept., State University of New York, Albany, NY, 12222
Y-T. Shy and S. Murarka, Materials Engineering Dept., RPI, Troy, NY, 12180

ABSTRACT

Interaction of thin film Cu and CoSi$_2$ has been studied using Rutherford backscattering spectroscopy and sheet resistance measurements. For temperatures $\leq$ 600° C., no measurable diffusion of Cu into CoSi$_2$ is observed.

As the micro-electronic feature size decreases toward a 0.25 μm scale, the higher conductivity of copper compared to aluminum makes it increasingly attractive as the metal for interconnections. Simultaneously, cobalt disilicide, because of its low resistivity (in the range of 10 - 15 $\mu\Omega$-cm) and advantages of self-alignment in processing and reliability, is emerging for application both as the chip level interconnection and the contact material to shallow junctions[1]. Thus in future applications, one can envision the use of copper as upper level connections directly on CoSi$_2$ which is used as contacts, gates and interconnections on the chip level. It is therefore important to learn about the compatibility of copper and CoSi$_2$. In this study we have investigated the interaction between copper and CoSi$_2$ films using sheet resistance measurement and Rutherford backscattering (RBS) techniques. For copper on CoSi$_2$, at least up to 600°C for anneals carried out in argon containing 3% hydrogen, no detectable interaction is observed.

The CoSi$_2$ samples were prepared either by sputtering $\approx$ 400 A of Co onto single-crystal Si wafers and reacting to form CoSi$_2$, or by cosputtering Co and Si in the desired Si/Co ratio onto single crystal Si or polycrystalline Si. All samples were then subjected to a rapid thermal anneal (RTA) at 900°C for 1 minute in AG Associates 210 annealer in Ar/3% H$_2$ gas mixture. In this process Co reacted with Si to form $\approx$ 1400-1500 A CoSi$_2$. Computer simulations of the RBS spectra indicate a Si/Co atomic ratio of 2 to 2.1 for the silicides formed by Co + 2Si = CoSi$_2$ reaction (Fig. 1) and 2.1 to 2.2 for the silicide formed by cosputtering. RBS analysis also shows that these CoSi$_2$ on Si films are stable up to at least 600°C. Temperature was measured by a thermocouple located next to the wafer; the real temperature of the wafer is estimated to be within $\pm$ 10°C of values quoted in this work. The anneal temperature of 900° C is reached within 5 seconds but the cool-down time is on the order of 20 - 30 minutes. Following the silicide anneal, the surfaces were cleaned using dilute-buffered hydrofluoric acid followed by a deionized water rinse and spin drying.

Mat. Res. Soc. Symp. Proc. Vol. 181. ©1990 Materials Research Society

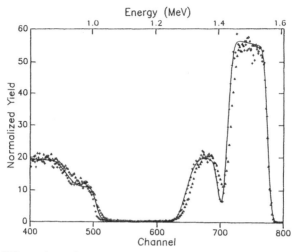

Figure 1 - RBS spectra of copper on reacted CoSi$_2$ on crystalline silicon. (+) - unannealed, (—) - computer simulation of unannealed spectrum, (△) - 600°C anneal. The simulation shows a Si/Co ratio of 2:1 for the first 1000 Å and decreasing Co concentration over the next 900 Å.

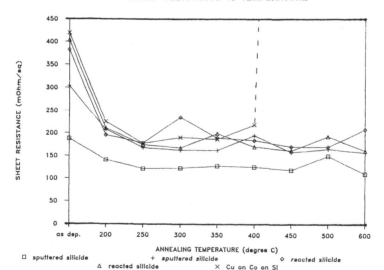

Figure 2 - Sheet Resistance vs Anneal Temperature - sheet resistance of the samples remains constant vs anneal temperatures indicating that Cu is not consumed by diffusion. For the sample that has Cu on Co film, the graph turns sharply upward at 400°C; the sheet resistance of this sample at 500°C is 4040 mΩ/square.

The CoSi$_2$ substrates then had $\approx$ 1250 A of Cu sputter-deposited onto the silicide layer. One piece of each wafer was reserved as a control both before and after the Cu deposition. It is estimated that the copper thickness varied across a sample or from sample to sample by $\pm$ 5%. These wafers were then cut into small pieces and the pieces subjected to 30 minute furnace anneals at temperatures ranging from 200°C to 600°C in 50°C intervals. All anneals were done in a gas consisting of 3% H$_2$, 97% Ar at a pressure of 1 atmosphere. An additional wafer was made by depositing the Cu directly on the Co metal film with no RTA done between layers. Hence, on this wafer no silicide was formed prior to copper deposition.

The sheet resistance was measured with a four point probe. In Figure 2 the sheet resistance is shown as a function of annealing temperature. We believe variations in layer thicknesses for different pieces of a sample account for the fluctuations in these results. The initial drop in resistance (at temperatures $\leq$ 200°C) shown in Figure 2 is known to be associated with the annealing of defects and release of trapped gases in the sputtered metallic films[2]. At higher temperatures all samples have the same resistance (within experimental error) except for the case of copper films on unreacted cobalt on silicon. In the latter case, resistance dramatically increases after 400°C anneal indicating copper is reacting with its substrate.

RBS, using 2 Mev He[+] ions, was used to study Cu-CoSi$_2$ interactions. Figure 1 compares samples with 1100-1200 A Cu on reacted CoSi$_2$ on crystalline silicon, both before furnace annealing and after a 600°C, 30 minute furnace anneal. The decrease in the dip between the Cu and Co peaks is not due to the difference in sample thickness. Such a decrease might be due to diffusion of Cu into CoSi$_2$ or increased roughness of the Cu film due to crystal growth during the anneal. However, the dip height does not correlate smoothly with annealing temperature. Therefore, in this case, and in all cases where no interactions were observed, the Cu film was selectively etched off the surface of the sample using dilute nitric acid; this acid does not etch the CoSi$_2$ at a significant rate. The etched surfaces were then carefully analyzed using RBS to see if any copper remained on the surface. By removing the metallic copper film, the sensitivity for detecting small amounts of copper which might diffuse into the CoSi$_2$ was enhanced by several orders of magnitude. Note: because Cu (mass = 64 amu) is only slightly heavier than Co (59 amu), their RBS signals tend to overlap. To test the sensitivity of this approach, we also made some measurements on samples incompletely etched in the nitric acid. From these measurements, we estimate that this method has a sensitivity of approximately 10^{15} Cu atoms/cm^2 within the first ≈ 500 Å of the surface.

Figure 3 shows an RBS spectrum for such an etched sample on a reacted silicide after 600°C/30 minute anneal. There is no evidence of copper on the surface of cobalt silicide. Similarly in Figure 4, which shows a RBS spectrum for a sample on cosputtered silicide made on polysilicon film. Again, there is no evidence of copper diffusion. The results of Figs. 1-4 clearly indicate that Cu and CoSi$_2$ do not significantly react below 600°C.

Figure 4 also shows a long tail on the low energy side of the Co peak. This feature was present on all samples (including unannealed ones) and is unaffected by the annealing. While we do not completely understand this feature, it is assumed to be due to the surface roughness of polycrystalline silicon and to grain boundary diffusion of Co into and through the polysilicon.

Finally, Figure 5 shows RBS spectra of the sample where copper was deposited on a cobalt film on silicon. In this case, reaction between copper and cobalt and silicon is detected even at 400°C and a complete catastrophic reaction is observed at 450°C. This is also seen in the resistance vs. temperature plots (Fig. 2). At 400°C - 450°C several interactions may be occurring at the same time. Cobalt is known to interact with silicon to form

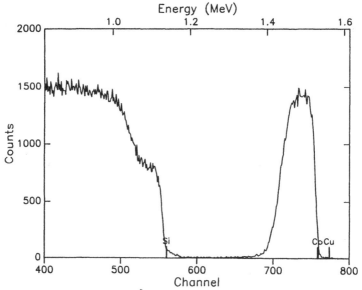

Figure 3 - RBS spectrum of 600°C annealed sample with reacted CoSi$_2$ on crystalline silicon after the Cu overlayer is etched off. Any Cu diffusion into the CoSi$_2$ layer should show at or below channel 778.

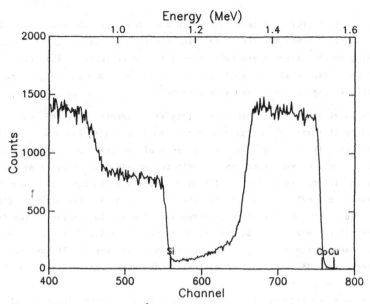

Figure 4 - RBS spectrum of 600°C annealed sample with cosputtered CoSi₂ on polycrystalline silicon after the Cu overlayer is etched off. Any Cu diffusion into the CoSi₂ layer should show at or below channel 778.

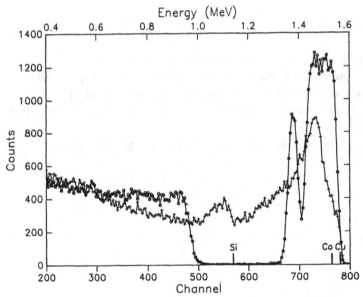

Figure 5 - RBS spectra of copper on cobalt film on crystalline silicon annealed at 400°C and 450°C. Element markers show the position of an element on the surface.

a cobalt rich silicide at temperatures in the range of 350-400°C. Also, since sputtered cobalt films are fine grain materials, copper may be diffusing quickly along grain boundaries into the silicon. The fact that cobalt can dissolve as much as 2 wt% of Cu at ≈ 305°C may be a way to explain the initial rise in the measured resistance[3].

In summary, both the sheet resistance measurements and RBS data show that up to at least 600°C copper does not react or interdiffuse with $CoSi_2$, and the $CoSi_2$ films are stable. The absence of interaction between copper and $CoSi_2$ could be very important if both copper and $CoSi_2$ find applications in silicon integrated circuits. The absence of Cu diffusion, in quantities detectable by RBS and sheet resistance techniques, cannot be simply attributed to the absence of a large density of grain boundaries in the $CoSi_2$ film on silicon where the possibility of epitaxy formation[4] is expected to lead to large grain size. This follows because the same result was obtained with copper on the cosputtered cobalt silicide (which has considerably lower grain size) on polysilicon films.

We are continuing to study copper-$CoSi_2$ interactions both at temperatures higher than 600°C and by using electrical methods which should be sensitive to the presence of very small quantities of Cu in silicon.

REFERENCES

1. S.P. Murarka, J. Vac. Sci. Technol. B4, 1325, (1986).

2. L.I. Maissel, in "Handbook of Thin Film Technology", Editors: L.I. Maissel and R. Glang, (McGraw-Hill, NY, 1970), p. 13-21.

3. M. Hansen, "Composition of Binary Alloys", (McGraw-Hill, NY, 1958), p. 469.

4. J.C. Bean and J.M. Poate, Appl. Phys. Lett. 37, 203 (1980)

Growth of monocrystalline $CoSi_2$ on $CoSi_2$ seeds in (100) Si

Karen Maex

Interuniversity Microelectronics Center (IMEC v.z.w.)
Kapeldreef 75, 3030 Leuven, Belgium

Abstract

Growth of monocrystalline $CoSi_2$ on top of (100) Si has been obtained by silicidation of deposited Co films on $Si/CoSi_2/Si$ heterostructures formed by ion implantation. The layers are of excellent crystalline quality and the $CoSi_2/Si$ interface is atomically flat. The required implanted dose can be reduced, since epitaxial silicidation is possible starting from aligned $CoSi_2$ precipitates in the Si.

Introduction

Since the introduction of metal silicides in IC fabrication technologies, the formation of $CoSi_2$ has received lots of attention. $CoSi_2$ is a very attractive silicide due its controlled formation by the reaction of Co thin films on Si and its good compatibility with MOS and bipolar technologies. Although the feasibility of $CoSi_2$ for use in full technologies has been demonstrated [1,2], its main disadvantage remains in its rather poor stability against high temperature treatments. Roughening of the polycrystalline silicide film occurs through coarsening of the grains. Eventually this can lead to penetration of underlying junctions. Therefore, a process which yields a better control of the metalsilicide/Si interface would open new technological possibilities.

Besides its technological advantages, $CoSi_2$ has the very interesting charateristic that it crystallizes in the CaF_2 structure with a lattice constant, which is only 1.2% smaller than that of Si. Epitaxial growth of $CoSi_2$ films on (111) Si has been reported widely, but only recently epitaxial growth has been achieved under severe MBE conditions [3] on (100) Si. With ion beam synthesis [4] monocrystalline $CoSi_2$ forms both in (100) and in (111) Si, buried under the Si surface. The growth of epitaxial $CoSi_2$ on (100) Si by direct reaction of a deposited Co film on Si was shown not to occur [5].

In this work the growth of $CoSi_2$ is studied by reaction of thin films of Co on a (100) Si substrate in which seeds of (100) $CoSi_2$ are present.

Experimental

The experiments were performed on 100mm wafers of (100) Si with a sheet resistance of 10-30 Ohm-cm. Co implantation was performed at an elevated temperature, i.e. at 400C, with doses ranging from 5 x 10^{16} to 2 x 10^{17} at/cm^2 with various energies. Co was deposited by DC- magnetron sputtering with a base pressure of 2 x 10^{-7} Torr. Silicidation was performed by rapid thermal processing in a N$_2$ ambient.

The sheet resistance of the samples was measured by 4 point probe (4PP). The silicide layers were monitored by cross-sectional SEM. Composition and structure of the layers were monitored by Rutherford backscattering (RBS), channeling, transmission electron microscopy (TEM) and X-ray diffraction (XRD).

Results and discussion

i. Seed consisting of a full buried layer

In a first set of experiments a full buried layer was synthesized in the Si by an implant of 2.5 x 10^{17} per cm^2 of Co at an energy of 200 keV. After heat treatment at 600C and subsequent step at 1000C, a 100nm thick buried layer was formed under approximately 100nm of Si. After deposition of 40nm of Co, silicidation of the top Si layer was obtained by a 2 step rapid thermal process at 550C followed by either 700C or 1000C step with an intermediate etch of the unreacted Co. Figure 1 shows random and channeled RBS spectra for the 2 samples. In fig. 1a the top Si layer has been completely converted to CoSi$_2$. The layer is, however, polycrystalline on top of the monocrystalline synthesized CoSi$_2$. After a heat treatment at 1000C for 10 s, as shown in fig.1b, an epitaxial CoSi$_2$ layer has been formed. The change of crystalline quality, from polycrystalline to monocrystalline is accompanied by a decrease of the sheet resistance of the sample with about 20%. The channeling data demonstrate the good structural quality of the layer oriented in the (100) direction with a X_{min} better than 6%. Cross-sectional TEM micrographs (fig. 2) show the original buried silicide after formation and the completed top epitaxial silicide after silicidation of the top Si layer. The thickness of the final layer is 220nm. No grain boundaries are present. Dislocations in the silicide layer can be observed and also some inclusions are visible. At present, the origin of these inclusions has not been identified, but they are most probably attributed to C or O impurities in the top Si layer after buried layer formation. It is interesting to note that the inclusions are not located at the original Si/CoSi$_2$ top interface, but midway in the reacted CoSi$_2$ layer. The high resolution micrograph (fig. 3), however, indicates that the CoSi$_2$/Si interface is atomically flat. XRD analysis (fig. 4) of a sample formed in a similar way reveals only (100) related peaks, thus

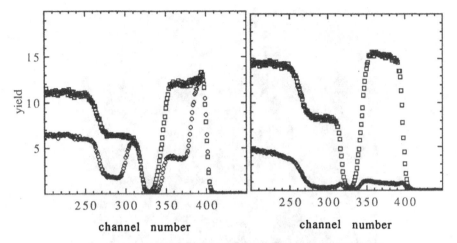

Fig.1: Random and channeled RBS spectra of a Si(100nm)/CoSi2(100nm)/Si
heterostructure after deposition of 38nm of Co and subsequent rapid thermal silicidation
a) at 700C for 30s, intermediate Co etch and 700C for 30s;
b) at 550C for 60s, intermediate Co etch and 1000C for 10s.

Fig.2: Cross-sectional TEM micrograph of a) the as formed heterostructure
and b) sample of fig.1b.

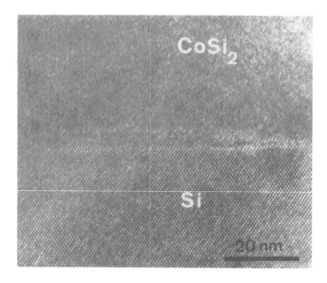

Fig.3: High resolution cross-sectional TEM of the CoSi$_2$/Si interface of the sample of fig.2b.

It should be stressed that the thickness of the deposited Co is critical. The amount of Si available for silicidation of the sputtered Co is limited, due to the presence of the buried silicide. From our experiments, it became clear that an overly thick Co layer yielded a Co rich silicide layer upon reaction. If the Co-rich silicide reaches the buried CoSi$_2$ layer, the structure of the layer is destroyed. In this case, after high temperature treatment more Si can be delivered by diffusion through the polycrystalline CoSi$_2$ to form a fully polycrystalline top layer.

ii. Seed consisting of precipitates

In order to make this technique, combining implantation and thin film reaction, more attractive, an experiment was carried out to reduce the required implantation dose and to make the thickness of the top Co layer with respect to the available Si in the top layer less critical. Therefore lower doses were implanted for the same thickness of deposited Co. Moreover, no effort was done with a high temperature treatment to make the synthesized CoSi$_2$ coalesce in a homogeneous layer. Instead, a distribution of CoSi$_2$ precipitates was introduced in the Si, over the complete implanted range. Figs 5 and 6 show RBS spectra for implanted Co doses of 8.5 x 10^{16} and 5 x 10^{16} per cm^2 of Co, respectively, after subsequent silicidation of 20nm Co. Again the channeled spectra demonstrate the alignment of the CoSi$_2$ with the (100) Si. From these results it can be concluded that a distribution of aligned CoSi$_2$ precipitates in (100) Si can act as a seed for epitaxial growth of CoSi$_2$ from deposited Co. The spacing between the

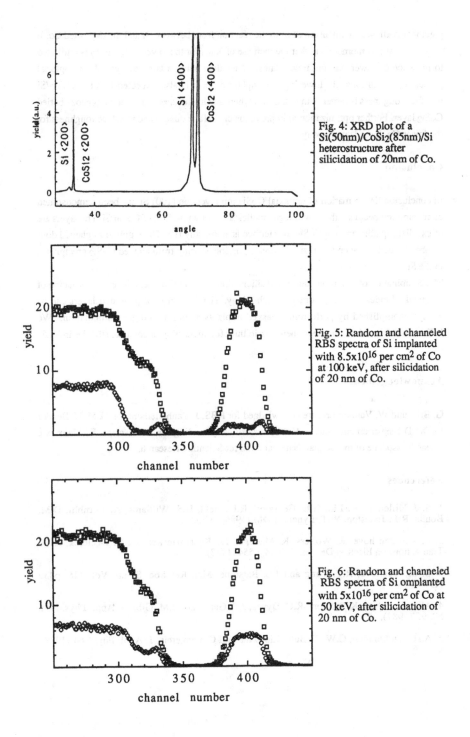

Fig. 4: XRD plot of a Si(50nm)/CoSi2(85nm)/Si heterostructure after silicidation of 20nm of Co.

Fig. 5: Random and channeled RBS spectra of Si implanted with 8.5×10^{16} per cm^2 of Co at 100 keV, after silicidation of 20 nm of Co.

Fig. 6: Random and channeled RBS spectra of Si omplanted with 5×10^{16} per cm^2 of Co at 50 keV, after silicidation of 20 nm of Co.

precipitates allows the diffusion of Si to the reaction front between Si and the deposited Co. It is not clear at this moment whether the increase of X_{min} for the lower dose (fig.6) is attributed to presence of fewer seeding precipitates, rather than being a consequence of non optimal processing conditions. By lowering the implant dose, it is also expected that the $CoSi_2/Si$ interface roughness is larger than in the case where seeding occurs from a homogeneous buried $CoSi_2$ layer. Further optimization to improve the crystalline quality and interface roughness for the lower Co doses is under way.

Conclusion

In conclusion the formation of epitaxial $CoSi_2$ top layers on (100) Si has been demonstrated combining ion beam synthesis and regular silicidation of a thin film of Co on Si. The layers are of excellent quality and the $CoSi_2/Si$ interface is atomically flat. The required implanted dose can be reduced, since epitaxial silicidation is possible starting from aligned $CoSi_2$ precipitates in the Si.

The combination of ion implantation and silicidation of a thin film metal layer for formation of epitaxial silicides is a very attractive technology. The formation of patterned layers can be easily accomplished by performing masked implants and/or self-aligned silicidation. We believe that this technology opens new possibilities for application of epitaxial silicides in VLSI

Acknowledgements

G. Brijs and W. Vandervorst are ackowledged for RBS, J. Vanhellemont for TEM, H. Bender for XRD measurements and L. Hobbs for critically reading the manuscript. K. Maex is a research associate of the Belgian National Fund for Scientific Research.

References

1. S. J. Hillenius, R. Liu, G.E. Georgiou, R.L. Field, D.S. Williams, A. Kornblit, D.M. Boulin, R.L. Johnston, W.T. Lynch, IEDM (1986), p 252.

2. L. Van den hove, R. Wolters, K. Maex, R. De Keersmaecker and G. Declerck, IEEE Transactions on Electron Devices, ED-34, p554 (1987).

3. S. M. Yasilove, R.T. Tung and J.L. Batstone, Mat. Res Soc. Symp. Vol. 116, p439 (1989).

4. Alice E. White, K.T. Short, R.C. Dynes, J.P. Garno and J.M. Gibson, Appl. Phys. Lett. 50, 95 (1987).

5. A.H. van Ommen, C.W. T. Bulle-Lieuwma and C. Langereis, J. Appl. Phys., 64 (1988), p 2706.

GROWTH AND CHARACTERIZATION OF EPITAXIAL CoSi$_2$-CONTACTS

C. Adamski, S. Meiser, D. Uffmann, L. Niewöhner and C. Schäffer, Institut für Halbleitertechnologie, Universität Hannover, Appelstr. 11A, D-3000 Hannover, Federal Republic of Germany

ABSTRACT

The growth of epitaxial CoSi$_2$ by means of SPE on heavily ion implanted Si(111) was investigated with LEED, RBS and TEM. After silicide formation, the dopant distribution in silicide and silicon was determined by means of SIMS. In a self-aligned process epitaxial CoSi$_2$/Si p$^+$ contacts have been produced. The specific contact resistance was found to be lower than for polycrystalline CoSi$_2$ contacts.

INTRODUCTION

With further scaling down of device dimensions to the submicrometer regime the contact resistance of source-drain contacts will have great influence on the device performance. In order to minimize the series resistance of MOS-transistors, new contact materials with low contact resistances will be required. For this purpose, metal silicides like TiSi$_2$ and CoSi$_2$ are of interest because of their low resistivity and good thermal stability. These materials are also suitable for a self aligned process in which the deposited metal only reacts to silicide where it is in contact with silicon. Using rapid thermal processing for silicide formation from metal films evaporated under high vacuum conditions, high quality contacts have been produced. For CoSi$_2$ contacts, the achieved contact resistance was as low as 1.5x10^{-7}Ωcm^2 [1].

Because of its small lattice mismatch (1.2%), CoSi$_2$ can be grown epitaxially on silicon. In the present work we investigated the quality of epitaxial CoSi$_2$ grown on heavily implanted Si(111) substrates. For boron doped substrates, the contact resistance was determined using a cross bridge kelvin resistor (CBKR). The results were compared with polycrystalline CoSi$_2$ contacts.

EXPERIMENTAL

To investigate the crystal quality of CoSi$_2$ films grown on heavily doped Si(111), unpatterned 3" wafers were used. The wafers were doped by ion implantation of arsenic (200keV) and boron (35keV) with doses varying from 1x10^{15}cm^{-2} to 7x10^{15}cm^{-2}. In the case of arsenic implantation, the doses exceeded the critical value where a completely amorphous silicon layer was formed [2]. The dopants were activated and the silicon was recrystallized by a furnace anneal at 975°C for 25 minutes. After having removed the stray oxide, the samples were loaded into the MBE. The wafers were cleaned in-situ by an annealing step at 850°C for 10 minutes. To improve the oxide desorption, the substrates were exposed to a weak silicon beam during thermal treatment. The required amount of cobalt to form a 70nm thick CoSi$_2$ layer was deposited on substrates at temperatures below 70°C. The silicide formation was carried out in-situ at 600°C for 5 minutes. This temperature was found to be sufficient to form CoSi$_2$ layers with acceptable crystal quality (RBS $\chi_{min} \approx 7.5\%$) on undoped Si(111) substrates. The pressure in the evaporation chamber remained below 5x10^{-8}Pa during the whole process.

For the determination of the specific contact resistance between epitaxial CoSi$_2$ and boron doped silicon, self-aligned CBKR structures were produced. Tungsten was used for interconnects.

CoSi₂ GROWTH ON ARSENIC DOPED Si(111)

After having cleaned the $1 \times 10^{15} \text{cm}^{-2}$ implanted substrate, a streaky 7x7 reconstruction with high intensity in the diffuse background could be observed using RHEED. This indicated a surface with random distributed steps and a high defect density. With increasing implantation doses, the intensity of the background increased too. With AES, contaminations on the surface were not detectable. After silicide formation, a 1x1 LEED pattern was visible with a characteristic energy dependence for a $CoSi_2$-S surface [3]. In comparision with layers grown on undoped Si(111), the threefold point symmetry of the 1x1 pattern was reduced.

RBS/channeling spectra of a $CoSi_2(70nm)/Si(111)$ (arsenic dose: $5 \times 10^{15} \text{cm}^{-2}$) sample are shown in Fig. 1. The spectra were taken by 1.5MeV He⁺ ions (channel 200 = 1.5MeV). The smeared cobalt signal indicated a non-uniform $CoSi_2$ thickness. In comparison with films grown on undoped substrates, the epitaxial quality was degraded. The minimum backscattering yield χ_{min} for $CoSi_2$ grown on undoped silicon was $\approx 7.5\%$ whereas the minimum backscattering yield for layers on arsenic doped substrates was $\approx 15\%$. This was in agreement with LEED measurements.

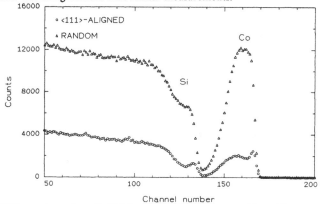

Fig. 1 RBS spectra of a $CoSi_2(70nm)/Si(111)$ structure. The silicon was arsenic implanted with a dose of $5 \times 10^5 \text{cm}^{-2}$. χ_{min} was determined to 15% for $CoSi_2$.

TEM cross-sections of this sample (Fig. 2) showed an interface between silicon and $CoSi_2$ with flat regions seperated by steps which were several nanometers high. This is a result of inhomogeneous diffusion of silicon and cobalt during silicide formation. Although the film thickness was not uniform, pinholes were not detected using TEM or SEM.

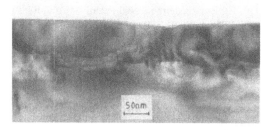

Fig. 2   TEM cross-section of a 70nm $CoSi_2$ layer on arsenic doped Si.

In order to determine the dopant distribution, a SIMS depthprofile was taken from a sample implanted with a dose of $1 \times 10^{15} \text{cm}^{-2}$ (Fig. 3). For arsenic, interface segregation or diffusion into the silicide could not be observed.

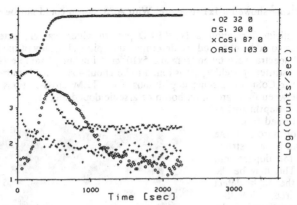

Fig. 3 SIMS depthprofile of the CoSi₂/Si structure. Diffusion of arsenic in the silicide was not detectable.

The resistivity of the CoSi₂ layers (17μΩcm) is higher than the resistivity of comparable films (15μΩcm) deposited on undoped substrates. This is in accordance to the higher defect density determined by LEED and RBS measurements.

CoSi₂ GROWTH ON BORON DOPED Si(111)

In contrast to arsenic implanted samples, the implantation dose used for boron is to low to produce an amorphous silicon layer. After the cleaning process, samples which were implanted with a dose of 5x10¹⁵cm⁻² showed a √3x√3R30° pattern. This structure is well known as a boron stabilized reconstruction of the Si(111) surface [4]. Even with AES, small amounts of boron could be found on the surface with a peak to peak ratio of B(179eV)/Si(92eV)≈1/100. Samples which were implanted with a boron

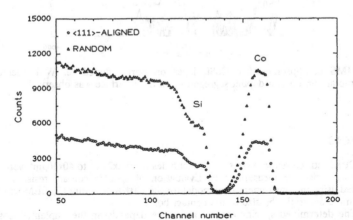

Fig. 4 RBS spectra of a CoSi₂(70nm)/Si structure. The Si was boron doped with a dose of 5x10¹⁵cm⁻². For CoSi₂ X_{min} was determined to 40%.

dose of $1x10^{15}cm^{-2}$ showed a 1x1 structure. With AES, boron could not be detected on the surface.

After silicide formation, a 1x1 LEED pattern which was characteristic for a $CoSi_2$-S surface could be observed. To determine the epitaxial quality of the $CoSi_2$ layer, RBS/channel spectra were taken from the $5x10^{15}cm^{-2}$ implanted sample (Fig. 4). The minimum backscattering yield χ_{min} was found to be about 40%. This is more than twice as much as for $CoSi_2$ on arsenic doped substrates. TEM cross-sections showed no differences between films grown on boron or arsenic doped Si(111).

SIMS depthprofiles (Fig. 5) turned out a segregation of boron at the interface and a strong diffusion of the dopant into the silicide. This may be the reason for the high defect density measured with RBS.

The resistivity of the $CoSi_2$ films was determined to be $17.8\mu\Omega cm$. This is corroberated by the higher χ_{min} in comparison with arsenic doped samples.

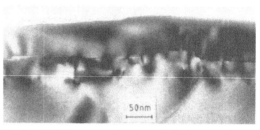

Fig. 5   TEM cross-section of a $CoSi_2$ layer on boron doped Si.

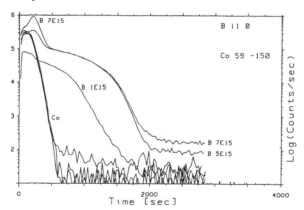

Fig. 6 SIMS depthprofile of a $CoSi_2$ layer on boron doped Si. With increasing implantion dose, a dopant segregation on the interface was observed.

CONTACTS

CBKR structures with sizes of contact holes from $2x2\mu m^2$ to $40x40\mu m^2$ were used for contact resistance measurements. Evaluation of specific contact resistance was performed using the one dimensional model for the CBKR structure [5]. SEM was used to determine accuratly the size of the contact holes.

The determined specific contact resistance depends on the implanted dose of boron (Fig.6). A great improvement of specific contact resistance was observed for a dose of $5x10^{15}cm^{-2}$ compared to a dose of $1x10^{15}cm^{-2}$. Sheet resistance measurements showed similar values for implanted doses of $5x10^{15}cm^{-2}$ and $7x10^{15}cm^{-2}$. According to

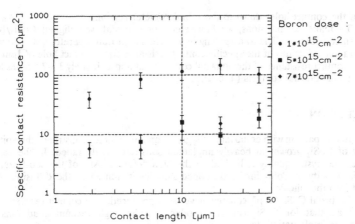

Fig. 7 Specific contact resistance in dependence of boron dose and contact dimensions.

the latter, a great enhancement of boron at the interface and in the silicide was found by SIMS (Fig.5). The doping profile in the bulk silicon was almost identical for both of the higher implanted doses. We assume that the boron atoms could not be fully activated for the highest applied dose. There was no significant improvement of contact resistance for a dose of $7 \times 10^{15} cm^{-2}$ compared to a dose of $5 \times 10^{15} cm$-2. A C(V) depthprofile of this sample showed a junction depth of 200nm below the silicide surface. The carrier concentration at the interface was above $1 \times 10^{20} cm^{-3}$ (Fig.7).

Boron dose $[10^{15} cm^{-2}]$	R_S $[\Omega/\ ]$	ρ_C $[\Omega\mu m^2]$	l_T $[\mu m]$
1	120-122	87	0.84
5	35-40	7.5	0.45
7	37-39	5.5	0.35

Tab. I Characteristic parameters for $5 \times 5 \mu m^2$ contacts. R_S is the sheet resistance of the diffusion area, ρ_C is the specific contact resistance and l_T the transfer length.

The evaluated specific contact resistance also depends on the size of the contact hole. The curve is typical for evaluations using the one dimensional model. For larger contact sizes, the specific contact resistance is overestimated due to current crowding effects [5]. The transfer length is a parameter which is used to estimate these current crowding effects. l_T is given by $l_T = \sqrt{\rho_C/R_S}$, where ρ_C is the specific contact resistance and R_S is the sheet resistance of the diffusion area. We determined the transfer length for the 5μm contacts (Tab.I). The values show that the specific contact resistance of our epitaxial contacts would be improved for

Method	$\rho_C/\Omega\mu m^2$
high vacuum dep., RTA	30
high vacuum dep.; RTA, [1]	15
MBE in-situ anneal	4

Tab. II Comparison of $1 \times 1 \mu m^2$ CoSi$_2$ contacts produced under various conditions.

sizes of the contact holes down to the submicrometer region.

To make comparisions, we fabricated non-epitaxial, self-aligned $CoSi_2$/boron doped silicon contacts. The transfer lengths were less than $1\mu m$. Therefore, we compared the extrapolated values of the specific contact resistance for a contact hole of $1\mu m$. The specific contact resistance of the epitaxial contacts was approximately 3.5-7.5 times lower than for nonepitaxial contacts (Tab.II).

CONCLUSION

In comparison with epitaxial $CoSi_2$ grown on unimplanted substrates, the epitaxial quality of $CoSi_2$ grown on heavily implanted substrates is degraded. In the case of arsenic, the crystal quality is limited by the high defect density of the substrate. The epitaxial quality of $CoSi_2$ for boron doped samples is limited by the diffusion of the dopant into the silicide.

Epitaxial $CoSi_2$/Si p^+ contacts have been produced. The contact resistance was lower than that for $CoSi_2$ contacts produced under high vacuum conditions and comparable implantation dose.

ACKNOWLEDGEMENTS

The authors would like to acknowledge the support of the Bundesministerium für Forschung und Technologie of the Federal Republic of Germany. We would also like to thank Dr. A. Mitwalsky of Siemens AG, Munich for the TEM work.

REFERENCES

1.   L. Van den Hove, R. Wolters, K. Maex, R.F. de Keersmaecker, G.J. Declerck, IEEE Trans. Electron Devices ED-34, 554, (1987).

2.   H. Ryssel, I. Ruge, Ion Implantation, (J. Wiley & Sons, 1986)

3.   R.T. Tung, F. Hellman, Mater. Res. Soc. Proc. 94, 65, (1987)

4.   R.L. Headrick, I.K. Robinson, E. Vlieg, L.C. Feldman, Phys. Rev. Lett. 63, 1253, (1989)

5.   W.M. Loh, S.E. Swirhun, T.A. Schreyer, R.M. Swanson, K.C. Saraswat, IEEE Trans. Electron Devices ED-34, 512, (1987)

INTEGRATED TITANIUM SILICIDE PROCESSING

JAIM NULMAN
Applied Materials Inc., 3050 Bowers Ave., Santa Clara, CA 95054

ABSTRACT

The processing of titanium silicide in a multichamber processing system is described. Three processes are included: wafer cleaning, Ti deposition, and annealing. The results are compared to wafers processed in a conventional way with exposure to ambient air between Ti deposition and the annealing step. TEM, RBS and sheet resistance measurements indicate that the films processed without exposure to ambient air have thicker and purer TiN layers as compared to the films with air exposure. Furthermore, integration allows for a wider processing window for the first annealing step. The use of reactive cleaning chemistry prior to Ti deposition results in a smooth silicon surface and therefore uniform silicidation as compared to inert cleaning technology, where redeposit of etched material occurs.

INTRODUCTION

As semiconductor device geometries continuously shrink to submicron dimensions, the junction ohmic contact and the interconnect resistance are some of the limiting factors in obtaining the desired high speed device performance. Silicidation of polysilicon gate electrodes and source drain ohmic contacts has become an accepted processes to reduce these limiting resistances values. One process which has received major attention is the self aligned silicide (SALICIDE) technology [1]. In the salicide technology, silicide is simultaneously formed on the source, drain and polysilicon gate electrodes, by the reaction of the deposited metal layer with the exposed silicon areas. Both refractory and near-noble metals have been used for applications in the salicide technology [2]. However, titanium has received the most attention due to the low resistivity of its silicide, 13 to 16 $\mu\Omega\cdot$cm, and its ability to reduce interfacial native oxide [3-5].

Figure 1 shows a schematic Ti SALICIDE process flow. After formation of the polysilicon electrode side walls, the exposed silicon surfaces are cleaned and a layer of Ti is deposited, typically 40 to 100 nm thick. then the wafers are transferred to the annealing reactor, most commonly a rapid thermal processor, where the wafer is annealed at a temperature in the range of 650 to 675 °C for a time of 10 to 40 seconds [6]. This first anneal allows Ti to react with the exposed silicon areas without interaction with the silicon dioxide. A selective etch removes unreacted Ti from oxide areas. However, for the Ti-Si system, Si is known to be the moving species during $TiSi_2$ formation. Therefore, in order to avoid silicon from diffusing into the Ti on the silicon dioxide areas, short processing times and the use of nitrogen atmosphere are necessary. This first anneal in nitrogen ambient also results in the formation of a passivating TiN layer [7-9]. After the selective etch, a second anneal at about 800 °C follows. This anneal allows for the conversion of the C49 $TiSi_2$ into its low resistivity C54 phase.

A limiting factor in the integrity of the titanium silicide process is the presence of titanium-oxygen compounds on the deposited film. These compounds are created by the exposure of the Ti coated wafers to ambient air during the transfer of the wafers from the deposition apparatus to the annealing reactor. The chemisorbed oxygen results in the formation of thin TiN_xO_{1-x}, instead of TiN, on the surface of the $TiSi_2$ layer [7], thus limiting the effectiveness of titanium nitridation over the silicon dioxide areas. Furthermore, some of this chemisorbed oxygen as well as oxygen generated from the interfacial native oxide will be present in the $TiSi_2$ layer resulting in chemical instabilities during high temperature annealing

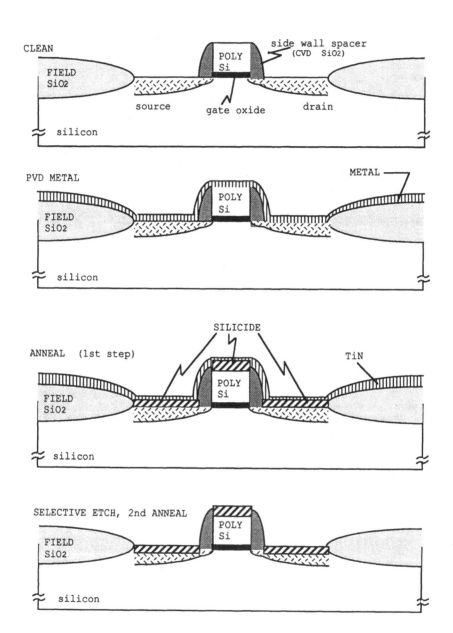

Fig. 1 Schematic process flow for a titanium SALICIDE process: cleaning of silicon surfaces, Ti deposition, first anneal step, selective etch of TiN and unreacted Ti, and second anneal step.

steps that might follow during the fabrication of semiconductor devices [10].

In order to properly implement a reliable $TiSi_2$ process in production, it is necessary to control both Ti interfaces: silicon and ambient. The elimination of native silicon dioxide insures the formation of a uniform silicide layer with no oxygen present, while the lack of exposure to ambient air improves the effectiveness of the nitridation process. In the following the effects of silicon surface cleaning and the ambient exposure of Ti in the kinetics of $TiSi_2$ are described.

INTEGRATED $TiSi_2$ PROCESSING

An integrated system for processing $TiSi_2$ includes a preclean chamber, a deposition chamber, and an annealing chamber connected through a central evacuated transfer chamber. Wafer loading into each of the chambers is performed at the transfer chamber base pressure, then each chamber is backfilled to the corresponding operating pressure. This process ensures the exposure of all surfaces to either vacuum or high purity gases only. Contrary to the conventional method where the surface of Ti is exposed to ambient air during transfer of wafers from the deposition to the annealing systems. Furthermore, in the integrated system, the annealing chamber is not exposed to ambient air, therefore ensuring the absence of oxygen and moisture and eliminating the need for high purge gas flows and extensive times.

Preclean

The conventional method of cleaning prior to Ti deposition uses Ar gas in an RF excited plasma. This method results not only in surface damage but also on device damage due to the relatively high DC bias, 500 to 1500 V. Furthermore, for source/drain and other device contact areas with dimensions in the submicron regime, redeposition of removed material becomes critical. The use of etching technologies where the gases are reactive, results in reaction by-products which are all volatile. For the present technology, reactive RF ion etching with fluorinated gases has been used to remove any oxides from the silicon areas where $TiSi_2$ is to be formed. Since reactive ion etch enhances the etch rate in comparison to inert ion etch, lower bias and shorter times can be applied for the same etch rate, thus eliminating device and surface damage effects.

Figure 2 shows SEM micrographs of 0.6 µm contacts on silicon after removal of the patterned 1.2 µm SiO_2 film and the formed $TiSi_2$ in buffered hydrofluoric acid. The $TiSi_2$ was formed by the deposition of about 50 nm Ti on the contact areas followed by annealing at 675°C for 30 seconds in nitrogen, selective etch and second annealing at 800 °C for 20 seconds. Fig. 2a corresponds to a wafer cleaned using inert Ar. Pitting of the silicon surface in the contact area is observed due to excess silicidation. In the non pitted areas silicidation was delayed by the possible redeposition of by-products of the inert etching process. Fig. 2b shows the case where reactive ion etching was used prior to Ti deposition, no pitting of the silicon in the contact area is observed, a uniform $TiSi_2$ was be formed in the contact areas, because all the etch by-products are volatile. In both cleaning processes, the same RF power and DC bias was used, typically 150 W and -140 V, but the etching times were adjusted in order to etch the same amount of silicon dioxide in both cases.

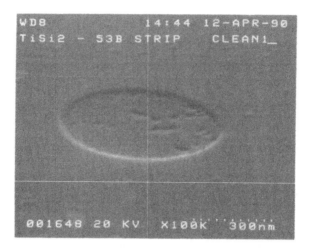

a)

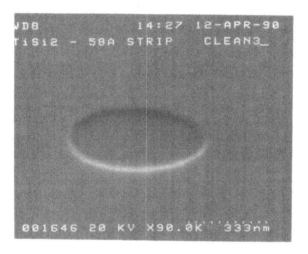

b)

Fig. 2 Scanning electron micrographs of 0.6 μm contacts in silicon after removing the patterned SiO_2 and formed $TiSi_2$. showing the effect of cleaning with a) inert gas and b) reactive gas.

Titanium silicide kinetics

The Ti silicidation kinetics, with and without air exposure between Ti deposition and annealing steps, was studied for wafers with reactive clean and 50 nm deposited Ti films. The ambient air exposure was about two hours. The triangle data points in Fig. 3 show the sheet resistance of the composite TiN/TiSi2 layer as function of the first anneal temperature for 20 seconds in nitrogen ambient at atmospheric pressure for wafers with and without air exposure. The sheet resistance for the wafers without air exposure is higher than the sheet resistance for the wafers with air exposure. Furthermore, the wafers without air exposure annealed between 700 and 750 °C do not show the sharp decrease in sheet resistance as observed on the wafers with air exposure. In the 650 to 700 °C regime, the slope of the silicidation kinetics is about 0.06 Ω/sqr./°C and 0.05 Ω/sqr./°C for wafers exposed and not exposed to ambient air respectively, indicating that the silicidation kinetics is slowed down. These effects are the result of a more effective nitridation in the case of wafers without exposure to ambient air. The formed TiN layer is thicker for the wafers not exposed to ambient air. The absence of chemisorbed oxygen, allows for nitrogen to diffuse deeper into the Ti, thus limiting the amount of C49 TiSi$_2$ that forms during the first anneal step. For wafers exposed

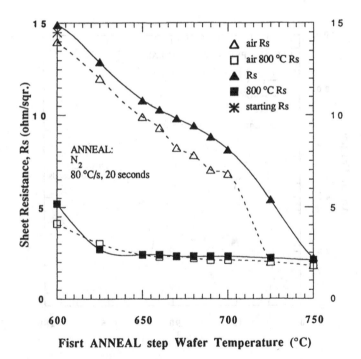

Fig. 3   Sheet resistance of Ti silicidation kinetics as function of the first anneal temperature for wafers with and without air exposure between Ti deposition and annealing.

to ambient air, the chemisorbed oxygen inhibits the nitrogen diffusion, and furthermore allows for some silicon to diffuse into the titanium oxygen compounds. These effects are confirmed by the atomic fraction vs. depth profiles, derived from Rutherford Backscattering (RBS) data, shown in Figs. 4 and 5 for the case of exposure and no exposure to ambient air respectively.

The RBS data in Fig. 4, show that the nitrogen atomic fraction is limited to about 22% and its profile correlates to the oxygen profile, at the same time some silicon with an atomic fraction of about 16% is observed in the top layer. In comparison to the wafer processed without exposure to air, Fig. 5, the surface atomic fraction for nitrogen is about 57%. No oxygen is observed, and the silicon stops at a depth were the nitrogen atomic fraction is about 37%. Furthermore, the data in Figs. 4 and 5 show less $TiSi_2$ for the case without exposure the air, because more Ti was converted to TiN. This RBS data is in agreement with the sheet resistance observations in Fig. 3. Further confirmation of these effects can be observed by the cross section TEM micrographs shown in Fig. 6. The wafer annealed without exposure to air, shows two distinctive layers: a fine grain TiN and a large grain C49 $TiSi_2$, Fig 6a. The wafer with air exposure shows a very thin top TiN layer, followed by an oxygen containing layer, and then the C49 $TiSi_2$.

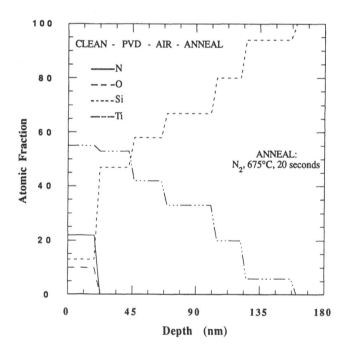

Fig. 4   Atomic fraction as function of depth derived from Rutherford Back Scattering data for a wafer exposed to ambient air between Ti deposition and annealing steps.

Figure 3 also shows the resulting sheet resistance of the C54 TiSi$_2$ as function of the first anneal temperature for both with and without air exposure before the first anneal step. As the temperature of the first annealing step increases, the resulting sheet resistance for the wafers with air exposure decreases compared to the case without air exposure. This is again in agreement with the RBS observations on more effective nitridation for non air exposed wafers.

All annealing steps to form titanium silicide has been done in atmospheric nitrogen pressure in order to provide for an effective formation of a TiN layer. The data described above indicates that without exposure to air the nitridation is more effective. Since in an integrated system, wafer transfer in and out of the annealing chamber is done under vacuum conditions, it is desired to operate under reduced nitrogen pressure in order to maximize throughput in an integrated system. Fig. 7 shows the sheet resistance of Ti after annealing, as function of the nitrogen vacuum level during annealing at 675 °C for 20 seconds. Also shown is the sheet resistance after the selective etch to remove the TiN layer as well as the difference between these two values.

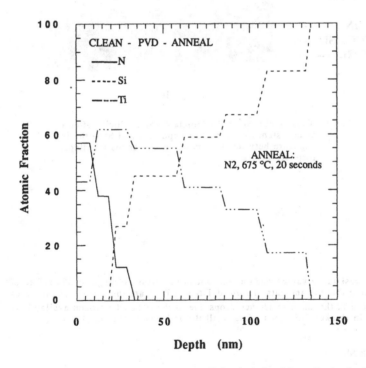

Fig. 5   Atomic fraction as function of depth derived from Rutherford Back Scattering data for a wafer not exposed to ambient air between Ti deposition and annealing steps.

**WITHOUT
AIR EXPOSURE**

TiN ------->

TiSi$_2$ ------>

<u> 90 nm </u>

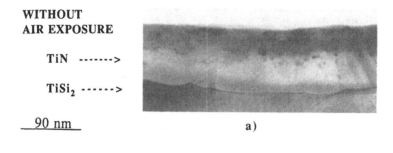

a)

**WITH AIR
EXPOSURE**

TiN -------->
TiO$_y$:Si$_x$ ---->
TiSi$_2$ ------->

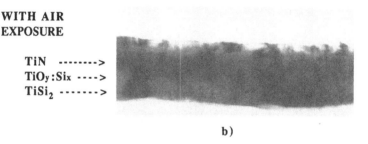

b)

Fig. 6   Cross sectional TEM micrographs of Ti silicide after a 675 °C anneal step. a) without air exposure, and b) with air exposure between Ti deposition and annealing steps, respectively.

The sheet resistance values remain constant in the pressure range of 500 mT to 800 T, indicating that the amounts of formed TiN and TiSi$_2$ are the same. For pressures below 500 mT, the sheet resistance drops due to the reduced nitrogen availability, resulting in thicker TiSi$_2$, until finally all the Ti is converted into TiSi$_2$.

CONCLUSIONS

The formation of titanium silicide without exposure to ambient air between the titanium deposition and annealing steps results in more effective TiN layer formation and wider processing window as compared to the conventional case, where wafers are exposed to ambient air. This integration further allows for an effective TiN/TiSi$_2$ layer formation under reduced nitrogen pressure. For wafers

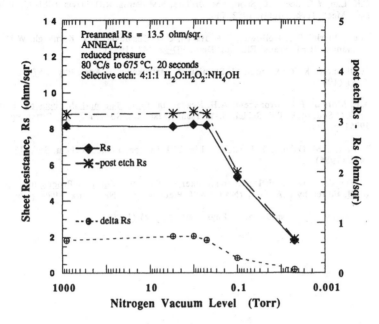

Fig. 7   Sheet resistance of composite TiN/TiSi$_2$ and TiSi$_2$
as a function of N$_2$ vacuum level.

exposed to ambient air the chemisorbed oxygen inhibits the effective nitridation of the Ti film. Silicon diffusion into the Ti to form the TiSi$_2$ stops at the point where the nitrogen atomic fraction is about 37%. For lower atomic fractions a TiN$_x$:Si layer is obtained.

ACKNOWLEDGMENTS

The author would like to thank the technical assistance and contributions from S. Tam, A. Tepman, R. Mosely, and H. Hanawa. S. Antonio and M. Dil assisted in sample preparation. The critical review of B. Cohen is also appreciated.

REFERENCES

1.   T. Shibata, K. Hieda, M. Sato, M. Konaka, R.L.M. Dang, H. Izuka, IEEE Int. Elect. Dev. Meet. Tech. Digest, 1981, 647.

2.   C.Y. Ting, IEEE Int. Elect. Dev. Meet. Tech. Digest, 1984, 110.

3.   C.Y.Ting, S.S. Iyer, C.M. Osburn, G.J. Hu, A.M. Schweighart, Electrochem. Soc. Meet. Extended Absratcts. 82-2, 224 (1982).

4. C.K. Lau, Y.C. See, D.B. Scott, J.M. Bridges, S.M. Perna, R.D. Davis, IEEE Int. Elect. Dev. Meet. Tech. Digest, 1982, 714.

5. M.E. Alperin, T.C. Holloway, R.A. Haken, C.D. Gosmeyer, R.V. Karnaugh, W.D. Parmantie, IEEE Trans. Electron Dev., ED-32, 141 (1985).

6. T. Okamoto, K. Tsukamoto, M. Shimizu, T. Matsukawa, J. Appl. Phys., 57, 5251 (1985).

7. A.E. Morgan, E.K. Broadbent, A.H. Reader, in Rapid Thermal PRocessing, Edited by. T.O Sedgwick, T.E. Seidel, B.-Y. Tsaur, (Mat. Res. Soc. Proc., 52, Pittsburgh, PA 1985), p. 279.

8. T. Brat, C.M. Osburn, T. Finstad, J. Liu, B. Ellington, J. Electrochem. Soc., 133, 1451 (1986).

9. L. Van den hove, R.F. De Keersmaecker, in Reduced Thermal Processing for ULSI, Edited by R.A. Levy (NATO/ASI, Plenum Press, New York, 1989) p. 53.

10. S.-i. Ogawa, T. Yoshida, Appl. Phys. Lett., 56, 725 (1990).

TEMPERATURE DEPENDENT CURRENT–VOLTAGE CHARACTERISTICS OF $TiSi_2/n^+/p$-Si SHALLOW JUNCTIONS

BHUPEN SHAH AND N. M. RAVINDRA

Microelectronics Research Center, New Jersey Institute of Technology, Newark, NJ 07102.

ABSTRACT

Modelling of temperature dependent current-voltage characteristics of $TiSi_2/n^+/p$-Si shallow junctions has been presented here. The formation of shallow pn junctions, by ion implantation of As^+ through Ti films evaporated on p-Si substrates has been performed in these experiments. The temperature dependent factors such as band gap narrowing, intrinsic carrier concentration, mobilities and diffusivities are considered in the model.

INTRODUCTION

As MOS devices are scaled down to submicron dimensions, junctions with a shallow depth and high surface concentration become necessary. With this reduction in device size, sheet resistance contributing to the RC time delay increases. Refractory metal silicides are of great interest for very large scale integrated circuit application as a contact material for shallow junctions. Among these silicides, $TiSi_2$ is particularly interesting, because of its low resistivity and its stability at high temperature treatments[1].

The silicidation, by ion-implantation through metal(ITM) into silicon substrate, has attracted attention because of the possibility of complete conversion of Ti into $TiSi_2$ by appropriate annealing conditions, resulting in smooth surface[2]. Current-voltage characteristics at different measurement temperatures, show that $TiSi_2$ contact on n^+/p-Si are nearly ohmic. Therefore the forward-bias current-voltage(I-V) characteristics correspond to that of a pn junction device. In this study, the factors affecting the I-V measurements at different temperatures for $TiSi_2/n^+/p$-Si have been analysed.

EXPERIMENTAL DETAILS

Single-crystal silicon wafers(p-type Czochralski) of $< 100 >$ orientation and $1-15\Omega - cm$ resistivity were used in these experiments. The cleaning procedure consisted of the conventional RCA technique followed by a HF dip and a through rinsing with deionized(DI) water. The Ti films, of 30nm thickness, were evaporated through metal mask (circular in geometry and 32 mils in diameter) onto the silicon wafers by e-beam evaporation in a vacuum of 10^{-7} torr. The deposition rate was maintained less than 1nm per sec in order to minimize the interaction between the metal and the underlying silicon substrate.

Ion beam mixing was done with a 200kV Varian 350D implanter using 100kV As^+ ions. It is estimated that, at this energy, the As^+ ions have a range R_p in silicon equal to 58nm. In order to minimize substrate heating during ion implantation, the current density was maintained below 1 microamp per cm^2. It is anticipated that this would correspond to an increase in the temperature of the wafer by less than $60°C$. A constant dose of 10^{16} ions/cm^2 was used throughout the experiments.

An AG Associates 210T rapid thermal annealer(RTA) was deployed to sinter the films. In these experiments, two sets of samples were prepared. The first one consisted of single step anneal, after ion implantation through metal. This set of samples were annealed in RTA at temperature of 900°C for 20 sec in the presence of argon. The second set of samples were subjected to two step anneal comprising of $500°C$ in Ar ambient for half an hour followed by RTA at temperature of $1000°C$ for 10 sec in the presence of argon. Electrical measurements of the temperature dependent current-voltage characteristics(I-V) were performed on both sets of samples using Keithley-236 source measure unit. The experiments were repeated with a HP 4145A semiconductor parameter analyzer, low temperature microprobe and programmable temperature controller.

RESULTS AND DISCUSSION

Current-voltage(I-V) characteristics of $TiSi_2/n^+/p$-Si devices for single step annealed samples are shown in fig-1a and 1b, for temperatures of $250°K$ and $300°K$ respectively. Representative I-V plots for two step annealed samples for temperatures of $250°K$, $300°K$, $350°K$ and $400°K$ are shown in fig-2a,2b,2c and 2d respectively. From figs.1a and 1b, it is seen that the single step annealed samples do not follow the pn junction I-V characteristics. A forward bias voltage from 0.0 to -3.0volts was applied to see the device behavior. It is seen that the device works as a resistor, i.e current increases linearly with applied voltage. From the I-V curves in figs 2a-2d, it is observed that two step annealed sample follows the simple equation of pn junction under forward bias;

$$I = I_s(e^{\frac{qV}{nkT}} - 1) \tag{1}$$

Where, I_s is the saturation current in amp, V is the applied voltage, n is the ideality factor, k is the Boltzmann constant and T is the temperature in degree Kelvin. From fig-2, it is clear that $TiSi_2$ makes nearly ohmic contact with the n^+/p-Si devices, after appropriate annealing time and conditions. These figures show that as the measurement temperature increases, current flowing through the device increases. The saturation current is temperature dependent, and is given by,

$$I_s = \frac{qAD_n n_i^2}{L_n N_A} \tag{2}$$

Where, A is the area of the device, D_n is the diffusivity of electron in the p-region, L_n is the diffusion length, n_i is the intrinsic carrier concentration and N_A is the substrate dopping. The intrinsic carrier concentration, n_i, is temperature dependent. It is given by[3],

$$n_i(T) = a_i T^3 exp(\frac{-E_{G0}}{kT}) \tag{3}$$

where

$$a_i = 4(2\pi k/h^2)^3 (m_n m_p)^{3/2} exp(-\alpha_G) \tag{4}$$

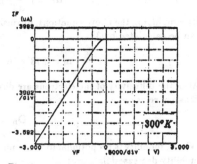

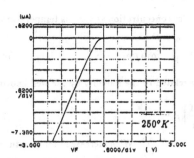

Fig-1a: I-V plot for $TiSi_2/n^+/p$-Si device for single step annealed sample

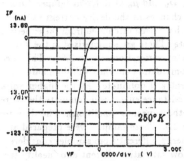

Fig-1b: I-V plot for $TiSi_2/n^+/p$-Si device for single step annealed sample

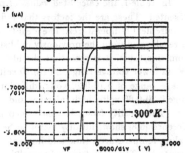

Fig-2a: I-V plot for $TiSi_2/n^+/p$-Si device for two step annealed samples

Fig-2b: I-V plot for $TiSi_2/n^+/p$-Si device for two step annealed samples

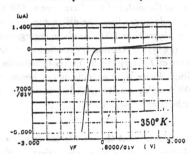

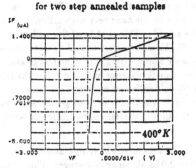

Fig-2c: I-V plot for $TiSi_2/n^+/p$-Si device for two step annealed samples

Fig-2d: I-V plot for $TiSi_2/n^+/p$-Si device for two step annealed samples

Where, α_G is an empirical constant characteristic of the semiconductor. It is given by the equation,

$$\alpha_G = (\frac{dE_G}{dT})(T - 300^\circ K)/kT \qquad (5)$$

By simplifying the above equations, intrinsic carrier concentration at various temperatures can be calculated by[3],

$$n_i = 3.73 \times 10^{16} T^{3/2} exp(-7014/T) \qquad (6)$$

Also, the diffusivity of the electrons are directly related to the mobility by the well known Einstein relation;

$$D_n = \left[\frac{kT}{q}\right] \mu_n \qquad (7)$$

Where, mobility is propotional to the temperature as $T^{-3/2}$. With increase in temperature, mobility decreases[4].

In Table-I, the calculated values of n_i, E_G, D_n and μ_n at different temperatures are presented. These are the factors that influence changes in the device current at different measurement temperatures. The ideality factors have been calculated from the experimental results and from the previously published data. These are summarised in the Table-II. From this table, it is seen that the ideality factor varies from 1.6 to 3.60. This means that at lower voltage range, 0.01-0.10 volts, recombination currents will dominate. In the voltage range of 0.10-0.20 volts, diffusion current will dominate. At the higher voltage range 0.20-0.60 volts, high injection and series resistance effects will dominate the current flow. The ideality factor is seen to depend on the annealing conditions. The data from references[6,8] shows ideality factor to be below 2.0. These samples were prepared at annealing temperatures, time and implantation energy different from samples considered in this study. The data from reference[7] shows low saturation current and ideality factor compared to all other data. This work pertains to $TiSi_2/p^+/n$-Si structure[7] and all the other data is based on the $TiSi_2/n^+/p$-Si structure. For the $TiSi_2/n^+/p$-Si structure, arsenic atoms have very high diffusion coefficent of the order of 10^{-13} cm^2/sec at 600$^\circ C$ in the silicide. This is at least five orders of magnitude larger than the diffusivity of arsenic in Si. Therefore, post annealing at temperatures higher than 600$^\circ C$ leads to segregation of arsenic atoms into $TiSi_2$. The loss of atoms from n^+/p-Si junction is expected to result in the schottky-barrier formation. For the structure of $TiSi_2/p^+/n$-Si, boron distribution in $TiSi_2$ is unaffected by post-silicidation annealing. There is no outdiffusion of boron from p^+/n-Si into $TiSi_2$ even at temperature of 900$^\circ C$. This results in an increased electrical activation of carrier concentration of boron without any loss[5,6,7,8].

At measurement temperature of 400$^\circ K$, it is found that the ideality factor is greater than 2.0 in all range of voltages. This means that recombination and high injection effects will dominate over the diffusion current through the device at high measurements temperatures. The saturation current I_s increases by three orders of magnitude as measurement temperature increases from 250$^\circ K$ to 400$^\circ K$. This current is proportional to the intrinsic carrier concentration, n_i. As the temperature increases, value of n_i increases.

In this work, it is observed that for two step annealed sample, complete conversion of Ti into $TiSi_2$ has occured and most of the dopants are electrically activated. Earlier

TABLE I

TEMPERATURE DEPENDENT PARAMETERS				
Temperature ($^\circ$ K)	n_i (cm^{-3})	E_g (eV)	D_n (cm^2/sec)	μ (cm^2/V-sec)
250	9.6×10^7	1.13	0.0646	3.0
300	1.45×10^{10}	1.12	0.117	4.5
350	4.8×10^{11}	1.108	0.012	4.0
400	7.23×10^{12}	1.097	0.12	3.5

TABLE II

ANALYSIS OF CURRENT-VOLTAGE CHARACTERISTICS					
Temperature ($^\circ$ K)	Voltage Range (V)	n	I_s (amp)	Area (cm^2)	J_s (amp-cm^{-2})
250	0 - 0.07 0.07 - 0.20 0.20 - 0.25 0.25 - 0.35 0.35 - 0.60	2.42 1.60 3.60 3.06 3.60	6.8×10^{-11}	5.18×10^{-3}	1.31×10^{-8}
300	0 - 0.10 0.10 - 0.20 0.20 - 0.275 0.275 - 0.60	1.09 1.92 2.90 3.30	1.8×10^{-10}	5.18×10^{-3}	3.47×10^{-8}
350	0 - 0.05 0.05 - 0.20 0.20 - 0.60	0.50 1.37 1.96	20×10^{-9}	5.18×10^{-3}	3.86×10^{-6}
400	0 - 0.20 0.20 - 0.275 0.275 - 0.35 0.35 - 0.60	2.26 2.78 2.26 3.00	53×10^{-9}	5.18×10^{-3}	1.023×10^{-5}
Ref[6] 300	0 - 0.20 0.20 - 0.40	1.19 1.85	2×10^{-9}	4.22×10^{-3}	4.73×10^{-7}
Ref[7] 300	0 - 0.10 0.10 - 0.40	0.64 1.06	1×10^{-12}	3.2×10^{-2}	3.125×10^{-11}
Ref[8] 300	0 - 0.20 0.20 - 0.40 0.40 - 0.60	1.63 1.35 1.67	8×10^{-11}	1×10^{-4}	8.0×10^{-7}
Ref[9] 300	0 - 0.15 0.15 - 0.40 0.40 - 0.60	2.40 2.50 2.80	5.6×10^{-9}	1×10^{-2}	5.6×10^{-7}

studies[2] of the structural properties of these samples, using transmission electron microscopy(TEM), have shown the formation of a uniform 40nm thick silicide for the As^+ implanted samples. No evidence of ion implantation induced damage is observed in the underlying silicon substrate. The TEM micrograph of the same sample(single step) sintered at $900°C$ for 20 sec.[2] shows that the interface between the silicide and the substrate is rough. The sheet resistance, R_{sh}, for the As^+ implanted samples were typically in the range of 58 ohms/square to 30 ohms/square. This decrease in R_{sh} was monitored as function of sintering temperature from $350° - 900°C$ for 10 minutes in argon[2].

CONCLUSION

Temperature dependent current-voltage characteristics of furnace and rapid thermally annealed $TiSi_2/n^+/p$-Si device structures, formed by ion implanting As^+ ions through Ti films on p-Si, have been reported in the above study. Analysis of experimentally measured current-voltage characteristics, in the temperature range of $250° - 400°K$, has been reported. In general, it is found that the saturation current increases with increasing measurement temperature. The ideality factor is found to have values above 2.0 in the lower voltage range   and higher voltage range   .This means that high level injection condition with significant recombination currents dominate over diffusion currents. A comparison with published data in the literature leads us to conclude that ideality factors are strongly influenced by process variables such as annealing temperatures, time and implantation energy.

ACKNOWLEDGMENTS

We acknowledge with thanks the assistance of V.Venkatramana and Y.Wang in performing I-V measurements.

References

[1]. S.P. Murarka, Silicides for VLSI Applications. Academic Press. 1983.

[2]. N. M. Ravindra, et al., Mal. Res. Soc. Symp. Proc., Reno, May 1988

[3]. Helmut F. Wolf, Semiconductors. Pergamon Press, Oxford. 1969.

[4]. S.M. Sze, Seniconductor Devices Physics and Technology, John Wiley & Sons, chapter-2 1985

[5]. K. Maex and R.P. Keersmaecker, Physica 129B, North-Holland, Amsterdam, 192-196, 1985

[6]. Jeng-Rern Yang and Juh Tzeng Lue, J. Appl. Phys. 65-3, pp-1039-1043, 1989.

[7]. M.Delfino, et al., J.Appl. Phys. 62-5, pp-1882-1886, 1987

[8]. Leonard Rubin, et al., J. IEEE Trans., 37, 183-190, 1990

[9]. N.M. Ravindra, Private Communication.

STRUCTURE OF THE Ti-SINGLE CRYSTAL Si INTERFACE

S.Ogawa*, T.Kouzaki**, T.Yoshida*, and R.Sinclair***
*Matsushita Electric Ind. Co., Ltd., Semiconductor Research Center, Moriguchi, Osaka, Japan
**Matsushita Technoresearch, Moriguchi, Osaka, Japan
***Dept. of Materials Science & Engineering, Stanford Univ., Stanford, CA

ABSTRACT

The Ti-single crystal Si interfaces, before and after annealing in argon, were examined by cross section high resolution transmission electron microscopy (HRTEM) combined for the first time with 2nmϕ probe for energy dispersive spectrometry (EDS). HRTEM shows that there is amorphous alloy formation at the Ti-Si interface. The thickness of the reacted layer is ~1.7nm for single crystal Si, independent of doping level and impurity species such as As and B, and is ~2.5nm for back sputter-amorphized Si. After annealing at 430°C for 30min, the thickness of the amorphous alloy increases up to ~11.5nm. High spatial resolution EDS microanalysis has been obtained. The results show that reliable compositions can be deduced at this level since some of the layers are only about 2nm thick. The amorphous alloy formed at the deposition step was found to be $Ti_{55}Si_{45}$. After annealing, the composition across the amorphous layer varied from about 70%Si near the substrate to about 30%Si close to the Ti interface. The substrate interface is atomically flat. Interpretation of the behavior in terms of the metastable Ti-Si phase diagram calculated by Holloway and Bormann will be discussed.

INTRODUCTION

Recently, the contact size in ULSI has been decreased well below sub-μm level, and yet ohmic and low resistivity properties are required. Ti is one of the best contact metals between single crystal Si and TiN diffusion barrier metal. Severe problems in high contact resistivity for p+Si substrates has been reported, the reason for which, however, has not been clear yet [1]. Holloway and Sinclair have demonstrated that a thin interdiffused layer exists at Ti-amorphous Si interface even in the as-deposited state [2]. Their result indicates that the Ti-Si contact really is the Ti-amorphous Ti-Si alloy-Si contact. Therefore, at the development step of ULSI these days, to obtain sub-μm contacts with excellent electric properties, it is strongly required to understand what happens at the Ti-single crystal Si interfaces during ULSI processing.

In this paper, using HRTEM combined with 2nmϕ EDS, we show that there is amorphous alloy formation at the Ti-single crystal Si interface. The thickness of the reacted layer depends on the crystallinity of the Si surface, but not on doping level or impurity species. After annealing, the thickness of the alloy increases, and it remains still amorphous. High spatial resolution EDS microanalysis has been obtained. The results show that reliable compositions can be deduced at this level since some of the layers are only about 2nm thick.

Mat. Res. Soc. Symp. Proc. Vol. 181. ©1990 Materials Research Society

EXPERIMENTAL PROCEDURE

After chemical cleaning of Si single crystal surface by dip etching in 20:1 NH_4F:HF, 35nm thick Ti films were sputter deposited at room temperature on various (100)Si substrates, including lightly doped p-type single crystal, Ar back-sputtered single crystal, and heavily doped n^+ and p^+ ($\sim 3\times10^{20}cm^{-3}$) single crystals. The crystallinity and morphology of Ti-Si interfaces, before and after 430°C, 30min annealing in argon, were examined by cross section HRTEM using JEOL4000FX microscope, and compositions near the interfaces were also measured by HRTEM supplemented with nominal diameter 2nm probe EDS using Akashi Beam Technology EM-002B microscope [3]. Compositions were calculated using k factor of Philips EDS computer software. Electron irradiation damage at Ti-Si interfaces has also been investigated during 20~120sec irradiation at 200keV accelerating voltage with 0.1nA electron beam current. The chosen annealing condition is widely used in ULSI processing as annealing after aluminum interconnect formation.

RESULTS AND DISCUSSION

A cross section HRTEM micrograph of as-deposited Ti-single crystal Si interface for which Ti was deposited on p^+ single crystal Si substrate without back-sputtering is shown in Fig.1. An amorphous layer is observed at the interface. EDS analysis makes it clear that it is a Ti-Si alloy as discussed later. The amorphous layer exsisted at most of the interface area, and occasionally, Ti lattice planes directly contacted the Si crystal in atomic scale as shown in the right part of Fig.1. This result shows that the reaction of Ti with Si crystal is somewhat inhomogeneous in the deposition step.

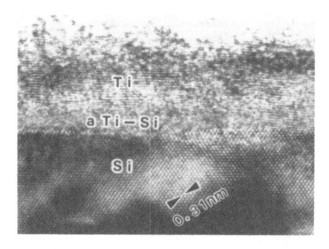

Fig.1. Cross section HRTEM micrograph of as-deposited Ti-single crystal Si interface for which Ti was deposited on a p^+ Si substrate without back-sputtering.

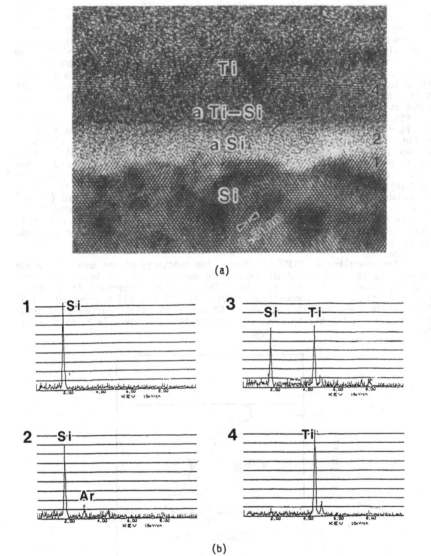

(a)

(b)

Fig.2.   (a) Cross section HRTEM micrograph and (b) EDS spectra (using 2nmφ
probe) from areas 1~4 of the Ti-Si interface for which Ti was sputter deposited
on a back-sputtered lightly doped p-Si substrate.   In the 2.0nm thick lighter
band(2), Si and Ar signals, and in the 2.5nm thick darker band(3), Si and Ti
signals were observed,   while Ar signal was not detected in the Si crystal(1)
adjacent to the lighter band and Si signal was not detected in the Ti layer(4)
adjacent to the darker band.

A dependency of the Si concentration in the as-deposited reacted layer on 2nmϕ electron beam irradiation time was examined. Positions to be analyzed were not changed for the increasing irradiation time. The composition was decided using Cliff-Lorimer k-factors established with a $TiSi_2$ standard[4]. The Si concentration X (X in $Ti_{100-x}Si_x$) was 27.5($\sigma=1.6$). No change was observed in the composition during the repeated EDS measurements at different occasions. Accuracy in composition determined by EDS seems to be within 10% in EM-002B system [5]. No damage or reaction was observed in the HRTEM observation during the electron beam irradiation. Hence, 100sec was chosen as the measurement time for the EDS observation.

Fig.2 shows cross section HRTEM micrograph and 2nmϕ EDS data of the Ti-Si interface for which Ti was sputter deposited on back-sputtered Si substrate. In Fig.2(a), HRTEM clearly shows that there were two amorphous layers in between Si lattice points and Ti lattice planes. One near the Si crystal, lighter band, was about 2nm thick, and the other darker band was about 2.5nm thick. EDS spectra from areas 1~4 in Fig.2(a), using 2nmϕ probe, were shown in Fig.2.(b). The lower lighter amorphous layer(2) was amorphized Si induced by Ar implantation during back-sputtering, and the upper darker layer(3) was amorphous Ti-Si interdiffused layer of approximately $Ti_{55}Si_{45}$ composition.

The results show that the compositions in the thin layer can be obtained exactly at this level. For instance, though some of the layers are only about 2nm thick, there is no Ti signal in the silicon immediately adjacent to the interface.

After annealing at 430°C for 30min, the thickness of the amorphous alloy increased up to ~11.5nm, and the growth did not depend on the impurity species or doping level in the Si crystals. The substrate interface is virtually atomically flat. The composition across the ~11.5nm thick amorphous Ti-Si alloy, which are determined by 2nmϕ EDS every 2nm step, varied from about 70%Si near the substrate to about 30%Si close to the Ti interface, as shown in Fig.3.

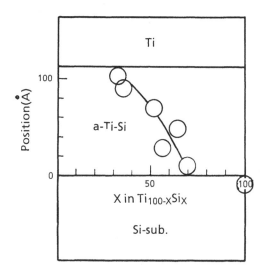

Fig.3. Si content across the ~11.6nm thick amorphous Ti-Si layer between Si substrate and Ti layer after annealing at 430°C for 30min in Ar atmosphere.

Table 1   The thickness(Å) of amorphous Ti-Si alloy
obtained from the HRTEM observation.   [Å]

Type	Crystallinity of Si	As deposited	430°C 30min
p	Amorphous	25 ± 3	
		17 ± 3	
p +	Single Crystal	17 ± 3	116 ± 25
n +		16 ± 8	113 ± 10

    The thickness of amorphous Ti-Si alloy obtained from the experiment is
shown in Table 1.   It is clarified that thickness of the as-deposited
amorphous layer depends on crystallinity of Si substrate surface, but not on
impurity species or doping level in the Si crystals and thickness of the
annealed amorphous layer also does not depend on high concentration impurity
species.   a   of
    Holloway and Bormann calculated the metastable Ti-Si phase diagram using
a CALPHAD method [6],[7],[8].   They indicated that, for instance, the single
phase field of the disordered phase of Ti-Si at 400°C extends from about 25
at.% Si to 65 at.% Si.   The calculation is consistent with our experimental
results.   Raaijmakers et. al. measured the composition of a 8nm thick Ti-Si
alloy which was formed between Ti and amorphous Si by Auger electron
spectroscopy sputter depth profile [9].   They found that it ranged from 73%Si
near the Si to 28%Si near the Ti interface.   Although ion mixing during the
Auger electron spectroscopy depth profiling might affect the compositions,
however, they obtained   quite similar results to those reported here.

CONCLUSION

    A thin amorphous Ti-Si interdiffused layer exists at the as-deposited
Ti/Si substrate interface, and after 430°C annealing, its thickness increases,
while remaining amorphous.   The thickness of the amorphous Ti-Si interdiffused
layer depends on the crystallinity of the Si-substrate surface, but does not
depend on impurities or doping level.   High spatial resolution 2nmφ EDS
analysis combined with HRTEM was also performed, showing that reliable
compositions can be obtained for such thin amorphous interdiffused layers.

ACKNOWLEDGEMENT

The authors wish to thank T.Yanaka (Akashi Beam Technology Company) for helpful EDS analysis and discussions, S.Nakamura for fruitful discussion and K.Tsuji for encouragement throughout this work.

REFERENCES

[1] J.Hui, S.Wong, and J.Moll, IEEE Electron Device Letters, vol.EDL-6, 479 (1985)

[2] K.Holloway and R.Sinclair, J. Appl. Phys. 61, 1339 (1987)

[3] T.Yanaka, K.Moriyama, and R.Buchanan, Mats. Res. Soc. Symp. Proc. vol.139 271 (1989)

[4] G.Cliff and G.Lorimer, Proc. 5th European Congress on Electron Microscopy, 140 (1972)

[5] T.Yanaka, Private communication

[6] K.Zoltzer and R.Bormann, J.Less Common Met. 140, 335.(1988)

[7] K.Holloway, "Interfacial Reactions in Metal-Silicon Multilayers", Ph.D.Thesis, Stanford University (1989)

[8] K.Holloway, P.Moine, J.Delage, R.Sinclair, and L.Capuano, Mats. Res. Soc. Symp. Proc. in press (1990)

[9] I.J.M.M.Raaijmakers, A.H.Reader, and P.H.Oosting, J. Appl. Phys. 63, 2790 (1988)

INITIAL PHASE FORMATION AND DISTRIBUTION
IN THE Pt-Ge$_x$Si$_{1-x}$ AND Cr-Ge$_x$Si$_{1-x}$ SYSTEMS

Q. Z. Hong and J. W. Mayer
Department of Materials Science and Engineering, Cornell University, Ithaca, NY 14853

ABSTRACT

Metal-GeSi reactions have been investigated in the Pt-GeSi and Cr-GeSi systems. Pt started to diffuse into and react with the alloys at 250 °C. The reacted region consisted of a uniform mixture of Pt$_2$Si and Pt$_2$Ge, with the same ratios of Ge to Si as those in the unreacted region. On the other hand, the Cr-GeSi reaction was induced by Ge motion at 375 °C. As a result a two-layer phase separation was observed. A Si-rich ternary layer was sandwiched between a germanide layer and the unreacted alloy.

INTRODUCTION

Recent progress in fabrication of epitaxial thin films on Si [1] has raised hopes of using Ge-Si in semiconductor devices. Inevitably, metal contact to Ge-Si will become a problem of concern. Metal-Si [2] and metal-Ge [3] binary reactions have been well investigated. However, only a few reports on metal-GeSi reactions have been documented [3-6]. In this experiment structural and morphological changes were studied for Pt, a near noble metal, and Cr, a refractory metal, in their reactions with amorphous Ge-Si alloys. It is well known that in Pt-Si reactions, Pt is the dominant moving species at low temperatures while Si becomes mobile at 425 °C in Cr-Si reactions. It is therefore of interest to investigate whether this difference still exists in the metal-GeSi ternary reactions and its impact on phase formation and distribution.

EXPERIMENTAL

Amorphous Ge$_x$Si$_{1-x}$ alloys were prepared on SiO$_2$ substrates in a dual e-gun evaporator with a base pressure better than 1×10^{-7} Torr. The alloy compositions used in this experiment varied from 25 percent of Ge to 75 percent of Ge. After the coevaporation, a thin Pt layer was deposited. In the case of Cr, the deposition sequence was reversed in order to prevent oxidation of Cr films in subsequent heat treatments, to provide a good adhesion between the Ge-rich alloys and the SiO$_2$ and to separate the signals of Cr and Ge on Rutherford backscattering spectroscopy (RBS). In some cases, Cr-Ge bilayers were also prepared to study the binary reactions which had not been reported. For the study of moving species, thin Ta and Ti films were chosen as markers in the Cr-Ge and Pt-GeSi systems, respectively. The as-prepared samples were then annealed in a vacuum furnace with a pressure kept around 1×10^{-7} Torr in the temperature range of 200 to 500 °C. RBS was used to monitor elemental profiles, and glancing angle x-ray and electron diffraction were applied to study phase transformation.

RESULTS

Pt-Ge$_x$Si$_{1-x}$ reactions

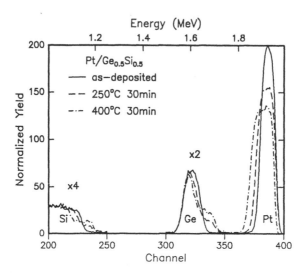

Fig.1 RBS spectra of a Pt-Ge$_{0.5}$Si$_{0.5}$ sample before and after a 30min annealing at 250 and 400 °C, respectively.

When annealed at 250 °C, Pt began to react with Ge-Si alloys, as indicated by the RBS spectra in Fig. 1. It shows that a half hour annealing at 250 °C has consumed the entire Pt layer on a Ge$_{0.5}$Si$_{0.5}$ alloy. Computer simulation revealed that the reacted layer consisted of equal amounts of Pt$_2$Si and Pt$_2$Ge. In other words, the ratio of Ge to Si in the reacted region remained the same as that in the as-deposited sample. Electron diffraction studies confirmed the existence of Pt$_2$Ge, a hexagonal phase, and Pt$_2$Si, a tetragonal phase, with no signs of ternary phase formation. Marker experiments showed that Pt was the dominant moving species in the reactions. Annealing at higher temperatures induced further reactions between the metal-rich silicide and germanide with the unreacted Ge-Si alloy. The RBS spectrum (Fig. 1) of Pt-Ge$_{0.5}$Si$_{0.5}$ after a 30-min annealing at 400 °C shows that a surface layer is formed on the alloy film with a composition close to that of 50% PtSi and 50% PtGe. The coexistence of PtSi and PtGe was again confirmed by electron diffraction. Similar experimental results were also obtained in the Pt-Ge$_{0.25}$Si$_{0.75}$ and Pt-Ge$_{0.75}$Si$_{0.25}$ samples.

Cr-Ge reactions

The Cr-Ge reactions started at 350 °C. When annealed above 375 °C, a uniform reacted region formed and grew with time. The composition of the reacted region was determined as Cr$_{4.3}$Ge$_3$. X ray diffraction patterns of the reacted sample consisted of reflection lines that could be indexed as those from the orthorhombic Cr$_4$Ge$_3$ compound. A marker experiment was also carried out to determine the identity of moving species during the initial phase formation. Fig. 2 shows RBS spectra of a Cr-Ta(3 Å)-Ge sample before and after a half hour annealing at 400 and 450 °C, respectively. The formation of Cr$_4$Ge$_3$ was again observed and the reaction kinetics was comparable to that without the marker.

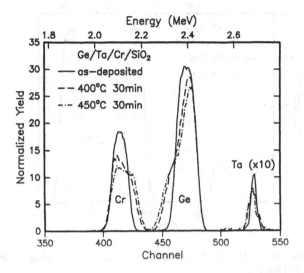

Fig. 2 RBS spectra of a Cr/Ta/Ge sample before and after annealing at 400 and 450 °C, respectively.

Therefore the presence of the marker did not change the reaction process significantly. A computer simulation of RBS spectrum of the annealed sample showed that the marker was embedded in the middle of the compound layer, indicating that both Cr and Ge moved with almost the same flux.

Cr-Ge$_x$Si$_{1-x}$ reactions

Compared with Pt, Cr reacted with Ge-Si alloys at higher temperatures, around 375 °C. Fig. 3 shows a series of RBS spectra of a Cr-Ge$_{50}$Si$_{50}$ sample annealed at various temperatures. A two layer structure started to develop at temperatures as low as 375 °C. The peak and dip of the Ge signals clearly indicates the formation of a Ge rich layer below the unreacted Cr and a Si-rich layer next to the unreacted Ge-Si alloy. A computer simulation of the spectra showed that the layer below the Cr was essentially free of Si and had a Cr to Ge ratio equal to 3:1. The composition of the Si rich layer was determined as Cr$_{1.5}$Ge$_{0.34}$Si$_{0.66}$. It was interesting to note that the two layers always formed simultaneously and continued to grow with increasing temperatures until all the free Cr was consumed. The fact that the compositions of the two layers remained unchanged during growth suggested each layer consisted of a compound or mixtures of compounds. Indeed, both glancing angle x-ray and electron diffraction indicated the existence of a Cr$_3$Ge compound and a ternary solid solution of Cr$_5$Ge$_3$ and Cr$_5$Si$_3$. By using Vegard's law, the composition of the solution was estimated to be Cr$_5$(Ge$_{0.4}$Si$_{0.6}$)$_3$, which was in good agreement with the backscattering results. The layer phase separation was also observed in the Cr-Ge$_{20}$Si$_{80}$ sample.

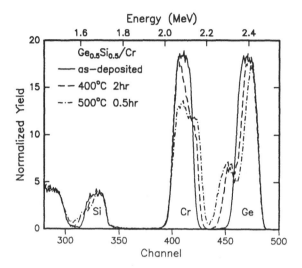

Fig. 3 RBS spectra of a Cr/Ge$_{0.5}$Si$_{0.5}$ sample before and after annealing at 400 and 500°C, respectively.

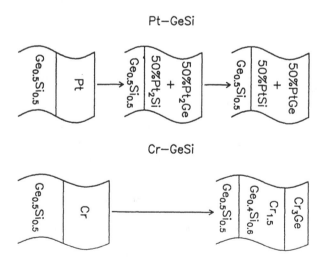

Fig. 4 Schematic diagram showing the reaction processes of Pt-Ge$_{0.5}$Si$_{0.5}$ and Cr-Ge$_{0.5}$Si$_{0.5}$.

DISCUSSION

Reactions of Pt and Cr with Ge-Si alloy have been investigated. Different reaction behaviors were observed in the two systems (Fig. 4). For Pt, the reaction produced a uniform mixed layer with no atomic redistribution of Ge and Si. Moreover, the phase formation sequence basically followed that of binary reactions of Pt with Si, and Pt with Ge. However, in the case of Cr, a Ge-rich layer was formed on top of a Si rich layer. Furthermore, the phases formed in the ternary reactions were different than that observed in the binary Cr-Ge reaction. We believe that the differences between the refractory metal and near-noble metal system is mainly due to the different moving species involved. In the Pt system, Pt diffuses into the Ge-Si alloy matrix and reacts separately with Ge and Si to form Pt_2Ge and Pt_2Si. Consequently no redistribution of Ge and Si occurs after reactions. In contrast, marker experiments prove that both Cr and Ge are quite mobile in germanides even at relatively low temperatures around 375 °C. Furthermore it is believed that the mobility of Si is low compared with Ge that has a lower melting point. Experimental evidence also shows that Si is essentially immobile below 450 °C in its reaction with Cr. Hence in the ternary reactions between Cr and Ge-Si alloys, Ge moves at low temperatures to react with Cr, leaving behind a Si-enriched alloy, while at the same time Cr transports through the germanide layer and reacts with the alloy to form a ternary solid solution , which has been reported by Rykova et. al. [7] . The fact that a more Cr-rich germanide , instead of the Cr_4Ge_3, is formed in the ternary reaction might be related to the limited supply of Ge. In order to react with Cr, the Ge has to first dissociate from the Ge-Si alloy and then transports through a Si rich compound layer.

CONCLUSION

A comparison study has been carried out on the initial phase formation and distribution in the Pt-Ge_xSi_{1-x} and Cr-Ge_xSi_{1-x} systems. Pt was found to be the dominant moving species, and the ratio of Ge to Si stayed unchanged after reactions. On the other hand, both Ge and Cr were mobile in the reaction, resulting in a two layer phase separation.

ACKNOWLEDGMENT

Research was supported in part by National Science Foundation (NSF) (John Hurt). Evaporation was carried out in the National Nanofabrication Facility, which is supported by NSF under Grant No. ECS-8619049, Cornell University and industrial affiliates.

REFERENCES

1. J. C. Bean, L. C. Feldman, A. T. Fiory, S. Nakahara, and I. K. Robinson, J. Vac. Sci. Technol. A 2, 436 (1984).
2. M.-A. Nicolet and S. S. Lau, in VLSI Electronics, Microstructure Science, edited by N. G. Enspruch and G. B. Larrabee (Academic, New York, 1983), Vol. 6, p. 330.
3. E. D. Marshall, C. S. Wu, C. S. Pai, D. M. Scott, and S. S. Lau, Mater. Res. Soc. Symp. Proc. 47, 161 (1985).

4. R. D. Thompson, K. N. Tu, J. Angillelo, S. Delage, and S. S. Iyer, J. Electrochem. Soc. 135, 3161 (1988).
5. Q. Z. Hong and J. W. Mayer, J. Appl. Phys. 66, 611 (1989).
6. O. Thomas, S. Delage, F. M. D'Heurle, and G. Scilla, Appl. Phys. Lett. 54, 228 (1989).
7. M. A. Rykova, A. V. Sabirzyanov, V. L. Zagryazhskii, and P. V. Gel'd, Inorg. Mat. 7, 835 (1971).

SIMULATION OF METALLIZATION FORMATION PROCESSES ON Si

V.V. Tokarev, A.N. Likholet and B.N. Zon
Institute of Solid State and Semiconductor Physics, the BSSR
Academy of Sciences, P.Brovka 17, Minsk 220726, USSR

INTRODUCTION

The processes occuring in thin films, in particular such
as silicide formation processes, are rather difficult to
control using analytical measurements . Such processes are
widely used for the formation of metallization layers and
usually require the precise control of film thickness, atomic
composition, atomic distribution profiles, phase composition
and electric resistance 1 . The absence of model concepts
of the process of silicide formation hinders the wide use of
such processes and makes them labour-consuming ones due to
the large number of control measurements.

This article presents the theoretical description of the
silicide formation processes upon solid state reaction of a
metal film with monocrystalline silicon.

THEORY

It is known that the kinetics of solid state reactions
is described by complex nonlinear boundary-value problems for
partial differential equations. However, recently for solid
state reactions there were developed effective methods of
solving diffusion equations 2 .

For the sake of definiteness we assume that silicide is
formed as a result of metal diffusion in silicon. Let us sup-
pose that before the process starts metal occupies the region
$x < 0$ and silicon $x > 0$. The following equation can describe
diffusion process

$$\frac{\partial n_A}{\partial t} = \frac{\partial}{\partial x}\left(D\,\frac{\partial n_A}{\partial x}\right) + S \qquad (1)$$

Here n_A is the concentration of free metal atoms, S is the
density of metal atom "drain", which is connected with the
possibility of silicide formation.

The boundary and the initial conditions for the eq.(1)
are of the common form:

$$n_A(0,t) = N_c \quad , \quad n_A(x,0) = 0, \quad x > 0 \qquad (2)$$

where N_o is the atomic density in a pure metal.

In order to calculate S let us consider silicide formation velocity. The concentration of free silicon atoms can be denoted as n and the concentration of the reaction product atoms in silicide as n . Then

$$\frac{\partial n_c}{\partial t} = k n_A n_B \tag{3}$$

where k is the reaction velocity constant. Consider the case of a stable formation of a one phase in a layer which is characteristic of a solid phase reaction. Note that the well-known mode of Wagner kinetics appears in the limit of high-speed reactions at $k \to \infty$.

Neglecting the possibility of heat release or absorption during the reaction we regard k as well as the diffusion coefficient D as not dependent on time.

It is clear that

$$S \equiv \left(\frac{\partial n_A}{\partial t}\right)_R = -\frac{\partial n_c}{\partial t} \tag{4}$$

where $\left(\frac{\partial n_A}{\partial t}\right)_R$ denotes the decrease in the number of metal atoms in some small area region, attributed to the course of the reaction.

The change in volume of a solid in the course of the reaction can be neglected and alongside with the eq.(3) we can write the relation:

$$n_B + n_c = n_{BO} \tag{5}$$

where n_{BO} is the concentration of atoms in a free Si. Eq.(5) determines the silicon density conservation law valid in the absence of silicon diffusion.

Substituting eq.(5) into eq.(3) we obtain:

$$\frac{\partial n_c}{\partial t} = k \left(n_{BO} - n_c\right) n_A$$

The solution of this equation under the condition $n_c(x,0)=0$ can be written in the form:

$$n_c(x,t) = n_{BO} \left\{ 1 - \exp\left[-n \int_c^t n_A(x,t') dt'\right] \right\} \tag{6}$$

Substituting this expression into eq.(4) and (1) we derive the equation for the distribution of metal atoms in the region $x > 0$:

$$\frac{\partial n_A}{\partial t} = \frac{\partial}{\partial x}\left(D \frac{\partial n_A}{\partial x}\right) - k n_{BO} n_A \exp\left[k \int_0^t n_A(x,t') dt'\right] \tag{7}$$

Solving in parallel the problem of the distribution of metal atoms in silicide for the known boundary motion velocity we obtain

$$n_{Me \to MeSi}(x,t) = \begin{cases} n_0\left[1 - \phi(x/\sqrt{4Dt})/\Phi(\lambda/\sqrt{4D})\right], & x < \lambda\sqrt{t} \\ 0 & x > \lambda\sqrt{t} \end{cases} \tag{7a}$$

The main attention in this article will be focused on the distribution of metal and silicon atoms in the silicide-Si transition region.

Eq.(7) is a nonlinear integro-differential equation and its analytical treatment is rather difficult. It can be substantially simplified if one assumes that the diffusion coefficient of metal in Si and silicide is the same. Further on we shall confine ourselves to this assumption and focus on the inquiring into the role of the finite value of k.

Thus, assuming D=const and introducing dimensionless variables

$$\frac{n_A}{N_0} = n; \quad \frac{n_{BC}}{N_0} = \nu; \quad tkN_0 = \tau; \quad x\sqrt{\frac{kN_0}{D}} = y \tag{8}$$

we rewrite the eq.(7) together with the additional conditions (2):

$$\frac{\partial n}{\partial \tau} = \frac{\partial^2 n}{\partial y^2} - \nu n \exp\left[-\int_{\tau}^{\tau} n(y,\tau')d\tau'\right]; \quad n(0,\tau)=1, n(y,0)=0 \tag{9}$$

We can now proceed from the integro-differential eq.(9) to the differential equation if we introduce a new sought-for quantity

$$f(y,\tau) = \int_{\tau}^{\tau} n(y,\tau')d\tau'$$

The equation for $f(y,\tau)$ is of the form of Kolmogorov-Petrovskij-Piskunov equation:

$$\frac{\partial f}{\partial \tau} = \frac{\partial^2 f}{\partial y^2} + \nu(e^{-f}-1); \quad f(y,0)=0, f(0,\tau)=\tau \tag{10}$$

The analytical solution of the boundary-value problem (10) can be obtained for 2 limiting cases.
(1). Region $f \ll 1$. The conditions of the physical realization of this region shall be considered below. Expanding e^{-f} in eq.(10) in series and leaving 2 terms we obtain a linear equation:

$$\frac{\partial f}{\partial \tau} = \frac{\partial^2 f}{\partial y^2} - \nu f \tag{11}$$

The solution of this equation with the additional conditions (10) takes the form

$$f_{(y,\tau)} = \frac{1}{2}\left\{ \left(\tau - \frac{y}{2\sqrt{\nu}}\right)e^{-\sqrt{\nu}y}\left[1 + \Phi\left(\sqrt{\nu\tau} - \frac{y}{2\sqrt{\tau}}\right)\right] + \left(\tau + \frac{y}{2\sqrt{\nu}}\right)e^{\sqrt{\nu}y}\left[1 - \Phi\left(\sqrt{\nu\tau} + \frac{y}{2\sqrt{\tau}}\right)\right]\right\} \quad (12)$$

Here Φ is an error integral.
Eq.(12) is simplified when

$$y \ll \sqrt{\tau} \ll 1 \tag{13}$$

In this case

$$f(y,\tau) \approx \tau - \sqrt{\frac{\tau}{\pi}}\, y \tag{14}$$

It is clear that the conditions of inequality (13) occur in the initial period of the reaction proceeding near the metal-Si interface. In this region metal atom density varies according to a linear law:

$$n(y,\tau) \approx 1 - \frac{y}{2\sqrt{\pi\tau}}, \tag{15}$$

Error integral is known to reach asymptotic value rather quickly. That is why when the conditions

$$y \gg \sqrt{\tau}; \quad y > 2\sqrt{\nu}\,\tau \tag{16}$$

are accomplished eq.(12) is also simplified:

$$f(y,\tau) \approx \frac{2\tau}{\sqrt{\pi}}\, e^{-\frac{y^2}{4\tau}}, \quad n(y,\tau) \approx \frac{y^2}{2\sqrt{\pi}\,\tau}\, e^{-\frac{y^2}{4\tau}} \tag{17}$$

The conditions of the inequality (16) and, hence, of the expression (17) are valid far from the silicide-Si interphase boundary, where metal atom density is low.
Further, in the region

$$y \gg \frac{1}{\sqrt{\nu}}, \quad y \gg \sqrt{\tau}, \quad y < 2\sqrt{\nu}\,\tau \tag{18}$$

we have

$$f(y,\tau) \approx \left(\tau - \frac{y}{2\sqrt{\nu}}\right)e^{-\sqrt{\nu}y}; \quad n(y,\tau) \approx e^{-\sqrt{\nu}y} \tag{19}$$

Region (18) is located closer to the silicide-Si interface, hence, metal atom density is decreasing exponentially, while in the region of the inequality (16) it decreases according to the Gaussian distribution.
(2). Region $f \gg 1$.

Here the exponent in the eq.(10) can be neglected and the boundary-value problem can be written in the following form:

$$f(y,\tau)=(\tau+\tfrac{y^2}{2})\left[1-\phi\left(\tfrac{y}{2\sqrt{\tau}}\right)\right]-\tfrac{y\sqrt{\tau}}{\sqrt{\pi}}e^{-\tfrac{y^2}{4\tau}}-1\int_0^\tau d\tau'\phi\left(\tfrac{y}{2\sqrt{\tau-\tau'}}\right) \qquad (20)$$

Eq.(20) in the stationary mode near the metal-silicide interface

$$\tau\gg 1 \; ; \; y^2\ll\tau \qquad (21)$$

can be rewritten in the form:

$$f(y,\tau)\approx\tau\left[1-\tfrac{2y}{\sqrt{\tau}}(1+\nu)\right] \; ; \; n(y,\tau)\approx 1-\tfrac{y}{\sqrt{\pi\tau}}(1+\nu) \qquad (22)$$

Here, similar to the expression (15) metal atom density is changing according to a linear law, however, the slopes of the linear functions at the process starting point and in the stationary mode are different.

Thus, n(y,τ) can be found near the metal-Si interface and in the region far from the silicide-Si interface when considering the linearized eq.(10). To find metal distribution at the silicide-Si interface one should solve a nonlinear eq.(10). For this purpose the method of least squares was used 3 . Function (y,) was taken in the form:

$$f(y,\tau)=\tau\exp\left[-\alpha(\tau)\tfrac{y}{\sqrt{\pi\tau}}\right]+\ln\left\{1+y^\lambda\left[\tfrac{y_0(\tau)-y}{\epsilon(\tau)}\right]\right\} \qquad (23)$$

$$\alpha(0)=1 \; ; \; y_0(0)=0 \; ; \; \epsilon(0)=0 \; ; \; \lambda\geq 2$$

It is clear that both additional conditions (10) are satisfied for the function (23). The value of $\alpha(\tau)$ determines the behaviour of the function for y→0. Proceeding from the formulae (14) and (22) the following expression was accepted for the function

$$\alpha(\tau)=\tfrac{1+2(1+\nu)\tau}{1+\tau} \qquad (24)$$

In this case in the stationary mode from the eq.(23) we obtain a modified Fermi-Dirac distribution for silicide density, $y_0(\tau)$ there at describes the silicide-Si interphase motion velocity and $\epsilon(\tau)$ - the spreading of this interphase boundary. Functions $y_0(\tau)$ and $\epsilon(\tau)$ are found numerically.

156

References

[1]. K.N. Tu and J.W. Mayer, in <u>Thin Films - Interdiffusion and Reactions</u>, edited by J.M. Poate, K.N. Tu and J.W. Mayer (John Wiley and Sons, 1978).

[2] L. Nanai, I. Hevesi, N.F. Bunkin, B.A. Zon, S.V. Lavrishev, B.S, Lukyanchuk and G.A. Shafeev, Appl. Phys.A <u>50</u>, 27, (1990).

[3] S.G. Mikhlin, in <u>Variatzionnye Metody v Matematicheskoj Fisike</u>, 2nd ed. (Nauka, Moscow, 1970).

PART IV

Silicides-Polysilicon Systems

THERMAL STABILITY OF CoSi$_2$ ON SINGLE CRYSTAL AND POLYCRYSTALLINE SILICON

J. R. Phillips, P. Revesz, J. O. Olowolafe, and J. W. Mayer
Department of Materials Science and Engineering
Cornell University, Ithaca NY 14853

ABSTRACT

The thermal stability of Co silicide on single crystal and polycrystalline Si has been investigated. Co films were evaporated onto (100) Si and undoped polycrystalline Si and annealed in vacuum. Resulting silicide films were examined using Rutherford backscattering spectroscopy, scanning electron microscopy, electron-induced x-ray spectroscopy, and sheet resistivity measurements. We find that CoSi$_2$ on single crystal (100) Si remains stable through 1000°C. In contact with undoped polycrystalline Si, intermixing begins at temperatures as low as 650°C for 30min annealing. The Co silicide and Si layers are intermixed after 750°C 30min annealing, giving islands of Si surrounded by silicide material, with both components extending from the surface down to the underlying oxide layer. The behavior of CoSi$_2$ contrasts with results reported for TiSi$_2$ which agglomerates on single crystal Si around 900°C but is stable on poly-crystalline silicon as high as 800°C. Resistivity measurements show that the Co silicide remained interconnected despite massive incursion by Si into the silicide layer.

INTRODUCTION

The disilicide phase of Co is of great interest as a possible replacement for TiSi$_2$ and other refractory metal silicides in integrated circuit metallization and gate contacts. TiSi$_2$ has been shown to undergo agglomeration on crystalline Si, [1,2,3] and hillock growth and other forms of degradation on polycrystalline Si [3,4]. CoSi$_2$ has one of the lowest resistivities of the silicides, and its cubic structure and lattice parameter make CoSi$_2$ attractive also for epitaxial applications. One advantage of the disilicide of Co over that of Ti stems from the relationship between moving species during silicide formation and the problem of short circuiting caused by silicide formation beyond the contact opening in self-aligned processes. Because Si is the moving species in TiSi$_2$ formation, lateral silicide formation may take place, with Si moving out through the growing silicide into the blanket metal layer beyond the contact window. In the case of Co silicides, Co, rather than Si, is the moving species at temperatures below about 700°C, precluding lateral silicide formation. The CoSi$_2$ phase has the cubic CaF$_2$ structure and small (1.2%) lattice mismatch with Si, allowing the formation of epitaxial films. CoSi$_2$ requires lower temperatures than TiSi$_2$ to achieve optimal resistivity. As device processing temperatures are reduced, the option to use CoSi$_2$ becomes

more promising. It is important therefore to know the
upper limits on CoSi$_2$ thermal stability. Here we report
that compared to TiSi$_2$, CoSi$_2$ is less stable on undoped
polycrystalline Si but more stable on single crystal
(100) Si.

EXPERIMENTAL

Silicide films were prepared by reacting electron-beam
evaporated Co films on polished (100) silicon wafers and on
low pressure chemical vapor deposited polycrystalline Si
films on oxide covered Si wafers. Polycrystalline Si film
thickness of 300nm and 500nm were used. The polycrystalline Si films were grown at 625°C, giving an average
columnar grain size of around 30-40nm, as determined by
transmission electron microscopy. E-beam evaporations were
in a vacuum of better than 6x10^{-7} Torr in an oil-free vacuum
system with a base pressure of better than 2x10^{-7} Torr.
Substrates were cleaned immediately before loading by
immersion in buffered HF. Samples were then cleaved and
annealed in vacuum of better than 3x10^{-7} Torr for 30min at
temperatures up to 1050°C. Rutherford backscattering
spectroscopy (RBS) was used for depth profiling, with
interpretation aided by simulation of laterally non-uniform
layers using the RUMP RBS data analysis program [5].
Scanning electron microscopy (SEM) was used for characterizing morphology, electron-induced x-ray energy dispersive
spectroscopy to identify elemental spatial distribution in SEM
images, and four-point probe measure- ments to determine the
effect of anneals on room temperature resistivities.

RESULTS

Fig. 1 contrasts Co/(100)Si and Co/polycrystalline Si,
showing RBS spectra for anneals to up to 1050°C and 950°C,
respectively. For anneals up to 1000°C, the CoSi$_2$ on
crystalline Si remains stable, with some intermixing beginning
at the interface only after the 30min 1050°C anneal. For the
polycrystalline Si case, intermixing has begun at the interface
after 30min at 650°C. After 750°C annealing, the Si signal has
increased near the surface and Co has moved deeper. For a
thicker (500nm) polycrystalline Si layer, the 950°C spectrum
shows the tendency for the silicide and silicon layers to
invert.
Due to the lateral nonuniformity in the sample films,
SEM micrographs were used to show that the RBS spectra indicate
intermixing. Micrographs showed small ($\approx$100nm) dark areas just
beginning to appear at the film surface after a 650°C 30min
anneal. After 750°C annealing, there were $\approx$300nm wide regions
of darker contrast material. Electron beam induced x-ray energy
dispersive spectroscopy confirmed that the dark regions are
silicon. Micrographs showing the two-phase appearance for CoSi$_2$
after annealing at 850°C are shown for Co on (Fig 2a) 300nm of
polycrystalline Si and (Fig. 2b) 500nm of polycrystalline Si.

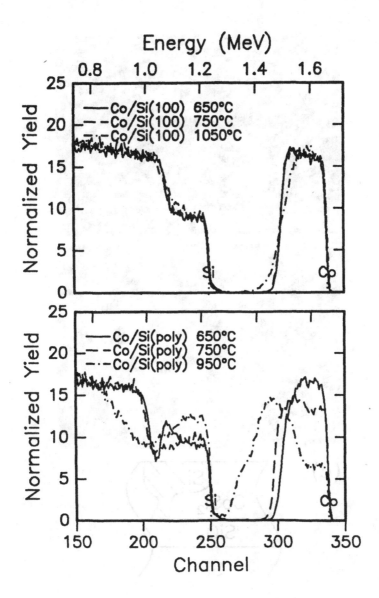

Fig. 1. 2.2 MeV He+ Rutherford backscattering spectra for Co silicide films prepared on (a) (100) Si and (b) 300nm and 500nm polycrystalline Si, and annealed for 30min in vacuum. Markers show the energies for Co and Si at the sample surface.

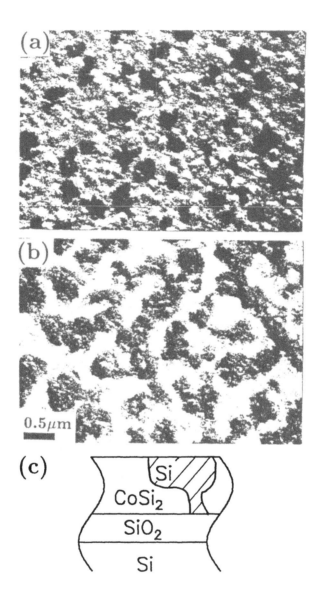

Fig. 2. SEM micrographs of Co after 850°C 30min anneal, on (a) 300nm polycrystalline Si (b) on 500nm polycrystalline Si. Dark areas were identified as Si. (c) Schematic of layer morphology for best RBS simulation for 500nm polycrystalline Si sample, showing tendency toward layer inversion.

Simulation of the wide-topped Si island morphology depicted in Fig. 2c produced a good fit to the RBS spectra for anneals of 950°C for the 500nm polycrystalline Si samples. For the single crystal substrates samples, even after annealing at 1050°C SEM micrographs showed no visible Si island formation, and x-ray spectroscopy scans across the $CoSi_2$ on the single crystal substrate showed uniform Co. Four point probe resistivity measurements are shown in Fig.3.

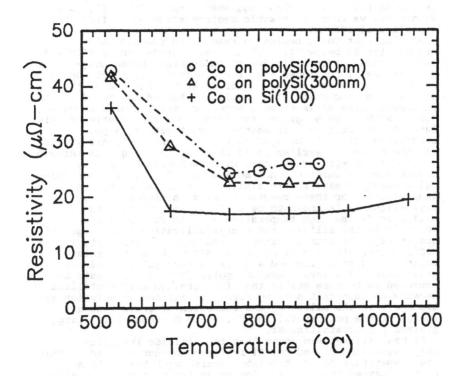

Fig. 3. Room temperature resistivity measurements for Co on single crystal Si, and on 200nm and 500nm polycrystalline Si films, versus annealing temperature. Conversions from sheet resistivities assumed a 220nm silicide film thickness for all samples.

Resistivity was estimated from four point probe data assuming identical silicide film thicknesses of 220nm for all samples. In general the silicide on polycrystalline Si films had higher resistivities. The good film conductivity in the poly-crystalline Si cases, despite the huge Si incursion into the silicide films, verifies the interconnectedness of the silicide phase shown in the SEM micrographs.

DISCUSSION

In this study, the $CoSi_2$ on single crystal Si was shown using RBS, SEM, and resistivity measurements to remain stable through 1000°C. $TiSi_2$ in contrast begins agglomeration at temperatures as low as 850°C, with clear agglomeration by 900°C [1,2,3]. With $CoSi_2$ we found no visible agglomeration even after 30min annealing at 1050°C.

We find that $CoSi_2$ begins to react with undoped poly-crystalline Si below 700°C with clear degradation by 750°C. An earlier study of Co silicides on phosphorous doped poly-crystalline Si concluded that the disilicide is stable up to 900°C [6]. Increasing amounts of Si at the surface of the silicide and incursion of the silicide toward the bottom of the polycrystalline Si layer were noted for temperatures above 900°C, with large Si grains extending from the underlying oxide layer to the surface. In another investigation, a poly-crystalline Si/$CoSi_2$ bilayer on Si was found to invert at 800°C (in the manner of earlier solid phase epitaxial growth through silicides) except for the case of phosphorous doped poly-crystalline Si which did not move at 900°C [7]. The morphol-ogies reported earlier are similar to those here, but the temperatures for these reactions are much higher. Poly-crystalline Si is known to be unstable in contact with silicides during high temperature anneals due to Si dis-solution in the silicide and recrystallization into larger Si grains [4]. In both instances cited above of stability in the $CoSi_2$/polycrystalline Si system at 900°C, high temperature P doping would have produced a large average polycrystalline Si grain size. The larger grained polycrystalline Si would be expected to be more stable than fine grained polycrystalline Si because an important driving force for silicon dissolution and transport in the silicide, the reduction of grain boundary energy of the polycrystalline Si, would be reduced for large-grained polycrystalline Si.

It has earlier been shown that for silicide stability on polycrystalline Si and for Si regrowth through silicides, that the temperature for Si transport scales with the silicide melting temperature [4,8]. One mechanism that may be respon-sible for this correlation is the increased ductility of the silicide near 0.6 of the melting temperature, allowing the silicide to deform [4,9]. Since melting temperatures are 1540°C for $TiSi_2$ and 1326°C for $CoSi_2$, we would expect the maximum stable temperature for $CoSi_2$ in contact with polycrystalline Si to be about 100°C to 150°C lower than for $TiSi_2$. This implies that large polycrystalline Si grain size is even more necessary for applications of $CoSi_2$.

CONCLUSIONS

We have shown that $CoSi_2$ in contact with single crystal (100) Si is stable to 1000°C, in contrast to $TiSi_2$ which has been reported to agglomerate by 900°C. On the other hand, $CoSi_2$ on undoped polycrystalline Si undergoes intermixing beginning around 700°C and becomes severe by 750°C, approaching inversion given sufficiently thick Si and higher annealing temperatures. The $CoSi_2$ thus is less stable on polycrystalline Si than $TiSi_2$, which is stable up to 800°C to 850°C. The lower melting temperature of $CoSi_2$ and greater ease of deformation may explain the lower stability of the $CoSi_2$ on polycrystalline Si compared to $TiSi_2$.

ACKNOWLEDGEMENT

This work was supported in part by the Semiconductor Research Corporation.

REFERENCES

1. C.Y. Ting, F.M. d'Heurle, S.S. Iyer, and P.M. Fryer, J. Electrochem. Soc. Abs. 85-1, 387 (1985).

2. P. Revesz, L.R. Zheng, L.S. Hung, and J.W. Mayer, Applied Phys. Lett. 48, 1591 (1986).

3. C.Y. Wong, L.K. Wang, P.A. McFarland, and C.Y. Ting, J. Appl. Phys. 60, 243 (1986).

4. L.R. Zheng, L.S. Hung, S.Q. Feng, P. Revesz, J.W. Mayer, and G. Miles, Appl. Phys. Lett. 48, 769 (1986).

5. L.R. Doolittle, Nucl. Instr. Meth. B9, 344 (1985).

6. S. Vaidya, S.P. Murarka, and T.T. Sheng, J. Appl. Phys. 58, 971 (1985).

7. S.P. Murarka, C.C. Chang, and A.C. Adams, J. Vac. Sci. Technol. B5, 865 (1987).

8. S.S. Lau, J.W. Mayer, and W. Tseng, in Handbook on Semiconductors, Vol. 3, ed. S.P. Keller, Ch.8, North-Holland, 1980.

9. L.R. Zheng, J.K. Phillips, P. Revesz, and J.W. Mayer, Nucl. Instru. Methods B 19/20, 598 (1987).

TiSi₂ THIN FILMS FORMED ON CRYSTALLINE AND AMORPHOUS SILICON

Z.G. XIAO[1], H. JIANG[1,4], J. HONEYCUTT[1], C.M. OSBURN[2,3], G. MCGUIRE[3]
AND G.A. ROZGONYI[1]

[1]North Carolina State University, Dept. of Materials Science and Engineering,
Raleigh, NC, 27695, USA
[2]North Carolina State University, Dept.of Electrical and Computer Engineering,
Raleigh, NC, 27695, USA
[3]Microelectronics Center of North Carolina, Research Triangle Park, NC 27709, USA
[4]The Royal Institute of Technology Solid State Electronics, Box 1298, S-16428, Kista,Sweden

ABSTRACT

TiSi₂ thin films were formed on crystalline and amorphous silicon substrates obtained by Ge⁺ and Ge⁺+B⁺ implantation and optional subsequent annealing. Transmission electron microscopy, X-ray diffraction and electrical resistivity analysis revealed that the silicide formed on amorphous Si has more tendency to have a C54 structure rather than the metastable C49 structure. Also, the grain size is smaller and the silicide/silicon interface is smoother for silicides formed on amorphous Si. Comparison between implanted and unimplanted, (100) and (111) Si substrates indicated that the origin of the differences can be attributed to the latent energy stored in amorphous silicon, which favors the silicide with fine grains and promotes the transformation to the C54 phase. Non-random distribution of planar defects in C49 grains has been observed by plan-view TEM. A proposal that these defects are transformation stress induced microtwins is presented.

INTRODUCTION

Because of their low resistivity and ease of formation by thermal reaction of Si with a deposited Co or Ti film, TiSi₂ and CoSi₂ films are a prime choice for contacts and interconnections in ULSI devices. The microstructure of these silicide thin films plays an important role in their electrical performance. An investigation of CoSi₂ has been presented recently [1] where it was demonstrated how the condition of the silicon substrate influenced the formation and microstructure of the resulting CoSi₂. In this report we examine the microstructure, crystal phase, interfacial roughness and resistivity of TiSi₂ films from both a fundamental and practical point of view.

Two polymorphs of TiSi₂ have been observed in thin silicide films on Si substrates [2]. One is the C49 phase with a base-centered orthorhombic crystal structure (a=3.62 Å, b=13.67 Å, c=3.60 Å) and high electrical resistivity (~65 μΩ-cm), while the other is the C54 phase, which appears in a face centered orthorhombic crystal structure (a=8.24 Å, b=4.78 Å, c=8.54 Å) and low resistivity (~15 μΩ-cm), making it the desired phase for device applications. The C49 phase usually occurs as a metastable phase during lower temperature annealing, while the formation of the C54 phase requires annealing above 650°C [2]. Previous workers pointed out that the C49 to C54 transition temperature is influenced by impurities. For example, Pt [3], Cu, Al, O, C [2] and some dopant elements [4] increase the transition temperature.

A smooth silicide/silicon interface is required for IC devices, especially for ULSI with shallow junctions, since a rough interface can induce excess leakage current and premature breakdown of devices [5]. Interfacial roughness has often been observed and attributed to residual native oxide on the surface of Si substrate [6] prior to film deposition. Although grain size may have some influence on interfacial roughness and thermal stability of silicide films, not much has been published on this. Another process variable to be considered is ion implantation. Since the Si substrate may be amorphous or crystalline following ion implantation depending on the implantation parameters and annealing, and since implantation is a frequent procedure in an IC device process, it is of interest to correlate the state of the Si substrate with the resulting $TiSi_2$ microstructure.

EXPERIMENTAL

The substrates used were n-type, (100) oriented, 4 inch diameter Si wafers. The samples were preamorphized with Ge^+(80 keV, $2x10^{15}$ ions/cm^2) and subsequently implanted with B^+(10 keV, $1x10^{15}$ ions/cm^2). Half of the samples were furnace annealed at 550°C for 30 min in Ar and rapid thermal annealed at 1050°C for 10 sec in Ar for recrystallization and activation of B before metallization, simulating a conventional process in device fabrication. The remaining wafers were kept in the as implanted state, and the silicide was subsequently formed in a concurrent process along with the implant anneal. A 30 nm Ti deposition was carried out in a resistively heated evaporation system with a base pressure of $1x10^{-5}$ Pa following a Ti pre-evaporation which gettered the oxygen in the system. The silicidation was performed at 650°C for 120 sec in N_2 in an AG Associates rapid thermal processing (RTP) system calibrated with a chromel/alumel thermocouple. In addition, Ge^+ implanted (100) and (111) Si wafers without a B^+ implant, and unimplanted Si reference wafers were also investigated. For these samples the thickness of Ti deposited was 45 nm, the preannealing condition was 550°C, 30 min, and the silicidation ambient was Ar.

Plan-view and cross-sectional transmission electron microscopy (X-TEM), X-ray diffraction (XRD), four point-probe sheet resistance measurements, and Auger electron spectroscopy (AES) were used to determine the resistivity, microstructure, crystalline phase and the chemical composition of the silicides.

RESULTS AND DISCUSSION

Figure 1 compares TEM plan-views of titanium silicide formed on Ge^+ and B^+ implanted Si without (concurrent) and with preannealing (conventional) before metallization. The grain size is about 30 to 80 nm for the concurrent process (Fig.1a), and from 100 to 300 nm in diameter for the conventional process (Fig.1b). Figure 2 is a cross-sectional TEM of the same samples which reveals that the $TiSi_2$/Si interface is much smoother for the concurrent process. The amplitudes of interfacial roughness are about 15 nm and 40 nm, respectively. For the silicide thicknesses of about 65 nm and 70 nm, sheet resistances of 3 and 12 Ω/sq were measured for the two processes, yielding electrical resistivities of approximately 20 and 84 $\mu\Omega$-cm for the two processes.

X-ray diffractometry results in Fig.3 show that the large differences in resistivity can be correlated with different crystal structures of $TiSi_2$: C49 for the conventional process and C54 for the concurrent process. The strongest peak for the conventional sample is a (131) reflection from the metastable C49 $TiSi_2$ phase (Fig.3b), while the strongest peak for concurrent sample is identified as (311) of C54 $TiSi_2$ (Fig.3a). When the conventional sample underwent an additional annealing at 850°C for 10 sec, its XRD spectrum changed to that shown in Fig.3c, yielding the expected conversion from C49 to the C54 $TiSi_2$ phase. This was accompanied by a resistivity reduction to about 19 $\mu\Omega$-cm.

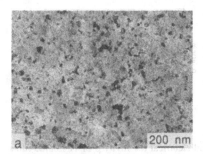 

Fig.1 TEM plan-views showing the TiSi₂ grain size following a 650°C,120 sec RTP on Ge⁺ and B⁺ implanted Si substrates which (a) were not preannealed, called concurrent process, and (b) were preannealed, called conventional process, prior to the silicidation reaction.

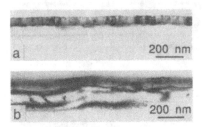 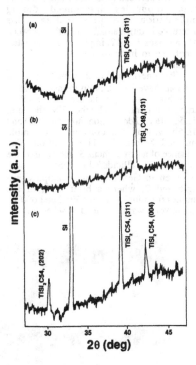

Fig.2 TEM cross-sectional view showing TiSi₂/Si interface roughness for the (a) concurrent and (b) conventional samples shown in plan-views in Fig. 1.

In order to explore the mechanism of the difference in the silicides formed via conventional and concurrent processes, a series of Ti silicide films have been generated on Ge⁺ implanted (100) and (111) Si. Figure 4 summarizes the results of phase composition and electrical resistance measurements. The ratio of XRD intensity from (311) of C54 TiSi₂ and from (131) of C49 TiSi₂ was taken as an indication of the relative quantity of the two phases, since these are the dominant reflections in Fig.3. It is evident that while C54 is the dominant phase in the samples without preannealing (process 1), C49 is essentially the only phase in the preannealed sample (process 2). The electrical resistivity obtained is consistent with the XRD results, if the standard resistivities of C54 and C49 phases are taken. This situation is true for both (100) and (111) Si. The results of an unimplanted sample are also included in Fig.4 (process 3). The

Fig.3 X-ray diffraction spectra of TiSi₂ formed during a 650°C, 120 sec RTP on: (a) Ge⁺ and B⁺ implanted Si; (b) sample with 550°C, 30 min preannealing before metallization; (c) sample with preannealing and 850°C, 10 sec additional annealing after silicidation.

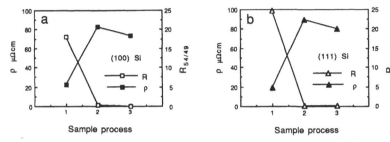

Fig.4 C54 to C49 phase composition ratio R, and electrical resistivity ρ of Ti silicide films formed during 650°C, 100 sec RTP on (100) and (111) Si. The wafers were Ge⁺ implanted without or with preannealing, and unimplanted, shown by 1,2,3 on the abscissa, respectively.

difference between processes 2 and 3 may be from impurities introduced during implantation and preannealing. This difference is much smaller than that between the processes 1 and 2, implying that the latter is not due to impurities introduced during preannealing.

The energy and dose of the implanted Ge⁺ ions have been chosen to insure that an amorphous Si layer is created. Such a structure is confirmed by the TEM shown in Fig.5a (after Ti deposition, no preanneal) and Fig.5b which is a conventional sample annealed before the Ti deposition. The influence of an amorphous Si substrate on the grain size and silicide/silicon interface was also seen in Ge⁺ implanted samples, see Fig.6 and 7, where a brief RTA of the samples (nominal 5 sec at 550°C and 600°C, respectively) was performed. A schematic is

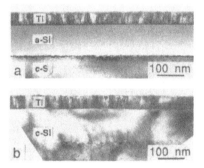

Fig.5 TEM cross-sectional view of Ge⁺ implanted (100) Si after Ti deposition, (a) without, and (b) with preannealing prior to Ti deposition.

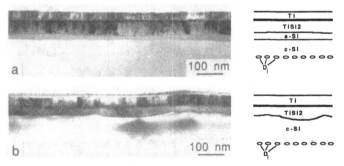

Fig.6 TEM cross-sectional view of samples following silicidation during 550°C, 5 sec RTP on Ge⁺ implanted (100) Si for (a) concurrent and (b) conventional samples. D_l represents the line of end-of-range dislocation loops corresponding to the a/c interface in Fig.5a.

attached to Fig.6 to identify the two thick layers formed on the top of Si. They are Ti (oxygen rich, close to Ti_2O) and C49 $TiSi_2$, as determined by XRD and AES (Fig.8) and confirmed by volume change calculations. AES did not show detectable O or Ge in the silicide. There is also a layer between the Ti and C49 $TiSi_2$, which may be TiSi and/or Ti_5Si_3.

There is little free energy difference between C49 and C54 $TiSi_2$, therefore the formation of C49 prior to the C54 phase in $TiSi_2$ is not fully understood [2,7]. A plan-view TEM microphotograph, see Fig.9, shows a single grain situated in the center

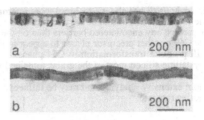

a 200 nm

b 200 nm

Fig.7 Ti silicide formed during 600°C, 5 sec RTP on Ge+ implanted (100) Si, without (a), and with (b) preannealing.

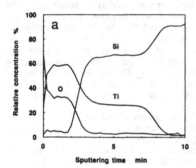

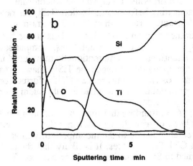

Fig.8 Depth profile from AES data, for Ge+ implanted, 550°C, 5 sec RTP annealed (100) Si, without (a), and with (b) preannealing.

surrounded by grains containing a high density of striations. From the special orientation distribution of these striations in the surrounding grains, it is evident that the striations are microtwins generated by stresses connected with the growth of the central grain. Similar striations, but randomly distributed in orientation, have been identified as stacking faults and are a characteristic feature of the C49 phase [2]. Besides Fig.9 we have seen striations with preferential orientation distribution due to stress also. So we propose that at the center in Fig.9 there is a growing C54 grain. The strain in phase transformation induced microtwins in the surrounding C49 grains and contributes to a certain extent to the barrier of the transition from the metastable C49 $TiSi_2$ to the equilibrium C54 $TiSi_2$. There is such a possibility, because the reaction of Ti with Si to form C49 $TiSi_2$ requires a smaller volume change than that which occurs for the C54 phase.

Fig.9 Non-random distribution of striations in C49 $TiSi_2$ grains.

The reason of the above mentioned dependence of silicide microstructure on the state of the Si substrate is believed to be related to the latent energy of the amorphous silicon. The amorphous Si has about 12 kJ/mol more internal energy than the crystalline Si [8], therefore, the reaction of Ti and Si at the interface will have a larger driving force on amorphous Si to overcome any encountered barriers than on the crystalline Si substrate. This energy will help the C49 or other precursor phases to appear in finer grains, resulting in fine grains of C54, and will promote transformation to C54 phase. Conversely, for the silicide film formed on the crystalline silicon most of the grains remained as C49 $TiSi_2$, which gives the high resistivity. The temperature range in which the effect of the substrate crystallinity is dominant requires more extensive studying and may be influenced by other factors such as impurities, ambient, heating rate, etc.

SUMMARY AND CONCLUSION

Amorphous and crystalline silicon substrates have been prepared by ion implantation and preannealing before metallization. $TiSi_2$ formed on amorphous Si has more tendency to have a C54 structure rather than C49. The grain size is smaller and the silicide/silicon interface is smoother than on crystalline Si. This is true for both Ge^++B^+ implanted and Ge^+ implanted Si and for (100) or (111) orientations. Comparison with unimplanted Si substrates indicates that any intentionally or unintentionally introduced impurities during implantation and preannealing did not appear to influence the $TiSi_2$ phase formation essentially. On this basis we propose that the microstructure difference of $TiSi_2$ formed on ion implanted, without and with preannealing before metallization, is situated in the latent energy of the amorphous layer of Si induced by the ion implantation.

A nonrandom distribution of striations in C49 $TiSi_2$ grains around a newly formed grain has been observed. A model that the observed striations are microtwins is presented. Their appearance is due to the incompatibility in volume and the transformation stress between C49 and C54 $TiSi_2$ phases. It is likely that the difference in volume reduction for C49 and C54 formed in Si-Ti reaction should be a reason for the easier generation of C49 prior to C54.

ACKNOWLEDGMENTS

The authors wish to acknowledge the NSF Engineering Research Center Program through the Center for Advanced Electronic Materials Processing (Grant CDR 8721505), MCNC, the John and Karin Engblom Foundation in Sweden, The Royal Institute of Technology in Sweden, and the Semiconductor Research Corporation for financial support.

REFERENCES

1. Z.G. Xiao, H. Jiang, J. Honeycutt, C.M. Osburn, G. McGuire, and G.A. Rozgonyi, J. Electrochem. Soc. 136, 296c (1989).
2. R. Beyers and R. Sinclair, J. Appl. Phys. 57, 5240 (1985).
3. F.M. d'Heurle, P. Gas, I. Engstrom, S. Nygren, M. Ostling, and C.S. Petersson, IBM Research Report, Solid State Pysics, RC 11151 (1987).
4. R. Beyers, D. Coulman, and P. Merchant, J.Appl. Phys. 61, 5110 (1987).
5. C.Y. Ting, F.M. d'Heurle, S.S. Iyer, and P.M. Fryer, J. Electrochem. Soc.133, 2621 (1986).
6. S.P. Murarka, M.H. Read, C.J. Doherty, and D.B. Fraser, J. Electrochem. Soc. 129, 293 (1982).
7. H.J.W. van Houtum and I.J.M.M. Raaijmakers, MRS Symposia Proc. 54, 37 (1986).
8. E.P. Donovan, F. Spaepen, D. Turnbull, J.M. Poate, and D.C. Jacobson, J. Appl. Phys. 57, 1795 (1985).

This paper also appears in Mat. Res. Soc. Symp. Proc. Vol 182

Geometrical effects and disintegration of narrow TiSi2/poly-Si lines

H. Norström, K. Maex and P. Vandenabeele

Interuniversity Microelectronics Center (IMEC v.z.w.),
Kapeldreef 75, 3030 Leuven, Belgium

Abstract

The geometrical shape and the thermal stability of the TiSi2/poly-Si interface on narrow lines has been studied. The examined line-widths varied between 0.8 μm and 1.5 μm. The thermal stability was found to strongly correlate to the actual line-width of the structures. At the onset of degradation, at and above 900 °C, narrow lines were observed to disintegrate at a much faster rate than wider ones. Cross-sectional microscopy (TEM and SEM) revealed the TiSi2/poly-Si interface to be curved inwards. The interface bowing was found to be more pronounced on narrow lines. It is suggested that the interface bowing results from a mechanical pinning of the TiSi2/poly-Si interface by the side-wall spacers.

Introduction

Reaction of thin films of titanium with mono- and poly-crystalline Si has recently been adopted by manufacturers of VLSI-circuits, to simultaneously lower the sheet resistance of poly-Si runners and to form self-aligned reacted contacts to source and drain areas of MOS devices. The use of polycide structures for interconnections requires very good stability upon high temperature treatments. A few studies have already addressed the limited temperature stability of the TiSi2/poly-Si system [1-5]. The major part of these investigations were, however, performed on large area structures suitable for C-V measurements and surface analysis.

Since silicides are applied on comparatively long and narrow poly-Si lines and in narrow openings on mono-Si, we have focussed our attention on the behavior of silicides on confined structures, ranging from 0.8 to 1.5 μm, and their thermal stability.

Experimental

Boron doped silicon wafers, (100)-orientated, with a diameter of 125 mm served as starting material. The wafers were thermally oxidized to get 300 nm of oxide prior to the deposition of 400 nm of poly-Si. In a few occasions, the mono-Si served as a basis layer during later silicidation.

Mat. Res. Soc. Symp. Proc. Vol. 181. ©1990 Materials Research Society

The poly-Si layer was patterned with a line-width reticle, using an I-line stepper. Subsequently, the poly-Si was dry-etched to define the integrated Van der Pauw and bridge-resistor structures. Approximately 400 nm TEOS was deposited on the wafers to allow for side-wall spacer formation, via etch-back by RIE.

Part of the structures were implanted with $2E15/cm^2$ of As or B. The dopants were activated by RTP. Then, 60 nm of titanium was sputter deposited. $TiSi_2$ was formed on all wafers by reacting the metal and Si in an RTP system at 730 °C. A second RTP-step at 850 °C was done after etching off unreacted metal. All wafers received 30 nm of oxide as an anneal cap, prior to further heat treatment in a furnace.

Results and discussion

(i) Electrical measurements

In previous work [6] we have reported on the thermal stability of $TiSi_2/poly-Si$ structures with narrow lines as a function of the dopants introduced into the poly-Si in the temperature range from 700 °C to 950 °C. Electrical information about the thermal stability of the bilayer was obtained from sheet resistance measurements on Van der Pauw structures in combination with line-width measurements of bridge resistors with a line width between 0.8 and 1.5 µm. To provide for statistical variation, about one hundred measurements were collected for each line-width and temperature setting. The thermal stability was observed to be strongly affected by the line width. At temperatures for onset of degradation (at and above 900 °C), narrow lines were found to disintegrate at a much higher rate than wider ones.

Figure 1, shows the percentage of failuring test sites as a function of the annealing temperature. In this particular case, the poly-Si was implanted with $2E15/cm^2$ of boron prior to silicidation.

Doping the poly-Si with high concentrations of arsenic or boron was found to improve the thermal stability, in agreement with previous findings by Lippens et al. [7].

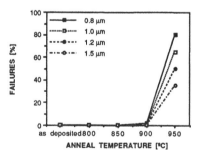

Fig.1: The number of failuring sites as a function of annealing temperature. The poly-Si basis layer was implanted with $2.E15/cm^2$ of boron prior to silicidation.

(ii) Physical characterization of narrow silicide runners

After electrical measurements the $TiSi_2$ film was etched off in diluted HF. At higher temperatures, chains of Si precipitates were observed in the middle of narrow lines. It was concluded that these precipitates are responsible for the increasing sheet resistance and finally lead to complete failure. At the same time another interesting phenomenon was observed, which will be discussed in more detail. Cross-sectional SEM- and TEM-micrographs of the polycide lines clearly revealed the $TiSi_2/poly$-Si interface to suffer from bowing. The effect was much more pronounced on narrow lines. Moreover, bowing was observed on both silicided poly-Si runners and on silicided mono-Si runners, figure 2a and 2b.

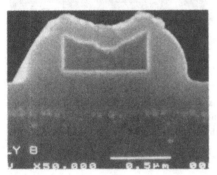

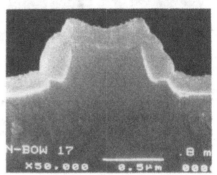

Fig. 2a: Cross-sectional SEM micrographs of silicided structures on poly-Si.

Fig. 2b: Cross-sectional SEM micrographs of silicided structures on mono-Si.

The cross-sectional TEM micrograph of fig. 3 shows an arsenic doped polycide runner. The $TiSi_2/poly$-Si interface is clearly seen to be bowed. No enlarged grain growth at the edges seems to occur as in the case of Phillips et al. (5). In their experiment, narrow $TiSi_2$ lines were formed on unpatterned poly-Si. They concluded that an increased grain growth at the silicide edge, together with a reduced stress was responsible for the observed interface bowing.

Fig.3: Cross-sectional TEM micrograph of $TiSi_2$ on boron doped poly-Si.

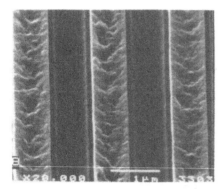 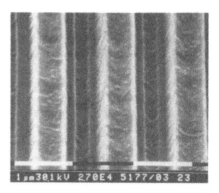

Fig.4a: Plan-view SEM micrographs of silicided stripes after removal of the TiSi$_2$ by wet etching on poly-Si.

Fig.4b: Plan-view SEM micrographs of silicided stripes after removal of the TiSi$_2$ by wet etching on mono-Si.

The plan view SEM micrographs (fig. 4a and 4b), recorded after etching off the silicide in diluted HF, confirm, both for silicided poly-Si and for silicided mono-Si, that no Si precipitation has occured, although the lines appear to be shaped concave at the center.

(iii) A model for interface bowing

Since grain growth of the poly-Si or Si precipitation at the line edges is not observed in our experiments using patterned Si lines, more experiments were carried out to elucidate this point. It is obvious from the TEM and SEM micrographs that the silicide appears to be pinned at the top of the spacer. It was therefore decided to overetch the spacer and monitor the silicide growth, before and after the reaction front reached the top of the spacer.

Therefore the spacer was partly removed by wet etching, whereupon 40 nm and 80 nm of Ti was deposited on the respective left and right part of the wafer. The cross-sectional micrographs, recorded after complete silicidation are shown in figs. 5a and 5b. Part of the spacer is seen to remain, but it is reduced in height. The unexpected shape of the spacers is due to the different etch-rates of the TEOS-oxide and the thermal oxide. For the thinner silicide, it is readily seen that there is no contact between the silicide and the spacer remnants and as such the silicide remains flat. For the thicker silicide, a completely different situation occurs. The evolution of the final shape is schematically illustrated in fig. 6a-c. When the horizontal interface levels with the spacer, the interface will eventually start to bow. The upper silicide is by that time already curved as a result of the reaction of the sidewall poly-Si with the covering Ti.

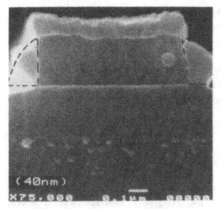 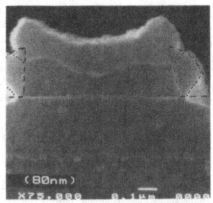

Fig.. 5a Cross-sectional SEM-micro-graphs of TiSi₂/poly-Si lines which were silicided after an etch-back of the spacers. The location of the spacers prior to etching is shown by the dashed profile. TiSi₂ formed via deposition of 40 nm of Ti on poly-Si.

Fig. 5b: Cross-sectional SEM-micro-graphs of TiSi₂/poly-Si lines which were silicided after an etch-back of the spacers. The location of the spacers prior to etching is shown by the dashed profile. TiSi₂ formed via deposition of 80 nm of Ti on poly-Si

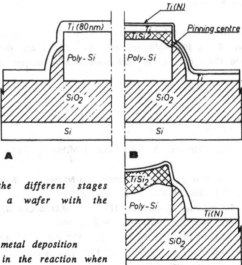

Fig. 6: Schematic of the different stages during silicidation of a wafer with the spacers etched back:

   a) directly after metal deposition
   b) at the stage in the reaction when the silicide front reaches the top of the spacers
   c) after completion of the reaction, but prior to selective etch

The model that the spacer top serves as a mechanical pinning center for the silicide is in agreement with the observed independence of the Si grain structure, the dependence on the line width and also the dependence on the silicide thickness.

Since more Si is taken up by a thicker deposited Ti film, a more severe bowing is expected. The mechanical pinning can be due to a small reaction of the Ti with the spacer material or more probably by the very rigid TiN top layer, which is formed on top of the silicide.

Conclusion

In addition to a disintegration of narrower $TiSi_2$/poly-Si runners at lower temperatures, an important bowing effect of the $TiSi_2$/poly interface was observed. All observations point to the idea that the top of the spacer edge serves as a mechanical pinning center. The reduction of the poly-Si thickness in the middle of narrow lines can have important consequences for the gate oxide breakdown. Moreover, a severe interface bowing on narrow lines may eventually yield a smaller effective thickness of the silicide and can therefore, result in a worse thermal stability.

Acknowledgements

The authors want to acknowledge the support of the IC processing facility at IMEC for the fabrication of the line-width test-vehicles. Thanks are also due to the following: R.Verbeeck and C.Alaerts for the SEM work and A.Romano and J.Vanhellemont for the TEM work.

References

1. C.Y.Ting, F.M.d'Heurle, S.S.Iyer and P.M.Fryer; J.Electrochem.Soc. 133, 1986, pp.2621

2. S.Nygren, M.Ostling, C.S.Petersson, H.Norstrom, K-H.Ryden, R.Buchta and C.Chatfield; Thin Solid Films, 168, 1989, pp.325

3. K.Shenai, P.A.Picante, G.A.Smith, N.Lewis, M.D.McConell, J.F.Norton, E.L.Hall and B.J.Baliga; Mat. Res. Soc. Proc.,106, 1988, pp.149

4. R.K.Shukla and J.S.Multani; Proceedings VMIC-87, 1987, pp.470

5. J.R.Phillips, L-R.Zheng and J.W.Mayer; Mat. Res. Soc. Proc., 106, 1988, pp.155

6. H.Norström, K.Maex and P.Vandenabeele; submitted to Am. Journal of Vacuum Sci. and Technol.

7. P.Lippens, K.Maex, L.Van den hove, R.De Keersmaecker, V.Probst, W.Koppenol and W.van der Weg; Journal de Physique, 49, 1988 pp.C4-191

THE EFFECT OF AMORPHOUS SILICON LAYER IN
PE-CVD TITANIUM POLYCIDE GATE DIELECTRICS

SHIH-CHANG CHEN, AKIHIRO SAKAMOTO, HIROYUKI TAMURA,
MASAKI YOSHIMARU AND MASAYOSHI INO
OKI Electric Industry Co. Ltd., VLSI R & D Center
550-1, Higashiasakawa, Hachioji, Tokyo, Japan 193

ABSTRACT

Titanium silicide ($TiSi_x$), used as polycide gate consists of $TiSi_{1.1}$ and amorphous silicon (a-Si), was deposited by Plasma Enhanced Chemical Vapor Deposition method (PE-CVD). The effect of a-Si layer in PE-CVD Ti polycide gate dielectrics has been studied. In order to evaluate the a-Si layer effect, three types of samples were prepared on gate SiO_2 film with following structures : a) a-Si / $TiSi_{1.1}$ / a-Si / phosphorus (P) doped poly-Si, b) a-Si / $TiSi_{1.1}$ / non-doped poly-Si / P doped poly-Si and c) a-Si / $TiSi_{1.1}$ / P doped poly-Si, respectively. Furthermore, in order to avoid the influence of native oxide existence at the interface, the pre-cleaning treatment was performed in-situ on the poly-Si film surface before $TiSi_{1.1}$ film deposition. The gate dielectric strengths of these samples indicate that the gate dielectric degradation in PE-CVD Ti polycide gate is greatly dependent on Si under layer crystallization. It is effective using a-Si film as the under layer in decreasing the gate dielectric degradation . This is due to the Ti oxide interlayer, formed at the interface of $TiSi_{2.0}$ and poly-Si films, which restrains the $TiSi_x$ local penetration.

INTRODUCTION

The delay-time in devices has been increasing as the VLSI's dimension has been scaling down. Refractory metal silicides are useful materials to solve this problem [1]. As $TiSi_x$ has the lowest resistivity in silicides, it has being considered as one of the superior electrode materials in advanced VLSI's [2].

Furthermore, since the PE-CVD $TiSi_x$ film can be formed in comparatively low deposition temperature and achieved the lowest resistivity at 650°C annealing [3], it has attracted much attention lately. Characteristics of deposition, thermal stability and interface reaction during annealing of PE-CVD $TiSi_x$ films has been reported [4,5]. Moreover, it has been reported that in Ti polycide gates, the existence of native oxide on poly-Si film surface causes the local $TiSi_x$ penetration and degrades the gate dielectrics [6]. But only a few papers have reported the electric characteristics dependence of PE-CVD $TiSi_x$ film on under layer characters. In this work, in order to evaluate the effect of a-Si layer, the dependence of gate dielectrics on Si under layer crystallization in PE-CVD Ti polycide gate has been studied.

EXPERIMENTAL PROCEDURES

The 6-inch P(100) Si substrates with 20nm thickness gate oxide film formed by thermal oxidation were employed. After

poly-Si film deposition at 620°C, the P diffusion was performed
into poly-Si film. At last, two kind of films were deposited.
One was prepared with TiSi1.1 and a-Si films continuous
deposition, and another was prepared with only TiSi1.1 film
deposition employing PE-CVD method, respectively.

In order to avoid the influence of native oxide in gate
dielectrics, the pre-cleaning treatment on poly-Si film surface
was performed in-situ before TiSi1.1 film deposition. And for
obtaining good film uniformity, the as-deposited TiSix film
composition was limited at 1.1.

Annealing was performed at 800°C in N2 ambient for 30
minutes. Analysis was performed by the gate dielectric strength
measurement, Auger Electron Spectroscopy (AES) analysis and
Scanning Electron Microscope (SEM) observation. The gate
dielectric strength is assumed with the forced voltage at 1uA
current flow.

EXPERIMENTAL RESULTS AND DISCUSSIONS

The gate dielectric strength of sample with or without 100
nm thick a-Si layer after anneal was shown in figure 1. In order
to avoid the native oxide influence, the pre-cleaning treatment
on poly-Si film surface in-situ before TiSi1.1 film deposition
was performed in all samples. In samples with a-Si layer, only
employing 150 nm thick poly-Si film is enough to obtain over 90
% of gate with intrinsic dielectric strength (> 8MV/cm). But in
sample without a-Si layers, even increased poly-Si film
thickness to 450 nm thick, the gate with intrinsic dielectric
strength is 0 %. This shows clearly that the gate dielectrics
depend greatly on the under layer. When comparing the film
characters between a-Si and P doped poly-Si films, it is
considered that the crystallization of the Si film is one of the
most important factors in the under layer.

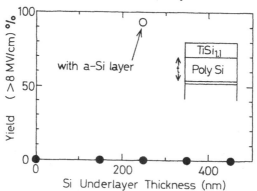

Fig.1 Gate intrinstic dielectric strength of Ti polycide
as a function of Si under layer thickness.

Therefore, one more sample was prepared, and employed
together with samples described above, to evaluate the Si under
layer crystallization dependence. These sample structures are
shown in figure 2. Sample a) and c) were prepared as used above,
sample b) was prepared with using 100 nm thick non-doped poly-Si
film instead of a-Si layer. All samples had the same total

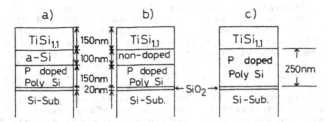

Fig.2 Sample preparation conditions in evaluating
the dependence of Si film crystallization.

thickness in under layer, and had annealed before the gate
breakdown voltage measurement performed.

The gate dielectric strength of these samples are shown in
figure 3. Sample a), b) and c) have gate intrinsic dielectric
strengths of 92%, 30%, and 0%, respectively. In the fact,
comparing grain size among these samples, there has a great
difference in grain size especially between a-Si film and P
doped poly-Si film. This indicates clearly that the gate
dielectric strength degraded as the Si under layer crystallized.

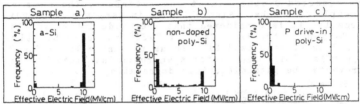

Fig.3 Effective electric field distribution variations
as a function of Si under layer characters.

Generally, the gate dielectrics is affected by the reaction
of Ti and Si, and also affected by the penetration of $TiSi_x$ [7].
Therefore, it is considered that the interface reaction was

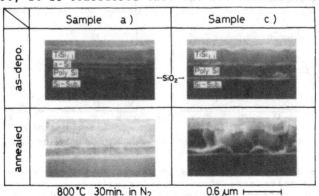

Fig.4 SEM cross-sectional observation of Ti
polycide gate with or without a-Si layer.

caused differently in these samples by different Si under layer crystallization. All samples were observed by SEM observation, and Sample a) and c) were shown in figure 4 with cross-sectional view. After anneal, in sample a), it has been observed that Ti and Si reaction was restrained at the interface of TiSi$_x$ and poly-Si films, and the interface is comparatively smooth. Furthermore, a new layer formed at the interface has been observed. But in sample c), the locally broken interface caused by the TiSi$_x$ penetration, and the penetration of TiSi$_x$ into gate oxide have been observed. The same interface reaction like this has also been observed in sample b). It shows clearly that the TiSi$_x$ penetrating behavior is affected by the crystallization of Si under layer and effects the gate dielectric degradation.

The penetration of TiSi$_x$ and the nature of the new interlayer as described above were examined by employing AES analysis. The results of sample a) and c) are shown in figure 5. When comparing the annealed spectra with the as-deposited spectra, in sample a), the Ti spectrum extended to the interface, but stopped in poly-Si layer. And oxygen (O) was detected at the interface. From these detected Ti, O and Si spectra, it is considered that this new layer is a Ti oxide layer containing a little Si. In sample c), the extension of Ti spectrum into the gate oxide region has been observed and there has no detected O spectrum at the interface. The same result like sample c) has been also obtained in sample b). According to this result, it is considered that the oxide interlayer is able to be formed at the interface of TiSi$_x$ and poly-Si films by using a less crystallized Si film as the under layer for PE-CVD TiSi$_{1.1}$ film. This oxide interlayer restrains TiSi$_x$ penetrating and prevents the gate dielectric degradation.

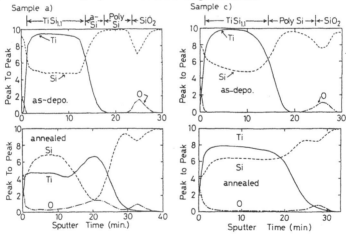

Fig.5 Auger depth profiles of PE-CVD Ti polycide gate with (left) or without (right) a-Si layer.

In order to evaluate the effect of this oxide interlayer in gate dielectrics more clear, samples with thick and thin oxide interlayer formed by selected process procedure have been prepared. These samples are shown with SEM cross sectional view in figure 6. The gate dielectric strength is shown in figure 7. The result shows that over 90% gate intrinsic dielectric

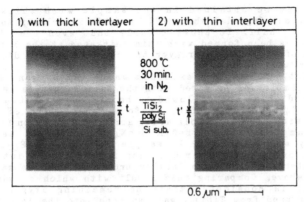

Fig.6 SEM cross-sectional view of samples
with thick or thin oxide interlayer.

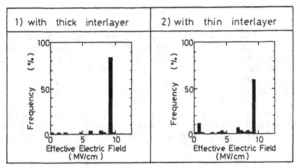

Fig.7 Effective electric field distribution difference
between with thick and thin oxide interlayer.

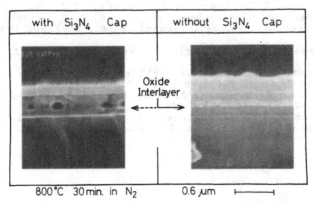

Fig.8 SEM cross-sectional observation of oxide interlayer
in samples with or without surface Si3N4 capped.

strength can be obtained in the sample with thick oxide interlayer, but degraded about 20% as oxide interlayer thickness decreased to half. It shows clearly that even the Ti oxide interlayer has been formed with thin thickness at the interface, the effect of Ti oxide interlayer to prevent the gate dielectric degradation will decrease.

At last, it has been examined where the oxygen diffused from. The sample was prepared with the same structure as sample a) described above and followed to be capped with 100nm thick PE-CVD Si_3N_4 film on surface before anneal. The SEM cross sectional view of this sample after anneal is shown in figure 8. Another sample without PE-CVD Si_3N_4 film capped is shown comparably at right hand. In sample with PE-CVD Si_3N_4 film capped, no oxide interlayer has been formed at the interface. And the interface is become locally broken as like as sample c) described above. Comparing this result with which of sample a) and c), it is considered that the remaining oxygen in the furnace diffused from $TiSi_{1.1}$ gate surface into the interface of $TiSi_x$ and poly-Si films during annealing, and formed Ti oxide interlayer while a-Si film consumed in silicidation.

CONCLUSIONS

In PE-CVD $TiSi_{1.1}$ polycide gate, employing a-Si film as the under layer for $TiSi_{1.1}$ film had much better gate dielectrics in comparing with non-doped poly-Si and P doped poly-Si films. It shows that the gate dielectrics is dependent on Si under layer crystallization. When used a less crystallized film as the under layer, the Ti oxide interlayer containing Si has been formed at the interface of $TiSi_x$ and poly-Si films after anneal. In case of employed a-Si under layer, the interlayer with enough thickness has been formed. This oxide interlayer restrains $TiSi_x$ penetrating and prevents the gate dielectric degradation.

REFERENCES

1)S.P. Murarka, Silicides for VLSI Applications, (ACADEMIC PRESS INC., New York, 1983), p.9.
2)C.Y. Ting, S.S. Iyer, C.M. Osburn, G.J. Hu and A.M. Schweighart, J. Electrochem. Soc. Abs. 82-2, 254 (1982)
3)D.G. Hemmes, J. Vac. Sci. Technol. B4(6), Nov/Dec, 1332-1335 (1986).
4)R.S. Rosler and G.M. Engle, J. Vac. Sci. Technol. B2(4), Oct/Dec, 733-737 (1984).
5)A.E.Morgan, W.T. Stacy, J.M. DeBlasi and T.Y. James Chen, J. Vac. Sci. Technol. B4(3), May/Jun, 723-731 (1986).
6)M.Tanielian, R. Layos and S. Blackstone, J. Electrochem. Soc. 132 (6), June, 1456-1460 (1985).
7)D. pramanik, S. Blackstone, M. Tanielian, R. Lajos and A. Brandes, J. Electrochem. Soc. Abs. 84-2, 524 (1982).

COBALT SILICIDE FORMATION ON POLYSILICON: DOPANT EFFECTS ON REACTION KINETICS AND SILICIDE PROPERTIES.

A.R. SITARAM * AND S.P. MURARKA
Materials Engineering Department/CIE, Rensselaer Polytechnic Institute, Troy, NY 12180.
*now with APRDL, Motorola Inc., 3501, Ed Bluestein Boulevard, Austin, TX 78721.

ABSTRACT

The concept of using cobalt disilicide as a self-aligned metallization scheme for gate/interconnection and contact formation is gaining wide acceptance. The silicide formation on the gate will require a Co-metal film interaction with the underlying polysilicon that may be doped heavily. We have investigated the silicide formation kinetics during this reaction. The effect of different dopants (B,P, and As) and their concentration on the Co-Si interaction and the dopant redistribution during such reactions were investigated. The results from these studies will be presented and discussed.

INTRODUCTION

Submicron and shallow junction technology has led to the lowering of process temperatures to a point where Group VIII metals can be considered for use in gate level, contact, and interconnect metallization. In addition to possessing the advantages of the refractory metal silicides, these silicides are self aligned, which make them highly attractive. Of the silicides of Ni, Co, Pt, and Pd, $CoSi_2$ is seen to offer numerous advantages which make it an ideal choice for this purpose [1,2].

In a typical scheme involving a silicide gate and interconnect pattern, the silicide would be formed on heavily doped polysilicon. Comparatively little data is available on the effect that high concentrations of dopants exert on the Co-Si interaction. Only Murarka and Lloyd [3] studied the reaction of cobalt with polysilicon that was doped heavily with phosphorus, and observed a retardation of the reaction. Most of the remaining work in this area has focussed mainly on the stability of the silicide on heavily doped silicon [4], its tendency to form compounds with the dopant [4,5], and on the main moving specie during the reaction [6,7]. Since the importance of understanding the effect of the dopants on silicide formation cannot be overstressed, this study was undertaken. Arsenic, boron, and phosphorus were used to dope the polysilicon and their influence on the silicide reaction was then monitored.

EXPERIMENTAL

Silicon wafers, (100), 75mm, n-type, and with a typical resistivity of 10 Ω-cm were subjected to an RCA clean [8]. This was followed by the growth of approximately 1300 Å of thermal oxide on the wafers. The next step was the deposition of around 5000 Å of undoped polysilicon at 625°C. This polysilicon was then diffusion doped with either phosphorus or boron in a diffusion furnace, using $POCl_3$ and BBr_3 sources, respectively. No in situ doping of the polysilicon was carried out. The diffusion temperatures ranged from 900 °C to 1000 °C, from 15 to 60 minutes. A different set of wafers had arsenic implanted into them. Since each Å of deposited cobalt would consume 3.6 Å of underlying silicon, the region damaged by the implant would be consumed by the forming silicide. Hence, no activation anneal was performed after the implant, in order to see whether an additional step could be avoided during actual processing. The details regarding the diffusions are presented in Table I. As was implanted into the polysilicon films at 30 and 100 keV, for doses of 3E15, 1E16, and 3E16 cm^{-2} at each energy. The range, straggle, and damage estimates due to the implants were obtained using TRIM [9]. Following the doping, the wafers were subjected to a dilute (50:1) HF dip for two minutes, and then an in situ backsputter etch in the sputter chamber which removed around 100 Å of material from the surface of the wafer. This ensured the removal of the native oxide and ensured a clean and reproducible interface. The thickness of the cobalt films ranged from 700 to 1400 Å, and are listed in Table I. The polysilicon films with the arsenic implants had 1400 Å of cobalt on them. Cobalt deposited on undoped polysilicon films was used as control during the experiment. Specimens from all these wafers were annealed in a furnace for 20 to 80 minutes at temperatures ranging from 400 to 700 °C. In an effort to draw some parallels between furnace and rapid thermal annealing, some samples underwent RTA for 15-60 s at temperatures ranging from 400 to 700 °C. The annealing ambient in all cases was Ar containing 3% H$_2$. Analysis of the extent of reaction in the samples was by sheet resistance measurement, RBS, and XRD. Since grain growth in polysilicon was observed during diffusion doping with both boron and phosphorus [10], TEM was used to determine the grain sizes of of the polysilicon films before and after doping with phosphorus or boron, in order to determine the effects of polysilicon grain size, if any, on the reaction.

RESULTS

The reaction of cobalt with a doped polysilicon substrates highlighted the fact that the dopants exerted considerable influence on the interaction. In the case of arsenic and phosphorus, this resulted in an overall retardation of the reaction, while in the case of cobalt on boron doped polysilicon, only the formation of CoSi$_2$ was retarded. While this was observed during furnace annealing, the RTA of cobalt on boron and phosphorus doped polysilicon resulted in a completely different dopant effect on the reaction. In all the cases, the concentration of the dopant played a significant part in the retardation of the reaction.

Figure 1 shows the variation of sheet resistance with temperature for cobalt reacting with polysilicon doped with phosphorus at 900 °C for 15, 30, and 45 minutes, and at 1000 °C for 30 minutes. An undoped polysilicon sample is used as control to show the Co-Si reaction proceeding in the absence of high dopant concentrations. It can be seen that an 80 minute furnace anneal results in complete CoSi formation in the case of the undoped polysilicon by 450 °C, with the subsequent decrease in R$_S$ caused by the formation of CoSi$_2$. In the case of doped polysilicon, the maximum R$_S$ is reached only by 550 °C for the specimens doped at 900 °C, and by 600 °C for those doped at 1000 °C for 30 minutes, thus highlighting the effect of dopant concentration. RBS spectra (Figure 2) revealed that for

Table I: Conditions during phosphorus and boron diffusion into polysilicon.

Dopant	Substrate	Temperature (°C)	Time (min)	R_s (Ω/sq.)	Grain Size (μm)	Co thickness (Å)
Boron	Poly-silicon	900	15	148.6	0.042	
			30	129.8	0.050	
			60	139.0	0.057	~900
		950	15	115.7	0.044	
			30	112.1	0.069	
			60	118.5	0.112	
Phosphorus	Poly-silicon	900	15	18.12	0.107	
			30	14.86	—	
			45	17.00	0.138	
		950	15	11.68	0.168	~800
			30	10.90	0.196	
			45	11.50	0.314	
		1000	30	10.02	0.471	
	(100)	950	30	8.58		
		1000	30	4.50		~1400
Control	Poly-silicon	–	–	–	–	~700
	(100)	–	–	–	–	~1400
	(111)	–	–	–	–	~1400

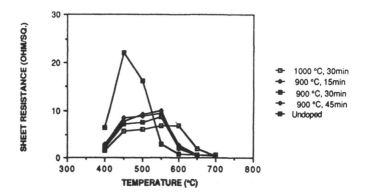

Figure 1. Variation of sheet resistance with temperature for cobalt on polysilicon, doped with phosphorus at 900 and 1000°C, following an 80 minute anneal.

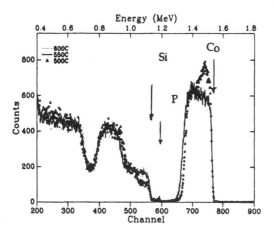

Figure 2. RBS spectra of cobalt on polysilicon, doped with phosphorus at 1000°C for 30 minutes, after furnace anneals of 80 minutes at 500, 550, and 600°C.

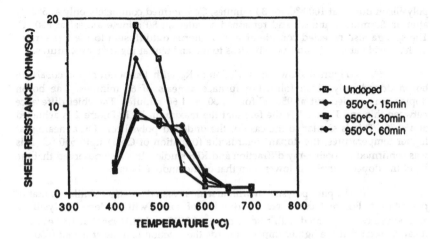

Figure 3. R_S vs. T following 60 minute furnace anneals of cobalt on boron doped polysilicon.

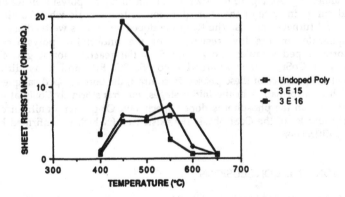

Figure 4. Rs vs. T for cobalt on As implanted (100 keV) polysilicon, furnace annealed for 60 minutes.

polysilicon doped at 1000 °C for 30 minutes, CoSi formed completely only at 550 °C after an 80 minute anneal, and remained so after an 80 minute anneal at 600 °C. The spectra also revealed considerable phosphorus outdiffusion to the top surface of the silicide at 450 °C, and possible loss to the ambient at higher temperatures.

Figure 3 shows the variation of R_S with temperature for cobalt on boron doped polysilicon following furnace anneals of 60 minutes. The boron doping was carried out at 950 °C for 15, 30, and 60 minutes. The chief difference between this and Figure 1 is the fact that the maximum R_S in Figure 2 is achieved at 450 °C. This is similar to the case of the undoped polysilicon. On annealing at higher temperatures, the dopant retards the formation of $CoSi_2$ upto 550 °C. This was confirmed by both x-ray diffraction and RBS studies. It is also observed that the R_S of the doped samples is lower than that of the undoped ones.

Doping polysilicon with arsenic revealed an effect similar to that of phosphorus. Figure 4 compares the reaction of cobalt with undoped polysilicon and polysilicon implanted with arsenic doses of 3E15 and 3E16 cm^{-2} at 100keV, and it can be seen that the lighter implant retards the reaction to a lesser extent (550 °C) than the 3E16 implant (600 °C) following furnace anneals of 60 minutes. On comparing the effect of implant energy on the reaction (Figure 5), it can be seen that the shallower implant retards $CoSi_2$ formation upto 550 °C, while the 100keV implant retards the reaction upto 600 °C. RBS results revealed that arsenic, like phosphorus, outdiffused to the top surface of the silicide and was lost to the annealing ambient. The deeper implants, however, tended to lose less arsenic while retarding the reaction to a greater degree. This is in agreement with the findings of Pai, et al [11].

RTA of cobalt on boron and phosphorus doped polysilicon revealed that the reaction during RTA differed considerably from that observed with furnace annealing. As Figure 6 shows, cobalt on undoped polysilicon follows the expected curve, in spite of the short duration of the anneals when compared to conventional furnace anneals. The R_S of the doped specimens were adjusted so as to compensate for the low resistivity of the underlying polysilicon. The phosphorus doped samples show a quick initial reaction forming CoSi. The formation of $CoSi_2$ is then retarded upto 650 °C. RBS and x-ray diffraction confirmed the presence of these phases. In contrast, the boron doped samples show a sluggish transformation in the initial stages, but transform to $CoSi_2$ at a lower temperature than the phosphorus doped specimens. Of greater significance is the extremely low R_S of the CoSi phase, the existence of which is confirmed by RBS and x-ray diffraction.

DISCUSSION AND CONCLUSIONS

Space restrictions limit the data that can be presented in this article, and for a fuller presentation of this work, the reader is referred elsewhere [12]. However, it is clear that phosphorus and arsenic, both n-type dopants, retard the reaction kinetics by affecting CoSi and $CoSi_2$ formation significantly. Boron, on the other hand, does not appear to affect the formation of CoSi to a great extent, but retards the formation of $CoSi_2$. It should be pointed out that CoSi formation occurs

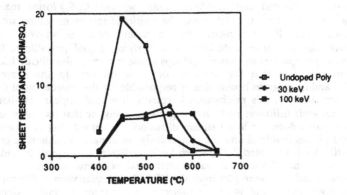

Figure 5. The variation of sheet resistance with temperature for cobalt on As implanted polysilicon (dose = 3E16 cm⁻²), following a 60 minute furnace anneal.

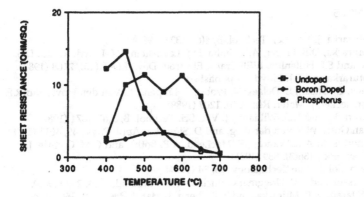

Figure 6. Sheet resistance variation during RTA of cobalt on phosphorus doped (1000°C, 30 minutes) and boron doped (950°C, 60 minutes) polysilicon.

predominantly by the diffusion of silicon atoms, whereas $CoSi_2$ forms mainly through the diffusion of cobalt [13]. Hence, the results of this investigation would appear to indicate that P and As retard silicon diffusion into Co_2Si, whereas boron retards cobalt diffusion from CoSi into the underlying doped polysilicon. Our observations of phosphorus and arsenic pileups at the surface of the silicide (CoSi) lead one to speculate that the presence of these dopants in the growing intermetallic, and in its grain boundaries, is responsible for the retardation of CoSi and $CoSi_2$ formation. If this mechanism is correct, it would appear that boron either does not stuff diffusion paths in the growing films, or that its presence in the diffusion paths does not affect silicon diffusion. On considering the melting points of the phases involved and the temperatures of anneal, considerable grain boundary diffusion is indicated [5,12], which lends credibility to the grain boundary plugging mechanism. However, actual experimental verification poses difficult problems. As regards to the results from the RTA, their explanation is difficult in light of the poor understanding of the subject. The apparent reversal of dopant behavior and the low resistivity CoSi phase formed, however, are extremely important.

Thus, it may be seen that the reaction kinetics of cobalt interacting with highly doped polysilicon depend on the type of dopant, its concentration, and its location in the polysilicon. N-type dopants retard both the formation of CoSi and $CoSi_2$, whereas boron retards only the formation of $CoSi_2$. A low resistivity CoSi phase is also observed during the reaction of cobalt with doped polysilicon.

ACKNOWLEDGEMENTS

The authors thank Dr. X-S. Guo of SUNY-Albany for RBS work, and Intel Corporation for partial financial support of this work. One of the authors (SPM) also acknowledges the financial support of SRC and SEMATECH Corporations.

REFERENCES

1. S.P. Murarka, J. Vac. Sci. Technol. B, 4(6), 1325 (1986).
2. S.P. Murarka, D.B. Fraser, A.K. Sinha, H.J. Levinstein, E.J. Lloyd, R. Liu, D.S. Williams, and S.J. Hillenius, IEEE Trans. Electron. Dev., ED-34(10), 2108 (1987).
3. S.P. Murarka and E.J. Lloyd, Unpublished.
4. K. Maex, G. Ghosh, L. Delaey, V. Probst, P. Lippens, L. Van den Hove, and R.F. De Keersmaecker, J. Mater. Res., 4(5), 1209 (1989).
5. S.P. Murarka and D.S. Williams, J. Vac. Sci. Technol. B, 5(6), 1674 (1987).
6. G.J. van Gurp, W.F. van der Weg, and D. Sigurd, J. Appl. Phys., 49, 4011 (1978).
7. R. Pretorius, M.A.E. Wandt, J.E. MacLeod, A.P. Botha, and C.M. Comrie, J. Electrochem. Soc., 136(3), 839 (198).
8. W. Kern, Solid State Technology, 15, 34 (1974).
9. J.P. Biersack and L.P. Hoggmark, Nucl. Instrum. Methods, 174, 257 (1980).
10. A.R. Sitaram, S.P. Murarka, and T.T. Sheng, J. Mater. Res., 5(2), 360 (1990).
11. C.S. Pai, F.A. Baiocchi, and D.S. Williams, J. Appl. Phys., 67(3), 1940 (1990).
12. A.R. Sitaram, Ph.D. thesis, Rensselaer Polytechnic Institute, Troy, NY, (1990).
13. F.M. d'Huerle and C.S. Petersson, Thin Solid Films, 128, 283 (1985).

MICROANALYSIS OF TUNGSTEN SILICIDE/POLYSILICON INTERFACE: EFFECTIVENESS OF IN SITU RIE CLEAN ON REMOVAL OF NATIVE OXIDE

Ronald S. Nowicki and Patrice Geraghty
Genus, Incorporated, Mt. View, CA 94043
and
David W. Harris and Gayle Lux
Charles Evans Associates, Redwood City, CA 94063

ABSTRACT

The presence of a thin (10-30Å) oxide ("native oxide") layer on a silicon surface prior to the deposition of another film on that surface can contribute to difficulties with subsequent device processing steps, e.g. contact metallization and high-temperature annealing or oxidation. Thus, the *in situ* process capability of "native oxide" removal affords an advantage over the conventional method of aqueous hydrofluoric acid cleaning prior to a film deposition step. This study describes such a technique, in which an *in situ* pre-deposition clean with C2F6 gas, using reactive ion etching (RIE) prior to tungsten silicide deposition, is employed. This technique allows post-silicide deposition high-temperature heat treatment and wet oxidation without loss of film adhesion or other obvious degradative effects. We also report the use of Secondary Ion Mass Spectrometry (SIMS) to show that this procedure has been effective in the removal of the oxide layer prior to silicide deposition. This study includes definition of the RIE etch parameters which provide acceptable etch selectivity of the oxide to silicon, and avoidance of excessive fluoropolymer formation on the silicon surface.

INTRODUCTION

The removal of the native oxide and cleaning of a silicon surface prior to contact of that surface with a CVD refractory silicide metallization has been necessary to minimize contaminants which can adversely affect subsequent device processing or electrical characteristics. For many years, the common method for such removal has been the use of aqueous, dilute hydrofluoric acid. However, this method can result in re-contamination of the surfaces being cleaned with metallic or other trace contaminants which can yield degradative effects in the electrical characteristics of the resultant devices [1]. Also, if device wafers are subsequently stored for any significant time in a non-inert ambient, the native oxide can regrow and yield similar adverse effects on device performance [2].

Therefore, an integrated processing tool which can provide removal of the native oxide in addition to maintenance of a relatively inert environment during transport and process sequencing prior to metallization, has an obvious advantage over the common, aqueous HF cleaning method.

The goals of our study were to remove this native oxide on a polysilicon surface with reasonable etch selectivity of oxide over polysilicon (ca. 2:1), deposit CVD tungsten silicide, and then examine the resultant polysilicon/silicide interface for native oxide removal using the SIMS analytical technique. A search of the literature revealed recent studies concerning removal and regrowth of native oxide [3-5]. Other early work has focused on the use of CF4/H2 as an etch gas for oxide removal [6]. Here, it was shown that the preferential etching of silicon oxide over polysilicon was a direct result of polymerization of the etch gas by-products on the silicon surface, which suppressed etching of the polysilicon. The resultant fluoropolymer has been thoroughly characterized by Oehrlein, et al. [7]. In our study, we chose not to use hydrogen to avoid any possible adverse consequences of the presence of this polymer on silicide film adhesion or subsequent processing.

Mat. Res. Soc. Symp. Proc. Vol. 181. ©1990 Materials Research Society

EXPERIMENTAL

For the plasma etch and deposition work, a cold-wall, low-pressure Genus CVD system (8700 Series) was modified to integrate with a Plasma-Therm dual-mode (RIE or planar etch) parallel-plate etcher. This "cluster" system provides a standard wafer transfer robot which can transfer the substrate in vacuo (<200 mT) from a cassette in a separate load lock into the etch chamber, then back to the cassette until transfer into the silicide deposition chamber. Undoped polysilicon was deposited by the CVD method onto only one-half of thermally-oxidized, 125 mm wafers using a deposition mask. The resultant wafers were then stored in a polypropylene container for several weeks, allowing the slow growth [2] of native oxide (ca. 10-15Å) before use. The etch gas used in this study was UHP C_2F_6. The flow rates varied between 2 and 21 SCCM, with the larger value being typical. Etch pressure was varied between 100 mT and 300 mT, as described herein. Etch power was varied between 75 and 200W RF (13.56 MHz) which yielded DC bias values between 40 and 400VDC at the 80 mm electrode spacing. Most of the etching was done in the RIE mode, wherein the substrate rested on the powered electrode. The DC bias was measured using an RF-filtered probe positioned near the substrate electrode. Typical base pressures prior to the etch step were in the millitorr range.

Tungsten silicide deposition parameters were typically: 570°C, 350 mT pressure, WF_6 flow ca. 20 SCCM, dichlorosilane flow ca. 300 SCCM, with argon as carrier gas at 25 SCCM.

SIMS depth profile analysis was done with a Cameca IMS-3F analyser using Cs+ primary ion bombardment with a net impact energy of 14.5 keV and negative secondary ion mass spectrometry. A nominal beam current of ca. 100 nA, rastered over an area of ca. 500 square micrometers, was used for the analyses.

RESULTS

I. Depth Profile Microanalysis by SIMS of RIE-Cleaned Structure

A half-polysilicon/half-oxide wafer was subjected to the RIE clean using C_2F_6 gas at 21 SCCM, 100 mT pressure and 70W power (80 V DC bias) for 60 sec. This treatment should have etched ca. 160Å of the oxide and 90Å of polysilicon, as determined by previous etch rate studies. The wafer was then transported to the silicide deposition chamber without exposure to atmosphere, where tungsten silicide was deposited from dichlorosilane and WF_6 at 570°C for 100 sec. to yield ca. 700Å of silicide. The wafer was then sectioned to allow SIMS depth profiling of the structure over the polysilicon. Figure 1(a) shows the resultant profile of this structure. Figure 1(b) shows the corresponding SIMS profile for a control wafer which had no preclean before silicide deposition. Comparison of Fig. 1(a) with Fig. 1(b) clearly shows the near-total extinction of the oxygen signal at the WSix/polysilicon interface (native oxide), on the polysilicon which had been plasma etched prior to silicide deposition.

Additional RIE studies were done with SF_6 and NF_3 gases. They yielded selectivities showing preferential etching of polysilicon over silicon oxide, and thus were excluded from further consideration as candidates for this integrated process.

II. Effects of High-temperature Annealing and Oxidation

Dilute, aqueous HF-cleaned (immersion until "beading") and RIE-cleaned samples of WSix (ca. 2000Å)/polysilicon (2000Å) on oxidized wafers (1000Å) were subjected to a 1000°C anneal for 30 min. while flowing nitrogen at ca. 10 l./min. Examination of the resultant structures showed no evidence of silicide film adhesion failure (tape test) over either the oxide or polysilicon surfaces. Additionally, pull tests with

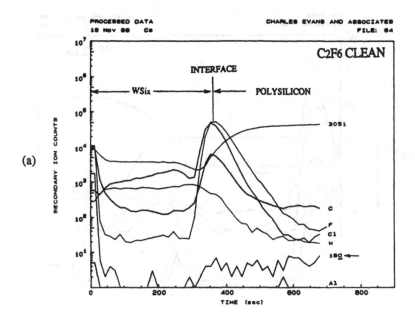

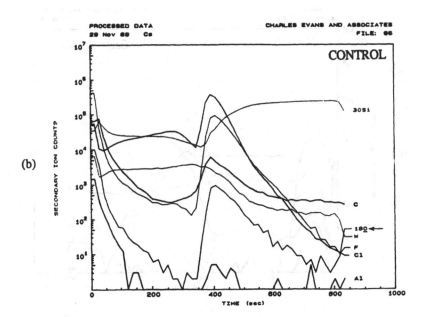

Fig. 1. SIMS depth profiles of (a) C2F6 cleaned, and (b) uncleaned poly with ca. 600 Å WSix overlayers showing O2 at WSix/poly interface.

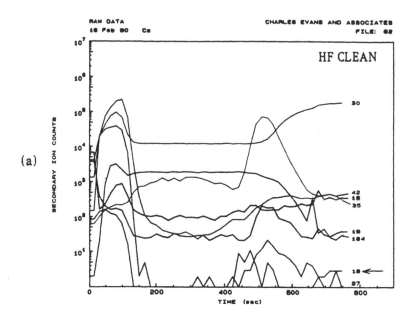

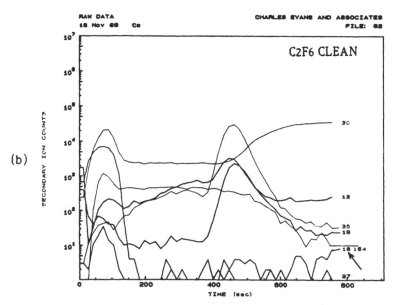

Fig. 2. SIMS depth profiles of annealed WSix films on polysilicon which had been precleaned by (a) aqueous HF, and (b) C2F6 RIE etch.

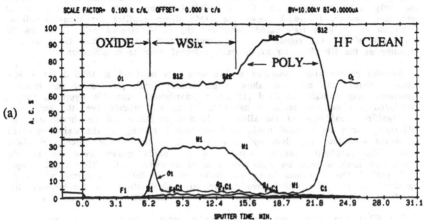

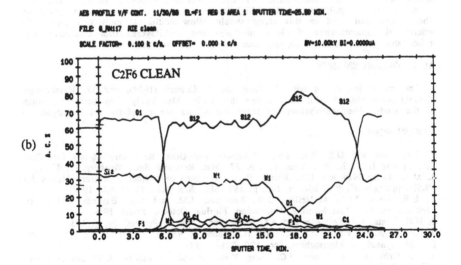

Fig. 3. AES depth profiles of (a) aqueous HF cleaned, and (b) C2F6 cleaned poly/WSix wafers which had been wet oxidized at 1050°C.

epoxied pins, using the Sebastian II tester, showed no significant loss of adhesion for silicide films deposited onto the RIE-cleaned samples in comparison to the HF-cleaned samples. Failure values of ca. 6600 psi were typical in all cases. Moreover, the only adhesion failure at the silicide/polysilicon interface occurred on an uncleaned sample. Figure 2 shows the SIMS depth profiles for annealed silicide films on (a) aqueous HF-cleaned poly, and (b) C2F6 cleaned poly. Based on the extinction of the O2 signal, one can conclude that the RIE clean is at least as effective as the HF clean for native oxide removal prior to metallization.

Samples of the above annealed wafers were wet oxidized at 1050°C for various times. Figures 3(a) and 3(b) show Auger (AES) depth profiles of the resultant structures from partially oxidized (15 min.) polysilicon. From this figure, it can be concluded that uniform oxidation has occurred on the silicide over the poly, with no significant oxidation of the silicide. Uniform oxidation did not occur over the WSix/polysilicon on thermal oxide, and substantial peeling of the oxide occurred, consistent with the study done by Saraswat, et al.[8] for similar silicide/polysilicon films. The resultant adhesion testing on the oxide grown over the silicide/polysilicon again showed no significant loss of adhesion to the RIE-cleaned sample, with typical failure values being ca. 6600 psi. All failures occurred within the epoxied test pin joint rather than at the silicide/polysilicon, or the oxide/silicide interfaces, suggestive of excellent film adhesion to the C2F6 etched surfaces.

The SIMS data raised some concern regarding what appeared to be the presence of a thin, fluorocarbon residue on the polysilicon surface after the C2F6 etch. However, ESCA analysis could not confirm the presence of such a polymer after silicide deposition.

CONCLUSIONS

The above data suggests that the RIE etching technique with C2F6 gas shows promise as a viable method of native oxide removal on polysilicon with acceptable etch selectivity of oxide over polysilicon. Tungsten silicide films subsequently deposited after the etch/clean, without breaking vacuum, exhibit excellent adhesion to the polysilicon after high-temperature annealing and wet oxidation The cluster tool used in this study should allow additional multi-step processing wherein the maintenance of clean interfaces and multilayered film structures, e.g. a bondable metal, such as copper, over the tungsten are required.

ACKNOWLEDGEMENTS

The many discussions with T. Gow and J. Coburn (IBM), and W. Harshbarger (Genus) were most helpful in directing this study. Ms. Sandy Otto performed much of the study. The contributions of these individuals are gratefully acknowledged.

REFERENCES

1. B.E. Deal and D.B. Kao, Proc. "Tungsten and Other Refractory Metals for VLSI Applications II", E.K. Broadbent, Ed., p. 27 (Mat. Research Soc., Pittsburgh, 1987).
2. M.A. Taubenblatt and C.R. Helms, Proc. "Tungsten and Other Refractory Metals for VLSI Applications", R.S. Blewer, Ed., p. 187 (Mat. Res. Soc., Pittsburgh, 1986).
3. B.E. Deal, M.A. McNeilly, D.B. Kao and J.M. deLarios, Extended Abstracts, Electrochem. Soc. Fall Meeting, Hollywood, Florida, October 15-20, 1989.
4. H.H. Busta, J.B. Ketterson and A.D. Feinerman, Ref. 2, Proceedings, p. 533.
5. P.A.M. van der Heide, M.J. Bean and H.J. Ronde, J. Vac. Sci. Technol. A7 1719 (1989)
6. L.M. Ephrath, J. Electrochem. Soc. 129, 2282 (1982).
7. G.S. Oehrlein, R.M. Tromp, J.C. Tsang, Y.H. Lee and E.J. Petrillo, J. Electrochem. Soc. 132, 1441 (1985).
8. K.C. Saraswat, R.S. Nowicki and J.F. Moulder, Appl. Phys. Lett. 41, 1127 (1982).

RAPID THERMAL ANNEALED TIW/TI CONTACT METALLIZATION FOR ADVANCED VLSI SI CIRCUITS

HENRY W. CHUNG AND AGNES T. YAO
Philips Research and Development Center, Philips Components-
Signetics, 811 E. Arques Av., Sunnyvale, CA 94088

ABSTRACT

Rapid thermal annealed (RTA) bilayers of TiW/Ti were evaluated
for contact metallization in advanced VLSI Si circuits. It
was found that RTA and a minimum thickness of Ti are necessary
to achieve consistently low p+ contact resistance. RTA has
little effect on n+ and polysilicon contact resistances. With
RTA, $TiSi_2$ is formed at the Ti/Si interface and a thin
nitrogen-containing layer is formed on the TiW surface. By
controlling RTA temperature and time, the interaction between
Si and TiW during RTA could be minimized while the gain factor
of p-channel MOSFETs was not degraded. Moreover, the leakage
currents of n+ and p+ contact chains did not increase after 30
minute anneals up to 525C.

INTRODUCTION

Thin films that provide low contact resistance and
barriers for Al to Si diffusion are critical for advanced VLSI
Si circuits. TiW has been successfully used as a diffusion
barrier for many years; however, direct contact of TiW to p+
diffusion produces high contact resistance [1][2]. $TiSi_2$,
formed from Ti and Si, is known to produce low contact
resistances to n+ Si, but is not an effective diffusion
barrier between Al and Si [3]. Hence, a new contact metallization scheme, taking advantage of both TiW and Ti silicide,
is shown. The scheme includes, prior to Al alloy deposition,
a sequential deposition of TiW/Ti followed by RTA in N_2. The
Al alloy/TiW/Ti sandwich is then patterned simultaneously. In
this paper, the effects of RTA temperature, time, and Ti
thickness on contact resistance, junction integrity and device
performance are reported.

EXPERIMENTAL

The evaluation of the contact metallization scheme was
done on a n-well, dual-metal CMOS process. (100) p-type Si
wafers of 8 - 12 ohm-cm were used. Polysilicon doping was
done in PH_3 to achieve a sheet resistance of 20 ohms/square.
As ion implantation of 7×10^{15} cm^{-2} for n+ junctions was done
at 120 keV. B ion implantation of 3×10^{15} cm^{-2} for p+
junctions was done at 40 keV. After B ion implantation, BPSG

was deposited and densified for 30 minutes at 920C in a wet ambient. After contact definition, P plug implantation at 50 keV with a dose of 3×10^{15} cm^{-2} was then performed for n-channel MOSFETs. Plug implant activation, which was also used to reflow the BPSG, was a 40 minute anneal at 920C in a N_2/O_2 mixture. The control wafers received 675 nm of Al-1%Si-1%Cu. On other wafers, 675 nm of Al-1%Cu was deposited on TiW (15 wt% Ti)/Ti films, which were sequentially deposited in a Varian S-gun sputter system. 100 nm of TiW films were used. RTA was done in an arc-lamp rapid thermal annealer. The RTA, unless otherwise specified, was 30 seconds at 620C in N_2.

Electrical measurements were made after completion of the dual metal process, which includes two anneals of 450C in forming gas.

RESULTS AND DISCUSSION

Table I shows the effects of AlSiCu only and AlCu/TiW/Ti without RTA, on the resistances of 1.5 x 1.5 um^2 Kelvin-type contacts for 20 and 35 nm Ti films. The contact resistance distributions of all three types of contacts are very tight. With TiW/Ti, lower n+ and polysilicon contact resistances resulted; however, p+ contact resistances were increased and became higher than n+ and polysilicon contact resistances. p+, n+ and polysilicon contact resistances were independent of Ti thicknesses of 20 or 35 nm. RTA was then evaluated to reduce p+ contact resistance.

Table I

Effects of TiW/Ti on
Contact Resistance

1.5 um Kelvin-Type Contact

Resistance in ohms

Thickness (nm)				
Ti		0	20	35
TiW		0	100	100
P+	Average	22.8	29.3	29.4
	Std. Dev.	1.4	1.9	1.8
N+	Average	29.4	11.3	10.6
	Std. Dev.	2.8	0.4	0.6
PS	Average	8.2	3.6	3.3
	Std. Dev.	0.8	0.2	0.2

Table II

Effects of RTA on Contact Resistance

1.5 um Kelvin-Type Contact

RTA: 30 Seconds @ 620C in Nitrogen

Resistance in ohms

RTA		NO	YES	NO	YES
Ti	(nm)	35	35	20	20
P+	Average	29.4	19.4	29.3	19.0
	Std. Dev.	1.8	2.3	1.9	1.1
N+	Average	11.0	10.6	11.3	10.7
	Std. Dev.	0.6	0.6	0.4	0.5
PS	Average	3.4	3.3	3.6	3.4
	Std. Dev.	0.2	0.2	0.2	0.2

Although RTA has little effect on n+ and polysilicon contact resistances, it significantly reduces the p+ contact resistance (Table II). More importantly, RTA produces a much tighter distribution of p+ contact resistances (Figure 1). Shown in Figure 2 are the tight resistance distributions of 1.25 x 1.25 um^2 p+, n+ and polysilicon contact chains that were achieved with RTA after TiW/Ti deposition.

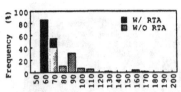

Figure 1. Distribution of P+ Contact Resistance in Contact Chains. Contact Size: 1.75 um x 1.75 um. Ti: 20 nm. TiW: 100 nm.

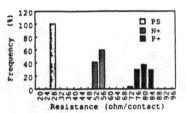

Figure 2. Resistance Distributions of 1.25 um x 1.25 um Contact Chains.

Auger Electron Spectroscopy (AES) was used to analyze the effects of RTA. 10 nm of Ti and 100 nm of TiW films were sequentially deposited on (100), p-type Si test wafers followed by RTA. Prior to RTA, there is a clean interface between Ti and TiW since the films were sequentially deposited without breaking vacuum (Figure 3). Oxygen was detected on the TiW surface. Ti silicide was formed during RTA (Figure 4). There was some interaction between Si and TiW during RTA as shown by a small dip in the Ti(1) and Ti(2) spectra and a small rise in the W spectrum at the TiW/Ti silicide interface. There also seems to be some interaction between TiW and the nitrogen ambient during RTA since the Ti(1) level, which is a combination of Ti and N, at the TiW surface was raised by RTA. The surface layer containing nitrogen and oxygen could enhance the diffusion barrier effectiveness of TiW. However, no Ti nitride was detected by low-angle XRD.

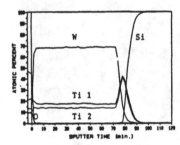

Figure 3. AES spectra of TiW/Ti/Si. Ti: 10 nm. TiW: 100 nm.

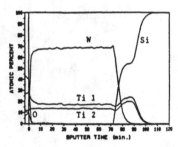

Figure 4. AES spectra after RTA. RTA: 30 sec. at 620C.

Uniform C-49 TiSi$_2$ of about 24 nm was formed during RTA (Figure 5). The TiSi$_2$/Si and TiW/TiSi$_2$ interfaces are free of defects. An interfacial layer of about 4 nm is present between TiSi$_2$ and TiW. A TEM dark field image of this area indicates that this interfacial layer is a Ti deficient TiW layer, consistent with the AES spectra. Figure 6 presents the TEM cross-section of a n+ contact with 20 nm of Ti, 100 nm of TiW and RTA. A $TiSi_2$ layer of about 29 nm has been formed.

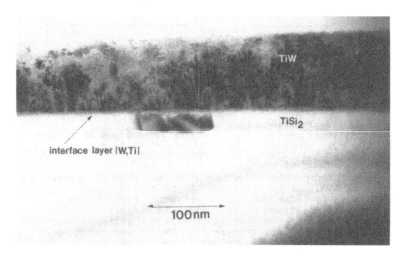

Figure 5   Rapid thermal annealed TiW/Ti/Si.   Ti = 10 nm.   TiW = 100 nm.   RTA condition: 30 seconds at 620C in N2.

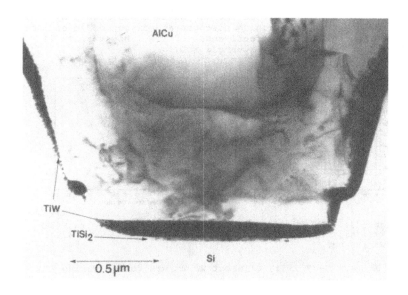

Figure 6 TEM cross-section of a 1.4 um x 1.4 um n+ contact with 10 nm of Ti, 100 nm of TiW and RTA.

Ti films thinner than 20 nm were evaluated for the purpose of minimizing Si and dopant consumption. The results are shown in Figure 7. It is apparent that n+ and polysilicon contact resistances are independent of Ti thickness (10 - 35 nm) and RTA time. However, p+ contact resistances increase when Ti films are thinner than 20 nm.

Table III

Effects of RTA Temperature & Time on Contact Resistance

1.5 um Kelvin-Type Contact

Resistance in ohms

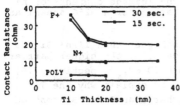

Figure 7. Contact Resistance vs. Ti Thickness. TiW: 100 nm. Al-1% Cu: 675 nm. Contact: 1.5 um x 1.5 um.

Temp. (C)	640	620	600	600
Time (sec.)	30	30	30	15
P+ Average	19.3	19.0	19.8	20.7
Std. Dev.	0.8	1.2	0.8	1.3
N+ Average	11.3	11.2	10.9	11.3
Std. Dev.	0.5	0.9	0.7	0.7
PS Average	4.0	3.9	3.7	3.9
Std. Dev.	0.3	0.3	0.3	0.3

Different RTA temperatures and times were evaluated on the (100) p-type test wafers with 20 nm of Ti and 100 nm of TiW. Table III and Figure 7 show that low and consistent p+, n+ and polysilicon contact resistances for 1.5 x 1.5 um^2 contacts could be achieved with a 15 second RTA even at a RTA temperature of 600C. Figure 8 shows the sheet resistances of Ti silicide formed at different RTA temperatures. The sheet resistances were measured after RTA and after the top TiW films were selectively removed. The sheet resistance of Ti silicide decreases with increasing RTA temperature.

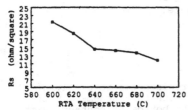

Figure 8. Sheet Resistance of Ti Silicide vs RTA temperature. Ti: 20 nm. TiW: 100 nm.

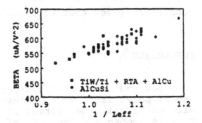

Figure 9. P-Channel Beta vs. 1 / Leff. Channel Width = 20 um.

In a self-aligned Ti silicide scheme, depletion of B in p+ junctions can result in high contact resistance at the $TiSi_2$/Si interface and degradation of the gain of p-channel MOSFETs [4]. The problem can be aggravated if a cap layer over the Ti causes further depletion of Si and dopants by curtailing the interaction between Ti and the nitrogen ambient [5]. These effects were evaluated. Figure 9 represents the dependence of the p-channel gain factor, beta, on the effective channel length. Data from control samples of AlSiCu only are also included. It is seen that formation of $TiSi_2$ in the contacts does not degrade the p-channel gain factor because low temperature RTA was used and no high temperature steps followed.

Samples were annealed at elevated temperatures for a reliability evaluation. There was no degradation of device parameters and contact resistances after 30 minute anneals at several temperatures up to 525C. The leakage currents of n+ and p+ contact chains did not increase after a 525C anneal at a reverse bias voltage of 5 V.

Al/n+ Si contact resistance is higher than Ti/n+ Si and $TiSi_2$/n+ Si contact resistances because of the higher barrier height difference, 0.72 vs 0.5 and 0.60 eV, respectively [6]. However, barrier height alone fails to explain why the $TiSi_2$/n+ Si contact resistance is not higher than Ti/n+ Si contact resistance since $TiSi_2$ has a higher barrier height to n+ Si.

The AlCuSi/p+ Si, Ti/p+ Si, and $TiSi_2$/p+ Si contact resistances are consistent with the respective barrier heights, 0.58, 0.61 and 0.56 eV [6] [7]. However, the nature of the interface also plays an important role in view of consistently low $TiSi_2$/p+ Si contact resistances (Table II, Figure 2 and Figure 6). In this study, the effects of contaminants and etch damage from dry etching to open contacts, as noted by R. Wolters, et al [2], were minimized by O_2 plasma clean and plug implant activation, respectively. It has also been found that AlSiCu/p+ Si contact resistances increase with higher P concentration in BPSG due to P doping of p+ junctions during BPSG reflow [8]. 450C anneals do not provide enough thermal energy for Ti to reduce the P contaminated surface layer; even with RTA, thin Ti films less than 15 nm, are insufficient. Silicidation with enough Ti reduces the native oxide, consumes the P contami nated surface layer, and forms an intimate $TiSi_2$/p+ Si interface for consistently low contact resistance.

CONCLUSION

Low and consistent contact resistances to p+ Si, n+ Si and n+ polysilicon were achieved with this scheme. Self-aligned Ti silicide, formed at the bottom of contacts with RTA, is smooth and free of defects. This scheme produces reliable contacts without degrading device performance.

Further, contacts produced with this scheme were shown to be stable up to 525C.

ACKNOWLEDGMENT

Thanks are extended to Signetics Fab 16 Operations for wafer processing, Brad Burrow for Auger analysis, Ken Ritz for TEM work, and Jan Reimer for the TEM photographs. The authors also thank Sheldon Lim, Hans Sigg, and Dominic Massetti for their support and discussion throughout this work.

REFERENCE

1. S.S. Cohen, M.J. Kim, B. Gorowitz, R. Saia, and T.F. McNelly, Appl. Phys. Lett. 45(4), 414, 1984.
2. R.A.M. Wolters and A.J.M. Nellissen, Proceedings of IEEE VLSI Multilevel Interconnection Conference, 351, 1987.
3. C.Y. Ting and M. Wittmer, J. Appl. Phys. 54(2), 937, 1983.
4. R.A. Haken, J. Vac. Sci. Technol. B, 3(6), 1657, 1985.
5. R. Kramer, Extended Abstract, Electrochem. Soc., Spring Meeting, 182, 1989.
6. S.M. Sze, Physics of Semiconductor Devices, John Wiley & Sons, 191.
7. J. Hui, S. Wong, and J. Moll, Electron Device Lett. 6(5), 479, 1985.
8. Agnes Yao, Philips Components internal report, 1988.

PART V

Non Au-Based
Ohmic Contacts to GaAs

CONTROLLED MODIFICATIONS IN THE ELECTRICAL PROPERTIES OF METAL/GaAs JUNCTIONS

M. EIZENBERG
Dept. of Materials Engineering and Solid State Institute,
Technion-Israel Institute of Technology, Haifa, Israel.

ABSTRACT

Controlled modifications in the electrical properties of metal/GaAs junctions were obtaind by a few different approaches. The first approach is based on modifications induced by solid state reactions occurring between the metal and GaAs substrate, resulting in compound formation and component redistribution. The characteristics of such contacts can further be modified when the contact metal is alloyed with another metal or with a dopant. The second approach is based on modifying the doping level of the near surface region of the GaAs. Here an enhancement of the barrier height was obtained by heavily counter doping the top GaAs region by recoil implantation of Mg from a Mg thin film irradiated by As^- ions. The correlations between the electrical properties of the junctions and the physical processes taking place using the above mentioned approaches are discussed.

INTRODUCTION

The Schottky barrier height on n-type GaAs is generally around 0.7 eV and is insensitive to the contact metal used [1]. This is attributed to the pinning of Fermi level at the metal/ GaAs interface, and has been explained by many models [2]. However, in many GaAs devices there is a need to modify the barrier. Thus, for example, the reduction of the effective barrier height is essential for the fabrication of ohmic contacts. On the other hand, in enhancement mode MESFETs an increase in the barrier height will allow a larger positive gate bias, thus permitting the design of circuits with an improved noise margin and a relaxed tolerance on threshold voltage uniformity.

Controlled modifications in the electrical properties of metal/GaAs junctions recently obtained by us by different approaches will be reviewed in the present work. The first approach is based on modifying the doping level of the near surface region of the GaAs by ion implantation. Ion implantation is normally utilized in contact technology to increase the doping level of the region beneath the contact in order to form, say, for an n-type substrate a heavily doped n^+ region that will result in a tunnelling ohmic contact [3]. We, however, produced counter doping by recoil implantation of Mg to form a thin p-doped region on top of an n-type GaAs substrate, which resulted in an enhancement of the effective barrier height [4]. The second approach to vary the electrical properties that is reviewed is based on modifications induced by solid state reactions occurring between the metallization and the GaAs substrate, which result in compound formation and component redistribution. In this frame our results for the interaction of a single element (Ni,Co and Ta) with GaAs will be compared to

Mat. Res. Soc. Symp. Proc. Vol. 181. ©1990 Materials Research Society

those obtained when two of the metals were alloyed (Ni-Ta) or
when a doping element (Ge) was added to the metal (Co).

ENHANCEMENT OF THE BARRIER HEIGHT BY RECOIL IMPLANTATION

    Enhancement of a metal/semiconductor barrier by counter
doping the near-surface region by ion implantation was first
demonstrated by Shanon [5] on a Si substrate. Only very
recently has this approach been applied also to GaAs by
Stanchina et al. [6] who implanted 5 and 10 keV Be$^+$ into n-GaAs
and following optimal heat-treatments observed an increase in
the barrier height of Au/GaAs by 50 meV. In our work [4]
shallow counter doping of the surface region was achieved by
replacing the conventional ion implantation with recoil
implantation. The p-type doping element, Mg, was introduced
into the n-type substrate by collisions between incident As$^+$
ions and Mg atoms of a thin deposited film of this material
which was subsequently chemically etched. Recoil implantation
was preferred over direct implantation because it provides an
extremely shallow Mg profile with maximum concentration at the
surface, a profile which is a prerequisite for the formation of
a "Shanon Diode".

    The primary ion beam parameters (60 keV, 1×10^{14} As$^+$/cm^2)
and the Mg layer thickness (400A°) were so chosen that the
maximum collision events occur at the Mg/GaAs interface (as
determined by Monte Carlo simulations). Following the
implantation the Mg layer was stripped by dipping in 25% HCl. A
depth profile of the recoiled Mg as determined by SIMS using a
8 keV ^{16}O$^-$ beam is given in Fig.1a. It shows the highest
concentration (1×10^{19} cm^{-3}) at the surface followed by an
exponential decay, reaching a value of 1×10^{17} cm^{-3} at a depth of
0.1μm.

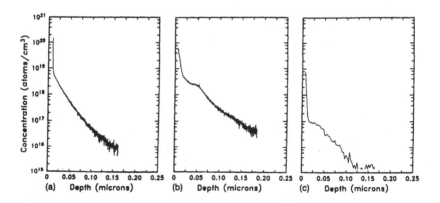

Fig.1. SIMS profiles of Mg recoil implanted in GaAs: (a) no
anneal, (b) RTA (AsH$_3$) 850°C/Os, and (c) 900°C/Os.

Rapid thermal annealing (RTA) in arsine ambient (2.5% AsH$_3$ in He) was performed to anneal the damage resulting from the As$^+$ implant and to electrically activate the Mg atoms. Schottky diodes were fabricated by depositing Ti-Pt-Au on the recoil implanted/annealed as well as on control unimplanted/unannealed or annealed samples by the lift-off technique. Current-voltage (I-V) measurements on the control wafers yielded a Schottky barrier height, Φ_b, of 0.7 eV and an identity factor, n, of 1.12 (see solid dots in Fig.2). Very similar values were obtained for the control wafers even after annealing up to 900°C (RTA-AsH$_3$). For the Mg-recoil implanted wafers, on the other hand, a definite enhancement in the barrier height dependent on the annealing temperature used was observed.

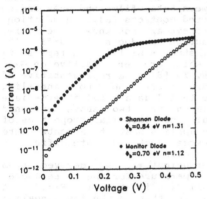

Fig.2. I-V characteristics of Ti-Pt-Au on n-GaAs control wafer (solid dots) and on recoil implanted wafer annealed at 850°Cs in AsH$_3$ (open dots).

The maximal enhancement in Φ_b was obtained after an arsine RTA at 850°C for nominal Os. The I-V characteristics (open dots in Fig.2) yielded a barrier height of 0.84 eV and an ideality factor of 1.3. The corresponding SIMS Mg profile (Fig.1b) showed some Mg redistribution compared to the unannealed state (Fig.1a), with a shoulder at a value of 2×10^{18} cm^{-3} appearing at a depth of 350A°. The I-V results are attributed to the formation of a "Shanon diode" under the contact metallization. Assuming a uniform acceptor distribution extending to a depth of 350A°, a barrier enhancement of 140 meV would require a p-type doping level of 7.5×10^{17} cm^{-3} in the surface region, on top of the 2×10^{17} cm^{-3} n-type doped subsrate.

A careful examination of the I-V characteristics of the recoil-implanted diodes shows at very low bias (<100 mV) some excess current above the extrapolation of the linear regime (the thermionic regime). This is attributed to excess generation-recombination current due to residual unannealed deep traps produced by the As$^+$ implant induced damage. This suggestion is supported by capacitance-voltage (C-V) measurements which yielded for these diodes (and not for the control wafers) anomalous large values of the flat-band voltage. The presence of residual damage was confirmed by channeling ion backscattering measurements which yielded a X_{min} value of 6.3% for the recoil implanted wafers annealed at 850°C compared to a value of 5.6% obtained for the control wafer. Cross-sectional transmission electron microscopy also showed the presence of small dislocation loops (150 A° mean diameter) in the surface region (≤500 A°).

Better annealing of the primary ion beam damage requires higher temperature heat-treatments, however the latter led to depletion of Mg at the surface due to its outdiffusion (see Fig.1c). The reduction of the acceptor level caused the "Shanon diode" effect to diminish.

INTERFACIAL REACTIONS OF THE CONTACT METALLIZATION WITH GaAs

a) Single element metallization

Unlike contacts to silicon, only limited information exists on the reaction mechanism between metal films and GaAs and on the electrical properties of reacted contacts [3]. In addition, ternary phase diagrams of such systems are not known, therefore, the prediction of metal/GaAs reaction products is difficult. Three metals were selected in our study - Ni [7] as a representatative of the near-noble metals which are very reactive on GaAs at relatively low temperatures, Ta [8] as a representative of the refractory metals which are inert up to annealing temperatures above 600°C, and Co [9,10] as a metal intermediate in its physical properties between the two above-mentioned groups. For each of these metals the electrical properties of the contacts as a function of the applied heat treatments were correlated with physical processes such as interdiffusion, compound formation and epitaxial growth.

The dependence of the barrier height ϕ_b, as derived from I-V measurements, on annealing temperature for the three above-mentioned metals deposited on (001) GaAs is shown in Fig.3. A common feature for the curves is that in the low annealing temperature regime there is a gradual increase from the as-deposited state (ϕ_b=0.76, 0.72, and 0.72 eV for Ni, Co, and Ta

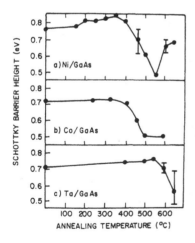

Fig.3. Schottky barrier height of studied metals as a function of annealing temperature. (a) Ni/GaAs, (b) Co/GaAs, and (c) Ta/GaAs.

respectively) up to a maximal value (ϕ_b=0.84, 0.74, and 0.77 eV, correspondingly) which was obtained after annealing at 350, 325 and 550°C, respectively. This slight increase in the barrier height was accompanied by a decrease in the ideality factor reaching its lowest value of 1.01-1.02, which indicates nearly ideal rectifying contacts, at the above mentioned temperatures. Following higher annealing temperatures a drastic fall in the barrier height followed by an increase in the ideality factor was noted. At this stage the apparent values of ϕ_b were measured by Norde's method [11] and were as low as -0.5 eV at the minimal point. On n⁺ GaAs substrates the annealing at this regime resulted in ohmic contacts, but with quite high values of specific contact resistivity.

The dependence of the electrical properties of the contacts on the applied heat-treatments can be correlated with the interfacial reactions between the metal and the GaAs substrate taking place. The various stages in the reaction progress are summarized in Table I. For Ni and Co the interaction with GaAs started at low temperatures (200 and 325°C, respectively) by indiffusion of the metal atoms toward the substrate. This is demonstrated in Fig.4, where a Ta marker deliberately evaporated on GaAs prior to the Ni deposition was found by Rutherford backscattering spectroscopy (RBS) to move toward the surface upon annealing [7]. A similar observation was noted also using an embedded inert marker (carbon layer) for the Co/GaAs system, see AES depth profile of Fig.5b [10]. The kinetics studies of the interfacial reaction at the first stage in this system [9] showed that the reaction was diffusion controlled with a relatively low activation energy of 0.6-0.7 eV.

Table I. Stages in the reaction progress of Ni, Co and Ta on GaAs (001).

	Ni	Co	Ta
1) No interfacial reaction	0-200°C	0-325°C	0-600°C
2) Formation of an epitaxial ternary phase M_xGaAs	200-350°C	325-400°C	-
3) Interfacial (non-stoichiometric) ternary phase + binary compounds at outer part	350-550°C	400-550°C	-
4) Existence of binary compounds only	≥ 600°C	≥600°C	>600°C

For both Co and Ni the first reaction product was a ternary phase in which the atomic ratio between Ga and As was maintained as in the substrate 1:1. For Co the ternary phase was identified as Co_2GaAs [9]. (Since this compound is isostructural (orthorhombic) with CoAs, it has very recently been denoted [12] as a solid solution of Ga in CoAs, but the common ternary nomenclature will be retained here to differentiate between the early stage reaction product and the final stage one). For Ni/GaAs the reaction product was Ni_xGaAs, where we have proposed x=2 [13,7], while there are reports that extend the range of x up to 3 [14] and even 4 [15]. For both metals discussed the ternary phase grows epitaxially on the (001) GaAs substrate. Thus, TEM analysis (Fig.6) enabled us to determine that for Co_2GaAs its (01$\bar{1}$) plane grew parallel to the (001) surface of GaAs, while the in-plane orientation was such that the [011] direction of Co_2GaAs was aligned either with the [110]

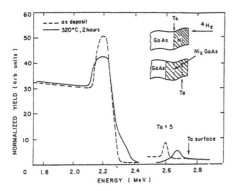

Fig.4. RBS spectra of Ni/(001) GaAs with Ta marker in the as-deposited state and after annealing at 320°C for 2 h.

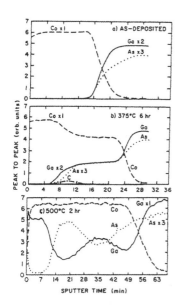

Fig.5. Auger depth profiles of Co/GaAs: (a) as-deposited state, (b) after heat treatment at 375°C, 6 h; (c) after heat treatment at 500°C, 2h.

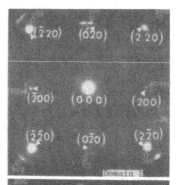

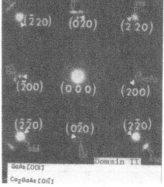

Fig.6. Electron diffraction from the Co₂GaAs ternary phase and indexing of the two complementary domains.

or the [110] direction of GaAs with a lattice mismatch of 1.15% [10]. The two possible perpendicular alignments are depicted in Fig.7. Since nucleation is in two possible

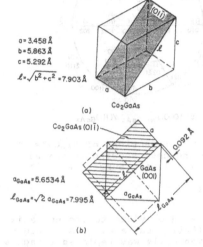

$a = 3.458 \text{Å}$
$b = 5.863 \text{Å}$
$c = 5.292 \text{Å}$
$\ell = \sqrt{b^2 + c^2} = 7.903 \text{Å}$

(a) Co_2GaAs

$Co_2GaAs (01\bar{1})$

$a_{GaAs} = 5.6534 \text{Å}$
$\ell_{GaAs} = \sqrt{2}\, a_{GaAs} = 7.995 \text{Å}$

GaAs (001)
a_{GaAs}
0.092Å
ℓ_{GaAs}

(b)

Fig.7. Schematic representation of the crystallographic structure and epitaxy of Co_2GaAs (a) the (011) plane of the ternary phase; (b) the epitaxy of Co_2GaAs(011) on GaAs(001).

orientations, the obtained microstructure consisted of elongated grains perpendicular to the substrate with both possible lateral alignments. For the Ni/(001)GaAs system the ternary phase Ni_2GaAs had a twinned structure and the epitaxial relationship between the hexagonal ternary phase and the substrate was:

$$(10\bar{1}1)Ni_2GaAs\,||\,(001)GaAs$$

and

$$[1210]Ni_2GaAs\,||\,[\bar{1}10]GaAs$$

Epitaxy was obtained also on a (111) oriented substrate of GaAs in the following manner:

$$(0001)Ni_2GaAs\,||\,(111)GaAs$$

$$[1\bar{2}10]Ni_2GaAs\,||\,[\bar{1}10]GaAs$$

These two epitaxial relationships are depicted in Fig.8.

The third stage in Table I is characterized by cnanges in the stoichiometry of the ternary phase which was still interfacing the GaAs substrate, but in addition binary compounds (MGa and MAs) appeared at the outer part of the metallization. This is demonstrated for the Co/GaAs system in the Auger depth profile after 500°C anneal, see Fig.5c, and was supported also by cross-sectional TEM analysis and by electron and X-ray diffraction. The metallization consisted at this stage of three major parts: an inner region of the ternary phase (not stoichiometric), an intermediate layer of CoAs, and an outer region of CoGa. This proves the metastability of the ternary

216

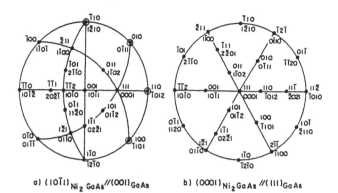

a) $(10\overline{1}1)_{Ni_2GaAs}//(001)_{GaAs}$ b) $(0001)_{Ni_2GaAs}//(111)_{GaAs}$

Fig.8. Superimposed stereographic projections of Ni_2GaAs
(four indices) and GaAs (three indices) showing epitaxial
relations on (a) (001) and (b) (111) GaAs substrates

phase, which was probably nucleated only due to its epitaxial
relationship with the GaAs (001) surface. At higher
temperatures it decomposed and eventually was replaced (stage 4
in Table I) by the thermodynamically stable binary compounds,
CoGa and CoAs. The latter were not intermixed, and their
vertical phase separation was attributed to two effects [10].
The first one was that when Co_2GaAs decomposed, the top
unreacted Co layer acted as a sink for Ga and As atoms to form
the stable binary compounds; the expected larger diffusivity of
Ga than that of As in the reacted layer should lead to the
observed phase separation. The second factor was that Co_2GaAs
is isostructural with CoAs, and therefore coherent nucleation of
CoAs on top of the ternary phase is energetically favored
(reduction of the surface energy term) encouraging thus the
appearance of CoAs adjacent to Co_2GaAs and below the CoGa phase.
    In the case of Ni/(001)GaAs at the third stage NiAs
precipitates were detected in addition to $Ni_{2-y}GaAs_{1-x}$, as can
be observed in the X-ray diffraction pattern of Fig.9c when
compared to 9b; NiGa was observed only at the fourth stage
(Fig.9d) when the ternary phase has completely disappeared. In
this system too we attributed the nucleation of the ternary
phase to epitaxy [7]. The importance of epitaxy was manifested
by the dependence of the stability of Ni_2GaAs on the crystallographic orientation of the substrate. In contrast to the above
mentioned instability of (001)GaAs, Ni_2GaAs was stable up to
600°C when formed on (111)GaAs, a situation which produced a
better epitaxial quality, as determined by channelling RBS and
TEM vertical cross-section analysis.
    In distinction to the cases of Ni and Co, the reaction
between Ta and GaAs did not take place up to annealing
temperature of 600°C, and therefore no ternary phase was formed.
Following higher temperature anneals the binary phases - TaAs,
adjacent to the substrate, and Ta_3Ga_2, at the outer part, were
observed [8].
    Based on the "metallurgical" processes described above one
can explain now the major features of the Schottky barrier

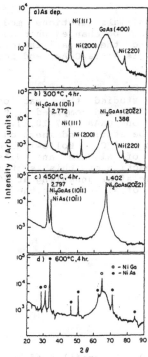

Fig.9. XRD patterns of as-deposited Ni on (001) GaAs (a), and after heat treatment for 4 h at:(b) 300°C,(c) 450°C, and (d) 600°C.

height dependence on annealing temperature as presented in Fig.3. The rectifying behavior (high Φ_b regime) corresponded either to a yet unreacted interface or to the case of a sharp and clean epitaxial interface between the ternary phase and the substrate. When the ternary phase changed stoichiometry while forming the binary phases, the contacts had very low effective barrier with high ideality factors. For the Ta/GaAs system this change in the electrical properties happened when the contact has eventually reacted, forming TaAs and Ta_2Ga_3. We have shown that the high temperature regime for all systems studied was characterized by interfacial reactions involving out-diffusion of Ga and As, which might have changed the GaAs stoichiometry beneath the contact leaving there a region full of defects and traps. In such a case other transport mechanisms such as tunnelling or generation-recombination and not thermionic emission dominate, which can explain the high current levels and low effective barriers. This is in contrast to the low-temperature regime where the formation of Co_2GaAs or Ni_2GaAs involved metal atom in-diffusion, maintaining thus the substrate's stoichiometry (Ga:As=1:1 in these compounds) and integrity. The change in the electrical properties after the high temperature anneals could have been also attributed to the presence on the GaAs surface of the binary phases, especially the metal-arsenide, replacing there the ternary phase. However, very recently Palmstrom et al [16] codeposited under UHV conditions CoGa and CoAs onto GaAs and measured barrier heights of 0.76 and 0.85 eV, respectively, ruling out, at least for Co/GaAs, the validity of the last hypothesis.

b) Bi-component metallization
     The electrical properties of metal/GaAs junctions modified
by   solid   state   reactions   can further be   changed when another
metal or a doping element are added   to   the   metallization.   A
demonstration   for   the   first   case   is Ni-Ta/GaAs, and for the
second one Co-Ge/GaAs.

     I. Ni-Ta/GaAs
     In the case of Ni-Ta/GaAs we studied the reaction of
$Ni_{60}Ta_{40}$ amorphous alloy film with GaAs and compared it to the
reactions of Ta/Ni and Ni/Ta bilayers on GaAs [17,18,8].   The
amorphous   alloy   film   was found to be stable and inert on GaAs
for annealing temperatures lower than   400°C.   Following heat-
treatment   at   450°C   the   formation   of Ni_2GaAs and NiAs was
detected. Phase separation in the reaction of this system   is
illustrated clearly in the AES profile after annealing at 500°C,
see Fig.10b. It reveals accumulation of Ni and Ga at the

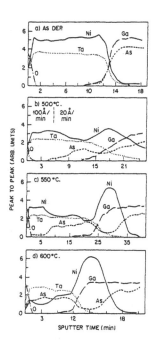

Fig.10. AES profiles of
$Ni_{60}Ta_{40}$/GaAs. (a) As-deposited
and annealed at (b) 500°C,
(c) 550°C, and (d) 600°C.

interface with GaAs and depletion of Ni and accumulation of As
in the adjacent region. By cross-sectional TEM we were able to
identify   that   the   inner   part   of   the   contact   consisted of
epitaxial Ni_2GaAs and β'-NiGa; a layer of TaAs was interposed
between these phases and the outer non-reacted Ni-Ta alloy.   A
progress in the interfacial reaction and phase separation can be
noted   at   550°C,   (Fig.10c),   and the process was completed at
600°C (Fig.10d), resulting in β'-NiGa at the interface and an
outer region of TaAs (with some Ni in it). The comparison with

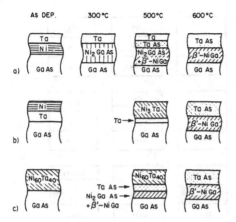

AS DEP.   300°C   500°C   600°C

Fig.11. Schematic representation of reactions of Ni-Ta films
on GaAs (a) Ta/Ni/GaAs, (b) Ni/Ta/GaAs, and (c) Ni₆₀Ta₄₀/GaAs.

the bilayer interactions, as schematically summarized in Fig.11
can explain the observed phase separation for the $Ni_{60}Ta_{40}$/GaAs
system. During the reaction Ni and As have opposite diffusion
directions. The Ni atoms leave the alloy and diffuse towards
the interface with GaAs to form Ni-Ga compound. The As atoms
released from the substrate as a result of this interaction
diffuse outwards into the metal film and react with Ta to form
TaAs. This trend is manifested very nicely in the case of
Ni/Ta/GaAs where after 600°C anneal component sequence reversal
is observed and the Ni compound is found beneath the TaAs.

The electrical properties of the three Ni-Ta configurations
on GaAs can be correlated with the metallurgical processes
discussed above. The temperature dependence of the Schottky
barrier height was very similar to that of Ni/GaAs already shown
in Fig.3a. This similarity is due to the formation of the Ni
phases at the interface with GaAs, namely Ni_2GaAs (100-350°C)
characterized by the rectifying contact properties, and β'-NiGa
(>500°C) forming almost on ohmic contact. It should be noted
that the epitaxial nature of both phases (especially β'-NiGa)
may be advantageous for contact preparation.

The obtained results suggest that amorphous Ni-Ta films,
which showed only very limited interactions at the interface,
may be utilized as extremely shallow reacted Schottky contacts.
Such contacts combine important properties such as high barrier
height (0.84 eV), almost zero contact depth, and stability up to
350°C.

II. Co-Ge/GaAs

The interfacial reactions between thin films of cobalt and
germanium and (001) oriented GaAs substrates were studied for
two configurations: Co/Ge/GaAs and Ge/Co/GaAs [19,20]. Ge was
added to the metallization in order to improve the contact
resistivity in the ohmic regime since it acts as a donor in
GaAs. The reactions occurred in several steps that are

220

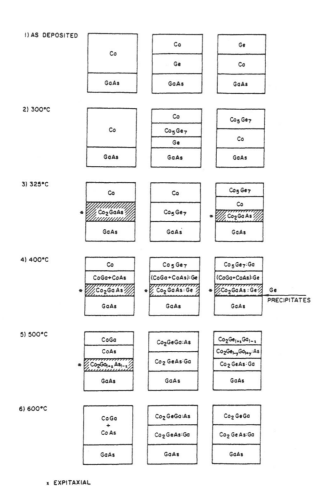

Fig.12. Schematic representation of phase
formation for the system of Co/GaAs,
Co/Ge/GaAs and Ge/Co/GaAs following heat
treatments at various temperatures (30 min.).

schematically shown (for 30 min. anneals) and compared with our
previous results for Co/GaAs [10] in Fig.12. At low temperatures, $250 \leq T < 325\,°C$, the only reaction that was monitored was the
formation of Co_5Ge_7 at the outer interrface, while the GaAs
substrate remained intact. The growth of Co_5Ge_7 was diffusion
limited with an activation energy of ~0.7 eV, see Fig.13. The
similarity with the behavior of the Co/GaAs system at the stage
of Co_2GaAs formation led us to suggest that here also the
reaction was controlled by Co atom in-diffusion across the
growing Co_5Ge_7 compound.

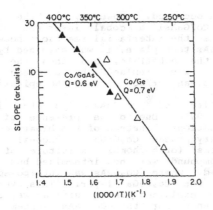

Fig.13. Arrhenius plots for the Co/GaAs and Co/Ge Systems.

Our previous study of the Co/GaAs system showed that the interfacial reaction between these two materials starts at 325°C by in-diffusion of the metal atoms into the substrate, resulting in the formation of epitaxial Co_2GaAs. And this was indeed observed at 325°C also for the Ge/Co/GaAs system where Co is directly contacting GaAs. For the other metallization (Co/Ge/GaAs), the appearance of epitaxial Co_2GaAs was delayed to 400°C, when all the Co needed for Co_5Ge_7 formation had been consumed, and the excess Co was transported to the substrate to form Co_2GaAs. (The thicknesses of the deposited layers, 50nm Co/50nm Ge, yield actually an atomic ratio of Co:Ge=2:1, leading to excess Co beyond what is needed for Co_5Ge_7 formation).

The instability of the ternary phase Co_2GaAs, discussed above for the Co/GaAs system, was manifested also in the complex metallization systems at 400°C, when on top of the epitaxial ternary phase polycrystalline CoGa and CoAs, the stable compounds of the Co-Ga-As system, were found. At this temperature for the Ge/Co/GaAs system an interesting components sequence reversal was observed, when coherent Ge precipitates were found on the GaAs surface beneath the layer of Co_2GaAs. Ge and GaAs are lattice matched, and therefore Ge can grow epitaxially on GaAs. Such a growth will reduce the surface-energy term in the total change in the free energy, and therefore is thermodynamically favored. A possible mechanism for the transport of Ge atoms to the GaAs surface is via diffusion during the grain boundaries of the earlier formed phases of Co_5Ge_7/Co_2GaAs. Alternatively one can suggest diffusion of Ge into the Co_2GaAs layer which might have an open structure, similarly to the recent model of Sands [21] for the cases of Si or Ge deposited on Pd or Ni on top of GaAs. The coherent Ge precipitates were obtained only for the Ge/Co/GaAs system and not for the other system of Co/Ge/GaAs, where actually this phenomenon should have been more expected, due to the unavoidable presence of some native oxide on the GaAs surface. This native oxide prevented the intimate contact between Ge and GaAs in the system of Co/Ge/GaAs, and therefore no epitaxial growth occurred. However, in the complementary metallization (Ge/Co/GaAs) a cobalt layer was interfacing the GaAs substrate. When the interfacial reaction between Co and GaAs to form Co_2GaAs took place, it was governed by indiffusion of Co atoms toward the substrate, and thus the contaminated Co/GaAs interface was replaced by a clean Co_2GaAs/GaAs interface. On this clean surface of GaAs the epitaxial growth of Ge could take place.

The instability of Co_2GaAs led to its complete decomposition at $T \geq 500\,°C$. Due to the presence of Ge in the discussed metallizations, instead of the binary compounds CoGa and CoAs, two ternary phases, Co_2GeGa (hexagonal structure of Co_2Ge) and Co_2GeAs (orthorhombic structure of CoAs) were obtained. These compounds were not intermixed but were rather spatially separated, where Co_2GeAs was interposed between the GaAs substrate and the Co_2GeGa layer. This vertical phase separation can be explained in a similar way used above to explain this phenomenon for the Co/GaAs system at elevated temperatures.

Current-voltage measurements of the Schottky barrier height reveal the existence of two different regimes. At annealing temperatures up to $400\,°C$ a nearly ideal rectifying behavior (thermionic emission mechanism) was observed, while higher heat-treatments resulted in contacts with very low effective barriers. This behavior is identical to that earlier discussed for Co solely deposited on GaAs (Fig.3b).

The electrical properties obtained can be correlated with modifications in the composition and structure at the metallization/semiconductor interface. Nearly ideal diodes were obtained following the $325\,°C$ anneal. The barrier height values were 0.73 eV and 0.69 eV for the Ge/Co/GaAs and Co/Ge/GaAs systems, respectively. At this temperature, at least for the Ge/Co/GaAs metallization, interfacial reaction between Co and the substrate had already occurred resulting in the formation of epitaxial Co_2GaAs. And indeed the value of 0.73 eV for the Ge/Co/GaAs system is very close to the above mentioned value of 0.74 eV for Co_2GaAs/GaAs. For the complementary system, Co/Ge/GaAs, we didn't see any macroscopic interfacial reaction with the GaAs substrate at this temperature; nevertheless, the fact that at $325\,°C$ the highest ϕ_b and lowest n values for this system were obtained, may serve as an indication to a limited interfacial reaction that could not be resolved with our analytical tools. The "good" rectifying behavior was maintained up to $400\,°C$ anneals, namely as long as Co_2GaAs was interfacing the substrate (See Fig.12).

In the high temperature regime ($T \geq 450\,°C$) a sharp decrease in the values of the barrier heights accompanied by an increase in the ideality factor was monitored. This is the temperature regime where Co_2GaAs was unstable and was not detected any more; it was replaced by two other ternaries: Co_2GeGa and Co_2GeAs, the latter interfacing directly the GaAs substrate. The change in the electrical properties may be hypothetically attributed to the Co_2GeAs phase which might have a low barrier with GaAs, similarly to the suggestion of Kuan et al [22], that the good ohmic behavior of Ni-Au-Ge/GaAs should be attributed to the presence of a very similar ternary compound, Ni_2GeAs. Alternatively, one can assume that in the high temperature contacts other transport mechanisms such as tunnelling or recombination and not thermionic emission dominate, and they are responsible for the high current level and apparent low barrier heights. The other mechanisms may be relevant since we have shown that in the high temperature regime the interfacial reactions involve outdiffusion of Ga and As which may change the GaAs stoichiometry beneath the contacts. The presence of interfacial Ge atoms at this stage, the diffusion of which into GaAs might be enhanced by the formed lattice defects, can lead to n$^+$ doping which will result in poor rectifying or even ohmic contacts, especially on n$^+$:GaAs substrates.

The ohmic behavior on n$^+$ substrates was examined by the TLM method. The lowest value for the specific contact resistivity in our study was obtained for the metallization where Ge was directly interfacing GaAs, namely Co/Ge/GaAs; it was 2.7x10^{-4} Λ.cm^2 following 500°C-30 min. anneal. This value was by a factor of 3 lower than that for the complementary metallization, and by two orders of magnitude lower than that for the Co metallization solely [19]. Thus, we may conclude that the presence of Ge played an important role in the ohmic contact formation. The fact that lower results were obtained for Co/Ge/GaAs than for Ge/Co/GaAs, in spite of the fact that for the latter system we observed interfacial epitaxial Ge grains, emphasizes that the dominant effect here is the ability to dope the GaAs with Ge atoms to form a tunnelling contact rather than the formation of a Ge/GaAs heterojunction. This conclusion agrees with the recent conclusion of ref.23 for the Ge/Pd/GaAs system, which replaced the previously suggested model of the epitaxial heterojunction for this system [24,25]. However, their suggested model of GaAs regrowth seems not to be applicable in our case [20]. It should be noted that the values for the contact resistivity reported here, are not necessarily the lowest possible ones for this metallization system, since only one set of layers thicknesses was studied (50nm/50nm) and only a very limited number of heat-treatments was applied. Further studies for other ratios between the Co and Ge layers and over a large range of heat-treatments are currently being carried out.

SUMMARY

Controlled modifications in the electrical properties of metal/GaAs junctions were obtained by a few different ways. Solid state interfacial reactions lead to the formation of compounds, the appearance of which at the semiconductor interface can change the electrical properties of the junctions. In addition, especially at elevated temperatures, the interfacial reactions involve also out-diffusion of Ga and As which changes the stoichiometry of the GaAs sub-surface region, leading to the appearance of traps, which result in high excess currents and apparent low barriers.

A second approach presented here is to modify the electrical properties by changing the GaAs sub-surface doping level. This was demonstrated in two different processes: 1) the counter doping of the top GaAs region (by Mg recoil implantation) which led to the formation of a "Shanon diode" and an enhanced effective barrier height, and 2) the increase in the doping level by addition of Ge to the metallization which resulted in an n$^+$/n:GaAs structure essential for the formation of an ohmic contact.

224

Acknowledgments

The work on Ni and Ti interactions with GaAs was done in collaboration with A. Lahav and Y. Komem and was partially supported by the Israeli Armament Development Authority. The study of Co and Co-Ge interactions was carried out in collaboration with M. Genut, and was partially supported by the Niedersachsischen Ministerium fur Wissenschaft und Kunst, W. Germany. The work on recoil implantation was carried out while the auther stayed at IBM, Watson Research Center, Yorktown Heights, and was done in collaboration with A.C. Callegari, D.K. Sadana, H.J. Hovel, and T.N. Jackson.

References

1. S.M. Sze, Physics of Semiconductor Devices, (Wiley, New York, 1981)
2. See e.g. W.E. Spicer, I. Lindau, P.R. Skeath, C.Y. Su, and P. Chye, Phys. Rev. Lett. 44, 420 (1980); L.J. Brillson, J. Vac. Sci. Technol. 15. 1378 (1978); J.M. Woodall and J.L. Freeouf, J. Vac. Sci.Technol. 21, 574 (1982); A.Zunger, Thin Solid Films 104, 310 (1983); J. Tersoff, Phys. Rev. Lett. 52, 465 (1984).
3. C.J. Palmstrom and D.V. Morgan, in Gallium Arsenide: Materials, Devices and Circuits, edited by M.J. Howes and D.V. Morgan (Wiley, New York, 1985).
4. M. Eizenberg, A.C. Callegari, D.K. Sadana, H.J. Hovel, and T.N. Jackson, Appl. Phys. Lett. 54, 1696 (1989).
5. J.M. Shanon, Solid State Electron. 19, 537 (1976).
6. W.E. Stanchina, M.D. Clark, K.V. Vaidyanathan, R.A. Jullens, and C.P. Crowell, J. Electrochem. Soc. 134, 967 (1987).
7. A. Lahav, M. Eizenberg, and Y. Komem, J. Appl. Phys. 60, 991 (1986).
8. A. Lahav, M. Eizenberg, and Y. Komem, J. Appl. Phys. 62, 1768(1987).
9. M. Genut and M. Eizenberg, Appl. Phys. Lett. 50, 1358 (1987).
10. M. Genut and M. Eizenberg, J. Appl. Phys, 66, 5456 (1989).
11. H. Norde, J. Appl. Phys. 50, 5052 (1979).
12. F.Y. Shiau, Y.A. Chang and L.J. Chen, J. Electronic Materials, 17, 433 (1988).
13. A. Lahav, M. Eizenberg, and Y. Komem in Layered Structures-Epitaxy and Interfaces, edited by J.M. Gibson and L.R. Dawson, (Mater. Res. Soc. Symp. Proc. 37, Pittsburgh, PA 1985) pp. 641-646.
14. T. Sands, V.G. Keramidas, J. Washburn, and R. Gronsky, Appl. Phys. Lett. 48, 402 (1986).
15. S.M. Chen, C.B. Carter, C.J. Palmstrom, and T. Ohashi in Thin Films - Interfaces and Phenomena, edited by P.J. Nemanich, P.S. Ho and S.S. Lau, (Mater. Res. Soc. Symp. Proc. 54, Pittsburgh, PA 1986) pp. 361-366.
16. C.J. Palmstrom, B,-O. Fimland, T. Sands, K.C. Garrison, and R.A. Bartynski, J. Appl. Phys. 65, 4753 (1989).
17. A. Lahav and M. Eizenberg, Appl. Phys. Lett. 45, 256 (1984).
18. A. Lahav and M. Eizenberg, Appl. Phys. Lett. 46, 430 (1985).
19. M. Genut and M. Eizenberg, Appl. Phys. Lett. 53, 672 (1988).
20. M. Genut and M. Eizenberg, J. Appl. Phys., in press.

21. T. Sands, Materials Science and Engineering $\underline{B1}$, 289 (1989).
22. T.S. Kuan, P.E. Batson, T.N. Jackson, H. Rupprecht, and E.L. Wilkie, J. Appl. Phys. $\underline{54}$, 6852 (1983).
23. L.C. Wang, B. Zhang, F. Fang, E.D. Marshall, S.S. Lau, T. Sands, and T.F. Kuech, J. Mater. Res. $\underline{3}$, 922 (1988).
24. E.D. Marshall, W.X. Chen, C.S. Wu, S.S. Lau, and T.F. Kuech, Appl. Phys. lett. $\underline{47}$, 298 (1985).
25. E.D. Marshal, B. Zhang, L.C. Wang, P.F. Jiao, W.X. Chen, T. Sawada, S.S. Lau, K.L. Kavanaugh, and T.F. Kuech, J. Appl. Phys. $\underline{62}$, 942 (1987).

RBS Analysis of Intermixing in Annealed Samples of Pt/Ti/III-V Semiconductors

W. SAVIN*, B. E. WEIR, A. KATZ, S. N. G. CHU, S. NAKAHARA, AND D. W. HARRIS**
AT&T Bell Laboratories, Murray Hill, NJ
*New Jersey Institute of Technology, Newark, NJ
**Charles Evans and Associates, Redwood City, CA

ABSTRACT

Low resistance Pt/Ti nonalloy contacts to GaAs, GaP, InAs, InGaAs, and InAlAs have been studied using Rutherford Backscattering. The samples were studied by comparing an as deposited structure to samples annealed at temperatures between 300°C and 600°C for 30 s. For all samples intermixing of the Pt and Ti was observed to start at 350°C. The Ti/III-V semiconductor interface mixing was strongest for As when it was present, but In and Ga also strongly intermixed for anneals of 450°C and above.

INTRODUCTION

In the design of ohmic contacts to III-V optielectric devices one has to consider the fact that they will often be operating at high current densities and under severe temperature conditions. One has to develop metallization schemes which (1) optimizes the contact's ohmic nature and yield the lowest resistance, (2) provides thermal stability over a wide temperature range, (3) form limited intermetallic reactions and have abrupt metal-semiconductor interface, (4) causes no excessive stress in the semiconductor and metal films, and (5) are compatible with the device manufacturing technology.

During recent years the technology used for contacts to p-InGaAs and n-InP were mainly Au-based alloys such as AuBe and for the former AuZn, and AuGe for the latter.[1,2] Subsequent to the metallization, the devices were subjected to anneals of 420°C for about 10 minutes and 350°C for about 5 minutes, respectively, in order to provide sufficient alloying of the AuBe or AuGe contacts. The main disadvantage of these contacts are: (1) the formation of an interfacial reacted region with about 2-3 times the thickness of the initial Au layer,[3] (2) nonuniform and spatial composition[4,5] of the contact leading to nonuniform current density flow through the contact, (3) lack of long-term metallurgical stability, (4) poor morphology and contact edge definition, and (5) not very low specific contact resistance, in the range of 7×10^{-5} to 6×10^{-6} Ω . cm^{-2}, depending on the doping level of the layers and other process conditions.[6]

The major disadvantage of the Au-based contacts is their reactive nature and their lack of long term stability, resulting from their relatively low eutectic melting temperatures. If one modifies the metallization scheme and process heating cycle one expects to produce more stable contacts. Using refractory metals in the metalization process created high quality nonalloyed contacts in n-InP,[7] p$^+$, InGaAsP[8–10] and p$^+$–InGaAs.[11,12] In addition, the use of rapid thermal processing (RTP) has been shown to produce improved ohmic contacts because of its accurate time and temperature control.

To understand the low resistance Pt/Ti nonalloyed contacts to III-V ternary and quaternary semiconductors recently reported in the literature[1,7] one has to understand the reactions that occur on annealing Pt/Ti metallizations of GaP, GaAs, InAs, InP and InGaAs systems. This is a first step toward understanding the low resistance contacts on InGaAs and InGaAsP. In this work we have characterized by Rutherford Backscattering (RBS) the reactions at the Pt/Ti/GaP, GaAs, InAs, InP and InGaAs systems which have been sintered by means of RTP at temperatures between 300°C and 600°C. Using the RBS utilities and manipulation package (RUMP)[13] simulations were done for each sample to determine if a systematic trend for interfacial layers formation could be detected.

EXPERIMENTAL PROCEDURE

Rutherford Backscattering spectra (RBS) at 170° were taken using 1.8 MeV ^{4}He$^+$ on Pt/Ti/GaP, Pt/Ti/GaAs, Pt/Ti/InAs, and Pt/Ti/InGaAs structures. In addition, RBS spectra was taken for Pt/Ti/InAlAs using 1.5 MeV ^{4}He. The sample thickness and beam energy were chosen to obtain clean spectra for the Pt and Ti layers, and sharp clean demarcations for the III-V semiconductor materials. For all samples the Pt and Ti layers thickness were about 60nm and 50nm.

For each III-V semiconductor material one samples was not annealed, the others were annealed at temperatures in the range of 300°C to 600°C for durations of 30s. The highest temperature provided such degraded structures that they were difficult to analyze and are not discussed. All spectra were collected at low counting rates and were reproducible. RUMP[13] simulations were done for each of the spectra to determine if a systematic trend for layer intermixing could be detected. The RUMP simulations fit very well in all cases to the spectra followed the data. Since we did the simulations for each annealing temperature, starting with the fit to the next lowest temperature, we were able to see the progressive evolution of interfacial intermixing. To help in the

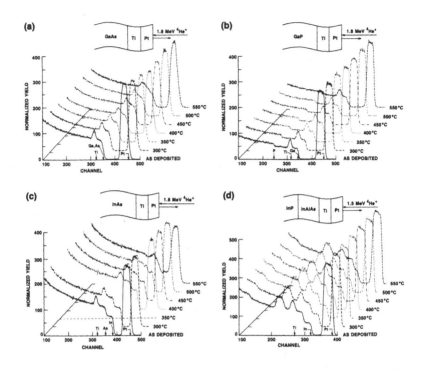

Figure 1:   RBS spectra for a) GaAs, b) GaP, c)InAs and d) InAlAs.

analysis, difference spectra between the unannealed standard and each annealed sample were used to help in tracking the migration and intermixing of the different elements. Examples of these spectra are shown in Figure 1.

RESULTS

RBS spectra for the as deposited sample and samples annealed for 30s at 300°C, 350°C, 400°C, 450°C, 500°C and 550°C show changes that can only be attributed to intermixing between the original layers. For all the studied Pt/Ti/GaAs samples the RBS results indicate that the as deposed Pt and Ti layers were not pure. This was previously verified by Auger electron spectroscopy (AES) depth profiles[11,16]. The RBS spectra for Pt/Ti/GaAs samples show Pt-Ti intermixing starting at 350°C, which agrees with other earlier reports.[15] The As undergoes mixing with the Ti at the same temperature. Both of these intermixed layers grow as the RTP temperature was increased to 550°C, where the Ga and As outdiffused widely and were detected in the Pt-Ti interfacial layer. The RBS spectra show that the Ga and As were intensively present where in the Ti layer, but there was always a pure Ti thin layer on top of it. The results of the RBS analysis are shown in Figure 2a.

AES confirms the presence of a significant Pt-Ti intermixed layer at the Pt/Ti interface region and a Ti-Ga-As intermixed layer at the Ti/GaAs interface after annealing at 400°C. It also confirms that the intermixed layers spread with increasing annealling temperature and at 550°C the Ga and As have mixed through the Ti layer and can be found in the original Pt layer. The AES depth profiles are shown in Figure 3. This phase evolution sequence was studied by transmission electron microscopy selected area diffractions patterns.[16]

For the Pt/Ti/GaP samples the RBS spectra and RUMP analysis are given in Figure 2b. They indicate that Pt-Ti mixing starts at sinitering temperature of 300°C and expend with increasing temperature. Only a limited interfacial intermixing is observed at the Ti/GaP interface at temperature lower than of 450°C, where the preliminar reaction between Ti and GaP is detected. As the temperature is elevated the intermixed layer grows in thickness and the P appears to migrate more rapidly than the Ga out into the adjacent Ti layer. At sintering temperatures of 500°C and above the P and Ga are detected throughout the entire Ti Layer.

The RBS spectra for Pt/Ti/InAs system are shown in Figure 2C. The Ti and Pt layers started mixing due to RTP at 300°C. This was also seen in the previous Pt/Ti/GaP samples. The amount of intermixing expanded as the sintering temperature was raised, but the Pt-Ti intermixing layer never exceeded a thickness of about 20 nm. As

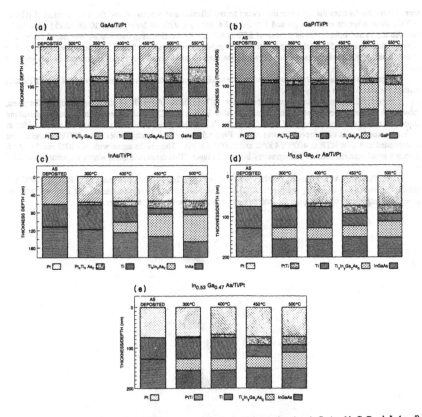

Figure 2: Schematic presentation of layer composition and thickness for the a) GaAs, b) GaP, c) InAs, d) InGaAs and e) InAlAs systems as obtained through RBS-RUMP analysis.

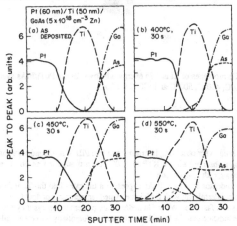

Figure 3: AES depth profiles of Pt/Ti/GaAs: a) as-deposited sample, and after RTP at b) 400°C, c) 450°C, and 550°C for 30s.

seen before, the As from the substrate underwent an outdiffusion and penetrated into the Ti as a result of RTP at 350°C. Both substrate elements (In and As) formed a ternary interfacial layer (about 10 nm thick) with the Ti between the Ti and InAs layers as a result of RTP at 400°C. As the sinitering temperature was raised the ternary layer got thicker and at 550°C the layered structure completely collapsed. At 550°C both substrate elements have migrated through the entire metallic cap layer to the contact surface and the Ti has penetrated deeply into the substrate.

Figure 4 shows a schematic presentation of an AES montage taken from the backview of the Pt/Ti/InAs contact, for the Ti and In energy windows.

Figure 2d shows the intermixing results for Pt/Ti/InGaAs. As with all the other samples discussed, substantial intermixing occurred at both interfaces after RTP at 400°C and this grew with increasing annealing temperature. We see a mixed layer of Ti and As at 300°C and this grew in thickness and had Ga and In added to the intermixed layer as the temperature was raised. Figure 5 shows AES depth profiles of the Pt/Ti/InGaAs for the as deposited and after RTP at 400°C, 450°C and 500°C for 30s. The results agree with our RBS results. AES detects a small concentration of In migraterey into the Pt layer. This concentration is within the error of the RBS results.

The Pt/Ti/InAlAs intermixing results are shown in Figure 2e. Pt-Ti mixing starts after RTP at 350°C and grows with increasing annealing temperature. The Ti/InAlAs interface shows no intermixing until we reach an annealing temperature of 450°C, but the intermixing grows with increasing temperature.

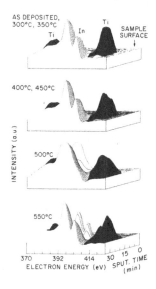

Figure 4:   AES depth profile montages of Ti and In energy windows for the Pt/Ti/InAs system as-deposited and after RTP at 350°C, 400°C, 500°C and 550°C for 30s.

CONCLUSION

RBS is a sensitive technique for observing intermixing in Pt/Ti/III-V semiconductor materials. At the Pt/Ti interface intermixing is always observed after a RTP of 350°C even though it was observed at 300°C in some systems. This intermixing grows with temperature.

At the Ti/III-V semiconductor interface, if As is present it migrates into the Ti at low RTP temperatures. In, when present, seems to start intermixing with the Ti at elevated temperatures of 450°C and above, but once free to move into the Ti it migrites at a fast rate. When Ga and/or P are present it appears they are more stable than the In and As. Our earlier results indicate that the In and As migrated preferentially into the metal contact layer.

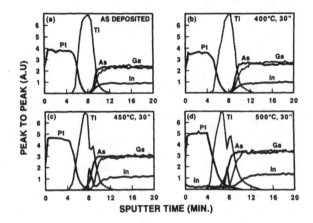

SPUTTER TIME (MIN.)

Figure 5: AES depth profiles of Pt/Ti/InGaAs a)as-deposited and after RTP at b) 400°C, c) 450°C, d) 500°C for 30s.

REFERENCES

1. A. Katz, P. M. Thomas, S. N. G. Chu, J. W. Lee and W. C. Dautremont-Smith, J. Appl. Phys. 66, 2056 (1989).

2. G. Bahir, J. L. Mertz, J. R. Abelson and T. W. Sigmon, J. Electronic Mat. 16, 257 (1987).

3. P. M. Thomas, W. C. Dautremont-Smith, J. Lopata, A. Appelbaum and R. L. Brown, Private Communication.

4. C. C. Cheng, T. T. Sheng, R. J. McCoy, S. Nakahara, V. G. Keramidas and F. Kermanis, J. Appl. Phys. 50, 7032 (1979).

5. J. M. Vandenberg, H. Temkin, R. A. Hamm and M. A. Digiuseppe, J. Appl. Phys. 53, 7385, (1982).

6. I. Camlibel, A. K. Chin, F. Ermanis, M. A. Digiuseppe, J. A. Lourenco and W. A. Bonner, J. Electrochem. Soc. 11, 2585 (1982).

7. A. Katz, B. E. Weir, S. N. G. Chu, P. M. Thomas, M. Soler, T. Boone and W. C. Dautremont-Smith, Appl. Phys. Lett., to be published.

8. A. Katz, W. C. Dautremont-Smith, P. M. Thomas, L. A. Koszi, J. W. Lee, V. G. Riggs, R. L. Brown, J. L. Zilko and A. Lahav, J. Appl. Phys. 65, 4319 (1989).

9. A. R. Goodwin, I. G. A. Davies, R. M. Ribb and K. H. Murphy, J. Lightwave Tech. 6, 1424 (1989).

10. A. Katz, P. M. Thomas, S. N. G. Chu, W. C. Dautremont-Smith, R. G. Sobers and S. G. Napholtz, J. Appl. Phys., 67, 884 (1990).

11. A. Katz, W. C. Dautremont-Smith, S. N. G. Chu, P. M. Thomas, L. H. Koszi, J. W. Lee, V. G. Riggs, R. L. Brown, S. G. Napholtz, J. L. Zilko and A. Lahau, Appl. Phys. Lett. 54, 2306 (1989).

12. A. Katz, B. E. Weir, D. Maher, P. M. Thomas, W. C. Dautremont-Smith, L. C. Kimerling, R. F. Karlickek, Jr. and J. D. Wynn, Appl. Phys. Lett. 55, 2220 (1989).

13. L. R. Doolittle, Vacl. Instr. ??? Meth. B9, 344 (1985).

14. A. Katz, W. C. Dautremont-Smith, P. M. Thomas, L. A. Koszi, J. W. Lee, V. G. Riggs, R. L. Brown, J. L. Zilko, J. Appl. Phys. 65(11), 4319 (1989).

15. A. K. Simha, T. E. Smith, M. H. Read and J. M. Poate, Solid-State Electron. 19, 409 (1976).

16. A. Katz, S. Nakahara, W. Savin and B. Weir, Submitted to J. Appl. Phys.

THERMALLY STABLE, LOW RESISTANCE
INDIUM-BASED OHMIC CONTACTS TO n AND p-TYPE GaAs

Masanori Murakami[*], P.-E. Hallali, W. H. Price, M. Norcott, N. Lustig, H.-J. Kim[**],
S. L. Wright, and D. LaTulipe

IBM T. J. Watson Research Center, Yorktown Heights, New York 10598, USA
Present address: ([*]) Dept. of Materials Science and Technology, Kyoto University, Sakyo-ku, Kyoto 606, Japan. ([**]) Gold Star Electronic Co., Seocho-Gu, Seoul, Korea

ABSTRACT

Recently, thermally stable, low resistance In-based ohmic contacts to n-type GaAs have been developed in our laboratories by depositing a small amount of In with refractory metals in a conventional evaporator, followed by rapid thermal annealing. By correlating the interfacial microstructure to the electrical properties, $In_xGa_{1-x}As$ phases grown epitaxially on the GaAs were found to be essential for reduction of the contact resistance (R_c). This low resistance was believed to be due to separation of the high barrier (ϕ_b) at the metal/GaAs contact into two low barriers at the metal/$In_xGa_{1-x}As$ and $In_xGa_{1-x}As$/GaAs interfaces. In this paper the effects of the In concentration (x) in the $In_xGa_{1-x}As$ phases and addition of dopants to the contact metal are presented. High In concentration is desirable to reduce the ϕ_b at the metal/$In_xGa_{1-x}As$ interface. Such contacts were prepared by sputter-depositing InAs with other contact elements, but the low R_c values were not obtained. The reason was explained to be due to an increase in the ϕ_b at the $In_xGa_{1-x}As$/GaAs interface due to the formation of misfit dislocations. However, addition of a small amount of Si to the contact metals reduced significantly the R_c value. This contact demonstrated excellent thermal stability: no deterioration was observed at 400°C for more than 100 hrs. In addition, the use of this Ni(Si)InW contact metal allowed us to fabricate the low resistance ohmic contacts by one-step (simultaneous) annealing for "implant-activation" and "ohmic contact formation", which simplifies significantly GaAs device fabrication process steps. For p-type ohmic contacts, low resistance contacts were fabricated by depositing the same NiInW contact material to p-type GaAs. This contact was also thermally stable during subsequent annealing at 400°C. Within our knowledge this is believed to be the first demonstration of low resistance, thermally stable ohmic contact fabrication using the same materials for both n and p-type GaAs.

INTRODUCTION

With increasing integration levels of GaAs devices, there is increasing demand for low resistance (R_c) ohmic contacts to both n and p-type GaAs which withstand high temperature thermal cycles without degradation. For 1μ-MESFETs (metal-semiconductor field-effect transistors) and HFETs (heterojunction FET) devices, n-type ohmic contacts with R_c less than 0.4 Ω-mm and p-type contacts with R_c less than 0.8 Ω-mm are required, respectively. Other important requirement for ohmic contacts is stability of the contacts at high temperatures during device fabrication. In order for GaAs technology to be a viable alternative to Si technology, the fabrication process for GaAs devices should be compatible with that of Si devices as much as possible. There are wide varieties of Si device fabrication processes among which the highest temperature cycle which the current devices must withstand is between room temperature and about 400°C. The ohmic contact properties should not deteriorate during this heat-treatment. Also, high temperature stability during device operation is required for GaAs devices. Since the energy gap of GaAs is wider than that of Si, the GaAs devices can operate at high temperature beyond the capability of the Si devices. If the maximum operation temperature which the Si devices can withstand is about 200°C, the GaAs devices can operate at temperature close to 400°C. This high temperature stable GaAs devices have two advantages: one is the devices can operate in high temperature environment such as sensor for car and jet engines, geothermal well-drills and nuclear reactors, and the other is that higher power can be applied to the devices to reduce the delay time. In order to fabricate high temperature stable GaAs devices, the active elements, such as Schottky, ohmic contacts and insulating layers, should be stable at the high temperature of ~ 400°C.

The thermally stable, low resistance NiInW ohmic contacts have been recently developed in our laboratories [1]. Development of this contact metallurgy was previously reviewed [2]. Using these contacts, 1μ-MESFETs (with one masking level) have been successfully fabricated [3] and the device performance was compared those prepared with AuGeNi ohmic contacts which have been extensively used in manufacturing GaAs devices. The performance of the as-fabricated devices was similar between the two, although the mean contact resistance ($R_c \simeq 0.3$ Ω-mm) [4] of the NiInW contacts was higher than that ($R_c \simeq 0.1$ Ω-mm) [5] of the AuGeNi contacts. However, the devices with the NiInW contacts were thermally stable during subsequent annealing at 400°C and no deterioration of the R_c values was observed after annealing at this temperature for more than 100 hours. On the other hand, the performance of the devices with the AuGeNi contacts deteriorated and the R_c values increased by a factor of ~3 after annealing at 400°C for 2 hours [6]. The use of the NiInW contacts in MESFETs devices have several advantages over the AuGeNi contacts: (a) TiAl(Cu) interconnect materials, which are universally used in the current Si technology, are able to be deposited directly onto the NiInW ohmic contacts without a diffusion barrier layer, (b) the devices have an increased flexibility in development of future packaging processes, because they withstand annealing at 500°C for 2 hours, and (c) the source and drain contacts have smooth surface, sharp contact edges, and shallow diffusion depth (~70 nm). These features make NiInW contacts attractive for future submicron devices. However, although the typical R_c values (~ 0.3 Ω -mm) of the NiInW contacts are acceptable for 1μ-MESFETs devices, a factor of ~ 2 reduction in the R_c values is required for submicron devices.

The purpose of the present paper is to review recent activities on NiInW ohmic contacts to n and p-type GaAs. For n-type GaAs the possibility of further reduction in R_c of NiInW contacts has been explored. The low R_c values of the In-based contacts which were formed by annealing is believed to be due to the formation of a thin $In_xGa_{1-x}As$ layer (with x ~ 0.4) at the metal/GaAs interfaces. This separates the high barrier the metal/n-GaAs into two low barriers (ϕ_a, ϕ_b), at metal/$In_xGa_{1-x}As$ and $In_xGa_{1-x}As$/GaAs, respectively. The corresponding energy band diagram is shown in Fig. 1(a). Since the $In_xGa_{1-x}As$ phases are grown epitaxially on the GaAs substrate, the barrier height ϕ_b is believed to be smaller and the barrier (ϕ_a) at the metal/$In_xGa_{1-x}As$ controls the R_c value. There are two approaches to reduce further the R_c value of the NiInW contacts. One is to reduce the ϕ_a value by increasing the In concentration (x) in the $In_xGa_{1-x}As$ layer, since the barrier height at the metal/$In_xGa_{1-x}As$ interface decreases with increasing x value [7]. The other is to add a small amount of dopant to the contact material during deposition and dope heavily at the $In_xGa_{1-x}As$ and GaAs interfaces during annealing.

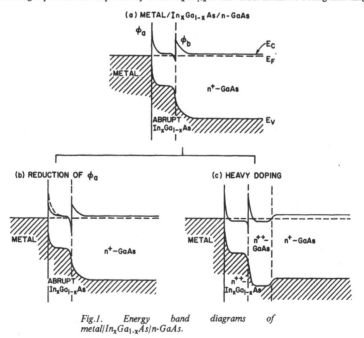

Fig.1. Energy band diagrams of $metal/In_xGa_{1-x}As/n-GaAs$.

The expected band diagrams are shown in Figs. 1(b) and 1(c), respectively. The contacts with high In concentration (x) in $In_xGa_{1-x}As$ layer were prepared by sputtering from an InAs target. Doping was achieved by incorporating through annealing a small amount of Si in the NiInW contacts. (Selection of Ni metals was due to formation of Ni_3In compounds with high T_m and W was used as a cap layer.) The details of the sample preparation procedures will be described in next section. Another active research area for NiInW ohmic contacts is the application to p-type GaAs so as to use the same metallurgy for both n and p-type ohmic contacts. This activity is also reviewed in this paper. In addition, since these contacts are formed by annealing at temperatures high enough to activate the implanted species, the ohmic contacts can be formed by simultaneous annealing for implant/activation and contact formation. Experiments to form the low resistance contacts were carried out for both n and p-type GaAs.

EXPERIMENTAL PROCEDURES

GaAs wafer preparation

For n-type GaAs wafers, conducting channels for the ohmic contact resistance measurements were formed on [100] oriented GaAs substrates by implanting $^{29}Si^+$ ions at a dose of 6.6 x 10^{13} atoms/cm^2 at 40 KeV through a photoresist stencil. After stripping the photoresist, the implanted species were activated by annealing at 800°C for 10 min in an arsine atmosphere except those used for one-step annealing experiments. A photoresist lift-off stencil defining the contacts metals for contact resistance measurements by the transmission line method (TLM) [8] was then applied. Prior to metal deposition the wafers were chemically cleaned using HCl solution.

For p-type contact formation, Be and F co-implanted GaAs wafers were used [9]. Be was chosen as an acceptor because of its relatively low diffusivity in GaAs. When it was implanted with high dose (> $1x10^{14}cm^{-2}$) into GaAs, only 10% of the implanted Be was activated. A high activation efficiency (35 to 48%) of Be implanted at 20 KeV with a dose of $1x10^{15}cm^{-2}$ has been obtained by co-implanting F with a dose in the range of $1x10^{15}cm^{-2}$ at energies of 10-20 KeV and annealing at 850°C for 5 sec by RTA with cap layers [10]. The sheet (channel) resistance of this GaAs wafer was very low (215 $\Omega/\square$) and was constant after annealing at ~ 900°C for 5 sec, and at 400°C for a long time, which are required for p-type NiInW ohmic contacts.

Contact resistance measurement and microstructure analysis

The contact resistances (R_c) and channel resistance (R_{ch}) were measured by the TLM [8]. Since the metal sheet resistance (R_s) of NiInW contacts with or without dopant addition is high (12-15 $\Omega/\square$), the contribution of the R_s to the measured R_c values was substracted using an equation given by Marlow and Das [11]. The validity of this equation in correcting the measured R_c values was experimentally demonstrated in AuGeNi [12] and NiInW contacts [3].

Metal and GaAs reaction and compound formation in the contact metal were studied by x-ray diffraction (XRD) and cross-sectional transmission electron microscopy (XTEM). XRD measurements were carried out by a computer-controlled Rigaku x-ray diffractometer using CuKα radiation, operated at 40 kV and 30 mA. The diffracted intensities were measured using a scintillation counter with the soller slits and a singly bent graphite monochromator. For cross-sectional TEM study, specimens were prepared by mechanically polishing and ion milling at liquid nitrogen temperature. A Philips 400T TEM/STEM, operated at 120KeV, was used to perform microanalysis at the interface. The compositions of Si in Ni films which were deposited onto carbon substrates were analysized by Rutherford backscattering (RBS).

EXPERIMENTAL RESULTS AND DISCUSSIONS

NiInW contacts to n-type GaAs

Ni/Ni-In/Ni/W ohmic contacts were prepared by depositing sequentially Ni(5nm), Ni-In(10nm), Ni(5nm), and W(40nm) on n-GaAs substrates where a sign "/" between two metals indicates the sequential deposition and a hyphen "-" indicates the co-deposition. Ni and W were evaporated by an electron beam, and In was evaporated by a RF induction heater. These contacts were coated with 50 nm thick Si_3N_4 cap layers. To improve adhesion of the Si_3N_4 layer to the GaAs substrate and the contact metal, the samples were heated to 300°C for about 30 min during the Si_3N_4 deposition. These samples were then annealed for ~ 1 sec at various temperatures, by rapid thermal annealing (RTA) in an Ar/H$_2$ atmosphere.

The total heating and cooling time is about 15 sec. The R_c values of the contacts are shown in Fig. 2 as a function of annealing temperatures. The R_c values of ~ 0.3 Ω-mm were obtained for the contacts annealed at temperatures in the range of 800 to 1000°C [1]. These contacts were thermally stable and no degradation of the R_c values was obtained after annealing at 400°C for more than 100 hrs and at 500°C for 20 hrs.

NiInAsW contacts to n-type GaAs

In order to prepare the contacts with high In concentration in the $In_xGa_{1-x}As$ layers, NiInAsW contacts were prepared by sputter-depositing InAs layers in an Ar atmosphere using an InAs target, with base pressure of 8×10^{-5} Pa [13]. The deposition rate was about 0.1 nm/sec and the resulting In to As ratio in the film was close to unity. The Ni and W layers were deposited in an e-beam evaporation system, which was pumped down to $\sim 8\times10^{-6}$ Pa before metal deposition. InAs/Ni/W and Ni/InAs/Ni/W contacts were thus prepared. The typical thicknesses of Ni, InAs, and W layers are 10, 30, and 40 nm, respectively. Samples were then capped by 50 nm thick Si_3N_4 layers, followed by rapid thermal annealing.

The electrical properties of the InAs/Ni/W and Ni/InAs/Ni/W contacts were measured after annealing the contacts at various temperatures. The current-voltage (I-V) curves of the as-deposited contacts showed Schottky behavior. For the InAs/Ni/W contacts, the ohmic behavior was first observed at 700°C. The R_c values decreased with increasing the annealing temperature (T) and the R_c of ~ 0.4 Ω-mm was obtained at the temperature range of 750-850°C as shown in Fig. 3. By annealing these contacts at higher temperatures, an increase in the R_c value was observed and typical U-shape dependence of R_c on T was obtained. A similar U-shape curve in the R_c vs T plots was observed in the Ni/InAs/Ni/W contacts, but the R_c values had less annealing temperature dependence. These contacts were isothermally annealed at 400°C and the R_c values of these contacts did not change even after annealing for more than 100 hrs.

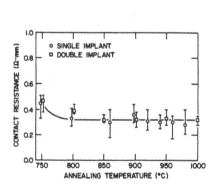

Fig.2. Contact resistances of NiInW ohmic contacts with single (circles) and double (squares) implant channels [1].

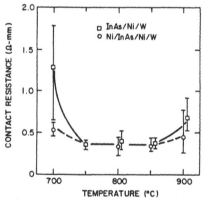

Fig.3. Contact resistance of InAs/Ni/W (squares) and Ni/InAs/Ni/W (circles) annealed at various temperatures [13].

The microstructures of the InAs/Ni/W contacts annealed at 300 °C for 30 min and at 800 °C for 1 sec were examined by XTEM and XRD. In the contact annealed at 300 °C, large polycrystalline InAs grains were observed. Some grains grew epitaxially on the GaAs substrate, but most grains had random crystal orientations. The average grain size of the InAs grains was measured to be approximately 20 nm. Ni reacted with the InAs layer at this temperature and a small volume fraction of $NiAs_2$ phases which have cubic structure were observed within the InAs grains. Indium islands, precipitated out from the InAs layer after Ni/InAs reaction, were observed close to the GaAs surfaces. The micrograph of the contact annealed at 800°C is shown in Fig. 4, where $In_xGa_{1-x}As$ phases are observed to cover about 80% of the GaAs surface. The In concentration (x) in the $In_xGa_{1-x}As$ phases was uniform and relatively high (x > 0.7). The $In_xGa_{1-x}As/GaAs$ interface is abrupt as indicated by Moire fringes at the interface. The misfit dislocations and twins which accommodated the lattice mismatch of $\sim 5\%$ (at x = 0.7) between

$In_xGa_{1-x}As$ and GaAs were also observed at the interfaces. The $NiAs_2$ compounds were found to transform to NiAs compounds with hexagonal structure and no "pure" In was detected in this contact.

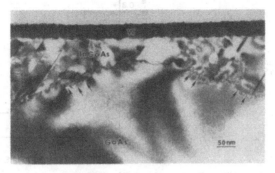

Fig.4. TEM micrograph of InAs/Ni/W contact annealed at 800°C [13].

The conventional NiInW contacts which form $In_xGa_{1-x}As$ layer with x = ~ 0.4 at the GaAs interface, provide routinely the R_c values of ~ 0.3 Ω-mm. Although, the NiInAsW contacts form $In_xGa_{1-x}As$ layer with x = 0.7, the R_c values did not reduce as seen in Fig. 3. The reason is believed to be formation of a high density of misfit dislocations the $In_xGa_{1-x}As$/GaAs interface as seen in Fig. 4. Woodall et al [14,15] found that the pinning of the GaAs Fermi level at interface between the heteroepitaxial layer and GaAs arose from the formation of misfit dislocations. Therefore, although "zero" barrier is expected at the metal/$In_xGa_{1-x}As$ interfaces, the barrier height at the $In_xGa_{1-x}As$/GaAs interface was high and the R_c values of the metal/$In_xGa_{1-x}As$/GaAs contacts were high. The expected energy band diagram is shown in Fig. 5. (Note that low resistance contacts were prepared by annealing InAs/GaAs contacts grown by MBE [16]. The difference between MBE grown and sputter-deposited ohmic contacts is believed to be due to doping levels and location of dislocations. However the exact reason is not known at the moment.) The only way to reduce the both barriers (ϕ_a, ϕ_b) is to prepare a compositionally graded $In_xGa_{1-x}As$ layer as an intermediate phase as demonstrated by MBE by Woodall et al [17].

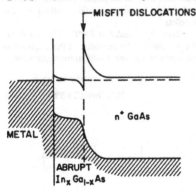

Fig.5. Energy band diagram of metal/$In_xGa_{1-x}As$/n-GaAs with misfit dislocation at the $In_xGa_{1-x}As$/GaAs interface.

Ni(Si)InW contacts to n-GaAs

In order to dope donors heavily at the GaAs and $In_xGa_{1-x}As$ interfaces, through the diffusion of donors from ohmic contact materials, Ni(Si) pellets were prepared by melting Ni and Si in an Ar atmosphere [18]. The Si compositions in the Ni(Si) pellets were in the range of 0.5 to 12 at.%. Ni(Si)/Ni(Si)-In/Ni(Si)/W contacts were prepared by evaporating Ni(Si) pellets instead of Ni pellets, and

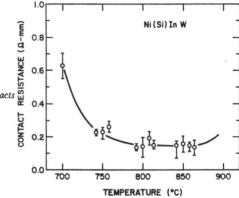

Fig.6. Contact resistances of Ni(Si)InW contacts prepared by Ni-5%Si pellets. [18].

this contact is denoted as Ni(Si)InW. (The other experimental procedures are the same as those used for NiInW contact formation). The nominal thickness of the first and third Ni(Si) layers was 5 nm, that of the second Ni(Si)-In layer was 10 nm, and the top W layer was 40 nm thick. After RTA annealing, some contacts were coated by Ti (5 nm) and Al-Cu (300 nm) overlayer which has a metal sheet resistance of less than 0.5 $\Omega/\square$. The thermal stability test was carried out by annealing the samples isothermally at 400°C in a furnace with an Ar/H$_2$ atmosphere.

The Ni(Si)InW contacts with various Si concentrations were prepared by annealing at temperatures (T) in the range of 750 to 900°C for ~ 2 sec. The dependence of the R_c values on the Si concentrations in the Ni(Si) pellets was studied for contacts annealed at 800-850°C. The R_c values decreased with increasing the Si concentrations in the Ni(Si) pellets, and reached the lowest value of ~ 0.1 Ω -mm at 5%Si, and increased to ~0.2 Ω-mm by further increasing the Si concentration. Figure 6 shows an example of the R_c vs T plots for the contacts prepared with Ni-5%Si pellets. The temperature dependence of the R_c values is similar to that observed in the NiInW contacts shown in Fig. 2. However, a factor of ~ two lower R_c values (0.1-0.2 Ω-mm) were obtained in the Ni(Si)InW contacts annealed at temperatures in the range of 800 to 850°C. Since the vapor pressure of Ni is higher than that of Si, the concentration of Si in the deposited films is expected to be smaller than that in the Ni(Si) pellets. The RBS analysis indicated that the Si concentrations in the films were about 5 times less than those in the pellets. However, although we cannot control accurately the Si concentration in the films by an evaporation method, we can reproduce the low contact resistances using the Ni-5%Si pellets.

The contacts prepared with Ni-5%Si pellets were isothermally annealed at 400°C. Figure 7 shows the resulting R_c values of the contacts. The R_c values did not change after annealing at this temperature for more than 100 hours. In contrast, the R_c values of the contacts with the TiAl(Cu) overlayer were

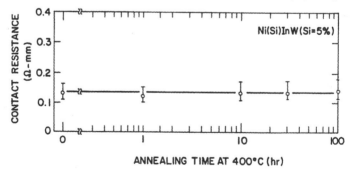

Fig.7. Contact resistances of Ni(Si)InW contact during annealing at 400°C [18].

constant for ~ 10 hours, but increased by 0.1 Ω-mm after annealing for 100 hours. X-ray diffraction was taken for the contact with the TiAl(Cu) overlayer before and after annealing at 400°C. Small interaction between the Ni(Si)InW contact and the TiAl(Cu) overlayer was observed after annealing for 1 hour, but no progress of the reaction was observed during subsequent annealing at this temperature. However, TEM cross-section examination showed formation of pin-holes at the In$_x$Ga$_{1-x}$As/GaAs interface before annealing at 400°C. The holes grew into small voids after annealing at 400°C for 100 hr which may be due to thermal stress, leading to an increase in the R$_c$ values for the contacts with the overlayers.

Since the optimum annealing temperature for NiInW ohmic contacts (with or without Si addition) are higher than 800°C, this temperature is high enough to activate the implanted species. Thus, the ohmic contacts can be formed by simultaneous (i.e., one-step) annealing for implant-activation and ohmic contact formation. Here, NiInW and Ni(Si)InW contacts were prepared by a one-step anneal, and the channel resistance (R$_{ch}$) and R$_c$ values were measured by TLM. The R$_c$ values are shown in Fig 8 for NiInW (squares) and Ni(Si)InW contacts (circles). For the NiInW contacts, Schottky like behavior was observed in the I-V curves when the contacts were annealed at temperatures below 800°C. The ohmic behavior was first observed at 850°C and the R$_{ch}$ values of ~ 600 Ω/□ and R$_c$ values of ~ 2 Ω-mm were measured. Upon annealing at higher temperatures, the R$_{ch}$ values decreased and the lowest value of ~ 300 Ω/□ was obtained at 950°C. This value is close to the typical R$_{ch}$ value (~ 260 Ω/□) obtained for the GaAs substrates in which implanted Si were activated by conventional cap annealing at 800°C for 10 min. Note that when SiF$^+$ in stead of ^{29}Si$^+$ were implanted at 60 KeV with dose of 3.3x10^{13}cm^{-2}, ohmic behavior was not obtained in the NiInW contacts by the present one-step anneal technique. The reason is not clear at the moment, but the implanted species strongly influenced the resistances of the contacts prepared by one-step anneal. Since Ni(Si)InW contacts prepared by Ni-5%Si pellets yielded the lowest R$_c$ values, this contacts were used for the one-step anneal experiment. The R$_{ch}$ values decreased with increasing the annealing temperatures and were close to those obtained in NiInW contacts when annealed at temperatures above 900°C. The R$_c$ values of about 0.8 Ω-mm was obtained at 750°C and decreased to ~ 0.2 Ω-mm after annealing at temperatures in the range of 800 to 950°C. Note that this value is close to the lowest value obtained by the conventional two-step anneal. The thermal stability of the Ni(Si)InW contacts, formed by annealing at 850°C, was also studied by annealing isothermally at 400°C. The R$_c$ values were constant even after annealing for 100 hours.

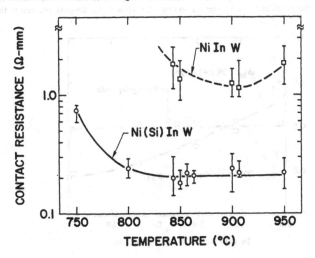

Fig.8. Contact resistances of NiInW (squares) and Ni(Si)InW contacts (circles) prepared by one-step annealing [18].

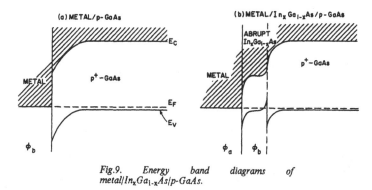

Fig.9. Energy band diagrams of metal/$In_xGa_{1-x}As$/p-GaAs.

NiInW contacts to p-type GaAs

Formation of NiInW ohmic contacts to p-type GaAs has been made [19]. Expectation of this In-based ohmic contact to p-GaAs was that the barrier ϕ_b at the metal/p-GaAs [shown in Fig. 9(a)] was reduced by separating it into two low barriers (ϕ_a, ϕ_b) (shown in Fig. 9(b)) at the metal/$In_xGa_{1-x}As$ and $In_xGa_{1-x}As$/GaAs, respectively, by forming $In_xGa_{1-x}As$ layer at the metal and GaAs interface, which was demonstrated in n-type GaAs. The barrier height at the metal/p-GaAs is close to 0.5 eV [20]. The barrier heights at the metal/p-$In_xGa_{1-x}As$ with various x values were not measured. However, these values are calculated assuming $E_g \simeq \phi_b(n) + \phi_b$ (p), and they are shown by open square in Fig. 10. It is noted that the ϕ_b values at the metal/p-$In_xGa_{1-x}As$ are constant for x values the range of 0 to 1. The height at the metal/p-$In_xGa_{1-x}As$ with x = 0.53 was measured by Veteran [21] and shown by a closed square. Note that the measured value is close to the calculated values, which indicates that the assumption, $E_g \simeq \phi_b(n) + \phi_b(p)$, is valid for the $In_xGa_{1-x}As$. From this figure, the reduction of barrier height is not expected by forming $In_xGa_{1-x}As$ layer at the metal/GaAs interfaces. The only way to reduce the R_c values to p-type GaAs is to adjust the metal/$In_xGa_{1-x}As$ interface to contact at close to the Be-peak position in the GaAs substrate.

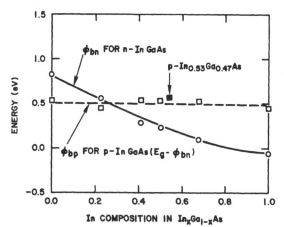

Fig.10. Barrier heights of metal/n-$In_xGa_{1-x}As$ **(circles) and** metal/p-$In_xGa_{1-x}As$ **(squares) as a function of In concentration (x) [7].**

NiInW contacts were deposited on the p-type GaAs in which Be and F were co-implanted [19]. The ohmic behavior was observed in the as-deposited sample although the R_c value was high due to contamination at the metal/GaAs interface. The contacts were annealed at temperature in the range of 300 to 900°C. The R_c values of these contacts were constant $\sim$ 1.4 Ω-mm. The result is consistent with the constant ϕ_b values with various In concentrations.

In order to reduce the R_c value of the NiInW contacts the GaAs wafers were etched so that the $In_xGa_{1-x}As$/GaAs interface contacts to the region with higher Be concentrations. The R_c values with various etched depths are shown in Fig. 11 where the upper horizontal axis shows the etched depth and the lower axis shows the corresponding hole concentrations read from the electrical profile. Note that the

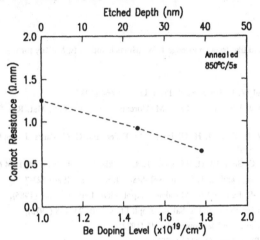

Fig.11. Contact resistances of NiInW contacts to p-GaAs as a function of Be concentrations [19].

R_c values decrease with increased etched depth. The lowest R_c value of 0.65 Ω-mm was obtained by etching at the depth of 40 nm, which is sufficiently low for 1μ-HFET devices. This contact was stable during annealing at 400°C for more than 30 hrs and is believed to be more thermally stable than conventional p-type ohmic contacts such as AuMn. In addition, the surface morphology was extremely smooth compared with these contacts.

SUMMARY

NiInW ohmic contacts are very attractive for variety of GaAs devices, because they provide low resistance ($\sim$ 0.3 Ω-mm), thermal stability during subsequent annealing at 400°C, shallow contact, and smooth surface. Recent activities for the NiInW ohmic contacts to both n and p-type GaAs were reviewed in this paper. Two approached were taken to reduce the contact resistance of the n-type NiInW contacts: one is heavy doping of donors by adding Si to NiInW and the other is reduction of the barrier height at metal/$In_xGa_{1-x}As$ interface by increasing the In concentration in the $In_xGa_{1-x}As$ layer. A factor of $\sim$ 3 reduction in R_c values was successfully achieved by adding a small amount of Si by evaporating Ni(Si) pellets. The R_c values of $\sim$ 0.1 Ω-mm were observed in the Ni(Si)InW contacts at temperatures in the range of 800-850°C. These values are comparable with those obtained routinely in conventional AuGeNi contacts. Although the contact resistance of the AuGeNi contact increased after annealing at 400°C, the resistance of the Ni(Si)InW contact was stable even after annealing at this temperature for more than 100 hrs. Moreover, the low resistance ($\sim$ 0.2 Ω-mm) Ni(Si)InW contacts were prepared by one-step (simultaneous) annealing for implant-activation and ohmic contact formation. However, the R_c values were not reduced by the second approach of increasing In concentration of the $In_xGa_{1-x}As$ layer formed at the metal/GaAs interfaces. This was explained to be due to an increase in the barrier height at the $In_xGa_{1-x}As$/GaAs interface by forming a high density of misfit dislocations at the interfaces, which was caused by large lattice mismatch between the $In_xGa_{1-x}As$ and GaAs.

The low resistance, thermally stable ohmic contacts to p-type GaAs were also prepared using NiInW metals. The contact resistance of 0.6 Ω-mm was obtained by annealing at $\sim$ 850°C and the contacts were stable during annealing at 400°C for more than 30 hrs. Note that the same ohmic contact metallurgy is applicable to both n and p-type ohmic contacts.

Our recent activities involve further reduction of the contact resistance of p-type ohmic contacts by adding a small amount of acceptors and evaluation of NiInW contacts for manufacturing integrated GaAs circuits.

Acknowledgement

The authors would like to acknowledge J. W. Mitchell and T. E. McKoy for technical assistance.

REFERENCES

1. M. Murakami, and W. H. Price, Appl. Phys. Lett. 51 664 (1987).

2. M. Murakami, H. J. Kim, W. H. Price, M. Norcott, and Y. C. Shih, Mat. Res. Soc. Symp. Proc. 148, 151 (1989).

3. M. Murakami, W. H. Price, J. H. Greiner, J. D. Feder, and C. C. Parks, J. Appl. Phys. 65, 3546 (1989).

4. M. Murakami, Y. C. Shih, W. H. Price, and E. L. Wilkie, J. Appl. Phys. 64, 1974 (1988).

5. A. Callegari, D. Lacey, and E. T. S. Pan, Sol.-Stat. Electron. 29 (1986) 523-7.

6. A. Callegari, E. T.-S. Pan, and M. Murakami, Appl. Phys. Lett. 46, 1141 (1985).

7. K. Kajiyama, Y. Mizushima and S. Sakata, Appl. Phys. Lett. 23, 458 (1973).

8. H. H. Berger, Solid-State Electronics 15, 145 (1972).

9. S. Adachi, Appl. Phys. Lett. 51, 1161 (1987).

10. P-E.Hallali, H.Baratte, F.Cardone, M.Norcott, F.Legoues and D.K.Sadana, Appl. Phys. Lett. (Submitted 1990).

11. G. S. Marlow and M. B. Das, Solid-State Electron. 25, 91 (1982).

12. M. Kamada, T. Suzuki, F. Nakamura, Y. Mori, and M. Arai, Appl. Phys. Lett. 49, 1263 (1986).

13. H.J. Kim, M. Murakami, S.L. Wright, M. Norcott, W.H. Price, and D. LaTulipe, IBM Res. Rep. RC15164 (1989).

14. J. M. Woodall, G. P. Pettit, T. N. Jackson, and C. Lanza, Phys. Rev. Lett. 51, 1783 (1983).

15. P. E. Batson, K. L. Kavanagh, J. M. Woodall, and J. W. Mayer, Phys. Rev. Lett. 57, 2729 (1986).

16. S.L. Wright, R.F. Marks, S. Tiwari, T.N. Jackson, and H. Baratte Appl. Phys. Lett. 49, 1545 (1986).

17. J. M. Woodall, J. L. Freeouf, G. D. Pettit, T. N. Jackson, and P. Kirchner, J. Vac. Sci. Technol. 19, 626 (1981).

18. M. Murakami, W. H. Price, M. Norcott, and P.-E. Hallali, IBM Res. Rep. RC15660 (1990).

19. P-E. Hallali, W. H. Price, M. Norcott and M. Murakami (unpublished)

20. W.E. Spicer, I. Lindau, P. Skeath, C.Y. Su, and P. Chye, Phys. Rev. Lett. 44, 420 (1980).

21. J. L. Veteran, D. P. Mullin, and D. I. Elder, Thin Solid Films 97, 187 (1982).

A Study of WInTe Ohmic Contact to n-Gaas

R. Dutta
AT&T Bell Laboratories, Princeton, New Jersey 08540

V.G. Lambrecht, M. Robbins
AT&T Bell Laboratories, Murray Hill, NJ 07974

Electrical, structural and diffusion characteristics of a solid phase reacted ohmic contact to n-GaAs are studied. Attempts were made to form a low band gap interfacial phase of InGaAs to reduce the barrier height at the conductor/semiconductor junction and thus yielding a low resistance, high reliablity contacts. The understanding of the interface is important from the point of view of device performance as well as device reliability. The contacts were fabricated by co-sputtering W, In and Te targets on n-GaAs with subsequent annealing. The as-deposited rectifying contacts became ohmic when annealed to 500°C, and showed a specific contact resistance of ~5x10⁻⁶ ohm cm². The Auger and Rutherford back scattering analysis of the interface revealed an InGaAs phase formation prior to the onset of ohmic conduction. The contacts were stable up to 500°C and the surface morphology was superior to presently used AuNiGe contacts. The contact pads were patterned by dry plasma etching without adversely affecting the GaAs substrate.

INTRODUCTION

There is an increasing demand for speed and bandwidth in telecommunication, in order to provide services with greater flexibility and of higher quality. Lightwave communication is bottlenecked, not by the fiber speed but by the speed of the processing and switching components. While attempts are being made to replace electronics altogether with optics, it is reasonable to assume that high speed electronics will fill the gap for quite sometime to come. In fact, in the near future, the state of telecommunication business will be driven by all the three following technologies,
1. Integrated circuit technology (signal processing)
2. Light wave technology (signal propagation)
3. Computer technology (signal switching)
A parallel development is needed in all three to meet telecommunication challenges of coming years. Performance, reliability, size and cost are the driving forces. Performance and reliability of high speed electronic components could be improved by:
1. minimization of parasitics
2. better heat dissipation
3. higher breakdown voltage & drain current
4. improved metallization, passivation and packaging
The fabrication of a stable ohmic contact

metallization to GaAs is inherently difficult since most metals do not form a thermally stable phase with GaAs. The reaction generally proceeds through formation of a metal-arsenide or a metal-gallium phase and hence a simple approach to ohmic contact formation based on metal work function and semiconductor band energy is not sufficient. It has been experimentally observed [1-3] that the Schottky barrier at a metal/GaAs interface falls within a very narrow energy range [0.7-0.9 ev] for metals with vastly different work functions. Thus, in order to make an ohmic contact to GaAs, either a tunneling junction or a low band gap interfacial phase needs to be engineered for easier charge transfer across the junction [4]. The extent of tunneling in a metal/GaAs junction, however, is also limited by the fact that the dopant atoms in a GaAs matrix can not be activated at a high doping level. With shrinking device dimensions, the quality and uniformity of the ohmic contacts are becoming increasingly important. The alloyed Au-Ge-Ni metallization, most widely used at the present time, involves melting and resolidification of the material and therefore demands a very carefully controlled process for reproducibility [5-7]. In addition, the lift-off technique which is essential to such a gold-based contact process places a limitation on the level of integration. Several attempts [8-14] have been made to replace Au-Ge-Ni by developing a non-gold ohmic contact with a non-alloyed, solid phase reacted interface.
The present study reports the ohmic contact properties of W-In based compound conductors

Mat. Res. Soc. Symp. Proc. Vol. 181. ©1990 Materials Research Society

to n-type GaAs. Such refractory contacts are expected to withstand the high temperatures, necessary to activate the ion implanted dopants, without deteriorating the lateral contact profile. A high temperature stable ohmic contact would also facilitate the adoption of a high temperature dielectric deposition process. Furthermore, a W based ohmic contact would not require a barrier layer to be used with Al interconnection metallization. Attempts are made to modify both the barrier width and the barrier height by forming a Te doped interface while at the same time fabricating a low band gap phase like GaInAs in the contact region. Tungsten is chosen for its dry etching compatibility while In is selected for its low diffusion constant [15] in GaAs.

EXPERIMENTAL

Semi-insulating (100) GaAs wafers with a VPE grown n-type epi-layer ($\sim 10^{18}$ cm^{-3}) approximately 100 nm thick were used in the present experiments. Substrates were prepared for deposition by etching in 5% NH$_4$OH solution for three minutes just prior to loading into the sputtering chamber. The compound contacts were deposited in a triple-target RF/DC magnetron sputtering machine which was cryo pumped to a base vacuum of $\sim 2 \times 10^{-7}$ torr. The Ar pressure during sputtering was held at 5 mtorr (~ 55 sccm) and the sputtering rate was typically 1 A/s. Ternary compounds of W, In and Te were co-deposited on substrates revolving above the targets at 10 rev/min. The films were typically 1000 A thick and were more than 80 at % W. In composition was kept below 20 at % while Te concentration in the films was around 2 at %.

The metallization pattern for the transmission line method (TLM) of contact resistance measurement was produced by standard photolithographic techniques. The negative lift-off patterns were removed using acetone in an adjustable power ultrasonic bath. The material was also patterned by subtractive etching of a positive TLM structure in a CF$_4$ plasma. The I-V measurements on transmission line patterns were made with a Wentworth prober and a HP 4145A Semiconductor Parameter Analyser connected to a HP 9000 personal computer. The samples were furnace annealed in the temperature range of 200°C- 500°C for ten minutes in an argon atmosphere.

The composition analysis, phase formation and preliminary diffusion characteristics were derived from Auger Electron Spectroscopy(AES) and Rutherford Backscattering (RBS) analysis. A 45 KV Cuα radiation was used for X-ray diffraction in a D spacing range of 0.98 to 8.83 A. A ~3.02 MeV He ion beam was used in RBS

to achieve a better resolution of W and In peaks.

RESULTS AND DISCUSSION

STRUCTURAL & COMPOSITIONAL:

The X-ray diffraction studies of the samples revealed the amorphous nature of the as-deposited films. The room temperature sample showed a broad amorphous W peak [fig 1] at $2\theta = 40°$ in

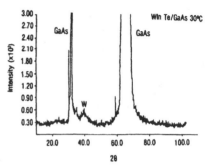

Fig 1. X ray diffraction of as deposited WInTe film

addition to the strong peaks corresponding to the (002), (004) and ß(004) reflections of the GaAs substrate. The samples annealed at 500°C [fig 2] showed a sharp, crystalline W peak at $2\theta = 40.4°$ representing a lattice spacing of 2.23

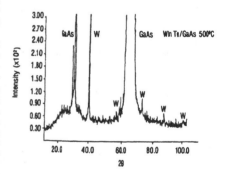

Fig 2. X ray diffraction of WInTe annealed to 500C

A. Also, the additional x-ray reflections in the spectrum of the annealed sample were attributed to a polycrystalline W matrix. The diffraction from the InGaAs interfacial phase was masked by the strong GaAs reflections. The presence of an InGaAs phase, however, was confirmed by RBS experiments.

The RBS spectrum of the as-deposited sample showed [fig 3] distinct peaks for W and In distributed uniformly in the metallization film. The RBS spectra of samples annealed at 500°C [dotted line] showed a decrease in In concentration in the WInTe film. The In concentration of the W-In-Te sample decreased from ~30 at% at room temperature to ~20 at% at 500 °C. The split peak of In at 500°C indicates the presence of In in two different environments, i) GaAs and ii) W matrix. The Auger Electron Spectroscopy (AES) of the samples [fig 4,5] confirms the reduction of In concentration in the metal film on annealing and its subsequent diffusion into the GaAs lattice. The fraction of

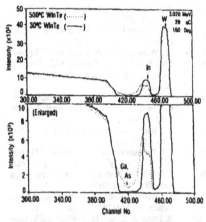

Fig 3. RBS spectra of as-deposited (solid) and 500C annealed (dotted) WInTe [normal and magnified]

Ga atoms replaced by In in the semiconductor matrix diffused out as indicated by AES.

ELECTRICAL:
I-V measurements were carried out on various W based compound contacts, annealed up to 500°C. Initially, attempts were made to make ohmic contacts by forming a tunnel junction through high doping of the interface. However, WTe$_x$ did not yield ohmic characteristics up to an annealing temperature of 500°C. Also, attempts to form a graded junction ohmic contact by In reacted phases without doping the interface did not yield low resistance contacts. A combination of interfacial doping and a graded junction was found to give ohmic contacts with low resistance below 500° C for W based materials. Doping was achieved by the incorporation of Te into GaAs at the annealing temperature while a low band gap phase of

GaInAs was formed by diffusion of In across the metal/semconductor interface. Thin films of W-Te on an interfacial layer of In resulted in an ohmic contact with a low specific contact

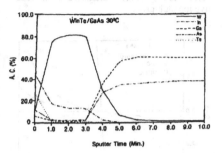

Fig 4. Auger spectroscopy of as deposited WInTe

resistance of 3×10^{-6} ohms cm^2. However, since an interfacial layer of In could affect the subtractive etching of the contacts, experiments were performed to develop a compound conductor of W-In with the dopant element. A low In concentration is preferred to preserve compatibility with the existing dry etching process for W films. Also, at high In concentration, the contacts lacked mechanical integrity. Thin films of W-In-Te with less than 20 at% of

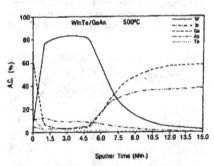

Fig 5. Auger spectroscopy of WInTe /GaAs samples annealed to 500C

In when annealed to 500°C displayed an ohmic behavior [fig 6] with an average specific contact resistance of 5×10^{-6} ohms cm^2 as calculated by TLM technique. The contact resistance decreases with increasing In concentration in the film. Below 20 at% of In, the morphology of the films looks excellent under optical microscopy and the mechnical integrity is found to be as good as W. Figure 7 shows that the contact morphology remained good upto 700°C, anneal, temperature high enough to degrade the GaAs substrate [fig 7b]. Preliminary experiments have shown the feasibility of dry etching these contacts in a CF_4 plasma. The parameters are being established for the reactive ion etch process.

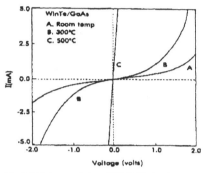

Fig 6. I-V characteristics of WInTe/GaAs, A. as deposited; B. annealed to 300C; C. annealed to 500C

CONCLUSIONS:

This work confirmed the formation of ohmic contacts to n-GaAs by compound conductors containing more than 80% tungsten. The contact process proceeds via the formation of an InGaAs phase at the interface and the doping of

Fig 7. Optical micrograph of WInTe pad annelaed to 700 C.

the interfacial layers. TLM patterns of this material have been subtractively etched by dry plasma processing. Results suggest that the optimization of the film composition may allow the specific contact resistance to be low-

ered below the present value. A contact resistance as low as 3×10^{-6} ohm cm^2 has been achieved in W-In based metallization systems.

Acknowledgements

We are grateful to Paul Sakach for the Auger data and Frank Baiocchi for RBS analysis. We would like to acknowledge the assistance of Bill Yost in various phases of this ongoing work. We would also like to thank Dr. L. Graham for helpful discussions.

References

1. C.A. Mead and W.G. Spitzer, Phys. Rev. A 134, 713 (1964).
2. W.E. Spicer, I. Lindau, P. Skeath and C.Y. Su, J. Vac. Sci. Tech. 17, 1019 (1980).
3. J.R. Waldrop, Appl. Phys. Lett. 44. 1002 (1984).
4. J.M. Woodall, J.L Freeouf, G.D. Pettit, T. Jackson and P. Kirchner, J. Vac. Sci. Tech. 19, 626 (1981).
5. T.S. Kuan, P.E. Batson, T.N. Jackson, H. Rupprecht and E.L. Wilkie, J. Appl. Phys. 54, 6952 (1983).
6. A. Illiadis and K.E. Singer, Solid State Commun. 49, 99 (1984).
7. C.J. Palmstron and D.V. Morgan, " Gallium Arsenide", Wiley, Chichester, Chap 6 (1985).
8. J.R. Waldrop and R.W. Grant, Appl. Phys. Lett 50(5), 250 (1987).
9. M. Murakami and W.H. Price, Appl. Phys. Lett. 51(9), 664(1987).
10. R. Dutta, A. Lahav, M. Robbins and V. Lambretch, AT&T Tech Report (1988).
11. Y. Yamane, Y. Takahashi, H. Ishii and M. Hirayama, Elec. Letters 23(8), 382 (1987).
12. E.D. Marshall et.el., J. Appl. Phys. 62(3), 942 (1987).
13. M. Murakami, W.H. Price, Yih-Cheng Shih and N. Braslau, J. Appl. Phys. 62(8), 3295 (1987).
14. R. Dutta, AT&T Tech Report (1988).
15. T. Otsuki, H. Aoki, H. Takagi and G. Kano, J Appl. Phys. 63(6), 2011 (1988).

Mn–In–Co, Mn–Pt and Mn–In–Pt Based Contacts to p–GaAs

T.S. Kalkur, Y.C. Lu[*] and , M. Rowe[*]
Microelectronics Research Laboratories,University of Colorado at Colorado Springs, Colorado Springs, CO 80933–7150.
[*] Department of Electrical and Computer Engineering, Rutgers University, P.O. Box 909, Piscataway, NJ 08855–0909

ABSTRACT

Mn–In–Co, Mn–In–Pt and Mn–Pt metallizations are used to form ohmic contact on Be–implanted rapid thermally annealed GaAs. The rapid thermal alloying of contact metallizations are performed in A.G. Associates Heat pulse system in nitrogen atmosphere in the temperature range of $350°C$ to $800°C$ for 5 seconds. The contacts were found to be ohmic at an annealing temperature of $450°C$. The In–Mn–Co metallization showed higher minimum contact resistivity (5×10^{-4} ohm.cm^2) than In–Mn–Pt metallization (1.5×10^{-5} ohm.cm^2) for an annealing temperature of $700°C$ and time 5 seconds. The surface morphologies of In–Mn–Pt metallizations were smooth even after alloying at $700°C$ for 5 seconds. The Auger analysis shows outdiffusion of Ga and As into the contact metallization and negligible indiffusion of In and Mn into GaAs.

INTRODUCTION

High performance ohmic contacts to p–GaAs are essential in the implementation of heterostructure bipolar transistors, junction FETs and optoelectronic devices (1–3). Although AuZnAu is the most conventional metallization used for the formation of ohmic contact to GaAs (4), recently there were efforts to replace gold from the contact metallization. This is because during the alloying process, Au forms a low melting point AuGa alloy with GaAs and this decreases the stability of contacts at high temperatures(5). In the case of contacts to n–type GaAs, encouraging results have been obtained in improving the stability of the contacts by using In–Ni–W and In–Mo–Ge metallizations (6,7). Although, In is a low melting point material, if a very thin layer of In is used, it can form InGaAs layer and the rest of Indium is used to form high temperature alloys with the constituents of contact metallization. Because of the fast diffusivity of Zn in GaAs, Au–Mn and Au–Mg based contact metallizations have been developed to improve the stability of contacts to p–GaAs (8,9). Recently, some research works have been reported in developing non–gold and non–zinc based contact metallizations to p–GaAs (10,11). In this paper we are reporting the results of In–Mn–Co, In–Mn–Pt and Mn–Pt based contact metallizations to p–GaAs.

SAMPLE PREPARATION

The samples used for the study were Be implanted semi–insulating GaAs as well as bulk Zn doped GaAs. The semi–insulating GaAs wafers were 2 degree off with respect to (100) orientation and Be implantation was done at an energy of 100 KeV at a dose of 8×10^{12} cm^{-2}. The wafers were rapid thermally annealed using the proximity technique and an activation efficiency of nearly 100% was obtained. Devices used for the measurement of contact resistivity were mesa isolated by etching in $H_2SO_4 : H_2O_2 : H_2O :: 1 : 10 : 10$ solution. The contact patterns for the electrical characterization of contact properties were defined by standard photolithography and lift–off techniques(12,13). The wafers were etched in a pre–evaporation etchant(15% NH_4OH) for 15 seconds before loading them to an evaporator. The metallization scheme used for the contact studies is shown in fig. 1. For In–Mn–Co contacts, In and Mn were coevaporated through resistive heating, and Co was deposited as an over layer by electron beam heating. During the coevaporation, the thickness of Mn and In were monitored independently using quartz crystal thickness monitors. The alloying of the contact metallization was performed in A.G. Associates Heat Pulse 410 system for various temperature and times in nitrogen atmosphere.

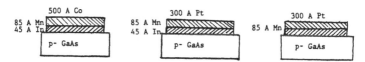

Fig. 1 Metallization Structure used for the study of Contacts to p–GaAs.

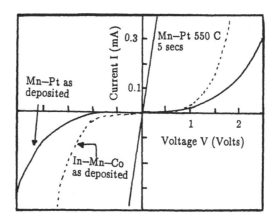

Fig.2 Variation of I–V characteristics with alloying temperature for In–Mn–Co and Mn–Pt contacts to p–GaAs.

RESULTS AND DISCUSSION

The current –voltage characteristics of as deposited In–Mn–Co metallization on Be implanted substrates is shown in fig. 2. With increase in annealing temperature to 450 C, the contact showed tendency to become ohmic. The contact became ohmic at an annealing temperature of 550 C. The variation of contact resistivity as determined by transmission line model (12), with alloying temperature for a time of 5 seconds is shown in fig. 3. The contact resistivity was found to decrease with increasing annealing temperature. The minimum contact resistivity of 5×10^{-4} ohm.cm^2 was obtained at an annealing temperatures in the range of 700–750°C. In the case of In–Mn–Pt contacts, the I–V characteristics for as deposited metallization was similar to that of In–Mn–Co. The variation of contact resisivity with alloying temperature is also shown in fig. 2. The contact resistivity was found to decrease with annealing temperature, and the minimum

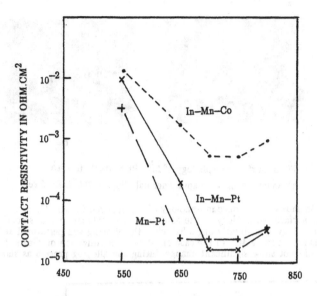

Fig.3 Variation of contact resistivity with alloying temperature for In–Mn–Co, Mn–Pt and In–Mn–Pt contacts to p–GaAs. Alloying time 5 secs.

contact resistivity obtained was 1.5×10^{-5} ohm.cm^2 and this is much less than the contact resistivity obtained for In–Mn–Co contacts. The I–V characteristics for as deposited Mn–Pt contacts are shown in fig 1.. The leakage curents for In–Mn contacts were found to be less than that for In–Mn–Pt contacts. This might be due to the absence of In in the contact metallization. The presence of Indium in In–Mn–Pt and In–Mn–Co gives rise to higher leakage current for as deposited metallization. This might be due to the modification of barrier height for contact metallization with GaAs due to the formation of InGaAs layer at the interface. But when the annealing temperature is increased beyond 650°C, the difference in contact resistivities for In–Mn–Pt and Mn–Pt was not significant. This shows that Mn plays dominant role in reducing the contact resistivity by forming a heavily doped p$^+$ layer under the contact.

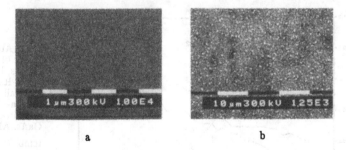

a   b

Fig.4 Surface morphology of In–Mn–Co contacts to GaAs.

a) As deposited b) Rapid Thermal alloying, 700°C for 5 secs.

a   b

Fig.5 Surface morphology of Mn–Pt contacts to GaAs

a) As deposited b) Rapid Thermal alloying 700°C for 5 secs.

Fig.4a shows the surface morphology of as deposited In–Mn–Co metallization on GaAs. The surface morphology of the as deposited metallization was smooth, but the surface morphology gets slightly degraded when the alloying was performed at 700 C for 5 seconds as shown in fig.4b. In the case on Mn–Pt metallization on GaAs, the surface morphlogy did not show significant change during the alloying process as shown in fig. 5a and 5b.

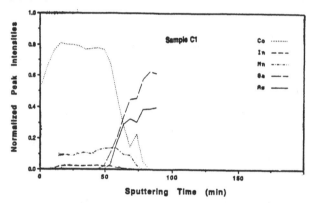

Fig.6 Auger profile of Co, In, Mn, Ga and As in As deposited In–Mn–Co metallization on GaAs.

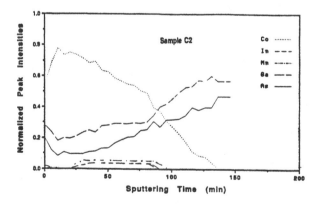

Fig.7 Auger profile of Co, Ga, As, In and Mn in Rapid thermally alloyed In–Mn–Co metallization on GaAs. Alloying temp. 750°C. Time 5 seconds

Fig.6 shows the Auger profile of In, Mn, Co, Ga and As in the as deposited In–Mn–Co metallization on GaAs. The profiles clearly show the top cobalt layer, followed by coevaporated indium and manganese layer. Fig. 7 shows the distribution of contact constituents due to rapid thermal alloying at 750 C for 5 seconds. The profiles clearly show the outdiffusion of Ga and As into the contact metallization and indiffusion of Co into GaAs. The profiles corresponding to Mn and In show very slight indiffusion into GaAs. The ohmic contact formation might be due to the combined action of reduction in barrier height of the metal–semiconductor contact with the formation of $In_xGa_{1-x}As$ (x = 0 to 1) and acceptor type doping action of Mn in GaAs. The outdiffusion of Ga into cobalt creates vacancies and a p^+ layer forms when these vacancies are occupied by Mn atoms.

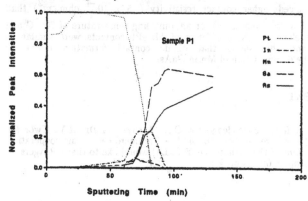

Fig.8 Auger Profile of Pt, Ga, As, In and Mn in as deposited In–Mn–Pt metallization on GaAs.

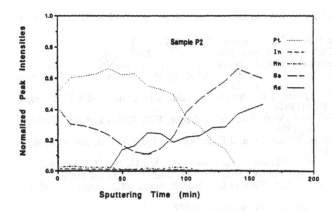

Fig.9 Auger profile of Pt, Ga, As, In and Mn in Rapid Thermally alloyed In–Mn–Pt metallization on GaAs. Alloying temperature 700°C and time 5 secs.

Fig. 8 shows the Auger profile for In, Mn, Pt, Ga and As in the as deposited In–Mn–Pt metallization. The profiles show In and manganese layer at the interface with Pt overlayer. Fig. 9 shows the profile for the metallization components for an alloying temperatue of 700°C and time 5 seconds. The profiles show indiffusion of platinum into GaAs and outdiffusion of Ga and As into the contact metallization. In and Mn were found to be redistributed in the contact metallization with negligble indiffusion into GaAs. The Mn–Pt metallization also showed similar redistribution due to the annealing processs. Minimum indiffusion of Mn into GaAs during alloying process shows that metallization ideally suitable for the implementation of contacts for heterostructure bipolar transistors.

CONCLUSIONS

Contact resistance and microstructural analysis were carried out for Mn–In–Co, Mn–In–Pt and Mn–Pt based contacts to Be implanted p–GaAs. The contact resistivity as determined by TLM method was found to depend on alloying temperature. The Mn–In–Co contacts showed higher contact resistivity (5×10^{-4} ohm.cm^2) than Mn–In–Pt contacts (1.5×10^{-5} ohm.cm^2) for an annealing temperature of 700°C and time 5 seconds. The electrical characteristics of Mn–Pt contacts were similar to In–Mn–Pt contacts and this shows that ohmic contact formation might be predominantly due to the doping action of Mn in GaAs.

ACKNOWLEDGEMENT

The authors gratefully acknowledge Dr. C.A. Araujo and Dr. R.Y. Kwor for their support. The authors are indebted to Dr. Lee Kammerdiner for his cooperation in experimental work. One of the authors (Dr.Y.C. Lu) would like to thank Rutgers Research Council for support of this project.

REFERENCES

1. M. Illegems, B. Schwartz, L.A. Koszi and R.C. Miller, Appl. Phys. Lett., 33(6), 29 (1978).
2. K. Lehovec, IEEE J. Solid State Circuits, SC–16, 797 (1979).
3. T. Izawa, T. Ishibashi and T. Sugeta, IEDM Tech. Dig. 328 (1985).
4. B.L. Sharma, Semiconductors and Semimetals, Academic Press, Vol.15, p.1 (1981).
5. J. Wood, Reliability and Degradation, edited by M.J. Howes and D.V. Morgan (John–Wiley and Sons. Ltd, 1981), p. 191.
6. M. Murakami, W.H. Price, Y.C. Shih, N. Brasalu, K.D. Childs and C.C. Parks, J. Appl. Phys., 62 (8), 3295 (1987).
7. M. Murakami, Y.C. Shih, W.H. Price, E.L. Wilkie, K.D. Childs and C.C. Parks, J. Appl. Phys., 64 (4), 1974 (1988).
8. C. Dubon–Chevalliar, M. Gauneau, J.F. Bresse, A. Israel and D. Ankri, J. Appl. Phys., 59 (11), 3783 (1986).
9. N.A. Papanicolaou and A. Christou, Electronics Lett., Vol.19, No. 11, 419 (1983).
10. T.S. Kalkur, Y.C. Lu and Robert Caracciolo, Mat. Res. Soc. 146, 437 (1989).
11. A. Katz, S.N.G. Chu, P.M. Thomas and W.C.D. Smith, Mat. Res. Soc. 146, 405 (1989).
12. H.H. Berger, Solid State Electronics, 15, 145 (1972).
13. Y.C. Lu, T.S. Kalkur and C. A. Araujo, Vol. 136, no.10, 3123 (1989).

SELECTIVE OXIDATION AND ETCHING OF REACTED Pt FILMS ON GaAs.

Eliezer Weiss,* Robert C. Keller, Margaret L. Kniffin, and C. R. Helms
Department of Electrical Engineering, Stanford University, Stanford, CA 94305.

ABSTRACT

The oxidation of prereacted Pt films on (100)-oriented n-GaAs substrates was studied in the temperature range between 550 and 750°C using Auger electron spectroscopy and Xe^+ ion profiling. The $GaPt/PtAs_2/GaAs$ structure formed during annealing in hydrogen was oxidized using a mixture of water vapor and hydrogen. The GaPt phase can be oxidized completely, whereas the inner $PtAs_2$ and GaAs interfaces are left unoxidized. The oxidation of the platinum-gallium phase is self limited by the diffusion of the Ga through the gallium oxide overlayer. The oxide can be etched off to leave a structure consisting only of platinum-arsenide on the GaAs substrate.

INTRODUCTION

The development of rectifying and ohmic contacts to GaAs devices which are stable at high temperatures, reliable and suitable for miniaturization is a challenging task. These requirements necessitate that metallic contacts to GaAs be in thermodynamical equilibrium with the substrate, which means they should react to completion during the processing treatments following deposition [1]. Unfortunately, these criteria are not easily met with the industry standard Au-Ge-based ohmic contacts to n-GaAs [2]. In addition most gate contacts, with the exception of WSi_x, are also unstable at temperatures required for implant activation. These difficulties have motivated numerous investigations of metallic contacts formed by solid-phase reactions as substitutes for alloy contacts [3]. In many cases the metal reacts with the GaAs substrate and forms intermetallic compounds upon heating [1,4]. The advantage of such a reaction is the removal of interfacial defects, introduced during the metal deposition and other processing steps. Although typical structures can be formed around 500°C, the electrical and metallurgical properties of such structures degrade at higher temperatures. In addition, at elevated temperatures Ga may be a fast diffuser or may phase separate into the liquid state. In these cases, good dimensional control, which is indispensable for miniaturization, will be difficult to achieve.

In this paper we report on a technique which shows promise for achieving stable contacts. This technique involves the selective oxidation of Ga in H_2O/H_2 mixtures simultaneous with the reaction of the metal and As to form a metal arsenide compound. In its simplest form the process can be represented by the reaction:

$$2mM + 2GaAs + 3H_2O \rightarrow Ga_2O_3 + 2M_mAs + 3H_2 \qquad (1)$$

Due to the large ΔG for the formation of Ga_2O_3 the partial pressure ratio of H_2O to H_2 can be adjusted so the equilibrium for reactions such as:

---

*On leave from SCD - Semi-Conductor Devices, D. N. Misgav 20179, Israel.

Mat. Res. Soc. Symp. Proc. Vol. 181. ©1990 Materials Research Society

$$2M_mAs + 6H_2O \rightleftharpoons As_2O_3 + M_{2m}O_3 + 6H_2 \qquad (2)$$

can be driven strongly to the left leaving the metal arsenide compound unperturbed. We note that this technique has been used successfully for the selective oxidation of Si in the presence of tungsten [5].

Reaction (1) can be broken into two: a reaction of the metal film with the GaAs substrate in an inert atmosphere:

$$(n+m)M + GaAs \rightarrow M_nGa + M_mAs \qquad (3)$$

and the oxidation of the intermetallic M_nGa in water vapor:

$$2M_nGa+3(1+n/m)H_2O+2(n/m)GaAs \rightarrow (1+n/m)Ga_2O_3+2(n/m)M_mAs+3(1+n/m)H_2 \qquad (4)$$

Finally the gallium oxide is etched off by an acid to form a single phase metal arsenide film:

$$Ga_2O_3 + M_mAs + 3H^+ \rightarrow 2Ga^{3+}(sol.) + M_mAs + 3OH^- \qquad (5)$$

It should be noted that such a scheme is not applicable to metals of groups IIIB, IV, and V because of their very high ΔG of oxidation.

We have chosen the Pt-GaAs system for our investigation of the removal of the Ga-rich phase. This is achieved by selective oxidation followed by chemical etching to remove the resultant oxides. Pt, which is both easily deposited as very thin continuous films (<100Å) and immune to oxidation, reacts with GaAs at low temperatures. Several studies of the Pt-GaAs interfacial reactions have been reported [1,4,6-10]. The interface reaction for nonoxidizing conditions leads to the formation of a bilayer structure $GaPt/PtAs_2/GaAs$. Pt/GaAs films heated in air behave similarily with two added features: a surface layer containing mostly Ga and O forms over the GaPt phase, and oxygen diffuses through the (reacted) Pt to form an oxygen-rich layer at the $PtAs_2/GaAs$ interface [7,8]. A weaker oxidant than air (a mixture of water vapor and hydrogen) is used in our case to completely oxidize the GaPt phase, while leaving the inner $PtAs_2$ and GaAs interfaces unoxidized.

EXPERIMENTAL

The GaAs substrates were 0.025 Ω-cm, Si doped (n-type), and had a (100) orientation (from Litton, Morris Plains, NJ). They were degreased in trichloroetane, acetone and 2-propanol, and etched in $1:10:NH_4OH:H_2O$ for 10 sec just prior to Pt deposition. Pt films of thicknesses 250 to 500Å were deposited by electron beam evaporation at pressures of ~10^{-7} Torr. The Pt films were first reacted at 450°C for 30 min in hydrogen (1SCFH) to form the bilayer structure. The $GaPt/PtAs_2/GaAs$ structures were oxidized using either water vapor in an Ar carrier (1SCFH) or a mixture of that and hydrogen. The water vapor pressure was regulated between 20 to 250 Torr by controlling the water bubbler temperature. The oxidation was carried out for various lengths of time at temperatures in the range 550-750°C. Auger spectroscopy coupled with Xe$^+$ ion sputtering and scanning electron microscopy were used to follow compositional and morphological changes in the samples.

ANNEALING AND OXIDATION

Producing a single-layer structure of PtAs$_2$ on GaAs was achieved in three steps: annealing the Pt film in hydrogen, oxidizing selectively the Ga-rich phase produced by the annealing (resulting in additional reaction of liberated Pt with the substrate), and finally removing the resultant gallium oxide. We shall first discuss the annealing and oxidizing steps. Auger profiles of the annealed samples [Fig. 1(a)] reveal the stratified GaPt/PtAs$_2$/GaAs structure proposed first by Sinha and Poate [6] and confirmed by others [9,10]. Even traces of air left in the furnace at early stages of the annealing (due to insufficient purging) caused the appearance of an oxygen-rich layer at the PtAs$_2$/GaAs interface [Fig. 1(a) and (b)]. In our work, the annealing at 450°C does not form islands of Ga-rich Pt-Ga, as found by Iwakuro and Kuroda [11], but rather continuous layers even for Pt films as thin as 250Å. In figure 1(a) the thin layer of oxygen on the surface of the structure is due to oxidation of reacted Ga by the ambient. If the oxide on the GaAs starting wafer is inefficiently stripped it is found in the reacted structure at the PtAs$_2$/GaAs interface and not in the GaPt/PtAs$_2$ interface, as reported by Sands et al. [10]. The oxygen peak found at both interfaces in earlier work [7,8] is due to diffusion of oxygen through the layers as will be discussed next.

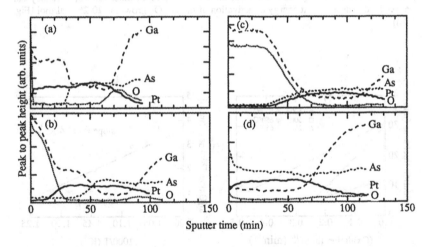

Fig.1: Auger depth profiles of 500Å Pt/GaAs films after various treatments: annealing in hydrogen at 450°C for 30 min (a); oxidizing the pre-reacted Pt/GaAs films by a H$_2$O(~20 Torr)/Ar stream at 650°C for 30 min (b) and 750°C for 15 min (c); and etching the sample depicted in (c) in hot (~70°C) concentrated HCl for 15 min (d). The oxygen peak at the PtAs$_2$/GaAs interface in (a) and (b) is due to traces of air left at the furnace during the annealing step.

The major product of oxidation of the pre-reacted Pt films by the H_2O/Ar stream is gallium oxide, as shown in Auger profiles of figure 1(b,c). The oxide resides on the surface of the structure. In contrast to the oxidized structure obtained by a single-step oxidation in air [7,8] we observed the structures: $Ga_2O_3/GaPt/PtAs_2/GaAs$ [Fig. 1(b)] or $Ga_2O_3/PtAs_2/GaAs$ [Fig. 1(c)] which are free of interfacial oxygen. This difference is due to our two-steps process and oxidation in water vapor rather than in air. The diluted oxygen ambient in the case of oxidation by air has a high enough oxidation potential to oxidize GaPt as well as $PtAs_2$ and GaAs. The water vapor, on the other hand, oxidizes only the GaPt phase:

$$2GaPt + 3H_2O \rightarrow 3Ga_2O_3 + 2Pt + 3H_2 \qquad (6)$$

The liberated Pt diffuses into the structure and reacts with As. The structures obtained by oxidation in a $H_2/H_2O/Ar$ stream are similar to those described above, however the gallium oxide film is much thinner for mixtures of $H_2:H_2O$ greater than 1:1. For our Pt contacts, the introduction of H_2 into the H_2O/Ar stream drives reaction (2) to the left as expected. Reaction (6) is also driven to the left, implying that the free energy for GaPt formation is comparable to that of Ga_2O_3.

Figure 2 depicts the results of the oxidation kinetic study. The gallium oxide thickness increases proportional to the square root of the oxidation time [Fig. 2(a)] suggesting a diffusion limited mechanism. We have found that the diffusion of oxygen-containing species is not the limiting step since a ~10-fold increase in the water vapor pressure (from ~20 to ~250 Torr) causes only a very small increase in the gallium oxide thickness. We conclude, therefore, that the limiting step is the diffusion of Ga atoms to the outer surface, were they can be oxidized. The apparent energy of activation of the Ga_2O_3 growth is 20-25 kcal/mol [Fig. 2(b)].

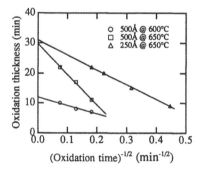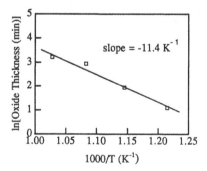

Fig. 2: (a) The dependence of the gallium oxide thickness on the oxidation time. (b) Arrhenius plot for oxidation of annealed 500Å Pt/GaAs films in $H_2O(\sim20$ Torr)/Ar stream in the temperature range 550-750°C. The oxidation time was 30 min except for 750°C where, in order to avoid severe tailing of the oxygen line, 15 min oxidation was employed which suffice to completely oxidize the GaPt phase [see Fig. 1(c)]. The oxide thickness is expressed in sputtering time yielding half of the oxygen plateau PPH.

ETCHING OFF THE OXIDE OVERLAYER

Having now formed a stable structure it is desirable to strip the surface oxide so that contact can be made. This is accomplished by wet etching in either concentrated HCl or concentrated HF. Both must be heated (to ~70°C or ~40°C, respectively) to dissolve efficiently and selectively the gallium oxide. The resulting structure is $PtAs_2/GaAs$ as is shown in figure 1(d). There is some Ga in the As-rich phase, probably as GaPt. However, if we assume the same sensitivity coefficients for Ga and As in both GaAs and $PtAs_2$ than the Ga to As ratio in the latter is only ~0.16. Arsenic is accumulated at the surface of the etched structure. Since there is only a small amount of oxygen at the surface and also arsenic oxides are highly soluble in both water and acids [12,13] we conclude that the accumulated species is elemental arsenic. Such accumulated As^0 was found earlier on GaAs surfaces etched in hot HCl [13,14]. The As^0 rich film at the surface can be dissolved in NH_4OH [13,14].

MORPHOLOGY

Auger profiles of samples oxidized at high temperatures or long periods show tailing of both the oxygen and the arsenic lines. It may be due to in-diffusion of oxygen and out-diffusion of arsenic. Also, non-continuous films may be suspected. However, we shall show below that this is only an artifact resulting from changes in the morphology of the gallium oxide film. The stratification remains and there is no lateral phase distribution in the inner interfaces.

The scanning electron microscope (SEM) image of the surface of an annealed 500Å Pt/GaAs film does not show a clear structure. In contrast 15 min oxidation at 750°C causes the appearance of grains with an average diameter of ~0.2μm together with bigger grains with ~1μm diameter. Energy dispersive spectroscopy (EDS) measurements performed on the bigger grains show a higher concentration of Ga relative to the rest of the area. They are probably Ga_2O_3 grains that cover the sample (oxygen could not be detected, since the apparatus was equipped with a Be window). However they can also be Ga droplets on top of the thin Ga_2O_3 film that cover the structure. Etching the oxide off yields a surface consisting of grains with an average diameter smaller than ~0.1μm. Clearly the $PtAs_2$ layer is a continuous one with smaller grains than the oxide.

The average oxide grain diameter is increased at longer oxidation periods indicating grain growth in the oxide. The grain growth occurs mainly in the gallium oxide layer and to a much lesser extent in the the other layers. Bigger grains at longer oxidation carried out at higher temperatures cause smearing of the various lines in the Auger profiles. Whereas even for samples showing a high degree of tailing, the profiles recorded after etching off the gallium oxide reveal relatively abrupt interfaces. The observed smearing is due to the increase in the oxide surface roughness caused by this grain growth. The bigger grains might have also different sputtering rate which will increase the grading of the interfaces.

CONCLUSIONS

We have demonstrated in the case of Pt/GaAs films that the gallium intermetallic compound can be eliminated by a three steps process. The Pt film is annealed in hydrogen, the GaPt phase is then oxidized in a H_2O/H_2 mixture, and the resultant gallium oxide layer is etched off in a hot acid yielding a single layer structure of $PtAs_2/GaAs$.

ACKNOWLEDGEMENTS

This work was supported by contract DAAL01-88-K-0828 funded by DARPA and the Joint Services Electronics Program contract DAAL03-88-C-0011. One of us (EW) would like to acknowledge partial support from SCD.

REFERENCES

[1] T. Sands, V.G. Keramidas, K.M. Yu, J. Washburn, and K. Krishnan, J. Appl. Phys. **62**, 2070 (1987).

[2] T.S. Kuan, P.E. Batson, T.N. Jackson, H. Rupprecht, and E.L. Wilkie, J. Appl. Phys. **54**, 6952 (1983).

[3] J.-C. Lin, K.J Schulz, K.-C. Hsieh, and Y.A. Chang, J. Electrochem. Soc. **136**, 3006 (1989) and references therein.

[4] R. Schmid-Fetzer, J. Electron. Mater. **17**, 193 (1988) and references therein.

[5] N. Kobyashi, S. Iwata, and N. Yamamoto, *Proc. IEEE Intern. Electron Dev. Meet.*, Dec. 1984, p. 122.

[6] A.K. Sinha and J.M. Poate, Appl. Phys. Lett., **23**, 666 (1973)

[7] C.C. Chang, S.P. Murarka, V. Kumar, and G. Quintana, J. Appl. Phys. **46**, 4237 (1975).

[8] V. Kumar, J. Phys. Chem. Solids **36**, 535 (1975).

[9] C· Fontaine, T. Okumura, and K.N. Tu, J. Appl. Phys. **54**, 1404 (1983).

[10] T. Sands, V.G. Keramidas, A.J. Yu, K-M. Yu, R. Gronsky, and J. Washburn J. Mater. Res. **2**, 262 (1987).

[11] H. Iwakuro and T. Kuroda, Jap. J. Appl. Phys. **28**, 223 (1989).

[12] *CRC Handbook of Chemistry and Physics*, 57th ed., edited by R. C. Weast (CRC, Boca Baton, FL, 1976-1977) p. 92.

[13] Z.H. Lu, C. Lagarde, E. Sacher, J.F. Currie, and A. Yalon, J. Vac. Sci. Technol. A **7**, 646 (1989).

[14] C.C. Chang, P.H. Citrin, and B. Schwartz, J. Vac. Sci. Technol. **14**, 943 (1977).

ALLOYING BEHAVIOR OF THE Ni/In/Ni/n-GaAs OHMIC CONTACT

Chia-Hong Jan, Doug Swenson and Y. Austin Chang
1509 University Avenue, Materials Science & Engineering Department,
University of Wisconsin - Madison, Madison, Wisconsin 53706

ABSTRACT

The interactions between Ni and Ni/In/Ni thin-films and GaAs were studied by SEM, SAM and AES. The presence of a molten phase was observed for both contacts after annealing at 820°C for 3 minutes. This behavior was rationalized in terms of the presence of a ternary eutectic reaction in the gallium-nickel-arsenic system. DTA confirmed the existence of the reaction:

$$L \longrightarrow NiGa + NiAs + GaAs$$

at 810°C. In the case of Ni/In/Ni, melting was thought to occur because of the segregation of indium metal to the contact surface and the subsequent melting of the nearly ternary interfacial region. Upon further cooling the formation of NiGa, NiAs, $In_xGa_{1-x}As$ and an unspecified compound Ni-In was believed to occur. The contact was shown to be either sintered or alloyed, depending upon processing conditions.

1. INTRODUCTION

Recently, a great deal of effort has been put forth to find a replacement for gold-germanium-based contacts to compound semiconductors. Gold-germanium contacts are referred to as alloyed contacts, meaning that contact formation is dependent upon the presence of a liquid phase. However, the existence in these contacts of low-melting temperature initial phases and reaction products makes them inherently unstable, even at moderate temperatures (e.g. 400 °C) required for device processing [1].

The trend has been toward the investigation of sintered contacts, i.e. contacts in which all reactions between contact material and semiconductor occur in the solid state. Most such contacts studied thus far have been based upon combinations of nickel or palladium with silicon, germanium, and indium [2-4]. Although sintered contacts require higher annealing temperatures and longer annealing times for activation, they have been shown to exhibit electrical stability and shape integrity under high-temperature processing conditions which are superior to those properties of gold-germanium contacts.

One particularly promising contact is based upon nickel and indium [5-7]. This contact has exhibited low contact resistance and excellent thermal stability. However, the mechanism of contact formation is not completely clear. Ohmic behavior may be observed in nickel-indium contacts to GaAs upon rapid thermal annealing (RTA) at 800 C for a few seconds. This implies that interfacial reactions are occuring at an extremely rapid rate, and is more consistent with a liquid phase reaction than a solid-state process. Therefore, an important question to be answered is whether nickel-indium contacts to GaAs are alloyed or sintered. Furthermore, if the contact is alloyed, the minimum alloying temperature and liquid phase present must be specified. These issues are addressed in this investigation.

2. EXPERIMENTAL PROCEDURE

GaAs wafers (Wacker Siltronic,(100)-oriented, Te-doped, n~$3.2 \times 10^{17}/cm^3$) were degreased in trichloroethylene and acetone. Native oxides were removed during a two-minute etch in a 50% HCl solution. Evaporation of high-purity nickel and dc magnetron sputtering of high-purity indium metal were carried out under a base pressure of 1×10^{-6} torr. Deposition configurations were as follows: Ni(500 Å)/GaAs and Ni(200 Å)/In(200 Å)/Ni(100 Å)/GaAs.

Samples were annealed in quartz tubes evacuated to 10^{-4} torr at 300,400,500,600,700 and 780°C for 1 hour and at 820°C for 1-3 minutes. Annealed samples were analyzed by scanning electron microscopy (SEM) using a JEOL JSM 35C microscope,and scanning Auger spectroscopy (SAM) and Auger electron spectroscopy (AES) with a Perkin Elmer Phi 600 SAM/SIMS/AES spectrometer.

Diffrerential thermal analysis (DTA) was used to determine sections of the gallium-nickel-arsenic phase diagram as a function of temperature. Samples were analyzed at many compositions along lines connecting stoichiometric GaAs and NiGa, NiGa and NiAs and NiAs and GaAs. Samples were prepared by mixing appropriate amounts of the master alloys

Mat. Res. Soc. Symp. Proc. Vol. 181. ©1990 Materials Research Society

NiGa,NiAs and GaAs in evacuated quartz tubes and homogenizing at high temperatures. There was no apparent interaction between the samples and the tubing. A Perkin-Elmer 1700 DTA system was used to determine the isoplethal sections GaAs-NiGa, NiGa-NiAs and NiAs-GaAs. All DTA data were taken in the heating mode to prevent uncertainty in reaction temperatures due to undercooling. Heating rates varied from 2°C/minute to 5°C/minute to 10°C/minute, and data were extrapolated to a rate of 0°C/minute to obtain the reaction temperatures. DTA results were confirmed by XRD experiments, using a Nicolet/STOE automated x-ray diffractometer.

Samples were prepared for electrical measurements by standard subtractive photo-lithography techniques. Measurements were made using a Keithley 617 electrometer/source at voltage steps of 50 mV. I-V measurements were performed on 240 μm frontside diodes with large area backside indium contacts. Contact resistivities were measured using the Cox-Strack method.

3. RESULTS AND DISCUSSION

Figures 1a,b and 2a,b show SEM micrographs of Ni/GaAs and Ni/In/Ni/GaAs contacts annealed at 780°C for 1 hour and 820°C for 3 minutes, respectively. These contacts show very similar microstructures. At 780°C, the surface morphologies are very rough and porous, an effect which is probably due in part to a mismatch of the thermal expansion coefficients between thin-film and substrate. At 820°C, however, it is apparent that a liquid phase has formed in both contacts. The presence of a liquid phase in either contact at this temperature cannot be explained in terms of any of the constituent binary phase diagrams except for those regions which are extremely nickel-poor and indium or gallium-rich. Because this melting phenomenon is observed in both contacts over the same temperature range, it may be inferred that this reaction is due to some ternary interaction in the gallium-nickel-arsenic system. This conclusion is further supported by Figures 3a-d, which show the results of an SAM linescan of a Ni/In/Ni contact annealed at 820°C for 3 minutes. It may be seen that indium,gallium and arsenic are present everywhere on the contact surface. However, there are many regions in which nickel is absent. These regions correspond to the area between the liquid droplets. Therefore, the molten phase is definitely due to the presence of nickel.

Before continuing with this discussion, it would be worthwhile to review the phase formation sequence in Ni/GaAs thin-films. It is well documented [8,9] that the first phase to form is a ternary compound Ni_3GaAs. Upon consumption of the elemental nickel and at higher temperatures, this phase decomposes into NiGa and NiAs. The final, three-phase equilibrium is NiGa + NiAs + GaAs. This result is confirmed by bulk phase diagram investigations [10].

Assuming that equilibrium conditions have been reached, the liquid phase must form by a reaction between NiGa, NiAs and GaAs, all of which have melting temperatures much higher than 780-820°C, the temperature range over which melting occurs. One possible explanation for the formation of a liquid phase in gallium-nickel-arsenic is the existence of a ternary eutectic point. Just as a relatively low-temperature liquid phase may exist between two high melting point phases in a binary system, so can a relatively low-temperature liquid phase exist between three high-temperature elements or compounds in a ternary system. In this case, the ternary eutectic point, if present, must lie within the composition region bounded by GaAs, NiGa and NiAs because these are the three phases presumably in equilibrium with each other prior to the formation of the liquid phase. One would therefore expect to find the following reaction in this system:

$$L ---> NiGa + NiAs + GaAs.$$

Figures 4a,b show the isoplethal section NiGa-NiAs as determined by DTA and the 600°C isotherm of the gallium-nickel-arsenic phase diagram, respectively. In Figure 4a, there exists an invariant reaction (i.e. a reaction which occurs at a single temperature and a single composition) at 810 °C, and NiGa, NiAs and GaAs coexist below this temperature. However, the minimum temperature in the single-phase liquid region is 820 °C Because the minimum single-phase liquid temperature and invariant reaction temperature do not coincide, the ternary eutectic point does not lie along this section. However, because the two temperatures are only 10°C apart, it is likely that the ternary eutectic point lies very close to this section.

The complete reaction sequence for Ni thin-film/GaAs couples may now be specified. This sequence is outlined in Figures 5a-e. As-deposited nickel reacts to form Ni_3GaAs. This phase decomposes to form NiGa and NiAs. This configuration is stable up to 810°C, at which point the NiGa, NiAs and GaAs melt to form a liquid phase. As is shown in Figure 5e, upon cooling to below 810°C NiGa, NiAs and GaAs will reappear. It is likely that some of the GaAs will

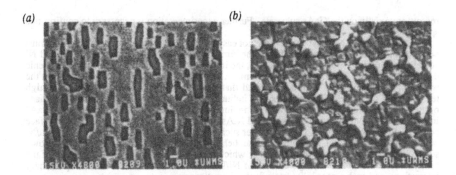

Figure 1. SEM micrographs of annealed Ni/GaAs contacts: (a) 780 °C, 1 hour, (b) 820 °C, 3 minutes.

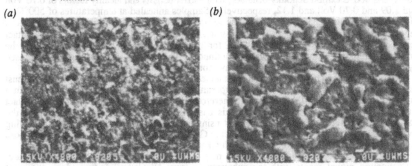

Figure 2. SEM micrographs of annealed Ni/In/Ni/GaAs contacts: (a) 780 °C, 1 hour, (b) 820 °C, 3 minutes.

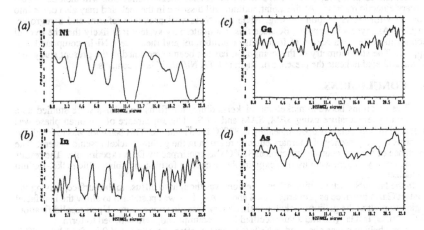

Figure 3. SAM linescan of a Ni/In/Ni/GaAs contact annealed at 820 °C for 3 minutes: (a) Nickel signal , (b) Indium signal , (c) Gallium signal and (d) Arsenic signal.

regrow epitaxially on the GaAs substrate. Further experimentation is necessary to confirm this speculation.

The results for the Ni/In/Ni contacts are not easy to interpret, for quaternary interactions must be taken into account. However, due to the similarity in behavior between Ni/In/Ni and Ni contacts, one may conclude that, in this case, the interaction between nickel, gallium and arsenic dominates the overall behavior of the system. A possible reason for this may be found in the work of Murakami et al. Murakami found that upon annealing a Ni/In/Ni contact at high temperatures, the nickel will react before the indium, and the indium will segregate to the surface of the contact. If the surface segregation of indium were rapid and complete, the contact/GaAs interface would consist entirely of NiGa and NiAs compounds on GaAs; in effect, the interface could be considered a ternary system. Under such circumstances, one would expect a Ni/In/Ni contact and a Ni contact to exhibit similar behavior. However, in this case, the regrown semiconducting layer would be $In_xGa_{1-x}As$, which is responsible for the ohmic behavior of the contact. Of course, the alloying behavior of Ni/In/Ni contacts could be related to quaternary interactions, and such a possibility bears further investigation.

Electrical measurements are consistent with the previous analysis. I-V and J-V plots are shown in Figures 6 a,b for Ni/In/Ni contacts at various temperatures. Samples annealed at 300°C and 400°C exhibit Schottky behavior, with barrier heights and ideality factors of 0.76 Volt and 1.09 and 0.70 Volt and 1.12, respectively. Samples annealed at temperatures of 500°C or higher show ohmic behavior. None of the samples had contact resistivities as low as those reported by Murakami, and the lowest resistivity was about 10^{-4} ohm-cm^2. The contact resistance of the sample annealed at 820°C for 3 minutes could not be measured because the droplet morphology of the contact was discontinuous. This implies that the tungsten layer used by Murakami actually plays the role of an active contact layer and is not just a protective cap.

Because ohmic behavior is observed upon annealing a Ni/In/Ni contact at 500 °C, there must be some $In_xGa_{1-x}As$ formation at this temperature and hence some penetration of indium through the contact toward the substrate. Therefore, the indium does not remain at the contact surface over long periods of time. In this case, there is no liquid phase present at the contact/substrate interface, and the contact is a sintered contact. Depending upon the annealing temperature and time, then, a Ni/In/Ni contact to GaAs may be either alloyed or sintered. Those high-temperature, RTA contacts studied by Murakami et al are most likely alloyed.

A reaction sequence for Ni/In/Ni contacts may now be postulated, as shown in Figures 7a-e. As-deposited nickel reacts with the substrate to form Ni_3GaAs. Indium does not react extensively with nickel or GaAs, but segregates to the surface. At higher temperatures, Ni_3GaAs will decompose into NiGa and NiAs. A limited amount of indium will diffuse through the contact and form $In_xGa_{1-x}As$. At 810°C, the interfacial, nickel-rich region will melt due to the ternary eutectic reaction. At this point, indium will dissolve in the melt and may also come into contact with GaAs. Upon cooling, NiGa and NiAs will probably reform, and a regrown, $In_xGa_{1-x}As$ layer will appear. Because this is a quaternary system it is likely that the overall composition lies in a region of four-phase equilibrium, and therefore a Ni-In compound will appear. The stoichiometry of this compound has not been determined in the present study, but Murakami et al indicate the presence of Ni_3In in RTA Ni/In/Ni/GaAs contacts.

4. CONCLUSIONS

The reactions between nickel and nickel-indium thin-films and GaAs were studied as a funciton of temperature using SEM, SAM and AES. The appearance of a molten phase was observed in both contacts upon annealing at 820°C for 3 minutes. This behavior was rationalized in terms of the presence of a ternary eutectic reaction in the gallium-nickel-arsenic system. The existence of a ternary eutectic reaction at 810°C was confirmed by DTA experiments. The results for the ternary system were used to explain ohmic contact formation in bimetallic nickel-indium metallizations.

In the Ni/In/Ni metallization scheme, indium was thought to diffuse to the surface, leaving the contact/GaAs interface as primarily nickel and GaAs. This was postulated to allow the interfacial region to melt at the gallium-nickel-arsenic ternary eutectic temperature. Upon resolidification, NiGa, NiAs, $In_xGa_{1-x}As$ and an unspecified Ni-In coumpound were believed to form.

Ohmic behavior was observed in Ni/In/Ni contacts after annealing at 500 °C for 1 hour. This result sugested that nickel-indium contacts could be either sintered or alloyed, depending upon processing conditions.

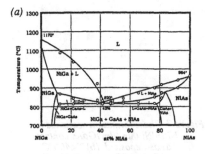

(a)

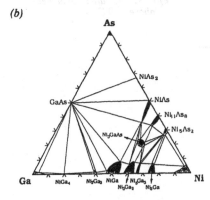

(b)

Figure 4. The Ga-Ni-As phase diagram:
(a) NiGa-NiAs isoplethal section ,
(b) 600°C isotherm (from ref. 10).

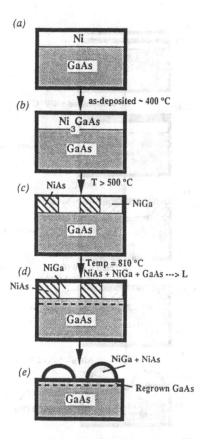

Figure 5. Phase formation sequence in Ni
thin-film/GaAs contacts:
(a) as-deposited , (b) T < 400 °C ,
(c) T > 500 °C, (d) T = 810 °C,
(e) T < 810 °C after annealed above 810 °C.

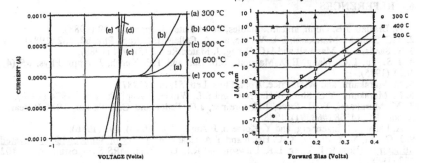

Figure 6. Electrical measurements on Ni/In/Ni contacts annealed at various temperatures
(a) I-V behavior for samples annealed at different temperatures for 1 hour.
(b) J-V behavior for samples annealed at 300 °C, 400 °C and 500 °C for 1 hour.

264

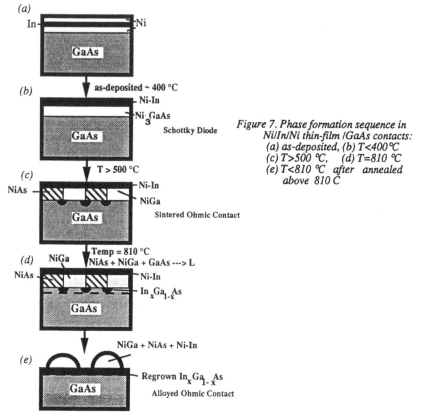

Figure 7. Phase formation sequence in Ni/In/Ni thin-film /GaAs contacts: (a) as-deposited, (b) T<400°C (c) T>500 °C, (d) T=810 °C (e) T<810 °C after annealed above 810 C

5. ACKNOWLEDGEMENT

The authors wish to thank the Department of Energy for their financial support through Grant No. DE_FG02-86ER452754.

6. REFERENCES

1. N. Braslau, J.B. Gunn, and J.L. Staples, Solid-State Electron. **10**,179 (1983).
2. Y.C. Shih, M. Murakami, E.L. Wilkie, and A. Callegari, J. Appl. Phy. **62**, 582 (1987).
3. T. Sands, E.D. Marshall and L.C. Wang, J. Mater. Res. **3** (5), 914 (1988).
4. L.S. Yu, L.C. Wang, E.D. Marshall, S.S. Lau and T.F. Kuech, J. Appl. Phys. **65** (4), 1621 (1985).
5. M. Murakami, and W.H. Price, Appl. Phys. Lett. **51**, 664 (1987).
6. M. Murakami, Y.C. Shih, W.H. Price, and E.L. Wilkie, J. Appl. Phys. **64**, 1974 (1988).
7. M. Murakami, W.H. Price, J.H. Greiner, J.D. Feder and C.C. Parker, IBM Research Report, RC 14158 (#63391), 1988.
8. A. Lahav, M. Eizenberg, and Y. Komem, J. Appl. Phys. **60** (3), 991 (1986).
9. J.-C. Lin, X.-Y. Zheng, K.-C. Hsieh and Y.A. Chang in <u>Epitaxy of Semiconductor Layered Structures</u>, (Eds: R.T. Tung, L.R. Dawson and R.L. Gunshor), MRS Symposium Proc., **102** 233 (1988).
10. X.-Y. Zheng, J.-C. Lin, D. Swenson, K.-C. Hsieh and Y.A. Chang, Mat. Sci. Engin., **B5**, 63 (1989).

INVESTIGATION OF THIN Pd-Ge LAYER FORMATION USING SYNCHROTRON VACUUM ULTRAVIOLET PHOTOEMISSION SPECTROSCOPY

P.L. MEISSNER*, J.C. BRAVMAN*, T. KENDELEWICZ**, K. MIYANO**, W.E. SPICER**, J.C. WOICIK***, C. BOULDIN***
*Stanford University, Department of Materials Science and Engineering, Stanford, CA
**Stanford University, Stanford Electronics Laboratory, Stanford CA
***National Institute of Standards and Technology, Semiconductor Electronics Division, Gaithersburg, MD

ABSTRACT

The formation of Pd-Ge layers was studied as a function of deposition and annealing using synchrotron Ultraviolet Photoemission Spectroscopy (UPS). Pd depositions ranging in thickness from 0.5 monolayers (ML) to 44 ML were examined in-situ on Ge (111) cleaved in ultra-high vacuum. The primary reaction components appear to be Pd_2Ge and PdGe. Comparison of bulk and surface sensitive Ge 3d core levels for even the highest coverages indicates that Ge segregates to the surface at room temperature. Such low temperature segregation suggests that Ge can diffuse via a rapid diffusion mechanism.

INTRODUCTION

Much of what is known about the formation of PdGe ohmic contacts to n-GaAs has been derived from studies which have focused either on bulk electrical measurements[1], or contact resistivity measurements together with TEM or sputter profile characterization of the interface [2,3]. While these analyses have yielded considerable information on the nature of the ohmic contact formation mechanism, relatively few quantitative details are available describing the relationship between microscopic reactions at the interface and the resulting effect on the barrier height at the interface. Models for ohmic contact formation typically do not consider changes in the barrier height at the GaAs interface[4]. However, some studies [5, 6] indicate that movement of the interface Fermi level and, therefore, changes in the interface barrier height, may play a role in forming ohmic contacts.

Photoemission spectroscopy is an effective tool for studying the relationship between microscopic interface chemistry and interface electronic behavior. In particular, UPS using synchrotron radiation, which can be precisely tuned to the desired incident photon energy, is a powerful tool to study such relationships. It is not surprising, however, that core level spectra for thin Pd-Ge multilayers on GaAs are extremely complex. Reactions are known to occur between Pd and Ge[7], Pd and GaAs[8,9] and, under certain circumstances, doping between Ge and GaAs[10]. Many of the reacted species produce only small chemical shifts in the core level photoemission spectra. It is very important, therefore, to determine the details of the component reactions involved in the Pd-Ge contact system. This study focuses on reactions between Pd and Ge.

EXPERIMENTAL

UPS spectra were taken for Pd coverages of 0 ML, 0.5 ML, 1 ML, 2 ML, 4 ML, and 14 ML evaporated in-situ on a Ge(111) surface cleaved ultra-high vacuum (UHV). (For this experiment, 1 ML corresponds to 1.08 Å of Pd.) The sample was then analyzed following in-situ anneals of 200°C for 15 minutes and 325°C for 15 minutes. In an effort to suppress the Ge signal, 30 ML of Pd was then deposited on the annealed layer, bringing the total deposited Pd thickness to 44 ML. Analysis of the sample followed this deposition and the final 200°C and 325°C anneals. Ge 3d bulk sensitive core levels were taken at a photon energy of 42 eV, while the corresponding surface sensitive core levels were examined using 80 eV photons. The base pressure in the UHV chamber for this experiment was 2×10^{-10} torr. The photoemission spectra were taken using the Vacuum Ultraviolet Beamline U16 at the National Synchrotron Light Source at Brookhaven National Laboratory. For phase identification, a 5000Å Pd film was evaporated with an E-beam source on a Buffer Oxide Etch treated Ge (111) surface. After an annealing sequence of 200°C for 15 minutes followed by 325°C for 15 minutes, an X-ray spectrum was taken using a Phillips XGR 3100 powder diffractometer.

Mat. Res. Soc. Symp. Proc. Vol. 181. ©1990 Materials Research Society

RESULTS

Figure1 shows Ge 3d bulk sensitive core levels together with their corresponding fitted components for selected Pd coverages. A description of the procedure and parameters for fitting these core levels can be found elsewhere[11]. While Ge 3d spectra for all coverages were fitted, those for higher coverages are shown because they are of most interest for this discussion. The bulk core levels have contributions from two primary components. Two components are also visible in surface sensitive spectra, as shown in Figure 2. Figure 3 summarizes Figures 1 and 2 by plotting the kinetic energy of each component against the experimental condition. A comparison of these figures shows that the core level intensity increases with annealing. The energy position of the reacted core level components, R1 and R2, remains relatively constant between anneals for both coverages. Both R1 and R2 are shifted with respect to the bulk Ge component of cleaved Ge (111), shown in Figure 4. Since both the surface and bulk spectra show shifts in these components with reaction, and since these shifts persist upon further annealing, R1 and R2 cannot be attributed to the underlying Ge.

The stability of the component energies can be explained by attributing them to two Pd-Ge phases. Charge transfer takes place as Pd reacts with Ge. As a result, if the reaction lowers the electron binding energy of one component, it should raise the binding energy of the other. The Pd 3d core level spectra moves towards lower kinetic energy (higher binding energy) with annealing, as shown in Figure 5. As a result, the higher kinetic energy component in the Ge 3d spectra can be attributed to a more Pd rich phase. X-ray diffraction data for 5000Å Pd films receiving the same annealing treatments, shown in Figure 6, suggest that the two dominant phases present are PdGe and Pd_2Ge, in agreement with other studies[7, 12, 13]. In the thick film case, PdGe and Pd_2Ge are the dominant reacted phases in the presence of a large concentration of unreacted Pd, as seen from the strong Pd peak in the diffraction pattern. Given that PdGe and Pd_2Ge are the two most Ge rich Pd germanides[14], they should be expected to remain the dominant phases under more Ge rich conditions. Because the Pd layers used for the photoemission analysis are so thin, the Pd-Ge reacted components R1 and R2 form in an environment which is more Ge rich than that of the film analyzed by X-ray diffraction. Therefore, it is reasonable to assign PdGe to R1 and Pd_2Ge to R2.

A particularly striking result can be seen by comparing the bulk and surface sensitive sensitive spectra taken for the 44 ML coverage (spectra c in Figures 1 and 2). From the kinetic energy of the photoelectrons, the escape depth for the bulk sensitive core level data can be determined to be approximately 12 Å, while that for the surface sensitive data is approximately 5 Å. (The effective sampling depth is about twice these values.) The fact that Ge 3d surface sensitive spectra is visible, while the bulk sensitive spectra falls below the background when 30 ML of Pd is deposited on the 14 ML reacted layer, indicates that Ge has segregated to the top of the Pd layer at room temperature. The surface sensitive spectra can again be divided into the same two Pd-Ge reacted components, suggesting that the surface segregated Ge reacts to form both PdGe and Pd_2Ge.

The surface sensitive spectra for the 14 ML coverage before and after annealing reveal that the surface of the reacted layer undergoes an evolution in the ratio between the two reacted components. Because it is sitting on a largely unreacted Pd layer, the surface segregated Ge is in a Pd rich environment. As the reaction proceeds, the layer becomes increasingly Ge rich as more Ge moves towards the surface. It is not surprising, therefore, that the relative intensity of the Pd_2Ge decreases between Figure 2a and Figure 2b. It is worth noting that the Pd_2Ge component persists for annealed layers with as little as 14 ML of deposited Pd. Because the layer is so thin, this layer is being annealed in an extremely Ge rich environment, where the PdGe phase is clearly more thermodynamically stable [14]. The rate of reaction to form Pd_2Ge in thick layers has been shown to be 12 times faster than the rate of PdGe formation[13]. This difference in the reaction rate may explain why both phases appear to be stable under these experimental conditions.

The segregation of Ge to the film surface at room temperature suggests that there is some rapid diffusion mechanism. Grain boundaries in the Pd-Ge reacted layer[2] and Pd film are one possible path for this rapid diffusion. Low temperature Ge transport through metal[15, 16], metal silicide[17], and metal germanide[3,4] layers has been known to occur. In particular, Ge transport from a Ge overlayer on PdGe to the GaAs interface is important to the

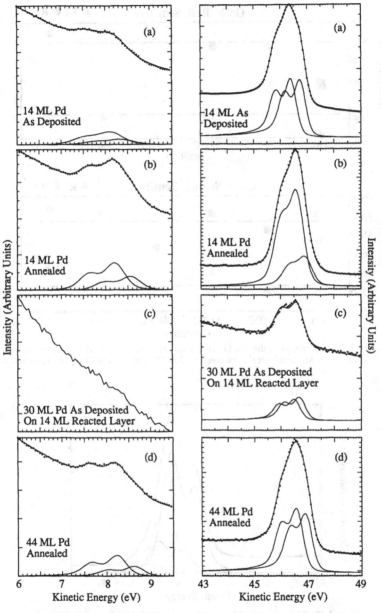

Figure 1: Bulk sensitive Ge 3d core level spectra and components for 14 and 44 ML and anneal (200°C for 15 min. + 300°C for 15 Min.).

Figure 2: Surface sensitive Ge 3d core level spectra and components for 14 and 44 ML and anneal (200°C for 15 min. + 300°C for 15 Min.).

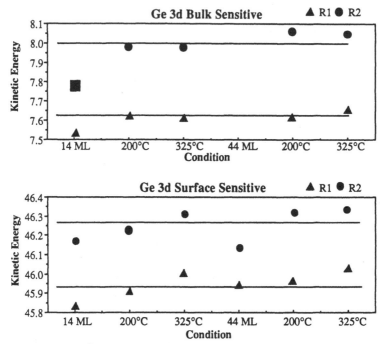

Figure 3: Kinetic energy of the two Ge core level components plotted against reaction condition for 14 ML and anneals, followed by 30 additional ML (44 ML total) and anneals.

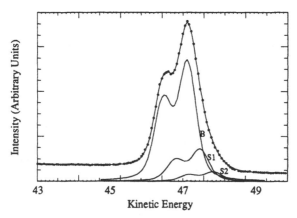

Figure 4: Ge 3d surface sensitive core level for Ge (111) cleaved in UHV. B corresponds to "bulk-like" Ge-Ge bonds while S1 and S2 correspond to changes in the Ge bonding at the surface. B can be compared with the the core level positions of the reacted components in Figure 2.

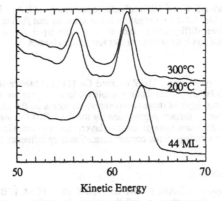

Figure 5: Pd 3d core levels for 30 ML
deposited on the 14 ML reacted Pd-Ge
layer, and anneals. The reaction shifts the
Pd core level towards lower kinetic
energy.

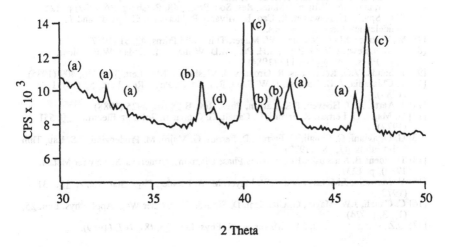

Figure 6: X-ray diffraction spectra for 5000Å Pd on Ge (111) annealed at 200°C for 15
minutes followed by 325°C for 15 minutes. The diffraction peaks are assigned as follows:
(a) PdGe (b) Pd_2Ge (c) unreacted Pd (d) Pd_5Ge_2

onset of ohmic behavior[3,4] for PdGe ohmic contacts on n-GaAs. It is not clear if the mechanism leading to Ge surface segregation is the same as that for Ge transport to a buried interface. It is, therefore, difficult to predict what role the rapid diffusion seen in this study plays in the overall PdGe to GaAs ohmic contact formation kinetics.

CONCLUSIONS

Reaction of thin Pd layers with a UHV cleaved Ge (111) substrate seems to result in the formation of PdGe and Pd_2Ge phases that are stable for low temperature anneals. As the layer thickness increases, a thin layer of these two compounds forms on the surface of the overlayer at room temperature due to surface segregation by the Ge. This Ge segregation most likely occurs by short circuit diffusion through the overlayer. Short circuit Ge diffusion through Pd and Pd-Ge reacted layers during ohmic contact formation may influence the overall kinetics of contact formation.

ACKNOWLEDGEMENTS

This work was supported DARPA (DAAL01-88-K-0828 PLM, JCB) and (N00014-89-J-1083 TK, KM WES). The authors gratefully acknowledge the assistance of Professor T. Rodin of Cornell University and Dr. K. Tsang of Brookhaven National Laboratory

REFERENCES

[1] A. K. Sinha, T.E. Smith, H.J. Levinstein, IEEE Transactions on Electron Devices 22 (5), 218 (1975).
[2] H.R. Grinolds, G.Y.Robinson, Solid St. Electron. 23, 973 (1980).
[3] E.D. Marshall, S.S. Lau, C.J. Palmstrøm, T. Sands, C.L. Schwartz, S.A. Shwarz, J.P. Harbison, and L.T. Florez in Chemistry and Defects in Semiconductor Heterostructures, edited by M. Kawabe, T.D. Sands, E.R. Weber, and R.S. Williams (Mater. Res. Soc. Proc. 148, Pittsburg, PA 1989) p. 163.
[4] L.S. Yu, L.C. Wang, E.D. Marshall, S.S. Lau, T.F. Kuech, J. Appl. Phys. 65 (4), 1621 (1989)
[5] R.W. Grant, J.R. Waldrop, J.Vac. Sci. Technol. B 5 (4), 1015 (1987); Chemistry and Defects in Semiconductor Heterostructures, edited by M. Kawabe, T.D. Sands, E.R. Weber, and R.S. Williams (Mater. Res. Soc. Proc. 148, Pittsburg, PA 1989) p. 125.
[6] W.E. Spicer, N. Newman, R. Cao, K. Miyano, P. Meissner, C. Spindt, and T. Kendelewicz, these proceedings.
[7] M. Wittmer, M.-A. Nicolte, J.W. Mayer, Thin Solid Films, 42, 51 (1977).
[8] T. Kendelewicz, W.G. Petro, S.H. Pan, M.D. Williams, I. Lindau, W.E. Spicer, Appl. Phys. Lett. 44 (1), 113 (1984).
[9] T. Sands, V.G. Keramidas, R. Gronsky, J. Washburn, Mat. Lett. 3 (9, 10), 409 (1985).
[10] P. Chiaradia, A.D. Katnani, H.W. Sang, R.S. Bauer, Phys. Rev. Lett., 52 (14), 1246 (1984).
[11] J. Aarts, A.-J. Hoeven, P.K. Larsen, Phys. Rev. B 38 (6), 3925 (1988).
[12] G. Majni, G. Ferrari, R. Ferrari, C. Canali, G. Ottaviani, Solid St. Electron., 20, 551 (1977).
[13] G. Ottaviani, C. Canali, G. Ferrari, R. Ferrari, G. Majni, M. Prudenziati, S.S. Lau, Thin Solid Films, 47, 187 (1977).
[14] Thaddeus B. Massalski, Binary Alloy Phase Diagrams, (American Society for Metals, 1986), p. 1237.
[15] V. Marrello, J.M. Caywood, J.W. Mayer, M.-A. Nicolet, Phys. Stat. Sol. A, 13, 531 (1972).
[16] C. Canali, J.W. Mayer, G. Ottaviani, D. Sigurd, W. van der Weg, Appl. Phys. Lett. 25, (1), 3, (1974).
[17] Q.Z. Hong, J.G. Zhu, J.W. Mayer, Appl. Phys. Lett. 55, (8), 747, (1989).

PART VI

Reliability and Degradation of Metal-III-V Systems

RELIABLE METALLIZATION FOR InP-BASED DEVICES AND OEIC'S

O. WADA AND O. UEDA
Fujitsu Laboratories Ltd., 10-1 Morinosato-Wakamiya, Atsugi 243-01, Japan

ABSTRACT

We describe techniques of reliable metallization in InP-based systems for application to discrete and opto-electronic integrated circuits (OEIC's). Strong metallurgical interaction between Au and InP-based compounds can cause serious contact degradation in light emitting diodes (LED's). By analyzing this interaction in detail, an improved thin Au/Zn/Au p-contact technique has been developed. The results are compared with Pt/Ti contacts, and it has shown that both provide sufficient reliability under temperature and current stresses in LED's. We then describe a metallization technique for flip-chip bonding of opto-electronic devices on other semiconductor chips for OEIC applications. An acceptable reaction barrier effect of Pt in AuSn/Pt/Ti metallization structure has been demonstrated and this structure has been used for a high-reliability, flip-chip integrated GaInAs/InP PIN photodiode/GaAs amplifier receiver circuit. We also discuss requirements for metallization for future monolithic OEIC's by taking up an example of metal-semiconductor-metal photodiodes in InP-based systems.

INTRODUCTION

III-V semiconductor opto-electronic devices such as light emitting diodes (LED's), lasers, and photodiodes have widely been used as key devices in optical communication systems. One of the most important requirements for these devices is high-reliability. For optical repeaters in a submarine transmission system, for example, no failure of any device within the component is allowable over a period of several decades. Metallization techniques are very imprtant in establishing fabrication technologies for devices with sufficiently high-performance and reliability. In general terms, the requirements for metallization for such devices include (1) acceptable electrical characteristics, such as low contact resitance for ohmic contacts and large barrier height for Schottky contacts, (2) high-uniformity over the entire wafer area, (3) high reproducibility in processes such as patterning and alloying, (4) high-stability under thermal and electrical stresses leading to overall high-reliability.

We focus in half of this paper on the reliability of p-type ohmic contacts for application to discrete devices, taking an example of InGaAsP LED's. Strong interaction of Au with InP-based compounds is studied in detail in the case of the most widely used Au/Zn p-type contact metallization. Then an improved technique for fabrication of high-reliability Au-based contacts is shown and a comparison with the Pt/Ti contact metallization system is made with regard to practical application.

Recently, the need for integrating optoelectronic devices with other electronic circuits has increased because of the demands of improving performance, functionality, and manufacturability for advanced optical components. Opto-electronic integrated circuit (OEIC) technology is extremely important in this regard, and many tehcnological challenges are currently being made [1]. One of the areas in which OEIC's are in strong demand is that of optical receivers. We describe in this paper, an approach called flip-chip opto-electronic integration, in which an opto-electronic device is flip-chip bonded to another chip, taking the example of PIN photodiode/amplifier receiver application. A reliable metallization structure using AuSn solder and a Pt/Ti reaction barrier is studied in detail. We also discuss the type of metallization required for monolithic OEIC's which will become increasingly important in the future. InP-based metal-semiconductor-metal (MSM) photodiodes using an improved Schottky barrier technique are demonstrated for such application.

CONTACTS FOR DISCRETE OPTO-ELECTRONIC DEVICES
—p-contacts in InGaAsP LED's—

Due to the larger Schottky barrier height for p-type InP-based materials compared with n-type, the formation of p-type ohmic contacts is more important. Also, since more severe electric field and current stresses are expected at p-contacts in many opto-electronic devices, the reliability of such contacts is of more practical significance. Au-based contacts have most widely been used for both n- and p-type materials. In this section, we look at the strong interaction of Au with InP and degradation of Au/Zn contacts in LED's, then study the improvement of reliability of Au-based contacts and compare them with Pt/Ti contacts.

Au-InP system

We have studied the thermal reaction of InP with an evaporated Au layer within a variety of temperature and duration ranges. In this experiment, undoped (001)-oriented InP wafers were used. The surface of the InP wafer was etched by 0.5% Br_2/CH_3OH solution prior to Au deposition. The evaporation of Au was carried out using a W filament heater under a vacuum pressure below 2×10^{-6} Torr. The Au-InP system was annealed at temperatures ranging from 345-500°C under a flow of purified nitrogen for a duration of 5 to 55 min.

Figure 1 illustrates schematic diagrams and SEM images of cross-sections of three types of Au/InP structures (970 nm-thick Au stripes were formed): i) as deposited; ii) annealed at 345-420°C for 5 min; iii) annealed at 450-460°C. Below 345°C no appreciable reaction was detected on the surface or in cross section. By increasing the temperature, a penetration of reacted alloy into the InP bulk became observable and the whole region below the Au stripe reacted at 380°C (see the SEM image lower left). Further increase of the temperature resulted in an extremely rapid reaction which often formed voids in the InP near the periphery of the Au stripe, presumably indicating the formation of volatile material. Annealing above a temperature of 450-460°C causes a drastic change in the structure. The Au layer completely reacts with InP to generate a nearly planar surface. The reacted region has been found to have structure walls which are comprised of planes close to the {111} and {001} crystallographic planes as shown Fig. 1 (see also the SEM image lower right). Observation of the surface geometry revealed also a preferential Au pattern spreading along the <110> and <1$\bar{1}$0> directions, consistent with the previous report [2].

An X-ray diffraction measurement using the Cu Kα line was performed to identify binary phases produced at various stages of annealing. The results indicate that at temperatures of 355-420°C, Au reacts with both In and P to form binary compounds, and the Au-In compound changes phase to form $AuIn_2$ (the Au-P compound no longer remains) at temperatures above 450°C (as schematically shown in Fig. 1). This is consistent with the microstructure results described above and in-depth composition measurements using SIMS and AES [3].

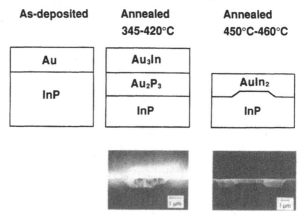

Fig. 1. Au-InP interaction during annealing

Au/Zn/Au contact

The Au penetration into InP is highly persistent; it continues until the whole Au layer is consumed to form alloys with In and P. This is suspected to occur in practical InGaAsP/InP DH lasers and LED's with Au-based alloy p-contacts, to accelerate the diode degradation. Our previous investigation of accelerated aging of InGaAsP/InP DH LED's, with Au/Zn/Au p-contact metallizations at a current density of approximately 8 kA/cm^2 revealed a thermally activated operating lifetime with an activation energy of 1.0 eV in the emission wavelength range of 1.15-1.5 μm [4,5]. This activation energy is different from the value of 2.31 eV determined for the Au-InP reaction and the LED life at moderate operating conditions is not necessarily related to the Au penetration into the InGaAsP and InP layers. However, when the LED is stressed by a large operating current, Au penetration through the p-InGaAsP contact layer into the p-InP cladding layer and even into the InGaAsP active layer is actually detected [6-8]. Figure 2 shows a schematic diagram of an InGaAsP/InP LED studied. The diode has a Au/Zn/Au alloyed p-contact with a thickness of 300 nm and a diameter of 40 μm. The emitting wavelength was 1.3 μm (see [4] for detailed structure). Figure 3 shows an example of the result of Au penetration into the active layer under heavy current application [9]. The diode was stressed at a current of 300 mA (24 kA/cm^2) with a current pulse width of 30 μs. During current application, a number of dark spot defects (DSD's) appeared in the electroluminescence (EL) image and these DSD's grew very rapidly to form a widespread dark region as shown in Fig. 3 (right). The in-depth composition profile were observed using a

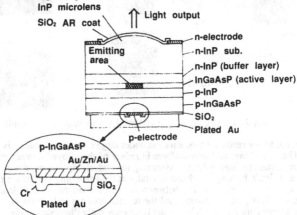

Fig. 2 Schematic diagram of structure of an InGaAsP/InP LED.

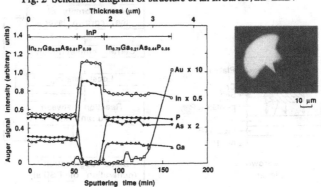

Fig.3 AES analysis of a degraded InGaAsP/InP LED by large current application.

microprobe AES system after removing the InP layer and the substrate by a selective etchant. The AES data taken with a primary electron probe diameter of 5 μm, within the moderately damaged region indicated by an arrow in the EL image, are also shown in Fig. 3 (left). The penetration of Au through the InGaAsP contact layer and the InP cladding layer into the InGaAsP active layer is clearly observed in this result, although its atomic content in the semiconductor bulk is one percent or less.

Structural analysis of the DSD's by TEM was also carried out, and the results are illustrated in Fig. 4. From these data, it has been established that the DSD's are associated with a) an amorphous area of the matrix, b) micrograins of the matrix crystal, and c) regions where the matrix crystal was alloyed with the metals of the electrode [10]. Furthermore, EDX analysis indicated that the DSD's or dark regions contained a large amount of Au [10]. These defects were considered to be generated by the alloy reaction between the matrix crystal and the metals of the electrode, especially Au.

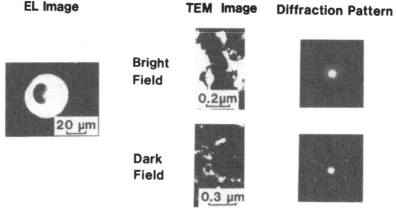

Fig. 4 TEM analysis of a degraded InGaAsP/InP LED by large current application.

Based on the above results, Au penetration under large current application is interpreted as follows. The non-planar and non-uniform interface of the alloyed contact results in an inhomogeneous contact resistance [11]. Assuming the underlying semiconductor bulk to be highly conductive, Joule heating is enhanced locally at the low-contact resistance parts of the interface. This causes strong reaction between Au and the matrix crystal, leading to Au penetration there. This process repeats until the reacted area expands to the entire emitting region forming a large dark region. This model is schematically illustrated in Fig. 5.

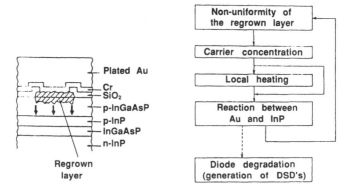

Fig. 5 Degradation mechanism of an InGaAsP/InP LED by large current application.

Improved contacts

By considering the results of degradation of LED's by large current application described previously, we have developed a new technique for producing high-reliability Au/Zn/Au contacts. Both contact structures fabricated by the conventional and the new metallization techniques are schematically shown in Fig. 6. The new technique includes the use of an ultra-thin Au/Zn/Au layer. In a conventional contact, this layer has a thickness of about 300 nm and is capped by a reaction barrier. The entire structure is covered with plated Au. In the ultra-thin process, the alloy layer thickness is typically 50 nm. As in the conventional contact, a reaction barrier is deposited through and opening in the SiO_2 protective coating. The advantage of the thin layer is that the Au is completely consumed during alloying, leaving no fresh Au available to penetrate the contact layer during operation. Fig. 7 illustrates threshold current for generation of DSD's by large current application in LED's with different contact structures. One can find first of all, that a substantial improvement in the threshold current can be achieved by using an ultra-thin Au/Zn/Au layer over a thick layer. Furthermore, we obtained higher threshold current by using a thicker InGaAsP contact layer, although the best results has been obtained for structures using a Pt/Ti layer. However, the contact resistance of Pt/Ti is one order of magnitude higher than that of Au/Zn/Au. These results lead us to the conclusion that both ultra-thin Au/Zn/Au and Pt/Ti contacts are practically useful, and that the former is better than the latter regarding contact resistance.

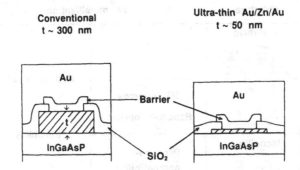

Fig. 6 Schematic diagram of an improved metallization technique for Au/Zn/Au contact to InGaAsP.

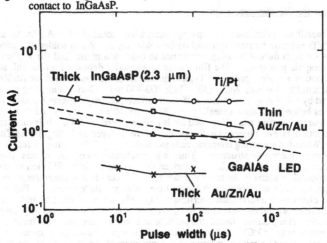

Fig. 7 Threshold current for DSD formation in LED's with different contact materials.

CONTACTS FOR OEIC'S
—Metallization for flip-chip integration—

Table 1 shows a variety of structures for integrating photodiodes with amplifier circuits. In conventional hybrid integration, parasitic reactances introduced by interconnect wiring degrades aspects of receiver performance such as speed and noise characteristics. Flip-chip and monolithic integration techniques are demanded to solve this problem. Monolithic integration is currently being investigated by many research groups [12-14], but more advancements in fabrication technology are needed due to the difficulty of integrating vastly different device structures in one chip. Requirements for metallization in monolithic integration include compatibility between optical and electronic devices in processing. On the other hand, flip-chip techniques would be more useful for urgent applications, provided reliable, reproducible flip-chip bonding techniques are developed. Techniques for metallurgical reaction-resistant microbump formation are important in this regard. We describe in this section the development of a reliable metallization structure using AuSn solder and a Pt/Ti reaction barrier for receiver applications.

Table 1 Structures for integrating photodiodes with amplifier circuits and requirements of metallization

Integration structure	Requirements of metallization	
Hybrid		**Reproducibility**
Flip-chip	**Microbumps** **Reaction barrier**	
Monolithic	**Process-compatibility**	**Reliability**

Metallization for flip-chip integration

The metallization structure for flip-chip integration consists of a Au/Zn/Au alloyed p-contact, Pt/Ti reaction-barrier layer and AuSn solder layer. AuSn solder was adopted for affinity to Au pads during bonding, resistance to oxidation in air and the applicability of photolithographic processes. The films were sequentially deposited on an InP substrate, after the contact alloying process. Typical thickness of the layers used are Au/Zn/Au: 50 nm, Ti: 50 nm, Pt: 300 nm, AuSn (30% Sn): 300-400 nm. During annealing the structure was covered by a SiN film to maintain sufficient surface planarity. Figure 8 shows AES data for samples before and after annealing at 305°C for 10 min. It is found that Sn and Au become separated during annealing and Sn becomes mixed with Pt. It is also noted that thermal reaction is not observed in the vicinity of the Ti layer including the Au/Zn/Au-InP interface. When the annealing temperature is too high (e. g., 450°C), stronger reaction occurs and deteriorates the whole structure. Thus, a thermal reaction occurs in this structure at moderate temperatures at the AuSn/Pt interface in a mode of inter-diffusion to produce a Pt-Sn compound. Essentially the same reaction seems to take place at lower temperatures (e.g. 190°C), too, although it takes a much longer time for appreciable reaction. Fig. 9 shows a plot of the effective interdiffusion coefficient, $D_{eff}=x^2/t$, where x and t stand for the Sn-Pt compound thickness and the annealing time, versus the reciprocal of annealing temperature, 1/T. Two distinct regions are found; the reaction at a temperature above the melting point of AuSn (approximately 280°C) is rather insensitive to temperature and that at lower temperature is distinct regions are found; the reaction at a temperature above the melting point of AuSn

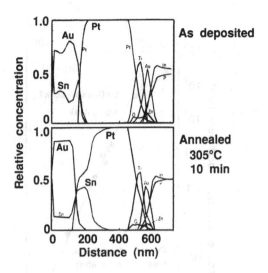

Fig. 8 AES analysis of metallization structure before and after annealing (305°C for 10 min)

(approximately 280°C) is rather insensitive to temperature and that at lower temperature is strongly dependent on the temperature. These results lead us to the conclusion that little interaction is expected during long-term operation of the device at lower temperatures, although a certain amount of interdiffusion would occur during the bonding process. The life time of the present metallization can be estimated from the time when half of the Pt layer reacts with Sn. This is in excess of 10^9 h for an operation temperature of 50°C, assuming a Pt thickness greater than 200 nm, indicating sufficient stability of this metallization structure.

Cross-sectional SEM observation of the structure indicates that the interfaces of the reacted layer are uniform. Furthermore, X-ray diffractometry was carried out to determine the alloy phases formed by the reaction. Before annealing, only peaks corresponding to pure Au, Pt, and AuSn phases are observed. After annealing, however, peaks corresponding to PtSn are present, and also an increase in Au peak intensity and an decrease in Pt peak intensity are found. This indicates that Sn and Pt react to form a PtSn phase to leave pure Au near the surface. No appreciable formation of other alloy phases of Pt and Sn such as $PtSn_4$, Pt_2Sn_3, and $PtSn_2$ has been found. As the annealing proceeds, i.e., the temperature increases, PtSn and pure Pt are consumed by the reaction. It has thus been confirmed that the Pt-Sn compound produced is PtSn.

We have applied this technique to fabrication of a high performance flip-chip integrated receiver in which an ultra-small diameter, microlensed GaInAs PIN photodiode and a low-capacitance GaAs MESFET amplifier are used [15]. Fig. 10 illustrates a schematic diagram of a cross-section of a flip-chip integrated photoreceiver. A small-diameter GaInAs photodiode having a monolithic microlens was flip-chip bonded on the semi-insulatind substrate of the GaAs amplifier. The PIN structure consists of a 3 μm-thick n-InP buffer layer, a 1.9 μm-thick n-GaInAs photo absorption layer, and a 1 μm-thick n-InP cap layer. The p-n juction was formed by 1 μm deep Zn diffusion within the cap layer. The photoabsorption area was delineated by chemical etching to form a mesa with a diameter of 17 μm. The mesa structure was passivated with a SiN film. The substrate was thinned to about 70 μm and a microlens with a radius of curvature of 55 μm was integrated by using Ar+ ion-beam etching technique. Then, a SiN film was deposited as the anti-reflection coating for 1.3 μm wavelength by plasma CVD. We have measured static characteristics of the photodiode. The dark current was less

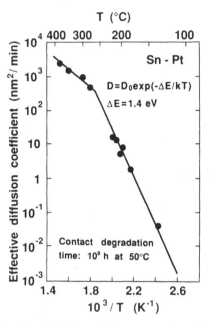

Fig. 9 Plot of effective interdiffusion coefficient versus the reciprocal of ambient temperature.

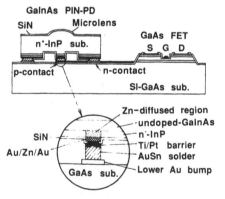

Fig. 10 Schematic cross-section of flip-chip integrated photoreceiver.

than 30 nA at -10V bias. Furthermore, the dark current did not change even after aging at 180°C for 3000h. Dynamic characteristics were also tested using an external equalizer/filter circuit, and a minimum received power of -27.4 dBm has been obtained for an error rate of 10^{-9} at a bit rate of 2Gb/s, NRZ which is, to the best of our knowledge, the best value for integrated PIN/amplifier receivers. These results have proven that flip-chip opto-electronic integration technique is useful in fabricating high-speed, high sensitivity photoreceivers.

CONTACTS FOR MONOLITHIC OEIC'S

In monolithic integration, a planar integrated structure is crucial to enable standard device processing techniques to be used for fine definition of device geometries. Also it is desirable to use processes compatible with both optical and electronic devices. In GaAs-based OEIC receivers, MSM photodiodes and MESFET's have been extensively used up to LSI scale [16]. In InP-based systems, however, low Schottky barrier heights in n-type material, which causes large dark current, have prohibited the development of MSM-based structures. We describe in the following an InP-based MSM photodiode using an improved Schottky barrier structure developed as a step towards future monolithic integration of receivers.

The structure of the GaInAs MSM-PD is shown in Fig. 11. The wafer used is a nominally undoped epitaxial wafer grown by pulse molecular beam epitaxy [17] on a semi-insulating InP substrate. The $Ga_{0.47}In_{0.53}As$ photoabsorption layer was 1.5 μm thick. To reduce the dark current, an undoped $Al_{0.48}In_{0.52}As$ barrier height enhancement layer (70 nm thick) was employed. Also, to minimize the effect of carrier trapping [18], which may be caused by a large band discontinuity at the AlInAs/GaInAs interface, a graded superlattice (200 nm thick) was incorporated between the AlInAs and GaInAs layers. This superlattice was composed of pairs of ultrathin (one to nine atomic layers) AlInAs and GaInAs layers, where the width ratio varies so that the effective band gap is graded linearly in depth as shown by dotted lines in Fig. 11. The same structure was used at the bottom interface for preventing the generation of a two-dimensional carrier gas there. The Al contacts had interdigited patterns (1.1 μm line and 1.4 μm space). A shallow mesa was formed at the periphery of the contacts to homogenize the field in the GaInAs layer. A photosensitive area with a size of approximately 20 μm x 20 μm was coated by a plasma-deposited SiN anti-reflection film. Au-based bonding pads were prepared on the substrate surface. This photodiode exhibited a dark current lower than 100 nA, an internal quantum efficiency of greater than 80% at a wavelength of 1.3 μm, and a capacitance of 40 fF, all at a bias voltage of 10 V. The response speed of this photodiode has been characterized by electro-optic sampling to exhibit a full width at half maximum of 14.7 ps.

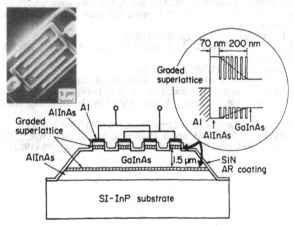

Fig. 11 Cross-section of GaInAs MSM photodiode using improved Schottky contacts incorporating an AlInAs/GaInAs graded superlattice.

CONCLUSIONS

We have reviewed the current status of reliable metallization techniques in InP-based systems for application to discrete devices and OEIC's. Based on a detailed study on the Au-InP reaction, a Au-based p-contact having sufficient reliability under both electrical and thermal stresses has been developed. LED lifetimes in excess of 10^9 hours have been measured at 20°C. For OEIC's, flip-chip metallization using AuSn solder and a Pt/Ti reaction barrier has been developed and high-thermal stability has been demonstrated in photodiode application. A planar structure InP-based MSM photo-diode has been developed by introducing an improved Schottky barrier contact structure. Planarity as well as process-compatibility between electronic devices, which can be fulfilled by this contact structure, will be very useful in future monolithic OEIC developments.

ACKNOWLEDGEMENTS

The authors would like to express their thanks to T. Misugi, O. Otsuki, and T. Sakurai for their encouragement . Special thanks go to A. Hobbs for valuable discussion.

REFERENCES

1. O. Wada, T. Sakurai, and T. Nakagami, IEEE J. Quantum Electron. QE-22, 805 (1986)
2. V. G. Keramidas, H. Temkin, and S. Mahajan, Proc. Int. Symp. on GaAs and Related compounds, Vienna, 1980, Inst. Phys. Conf. Ser. 56, 293 (1981).
3. O. Wada, J. Appl. Phys. 57, 1901 (1985).
4. S. Yamakoshi, M. Abe, O. Wada, S. Komiya, and T. Sakurai, IEEE J. Quantum Electron. QE-17, 167 (1981).
5. O. Wada, S. Yamakoshi, H. Hamaguchi, T. Sanada, Y. Nishitani, and T. Sakurai, IEEE J. Quantum Electron. QE-18, 368 (1982).
6. O. Ueda, S. Yamakoshi, T. Sanada, I. Umebu, and T. Kotani, J. Appl. Phys. 53, 7385 (1972).
7. A. K. Chin, C. S. Zipfel, F. Ermanis, L. Marchut, I. Camlibel, M. A. DiGiuseppe, and B. R. Chin, IEEE Trans. Electron Devices ED-30, 304 (1983).
8. O. Ueda, H. Imai, A. Yamaguchi, S. Komiya, I. Umebu, and T. Kotani, J. Appl. Phys. 55, 665 (1984).
9. O. Wada, T. Sanada, S. Yamakoshi, M. Abe, and T. Sakurai (unpublished).
10. O. Ueda, S. Yamakoshi, T. Sanada, I. Umebu, T. Kotani, and O. Hasegawa, J. Appl. Phys. 53, 9170 (1982).
11. N. Braslau, J. Vac. Sci. Technol. 19, 803 (1981).
12 O. Wada, H. Nobuhara, T. Sanada, M. Kuno, T. Fujii, and T. Sakurai, IEEE J. Lightwave Technol. LT-7, 186 (1989).
13. J. D. Crow, Tech. Dig. IOOC'89, Kobe, p. 86.
14. N. Suzuki, H. Furuyama, Y. Hirayama, M. Morinaga, K. Eguchi, M. Fushibe, M. Funamizu, and M. Nakamura, Electron. Lett. 24, 467 (1988).
15. Hamaguchi, M. Makiuchi, T. Kumai, O. Aoki, and O. Wada, Tech. Dig. 2nd OEC '88, paper 4C2-3.
16. J. D. Crow, Tech. Dig. Optical Fiber Communication (OFC) 1989, p. 83.
17. T. Fujii, Y, Nakata, Y. Sugiyama, and S. Hiyamizu, Jpn. J. Appl. Phys. 25, L254 (1986).
18. S. R. Forest, O. K. Kim, and R. G. Smith, Appl. Phys. Lettt. 41, 95 (1982).

ROLE OF INTERFACE-STATES IN THE REVERSE BIAS AGING
OF GaAs SCHOTTKY BARRIERS

K.A. CHRISTIANSON
Electrical Engineering Department
University of Maine, Orono, ME 04469

ABSTRACT

Forward bias capacitance has been used to examine the Au/W/GaAs and Au/Pt/Ti/GaAs Schottky barriers present in power microwave MESFET devices to see if interface-state generation plays any role in the previously reported reverse bias barrier height aging process. If a constant carrier capture cross-section is assumed, forward bias capacitance has shown that for samples strongly susceptible to aging (i.e. the Au/W/GaAs samples in this study) interface-state generation is taking place during the aging process. The validity of the constant capture cross section assumption has been tested by examining the I-V properties. For those samples whose reverse I-V properties were not dominated by thermionic-field emission, similar increases in interface-state densities were evaluated from the I-V characteristics for the degraded samples.

INTRODUCTION

A recent series of measurements have shown that the barrier height of Schottky diodes on GaAs may change under long term biasing conditions. The effect has been most pronounced for Ag/GaAs and Au/W/GaAs diodes, but has also been observed in Au/GaAs and Au/Pt/Ti/GaAs diodes [1,2]. The aging has been found to occur under reverse bias conditions with a logarithmic dependence on time. The major objective of the present work was to determine the role of interface-states in this aging process. A variation of admittance spectroscopy, due to Wu and Yang [3], was used. In this variation the Schottky diode forward bias capacitance has been modeled as the modulation of the effective barrier height by the interface charge. It is also possible in certain cases to evaluate for interface-state characteristics from the current-voltage characteristics, in a manner recently detailed by Horvath [4]. In this technique, interface-state energy distributions may be evaluated if it is possible to estimate the contribution of the interface-states to the slope of the forward and reverse I-V characteristics. This method of interface-state evaluation was used to check the major assumptions needed for the forward bias capacitance technique.

EXPERIMENTAL PROCEDURE

Samples

The same two groups of samples previously evaluated in the barrier height aging study were used in this study. The first group were GaAs power MESFETs, designed for two watts output at 7.5 GHz. They have 24 gate fingers, each 200 μm wide by 1 μm width, for a total gate width of 4.8 mm. The metalization system for these was Au/W/GaAs. The second group of samples were also GaAs power MESFETs, however this group of samples was designed for two watts output at 4 GHz. Their structure has six gate fingers, each 1 μm length, for a total gate width of 1.7 mm. The metalization system for these is Au/Pt/Ti/GaAs. Further details about these samples may be found in reference [2].

The use of MESFETs instead of simple Schottky barriers has complicated the investigation. With the MESFET the gate-source, gate-drain, and gate-source and drain connected together can all be examined. For this investigation the gate-source diode of the Au/W/GaAs devices were examined, while the gate-drain diode of the Au/Pt/Ti/GaAs devices were examined. These connections were chosen since the devices were previously seen to exhibit the greatest amount of aging when connected in this manner [2].

Measurement Procedures

Commercial LCR bridges were not suitable for the measurement of the forward bias capacitance of the samples because the capacitance is of relatively small magnitude compared to the conductance. If the phase accuracy of the bridge is less than perfect, the capacitance signal will be overwhelmed by the conductance signal and the bridge will give erroneous results. To overcome this problem the measurement system of Wu et al. [5] was implemented in a modified version [6].

The samples I-V characteristics were examined in an HP 4062A semiconductor parameter analyzer programmed to automatically age and then calculate barrier height and ideality factor from measured I-V properties, assuming thermionic emission applies.

EXPERIMENTAL RESULTS AND DISCUSSION

Forward Bias Capacitance Results

Figure 1 shows the forward bias capacitance results for a typical Au/W/GaAs sample both before and after aging with V_{gs} = -10 V for 12 hours. As can be seen after aging the peak height has increased slightly, and voltage at which the peak occurs has shifted from 0.36 V to 0.30 V. Analysis of the interface-state density was done by curve fitting the expression developed by Wu and Yang [3],

$$\omega C = \frac{q^2 I}{kT \left(1 + \frac{qR_s I}{kT} \right)} \frac{N_S\left(E_f^s\right)}{C_i \Delta V}$$

$$x \frac{\omega(1-f) \, 4 \, \sigma_n j / q}{\left(4 \, \sigma_n J / q + 1 / \tau_m \right)^2 + \omega^2} . \tag{1}$$

In this expression I is the diode current, R_s is the diode series resistance, C_i is the interface specific capacitance, ΔV is the probing voltage, ω is the angular frequency, f is the interface-state occupation factor, σ_n is the capture cross-section, j is the ac current density, J is the dc current density, τ_m is the metal interface relaxation time, and

$$N_S\left(E_f^s\right) = \int_{E_f^M}^{E_f^S} N_{SS}(E) \, dE = \overline{N}_{SS} \, q \, V_A \tag{2}$$

where N_{SS} is the average interface-state density evaluated at the position of the Fermi-level and V_A is the applied voltage.

In order to allow for comparison with the experimental data, it is desirable to locate any extrema of this expression. Taking the derivative of this expression with respect to the voltage is rather difficult, since any voltage dependence is

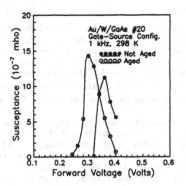

Fig. 1. Susceptance versus forward voltage for a Au/W/GaAs Schottky barrier in the not aged condition and after aging at -10 V for 12 hours. Measurement frequency was 1 kHz. Line is eye aid only.

implicit in the current terms. However, if we look at the expression as a function of angular frequency we see that it is of the form:

$$B \propto \frac{\omega}{a^2 + \omega^2}$$

(3)

which has a maximum at $a = \omega$. Thus it is possible to define an expression involving the capture cross section, the dc current density, the time constant of the relaxation of the carriers into the metal, and the maximum of the capacitance response as a function of frequency.

At this point Wu and Yang [3] took a limiting low frequency case in order to solve for the capture cross section and the relaxation constant independently. The assumptions involved in this step do not apply to the samples evaluated in this study, because the sample capacitance of the diodes looked at in this study are much smaller than that of Wu and Yang's (approximately 10 pF versus a few nF), and thus no data could be taken at the portion of the frequency spectrum where these assumptions apply.

Instead of Wu and Yang's procedure, an assumption as to the value of the capture cross section was made. It is believed that the interface-states present and being made by the aging process are responsible for making the barrier height decrease. This implies that these states have a positive charge associated with them, according to two different theories of the relationship between barrier height and charge at the interface [7,8]. Interface donor states located in the upper half of the bandgap will give such behavior, if we assume that the majority of them are empty and thus are in their positive charge state (note that as part of the evaluation of interface-state densities from forward bias capacitance the occupation factor for these states was calculated to be less than 25% [6]). Such donor states typically have a capture cross section of approximately 10^{-14} cm^{-2} [9], and this value will be used for calculations involving both the Au/W/GaAs and the Au/Pt/Ti/GaAs samples.

Note that the rest of this work thus assumes that the capture cross section does not change with aging. This may not be true, in fact for Si MOS transistors subjected to hot-carrier aging a recent study has indicated that the cross section varies between aged and non-aged samples [9]. With the samples and instrumentation used in this study, however, it was not possible to distinguish between concentration variations and cross section variations. In order to allow for an evaluation of the experimental data it was thus assumed that the concentration of the interface-states is the independent parameter.

In order to evaluate for the interface-state density it is also necessary to have a value of the interface layer specific capacitance, C_i. For these samples it is assumed that the interface is composed of GaO, which has an relative dielectric constant of 3.5 [10]. Also needed is a value of δ, the interfacial layer thickness. For this grouping of samples it was assumed this value is 25 Å, which is consistent with Auger results taken for this group of samples [2].

With these assumptions in mind, and with a measurement which showed the susceptance peak as a function of frequency does not change with aging, but instead stays constant at 30 kHz, the interface-state densities were calculated. The not aged average (N_{SS}) interface-state density was calculated to be 5-10 x 10^{12} cm^{-2} eV^{-1}, which changed to 3-4 x 10^{13} cm^{-2} eV^{-1} after aging for 12 hours with V_{gs} = -10 V.

The same procedure was repeated with the Au/Pt/Ti/GaAs samples with the assumption of a thinner oxide (10 Å), which is typical of good technology at the time these samples were manufactured [11]. Evaluation of the forward bias capacitance before and after aging of these samples with V_{ds} = -10 V for 12 hours showed no change, as Figure 2 illustrates. These was also no change in the frequency at which the maximum capacitance occurs, as well as no change in the forward bias conductance after aging. From these results the average interface-state density was calculated to be approximately 2 x 10^{12} cm^{-2} eV^{-1} for both the not aged and aged samples.

Fig. 2. Susceptance versus forward voltage for a Au/Pt/Ti/GaAs Schottky barrier in the not aged condition and after aging at -10 V for 12 hours. Measurement frequency was 1 kHz. Line is eye aid only.

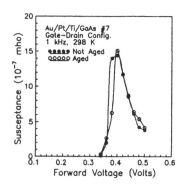

I-V Results

The validity of the constant capture cross section assumption is strengthened by the measurement of interface-state densities based upon an evaluation of the forward and reverse I-V characteristics. The reverse current flow of the Au/W/GaAs diodes could not be explained in terms of thermionic (T) emission, including the use of image force barrier height lowering. However, at room temperature the reverse current could be successfully modeled using thermionic-field (TF) emission [12,13] at reverse voltages greater than two volts with the following additional assumptions. First, there is a slight lowering of the barrier height due to the edge effects of the MESFET structure. Second, a bias dependent component of the barrier height was included to account for the movement of the depletion region across the channel and towards the channel-substrate interface. Figure 3 shows a comparison of experiment with modified TF theory for sample #38 for reverse voltages in the range of 2 to 10 Volts.

At voltages below two volts, the I-V characteristics were found to be partially determined by the contribution of the interfacial layer to the barrier height lowering. This dependence allowed for an evaluation of the interface-state density as per Horvath [4]. The appropriate expression for this evaluation is from

Horvath's model A, where the interface-state density may be determined from:

$$D_s = \frac{\varepsilon_O}{n_i q} \left(\frac{\varepsilon_s}{w_R} \left(\frac{q}{s_i kT} - 1 \right) \left(n_i - 1 \right) - \frac{\varepsilon_s}{w_F} \right)$$

(4)

where D_s is the interface-state density, n_i is the ideality factor in the forward direction due to interface states, w_R and w_F are the widths of the depletion regions under reverse and forward bias conditions, s_i is the contribution of the interface-states to the slope of the reverse bias characteristics, and with all other terms having their usual meaning.

The major difficulty in this evaluation was obtaining a value for s_i. As has been mentioned, for reverse voltages greater than 2 volts TF emission controls. Thus a reverse voltage range of 3 to 4 volts was used to establish s_{TF}, i.e. the slope due to thermionic-field emission. For reverse voltages less than one volt is believed that not all the interface-states may be filled, thus the reverse voltage range of one to two volts was used to establish s_{Total}, i.e. the sum of s_i and s_{TF}.

For this calculation n was assumed to be the same as n_i, i.e. the non-ideality of the forward characteristics was assumed to be due to the presence of interface-states. After evaluating w_R at a reverse voltage of 1.5 V, w_F at 0.3 V, then substitution into equation (4) yielded an average interface-state density of approximately 10^{13} cm^{-2} eV^{-1}. After aging with a gate to source voltage of -10 volts for 12 hours, an average interface-state density of 3×10^{13} cm^{-2} eV^{-1} was then evaluated. It was not possible to evaluate the Au/Pt/Ti/GaAs samples in this manner since their reverse I-V characteristics were found to be dominated by thermionic-field emission over the entire reverse voltage range examined.

Fig. 3. Reverse I-V characteristics of Au/W/GaAs sample #38. The slope between reverse voltages of 1 to 2 was evaluated as described in the text to determine the role of the interface-states in the aging process. The sample was measured between the gate and source.

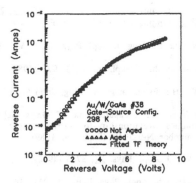

Discussion and Conclusions

In this work the reasons why the barrier height of certain GaAs Schottky barriers decrease after being held for a long time at reverse bias have been examined. In particular we have looked for the influence of interface-states in this process. The experimental method used was forward bias capacitance, as developed by Wu and Yang [3]. Several key assumptions had to be made in order to use this method. The most important of these involved assuming donor type interface-states were involved which had a constant capture cross-section throughout the measurement cycle, including aging. Assumptions also had to be made concerning the nature of the interface involved with each of the type of samples.

With these assumptions in mind the average interface-state concentration of the Au/Pt/Ti/GaAs diodes was found to be in the low 10^{12} cm^{-2} eV^{-1} range. This concentration of interface-states is probably not enough to account for the Fermi

288

level pinning of these samples, and other reasons, perhaps fixed charge and/or deep traps, are responsible. As has been previously reported and repeated here these samples do not show significant aging characteristics.

In marked contrast are the Au/W/GaAs samples. Again with the same type of assumptions, the average interface-state density was found to be in the mid to upper 10^{12} cm^{-2} eV^{-1} range. Once again, particularily in the lower end of this range, there are not enough interface-states to explain the Fermi-level pinning behavior. After aging these diodes in reverse bias the observed interface-state concentation has increased to the low to mid 10^{13} cm^{-2} eV^{-1} range, indicating interface-state generation has taken place. The pinning mechanism has apparently changed, and the generated interface-states now play at least some role in determining the barrier height.

ACKNOWLEDGEMENTS

The forward bias capacitance measurements were made while the author was a visitor at RADC/RBRP during July and August of 1989, under task number N-9-5940. There I would like to thank F. Rubinski, S. Dragoner, and M. Walter for their help. Early barrier height stability studies were funded by UES/AFOSR under contract number F49620-87-R-004. Continuation of that work has been sponsored under UES/AFOSR contract number F49620-88-C-0053.

REFERENCES

1. A. Miret, N. Newman, E.R. Weber, Z. Lilientiel-Weber, J. Washburn, and W.E. Spicer, J. Appl. Phys. 63, 2206 (1988).

2. K.A. Christianson, in Proceedings of the 27th Annual International Reliability Physics Symposium, p. 65 (IEEE Press, 1989).

3. X. Wu and E. Yang, J. Appl. Phys. 65, 3560 (1989).

4. Zs. J. Horvath, J. Appl. Phys. 63, 976 (1988).

5. X. Wu, H.L Evans, and E.S. Yang, Solid State Elect. 31, 167 (1988).

6. "Interface-State Measurements of GaAs Schottky Barriers Using Admittance Techniques: Relationship to Barrier Height Instability," final report for RADC task number N-9-5940 (Sept. 15, 1989).

7. C.Y. Wu, J. Appl. Phys. 51, 3786 (1980).

8. Z. J. Horvath, Appl. Phys. Lett. 54, 931 (1989).

9. D. Vuillaume, R. Bouchakour, M. Jourdain, and J.C. Bourgoin, Appl. Phys. Lett. 55, 153 (1989).

10. B.R. Pruniaux and A.C. Adams, J. Appl. Phys. 43, 1980 (1972).

11. E. Calleja and J. Piqueras, Electron. Lett. 17, 37 (1981).

12. F.A. Padovani and R. Stratton, Solid State Elect. 9, 695 (1966).

13. Zs. J. Horvath, J. Appl. Phys. 64, 6780 (1988).

RELIABILITY OF OHMIC CONTACTS FOR AlGaAs/GaAs HBTs

G.S. JACKSON*, E.TONG*, P. SALEDAS*, T.E. KAZIOR*, R. SPRAGUE*,
R.C. BROOKS**, AND K.C. HSIEH***
*Raytheon Co., Research Div., 131 Spring St., Lexington, MA 02173
**now at: Westinghouse Electric Co., Electronic Systems Group, P.O. Box 1521, Baltimore, MD 21203
***University of Illinois, Dept. of Elect. Eng., 1406 W. Green St., Urbana, IL 61801

ABSTRACT

The reliability of ohmic contacts to thin, heavily doped layers of GaAs is investigated. Pd/Ge/Au contacts to n-type GaAs display excellent electrical stability over extended periods of thermal stress. The contact resistance stays below 0.5Ω-mm during a 2500h, 280°C bake. Reactive ion beam assisted evaporation of Ti with N forms TiN which is introduced as a barrier layer in Pt/TiN/Ti/Au contacts to a thin p^+ layer. The TiN layer allows greater process latitude in the sintering process and improves long term stability of the ohmic contact. The microstructure of the p-type contacts is examined with TEM and Auger profiling at different instances of the 2500h, 280°C bake and compared to the contact resistance measurements.

INTRODUCTION

Reliability of heterostructure bipolar transistor (HBT) devices and of the fabrication process depends critically upon the ohmic contacts. In particular the base contact is placed on a very thin (~1000Å) heavily doped layer. A reliable process requires broad process latitude allowing optimization of contact formation, including the alloying or sintering cycle, without shorting the underlying p-n junction beneath the base contact. The issue of long term reliability under the fields and temperatures associated with device operation is related. Both involve interdiffusion of the contact metallurgy and the underlying semiconductor.

We report the development of ohmic contacts which simplify the HBT process techniques and exhibit good aging qualities. The contacts must exhibit low resistance and maintain this property for long periods. Preferably the same sintering conditions for both n- and p-type contacts will be used. For n-type contacts Pd/Ge/Au exhibits low contact resistance and excellent long term stability for FETs [1]. This metallurgy has been chosen for the n-type contacts. For p-type contacts the use of TiN as a diffusion barrier in a Pt/barrier layer/Au sandwich is investigated [2]. The TiN is deposited by reactive ion beam assisted evaporation (RIBAE). Thus it is compatible with a photoresist lift-off process utilizing electron-beam evaporation [3].

EXPERIMENT

Three wafers, which duplicate the base and collector layers of an emitter-up HBT, are processed. The structure, grown by MBE on semi-insulating GaAs, is 5000Å of n^+ GaAs ($3x10^{18}$ cm^{-3}), 5000Å of n GaAs ($5x10^{16}$ cm^{-3}), finishing with 1000Å p^+ GaAs ($5x10^{19}$ cm^{-3}). The n-type dopant is Si and the p-type dopant is Be. The process sequence starts with a mesa etch to isolate the devices, a collector recess etch to the n^+ layer and metallization, and finally base metallization on the p^+ GaAs. The photolithography patterns includes a transmission line model (TLM) [4] contact resistance test structure and a diode structure. The area of the diode is $1.0 x10^5$ μm^2. Contacts to n-type material are Pd/Ge/Au (30nm/40nm/600nm). The p-type contacts are Pt/Ti/Au (20nm/100nm/150nm) on the control wafer A and Pt/TiN/Ti/Au (20nm/50nm/10nm/150nm) on the remaining two wafers B and C. All contact patterns are created by lift-off of e-beam evaporated metal. The evaporations are performed in a Veeco evaporator equipped with a Commonwealth Scientific Kaufman type ion source. The thicknesses of the layers are monitored during deposition via an Inficon crystal monitor. The conditions for RIBAE of the TiN, in which the Ti is evaporated onto the sample simultaneously with a flux of ionized N, are: partial pressure: 2×10^{-4} Torr; beam voltage 250V; beam current 25mA.

A variety of alloying conditions are investigated. The p-type contacts on wafers A and B are alloyed in N_2 in an AG Associates Heatpulse rapid thermal processing (RTP) system. Pd/ Ge/Au sintered under the RTP conditions used for the p-contacts is too resistive ($\sim 10\mu\Omega\text{-cm}^2$), so the n-contacts on these wafers are heated in a furnace (peak temperature of 410^0C) prior to the deposition of the base metal. The RTP temperature profile includes a low temperature soak, and a fast ramp to the peak temperature Ts at which the sample is held for 10s. Four different Ts are used: 375, 400, 425 and 450^0C. For each Ts, a quarter piece of wafer A and of B are sintered during the same run, transported on a carrier wafer of GaAs. The temperature is monitored with a thermocouple in contact with the carrier wafer.

The sintering of both type contacts on wafer C is conducted in a furnace under H_2 atmosphere. The sample, on a quartz boat, is placed in a zone held at 465^0C. A thermocouple attatched to the boat monitors the temperature. The wafer is removed 30s after a value Ts' is reached. The same four values, from 375 to 450^0C, are used for Ts' during furnace alloys. Both n- and p-type contacts are sintered in a single treatment.

All the samples undergo thermal stress testing (aging) for an extended period. The samples are placed on a hot plate at 280^0C in room air. Electrical properties, morphology, and microstruture are examined prior to and during the aging test. The contact resistance is determined using the TLM pattern and a four point probe measurement. The spacings between contacts, nominally 5, 10, 15 and 20 μm, are verified on each wafer with an optical microscope. The average value of six measurements per sample is reported. Diode current-voltage (I-V) curves are also monitored. Morphology of the contacts is examined with phase-contrast optical micrographs. Auger electron spectroscopy (AES) and transmission electron microscopy (TEM) are used to study the microstructure of the contacts versus time.

RESULTS and DISCUSSION

The n-type contacts have a broad process window. The Pd/Ge/Au contacts exhibit low contact resistance (Ro <1 Ω-mm) for all the wafers under all annealing conditions of this study. On wafer C with furnace sintering, Ro<0.5 Ω-mm, and for wafers A and B with the combination of furnace and RTP treatment, 0.4<Ro<0.8 Ω-mm. This is consistent with other work investigating the particular Pd/Ge/Au metallurgy used here [1].

During the aging test the samples are removed from the hotplate and measured electrically. Figure 1 displays Ro and the specific contact resistance Rc of the quarter of wafer C, furnace sintered with Ts'=425^0C, as a function of aging time. There is a slight dip in both Ro and Rc initially. The values then rise gradually. Ro changes from an as-sintered value of 0.3Ω-mm to 0.4Ω-mm after aging ~2500h. The change in Rc is larger because in the TLM formulation Rc is proportional to Ro^2. Differences in the sheet resistance under the contact and between contacts are not accounted for in these measurements [5]. Similar results are obtained for the RTP treated wafers A and B; however the absolute values of Ro and Rc are higher (Ro$\approx$0.6Ω-mm, 2<Rc<3 $\mu\Omega$-cm^2). The morphology of these contacts is unchanged during the aging test. This is an excellent contact metallurgy that exhibits good long term stability for AlGaAs/GaAs HBTs and for GaAs-based FETs.

The contact resistance of the p-type contacts with the TiN barrier layer, for either type of sintering procedure, is lower than the value obtained with just a Ti layer. This is illustrated in figure 2 which displays Ro versus Ts or Ts', depending on the wafer. The values for wafers B and C are similar and all <0.1 Ω-mm. For wafer A, with just the Ti layer, 0.4 <Ro<0.9-mm. The same qualitative pattern appears in Rc. But, again, the difference between the values for B and C, Rc$\approx$0.3$\mu\Omega$-cm^2, and for A, Rc$\approx$20$\mu\Omega$-cm^2, is larger.

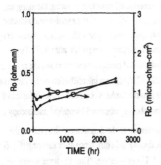

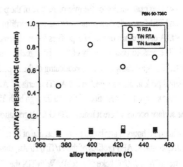

Figure 1. Contact resistance Ro and specific contact resistance Rc versus time under thermal stress (280°C) for n-type Pd/Ge/Au contacts sintered in a furnace at ~425°C.

Figure 2. Contact resistance Ro versus sintering temperature for (a) Pt/Ti/Au in RTP, (b) Pt/TiN/Ti/Au in RTP, and (c) Pt/TiN/Ti/Au in furnace.

The electrical characteristics of the p-type contacts vary significantly during the aging test depending upon the metallurgy and the sintering conditions. On wafer A, Ro of the Pt/Ti/Au contacts increases significantly during the first few hundred hours of the test. Then the value decays back towards the initial value after a few thousand hours. This behavior is illustrated in figure 3. The four curves show the behavoir of Ro for the four different sintering conditions over the course of the aging test. In no case does Ro return to the initial value within the period of the test. The sample sintered at 425°C changes the least over the trial; however the sample sintered at 375°C has the lowest value of Ro at the beginning and end of the trial. The Pt/TiN/Ti/Au contacts are stable in comparison to the Pt/Ti/Au contacts. Figure 4 is a plot of Ro versus time for wafer C which was furnace sintered. The vertical scale is expanded in comparison to figure 3. Similar results are obtained for contacts on wafer B, RTP; however they degrade more over the aging test than those on wafer C. For all RTP sintering conditions, Ro>0.1 Ω-mm after 2500h. Also the lowest contact resistance is obtained with the lowest Ts. With furnace sintering, less change in Ro is observed. When Ts'=425 or 450°C, there is only a slight difference in the intial and final values of Ro. For all samples the only detectable change in the diode I-V characteristics after aging is due to the increased contact resistance. No degradation in the reverse-bias I-V is observed even with the Pt/Ti/Au.

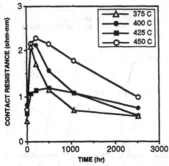

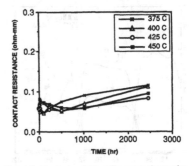

Figure 3. Contact resistance Ro versus time under thermal stress (280°C) for Pt/Ti/Au contacts sintered in RTP at (a) 375°C, (b) 400°C, (c) 425°C, and (d) 450°C.

Figure 4. Contact resistance Ro versus time under thermal stress (280°C) for Pt/TiN/Ti/Au contacts sintered in the furnace at (a) 375°C, (b) 400°C, (c) 425°C, and (d) 450°C.

Detailed examination of the microstructure of the p-type contacts is conducted prior to and during the aging test. AES and TEM provide data demonstrating the interdiffusion of the various constituents in the metal-semiconductor contact. A comparison of the AES profiles for as-sintered and the 1070h aged p-contact on wafer A is presented in figure 5. The peak-to-peak intensity of the AES signal versus sputtering time is plotted. The profile after RTP, Ts= 425°C, (fig. 5(a)) shows a slight broadening of the Pt peak from the as-deposited profile (not shown) and the presence of a small peak in the As signal at the Au-Ti interface. After 1070h at 280°C significant interdiffusion of the various element has occurred. Penetration of the Au through both Ti and Pt layers into the GaAs is observed. Also diffusion of Ti to the surface occurs where it reacts with O_2. Degradation of surface morphology observed by optical microscopy after a few hundred hours is attributed to the presence of Ti at the surface. In contrast to the degradation observed on the Pt/Ti/Au contacts, AES profiles of the Pt/TiN/Ti/ Au, shown in figure 6, display little difference after sintering (fig. 6(a)) and after 1070h, 280°C bake (fig. 6(b)). With TiN the Ti1 and the N AES lines interfere. Thus the layer where the Ti1 line is stronger than the Ti2 line contains a large concentration of N (compare the Ti layer of figure 5 with a stronger Ti2 signal). The TiN-Ti interface observable in the as-deposited profile is washed out after sintering. The Pt reacts to form a Pt-Ga compound [6], but the TiN successfully blocks any interdiffusion of Au and GaAs, even after the 1070h at 280°C.

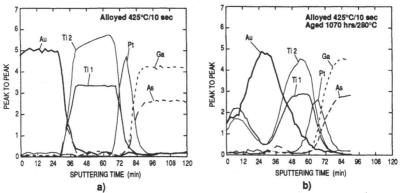

Figure 5. Auger sputter profile of Pt/Ti/Au contacts (a) as-sintered in RTP at $425^{o}C$ and (b) after 1070h at $280^{o}C$.

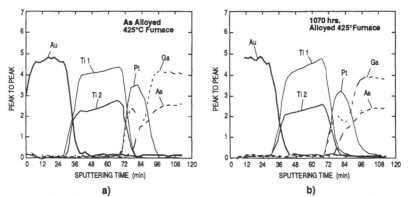

Figure 6. Auger sputter profile of Pt/TiN/Ti/Au contacts (a) as-sintered in furnace at $425^{o}C$ and (b) after 1070h at $280^{o}C$.

The effects of sintering and aging on the RTP treated contacts also are examined with TEM. Bright-field micrographs (diffraction conditions: g=004 for GaAs) of the as-deposited and the aged (~2500h at 280°C) Pt/Ti/Au contact, figure 7, illustrate the aging properties. The sintering conditions for the contact pictured in 7(b) are Ts=425°C. The Au and Ti layers of the as-deposited contact are clearly distinguishable in figure 7(a). The Ti layer measures 1000Å in the micrograph, as it should. The Au layer is 1700Å, slightly thicker than the 1500Å monitored during deposition. The interface between the Pt and the GaAs is not easily identifiable in this bright-field micrograph. In dark-field images (not shown) the thickness of the as-deposited Pt layer is apparent and agrees with the deposited value of 200Å. Another feature of the metals which is more easily determined with the dark-field image is the grain size of the Ti. The largest grains are 400 to 500Å. The sintered and aged contact of figure 7(b) displays the changes obseved by the AES analysis. The AES profiles of the Pt/Ti/Au, with the same sintering conditions, before and after aging (1070h at 280°C) are presented in figure 5. The Au and the Ti layers after aging are no longer well differentiated in the micrograph. The original Pt-GaAs interface is visible within the Pt-GaAs alloyed layer which is clearly identifiable. This alloyed layer also incorporates some Ti. Chemical indentifications are made via energy dispersive x-ray spectra (EDAX) measured in the TEM. The other particularly interesting feature of the sintered and aged contact in 7(b) is the nodule protruding into the GaAs from the contact. These nodules are precipitates of Pt, Ga and As. The nodule penetrates ~1000Å into the GaAs. It extends through most of the p$^+$ layer and is a potential reliability problem.

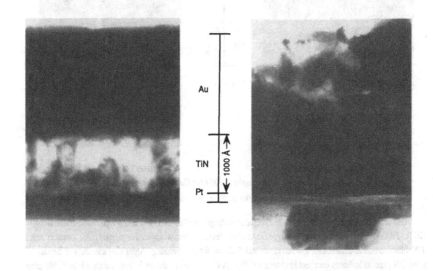

As-deposited   Sintered and aged 2500h

Figure 7. Transmission electron micrographs (g=004 for GaAs substrate) of the Pt/Ti/Au layers (a) as-deposited, and (b) after RTP sintering and aging 2500h at 280°C.

The comparison of the as-deposited to the sintered and aged Pt/TiN/Ti/Au contacts of wafer B presented in figure 8 confirms the AES and electrical data. The Pt, the TiN/Ti, and the Au layers of the contact are clearly delineated in the TEM micrograph of the as-deposited metals (figure 8(a)). After sintering (425°C, 10s) and aging (~2500h at 280°C) the Pt layer and the GaAs interdiffuse as shown in 8(b). The TEM data suggest that this Pt-GaAs layer is about 2.5 times the thickness of the as-deposited Pt layer (200Å). Data from AES indicate an increase in the full width at half maximum of about 1.5 times for the Pt, assuming equal sputtering rates for the Pt and Pt-GaAs layers. The thicknesses of the other layers in the contact remain unchanged after sintering and aging. The largest TiN grains measured in dark-field micrographs are 150 to 200Å. These are smaller than the Ti grains describe previously. The grain size is controlled by growth kinetics during deposition, which we have not investigated. The contact remains planar with no indication of the spiking behavior observed in the Pt/Ti/Au (figure 7(b)). EDAX data verify that only Pt, Ga and As are present in the interfacial layer. There is some mixing of Pt and Ga into the TiN layer near the Pt/TiN interface, but no As is detected. Also no N is detected because the thin C-based polymer window of the x-ray detector strongly absorbs the N signal. The electron diffraction patterns of the TiN and the Ti polycrystalline layers are different. The TiN diffraction pattern is sharper, and shows a preferred orientation with four-fold symmetry. This is expected since TiN has a cubic structure[7]. Ti has a hexagonal lattice. The data demonstrate the ability of the TiN layer to eliminate the interdiffusion of Au and GaAs maintaining a planar interface for reliable contacts.

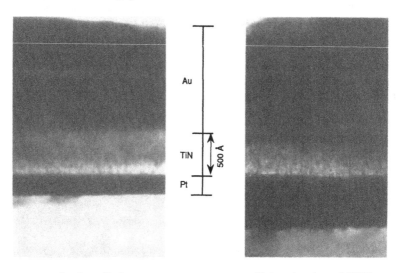

As-deposited   Sintered and aged 2500h

Figure 8. Transmission electron micrographs (g=004 for GaAs substrate) of the Pt/TiN/ Ti/Au layers (a) as-deposited, and (b) after RTP sintering and aging 2500h at 280°C.

SUMMARY

This work demonstrates advantages of sintered metallurgies applied to both n- and p-type ohmic contacts for HBTs. Pd/Ge/Au is a stable contact to n-type GaAs with a broad process window. For p-type GaAs, contacts with a TiN diffusion barrier demonstrate low resistance with RTP and furnace sintering. Aging characteristics of furnace sintered TiN material indicates excellent long-term stability. A single sinter treatment for both contacts is available using the furnace. Additional work is needed to determine conditions for a single RTP sinter treatment for both contacts. These contacts have been incorporated into power HBTs demonstrating excellent characteristics[8]. Long term reliablity studies of the devices are beginning.

REFERENCES

1. T.E. Kazior and R.C. Brooks, submitted to J. of ElectroChem. Soc.
2. H.P. Kattelus, J.L. Tandon, A.H. Hamdi and M-A. Nicolet in Layered Structures, Epitaxy, and Interfaces, edited by J.M. Gibson and L.R. Dawson (Mater. Res. Soc. Symp. Proc. 37, Pitts. PA 1985) pp. 589-594.
3. T.E. Kazior and R.C. Brooks, (presented at MRS 1990 Spring Meeting).
4. H.H. Berger, Solid State Electron. 15, 145 (1972).
5. G.K. Reeves and H.B. Harrison, Elect. Dev. Lett. EDL-3, 111 (1982).
6. T. Sands, V.G. Keramidas, A.J. Yu, K-M. Yu, R. Gronsky, and J. Washburn, J. Mater. Res. 2, 262 (1987).
7. The crystallography of the TiN has been verified by x-ray diffraction studies of thick layers.
8. M. Adlerstein, G. Flynn, W. Hoke, J. Huang, G. Jackson, P. Lemonias, R. Marjarone, E. Tong, and M. Zaitlin, submitted to Elect. Dev. Lett.

This paper also appears in Mat. Res. Soc. Symp. Proc. Vol 184

Thermal Stability of Al-Pt Thin Films/GaAs for Self-Aligning Gate Contacts

G.D. Wilk*, B. Blanpain*, J.O. Olowolafe*, J.W. Mayer* and L.R. Zheng**
*Department of Materials Science and Engineering,
Bard Hall, Cornell University, Ithaca, NY 14853
**Corporate Research Laboratories,
Eastman Kodak Company, Rochester, NY 14650-2011

Abstract

Several compositions of Al-Pt thin films have been co-evaporated on GaAs substrates to study the stability of the alloys at high-temperature anneals. The Al concentration in the alloys ranges from 45 at.% to 70 at.%, and we show that the films meet thermal stability requirements imposed by GaAs self-aligning gate technology for compositions between AlPt and Al_2Pt.

Introduction

Self-aligning gates offer many advantages in the manufacturing process of MESFETs (metal-semiconductor field effect transistors), and much research has been done on different metallizations to find a suitable gate material. The major difficulty with self-aligning gate contacts on GaAs substrates is the necessity for the metal to withstand a post-implantation anneal of 900°C for 20 seconds, and adequate adhesion between the metal and substrate must also be maintained during the process. Metallizations thus far researched have been based on refractory metals, alloys of refractory metals or alloys of refractory metals with Si. Although some of these materials show desirable qualities, they do not exhibit reproducible thermal stability at the required temperature, and the search is still open for a better gate material [1]. Sands et al. [2] has shown AlNi to be epitaxial on GaAs, and found it to be significantly more stable than noble metals, which react at very low temperatures [3,4]. Although the AlNi films exhibit increased stability over other materials, they cannot endure the post-implantation anneal. We selected the Al-Pt system because it is thermodynamically more stable than Al-Ni alloys (ΔH_f(AlNi) = -59 kJ/gmatom, ΔH_f(AlPt) = -100 kJ/gmatom) [5]. We have determined a composition region where the Al-Pt alloys meet the thermal stability requirements imposed by self-aligning gate technology on GaAs MESFETs.

Experimental

Samples were prepared on semi-insulating GaAs substrates, coated on the back with reactive sputtered Si_3N_4 (50 nm thick). We simultaneously co-evaporated the films on to SiO_2 and GaAs substrates, using SiO_2 as control samples and to facilitate the Rutherford Backscattering (RBS) analysis of the alloy composition. Before the deposition, the GaAs substrates were cleaned in a $HCl:H_2O$ (1:1) for 5 minutes, rinsed in H_2O for 1 minute, etched in $NH_4OH:H_2O$ (1:10) for 15 seconds and finally dried with nitrogen. The Al-Pt thin films were co-evaporated in a cryo-pumped dual electron-gun deposition system with a base pressure below 1×10^{-7} Torr. The deposition rate for Pt was fixed at 0.15 nm/sec and the Al deposition rate was adjusted according to the desired composition. Four different compositions were deposited ranging from Al concentrations of 45 at.% to 70 at.%. The samples were annealed in a RTA oven under flowing nitrogen at temperatures between 500°C and 1000°C for 20 seconds. The phase transformations and the reactions with the substrate were analyzed with RBS, Auger Electron Spectroscopy (AES) and x-ray diffraction. The Auger depth profiling was used to monitor the Al movement and oxygen contamination.

Results

We will present in particular the behavior of two different alloys of the Al-Pt, alloy D (70 at.% Al) and alloy C (64 at.% Al) on GaAs and SiO_2 substrates (see Fig. 1). The x-ray diffraction pattern of the as-deposited alloy D showed two broad peaks, indicating an initial amorphous phase. At low-temperature anneals, the Al-Pt phase transformed into two equilibrium phases, $Al_{21}Pt_8$ and Al_2Pt. At 900°C, the $Al_{21}Pt_8$ completely transformed into Al_2Pt, as confirmed by the x-ray diffraction pattern at this temperature. The RBS spectrum for samples annealed at 900°C clearly showed reaction between the substrate and the Al-Pt thin film, while AES confirms the presence of As. Figure 2 shows the RBS spectrum of alloy D as-deposited and annealed at 1000°C. The leading edge of the Pt peak has receded significantly compared to the as-deposited spectrum, which is most likely caused by the formation of a surface Al-oxide layer during the Rapid Thermal Annealing (RTA) process. AES confirms the presence of oxygen at the surface of the alloy for anneals at 1000°C. The growth of this oxide layer may also explain the entrapment of As at the surface of the film. Figure 2 also indicates that a significant amount of As has diffused into the alloy, which led to the formation of two phases, AlAs and Al₃Pt₂, as confirmed by the x-ray diffraction pattern in Fig. 3. The diffusion of As into the thin film at 900°C and subsequent deterioration of the metal-substrate interface indicates that alloy D does not meet thermal stability requirements.

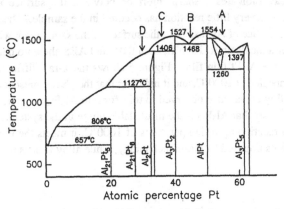

Fig. 1: Partial phase diagram of the Al-Pt system, with the four co-evaporated alloys labelled A(45 at.% Al), B(55 at.% Al), C(64 at.% Al), D(70 at.% Al)

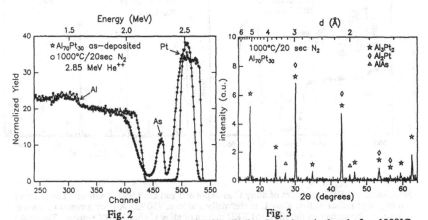

Fig. 2

Figs. 2&3: (2) RBS spectrum (sample tilt 60°) of alloy D as-deposited and after 1000°C anneal. Out-diffusing As from subsrtate has formed surface layer of an As compound in alloy, and an Al-oxide surface layer formed during RTA process. (3) XRD pattern of alloy D after 1000°C anneal. Reaction with out-diffusing As led to formation of AlAs and Al₃Pt₂.

Alloy C is more stable than alloy D (see Fig. 1). The as-deposited sample shows an x-ray diffraction pattern characteristic of the amorphous Al-Pt alloy previously reported by Legresy *et al.* [6]. The two broad peaks sharpen and new peaks appear as the amorphous phase transforms at low-temperature anneals to Al_2Pt and Al_3Pt_2 equilibrium phases. These phases were stable until anneals near 1000°C/20 seconds on GaAs substrates. Figure 4 shows the as-deposited sample and the sample annealed at 950°C/20 seconds. The steep trailing edge of the Pt peak indicates a sharp interface between the surface alloy and the substrate, thus very little reaction has occured in the sample. The spectrum also shows that a trace of As resides at the surface of the sample, not within the metal layer. For anneals up to 1000°C, both RBS and AES show no detectable Ga or As inside the Al-Pt thin film. Figure 5 shows the x-ray diffraction pattern of alloy C annealed at 1000°C, and it indicates that the phases present are Al_3Pt_2 and Al_2Pt. Alloy D was also observed to transform to Al_3Pt_2 for anneals at 1000°C, which suggests that Al_3Pt_2 is the most stable phase in the system. Since alloy C shows no reaction up to temperatures of 1000°C, it meets the thermal stability requirements imposed by GaAs technology, while alloy D does not.

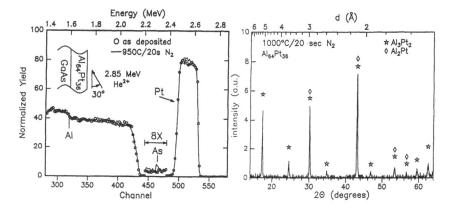

Fig. 4   Fig. 5

Figs. 4&5: (4) RBS spectrum of alloy C as-deposited and after 950°C anneal. The steep Pt trailing edge indicates that alloy C is thermally stable at 950°C. A trace amount of As has diffused to the surface of the alloy. (5) XRD pattern of alloy C after 1000°C anneal. The peaks correspond to Al_2Pt and Al_3Pt_2, the same Al-Pt phases found in alloy D at 1000°C (Fig. 3).

The Al-Pt/GaAs system is summarized in Fig. 1. Alloy A (45 at.% Al) failed after annealing at about 600°C, where Ga diffused into the alloy layer, transforming Al_3Pt_5 into (Al,Ga)Pt. This transformation yielded excess Pt, which was found to react with As to form a $PtAs_2$ layer between the alloy and the substrate. Alloy B failed at 950°C because of Ga out-diffusion, which transformed AlPt into $(Al,Ga)_3Pt_2$. This failure temperature, however, is high enough to meet thermal stability requirements. Thus, both alloys B and C may be considered thermally stable for self-aligning gate technology.

Discussion

Each of the alloys transformed to a final equilibrium Al_3Pt_2 compound. Thus, Al_3Pt_2 is the most stable phase in the systems studied. We believe that the thermodynamic properties of the Al-Pt alloy are responsible for the observed increase in stability near the Al_3Pt_2 composition. A possible explanation for the observed stability of Al-Pt/GaAs stems from the thermodynamics of the system. We propose that since Al and Ga have similar chemical properties, the Ga chemical potential in the Al-Pt alloy strongly depends on Al concentration. Thus, alloy compositions which are Al-deficient compared to Al_3Pt_2 have a lower Ga chemical potential than the substrate, and Ga diffuses into the thin film to cause failure. For alloys are richer in Al than Al_3Pt_2, the chemical potential is increased, preventing Ga diffusion. Hence, As out-diffusion is observed first, and the kinetics of the As out-diffusion may be responsible for the thermal instability in these samples. These two competing failure mechanisms result in maximum thermal stability at the composition Al_3Pt_2.

Conclusion

The equilibrium phase Al_3Pt_2 is the most stable phase of the Al-Pt/GaAs systems studied. It is believed that either Ga or As out-diffusion is primarily responsible for failure of the alloys, depending on the initial concentration of Al. The composition range between AlPt and Al_2Pt appears to meet the thermal stability requirements for self-aligning gate contacts on GaAs.

References

[1]   S.P. Kwok, J. Vac. Sci. Techn. B, 1383(1986).

[2]   T. Sands, J.P. Harbison, W.K. Chan, S.A. Schwarz, C.C. Chang, C.J.   Palmstrom
      and V.G. Keramidas, Appl. Phys. Lett. 52, 1261(1988). and T. Sands,   W.K.
      Chan, C.C. Chang, E.W. Chase and V.G. Keramidas,
      Appl. Phys. Lett. 52, 1338(1988).

[3]   J.O. Olowolafe, P.S. Ho, H.J. Hovel, J.E. Lewis and J.M. Woodall, J. Appl.
      Phys. 50, 955(1979).

[4]   T. Sands, V.G. Keramidas, A.J. Yu, K.-M. Yu, R. Gronsky and J. Washburn,
      J. Mater. Res. 2, 262(1987).

[5]   R. Hultgren, P.D. Desai, D.T. Hawkins, M. Gleiser and K.K. Kelley, Selected values
      of thermodynamic proerties of binary alloys, (American Physical Society), Ohio, 1973.

[6]   J.-M. Legresy, B. Blanpain and J.W. Mayer, J. Mat. Res. 3, 884(1988).

Requirements of Electrical Contacts to Photovoltaic Solar Cells

T.A. Gessert and T.J. Coutts
Solar Energy Research Institute, 1617 Cole Blvd., Golden, CO 80401 USA.

ABSTRACT

The importance of contacts to photovoltaic solar cells is often underrated mainly because the required values of specific contact resistance and metal resistivity are often thought to be relatively modest compared with those associated with very large scale integration (VLSI) applications. However, due to the adverse environmental conditions experienced by solar cells, and since many of the more efficient cells are economically advantageous only when operated under solar concentration, the requirements for solar cell contacts are sometimes more severe. For example, at one-sun operation, the upper limit in specific contact resistance is usually taken to be 10^{-2} Ω-cm^2. However, at several hundred suns, this value should be reduced to less than 10^{-4} Ω-cm^2. Additionally, since grid line fabrication often relies on economical plating processes, porosity and contamination issues can be expected to cause reliability and stability problems once the device is fabricated. It is shown that, in practice, these metal resistivity issues can be much more important than issues relating to specific contact resistance and that the problem is similar to that of providing stable, low resistance interconnects in VLSI. This paper is concerned with the design and fabrication of collector grids on the front of the solar cells and, although the discussion is fairly general, it will center on the particular material indium phosphide. This III-V material is currently of great importance for space application because of its resistance to the damaging radiation experienced in space.

INTRODUCTION

A solar cell exploits the photovoltaic (PV) effect and delivers electrical power to an external load when illuminated by a suitable light source. Since sunlight is a relatively dilute source of energy at the surface of the earth, the electrical output per unit area of a PV device is quite small; thus large solar cell areas are normally required to generate significant amounts of power. At present, the maximum reported conversion efficiency for a Si based solar cell is slightly over 23% [1]. Although Si based solar cells are becoming increasingly cost effective for many power generation applications, the most efficient solar cells (although not the most commonly used), with efficiencies approaching 35%, are those based on the III-V binary, ternary and quaternary alloys involving GaAs and InP [2]. However, these materials are expensive, when compared with Si, and thus the relative cost of generating the energy economically becomes a critical issue. Often, cost effectiveness will only be possible when the cells are operated under optical concentration. This has a two-fold advantage in that it not only minimizes the amount of the costly materials, but also, due to the electrical characteristics of the PV device, causes the junction efficiency actually to *increase* with concentration. As an example of this, it has been previously shown that it is theoretically possible to achieve an efficiency of greater than 40% by using a tandem cell based on a GaAs device mechanically stacked (i.e. not grown monolithically) on top of a lower cell of InGaAs or InGaAsP and operating this at a concentration of 1000 suns [3].

For both one-sun and concentration designs, inherent junction and optical losses limit achievable efficiency. However, a significant part of the available power is consumed through additional series resistance losses associated with the electrical contacts to the cell. These losses not only limit the amount of collectable power available from a typical one-sun cell, but also, since they increase with the square of the solar cell current, actually determine the upper limit to the effective concentration ratio of concentrator solar cells.

For the materials mentioned above, the achievement of the relatively low values of the series resistance required for concentrator operation is by no means a simple matter, particularly when it is remembered that the grid line contacts must have all the qualities required of VLSI contacts (i.e., reliability, ease of manufacture, durability, high temperature stability, etc.). It must also be noted that, unlike VLSI, these contacts are exposed to degrading environmental conditions such as moisture, temperature extremes, particulates and ultra-violet irradiation. Although the required values of series resistance have easily been achieved under laboratory conditions for heavily doped GaAs, this is not the case for InP since there are often restrictions on the thermal treatments which may be used. The most efficient cells are actually very shallow homojunctions, with an n-type emitter substantially less than 100 nm in thickness, and these junctions can easily be shorted by diffusion of the metallization during sintering. InP is an important solar cell material for space application because of its radiation hardness [4], and it has been of major interest to NASA for some years. If the cost of the device could be decreased, perhaps through techniques such as heteroepitaxy, then these InP based solar cells could also have considerable terrestrial applications.

The purposes of this paper are to discuss some of the requirements necessary for good quality contacts to the above mentioned III-V based solar cell devices, to discuss some usual (but often overlooked) stability problems, and finally to discuss some aspects influencing the design of solar cell grids and how the design can be used to overcome several reliability/stability concerns.

EXPERIMENTAL

Although the following discussion on solar cell grid design is completely general, these modeling methods are given more clarity when presented with respect to a specific group of solar cell devices. With this in mind, the following section gives the experimental details of one such device - the InP shallow homojunction.

The devices are grown on p^+-InP substrates, which are (100) oriented, Zn doped to about 4×10^{18} cm^{-3} and supplied in a mechanically polished and chemically etched condition with near perfect surfaces. After the wafers are cleaved into convenient sized substrates, the latter are loaded into the growth system without any further cleaning. Growth is performed using atmospheric pressure metal organic vapor phase epitaxy (APMOVPE) [5]. The device itself consists of a p^+ buffer layer doped to $\sim 4 \times 10^{18}$ cm^{-3}, a p-type base layer doped to $\sim 10^{17}$ cm^{-3} and a thin n^+ emitter layer of ~ 300 Å in thickness doped to $\sim 4 \times 10^{18}$ cm^{-3}

Ohmic back contact to the p^+-substrate can be achieved in a number of ways, one of which is by evaporating a thin film contact of either Au:Be or Au:Zn followed by contact sintering at a temperature of $\sim 380°C$. After this, a layer of high purity Au of about 2 µm thickness is electroplated onto the contacting layer. This method and others are explained in greater detail elsewhere [6]. Front grid contact patterns are formed photolithographically. It is the fabrication of these front grids which will be the main topic of discussion of this paper. The final step is to define electrical isolation trenches between the individual cells photolithographically and chemically etch these trenches with concentrated HCl.

In addition to the fabrication of actual solar cells, other analyses included scanning electron microscopy (SEM), resistivity studies of plated and evaporated Au deposits, contact resistance studies and adhesion studies of several metals onto InP.

METHOD

Before discussing the details of how certain metallizations can affect the performance and stability of a solar cell, it is necessary to establish the background physics. The current-voltage characteristics of an *ideal* solar cell junction (i.e., one in which all resistance terms have been neglected) are described by the basic diode equation with a light generation component:

$$J = J_0 [exp(eV/nkT) - 1] - J_g \qquad (1)$$

where J_0 is the reverse saturation current density, J_g is the minority carrier light generated current density, n is the diode ideality factor, and the other symbols are as conventionally used. For a good cell, it is reasonable to put J_g equal to the measured short circuit current density (J_{SC}). In this case, the open circuit voltage (V_{OC}) of the device is:

$$V_{OC} = nkT/e \ ln[(J_{SC}/J_0) + 1]. \qquad (2)$$

On the current-voltage characteristic, there is a point at which the power that may be supplied to an external load reaches a maximum. The coordinates of this point are known as the maximum power point current and voltage (J_{mpp} & V_{mpp}). The output power at this point, divided by the product of V_{OC} and J_{SC}, is the fill-factor (FF) of the solar cell, and, for typical good quality cells, is equal to ~80-85%. Since the power generated by a PV solar cell will be dependent not only upon the total amount of input power, but also the specific spectral content, for cell calibration purposes, the incident illumination (for terrestrial applications) is standardized to both 1000 W m^{-2} and a specific spectral content [7]. The final conversion efficiency of the cell is equal to the product of the above three junction parameters (V_{OC}, J_{SC} & FF) divided by the standardized incident energy.

In the presence of a significant series resistance (R_S), the voltage measured across the external terminals of the cell is reduced, and the ideal diode equation (Eq. 1) is modified, resulting in:

$$J = J_0\{exp[e(V + JR_S)/nkT] - 1\} - J_g \qquad (3)$$

The five main areas where series resistance can manifest itself in a solar cell are: 1) Resistance associated with the back contact [sheet resistance of the metal and contact resistance], 2) Resistance associated with current flow through the bulk of the solar cell substrate, 3) Resistance associated with current flow in the top emitter region of the cell, 4) Contact resistance between the grid line and the emitter, and (5) Resistance associated with current flow in the grid lines. Although any one of these five areas can dominate the the total R_S, in this paper the main discussion will be limited only to areas (3), (4) and (5), since it is these losses that can be reduced by proper metallization and grid design techniques.

The methodology of designing grids for solar cells is well established (see, for example, Ref. 8), but it is necessary to indicate the approach used here in order to establish the relevant points. First, to minimize series resistance losses in the emitter, the grid design must account for the, often greater than desired, emitter sheet resistance by placing many narrowly spaced grid lines on the surface. Having many grid lines on the surface is also consistent with the need to provide sufficient grid metal cross-section in order to avoid excessive series resistance losses in the grid lines. However, having many grid lines on the surface conflicts with the requirement of minimum cell obscuration. To balance these conflicting criteria, standard formalisms for the component power losses have been established and a computer program developed for optimization of the grid design.

This computer program is similar to those developed by others, but includes some additional capabilities. [9] [10] The first of these assesses the effect of current crowding, which, although usually associated with the fine line contacts of VLSI, is also relevant to the design of certain types of solar cells. The second involves the ability to calculate the total and component power losses of non-optimum designs - a highly useful tool when assessing the effects on non-ideal metallizations. The program, however, does employ the assumption that no current flows directly from the emitter to the bus-bar (i.e., one-dimensional current flow, see Figure 1) and is currently configured for use only with rectangular cells. Following is the mathematical formalism used to compute the fractional power losses, with each of the individual component power loss terms indicated.

$$\Delta P/P = (J_{mpp} \, S \, R_F \, B^2)/4 \, W_F \, V_{mpp} \qquad \text{(Power loss in tapered grid fingers)} \quad (4)$$

$$+ \; (J_{mpp} \, B \, R_B \, A^2)/4 \, W_B \, V_{mpp} \qquad \text{(Power loss in tapered bus-bar)} \quad (5)$$

$$+ \; (J_{mpp} \, S \, [r_c R_E]^{1/2} \, \coth[W_F/L_T])/2 \, V_{mpp} \quad \text{(Power loss due to contact resistance)} \quad (6)$$

$$+ \; (J_{mpp} \, S^2 \, R_E)/12 \, V_{mpp} \qquad \text{(Power loss due to emitter)} \quad (7)$$

$$+ \; (W_F/S + W_B/B) \qquad \text{(Power loss due to the finger and bus-bar shadowing)} \quad (8)$$

V_{mpp}	=	Voltage at Maximum Power Point (mV)
J_{mpp}	=	Current Density at Maximum Power Point (mA-cm^{-2})
r_c	=	Specific contact resistance (Ω-cm^2)
R_E	=	Sheet resistance of the emitter (Ω/sq)
R_B	=	Sheet resistance of the bus-bar (Ω/sq)
R_F	=	Sheet resistance of the grid fingers (Ω/sq)
W_F	=	Width of the grid fingers (cm)
A	=	Length of the cell parallel to bus bar (cm)
B	=	Half-width of the cell parallel to fingers (cm)
L_T	=	Calculated Transfer length, $L_T = (r_c/R_s)^{1/2}$ (cm) [11]
W_B	=	Calculated width of the bus-bar (cm)
S	=	Calculated spacings between the finger centers (cm)

After specifying the first nine parameters, the program performs a user interactive, two-step process, first minimizing the optical and electrical losses due to the single bus-bar and calculating the optimum bus-bar width, and second, iterating to the optimum value of the spacing between the fingers and calculating the total and individual component losses. After the optimum cell is designed, the program can then calculate the power losses associated with non-optimum configurations.

As mentioned earlier, operation under concentration is an important option for the more expensive materials such as GaAs and InP. In concentrator solar cells, the main reason for concentration is to acquire an economic advantage, and thus the cost of the additional equipment (concentrating optics, frames, alignment motors and electronics, etc.) must be balanced against the benefit of additional power output due to the greater illumination. Under these constraints, the cell in normally redesignd as follows: Since there will be a considerable amount of unused area beneath the optics (99.9% at 1000 suns concentration), the active area of the concentrator cell includes only that area which is actually illuminated by the equipment. Also, the bus-bar is increased in width and moved outside the illumination area; the associated shadow and resistance losses are removed from the actual solar cell and, thus, from the calculation of efficiency. When this is done, the economics indicate that a concentrator solar cell should have much less area than a typical 1-sun device. Therefore, the model concentrator solar cell used here has a smaller total area than the model one-sun cell. The final difference is, since the emitter is normally much thicker in a concentrator cell (decrease emitter resistance losses, as will be discussed), the model concentrator cell is assigned an emitter sheet resistance of 100 Ω/sq.

In presenting concentrator cells data, it is informative to calculate the efficiency of the modeled cell and compare it with the ideal junction characteristics (Eq. 1). For this, the ideal junction conversion efficiency must be calculated for each concentration ratio; thus, it is necessary to use a consistent set of one sun values of V_{oc}, J_{sc} and J_0 for Eqs. 1-3 from which the values of V_{mpp} and J_{mpp} can be determined (See Table I). For the present work, this was done by first multiplying both sides of Eq. 2 by voltage, differentiating the resulting equation for device power and equating it to zero. This results in a transcendental equation from which the value of V_{mpp} may be derived numerically. This value is then substituted into the original current/voltage equation to obtain J_{mpp}. Although the V_{mpp} and J_{mpp} are changed in the presence of series resistance, this is a relatively small effect for good quality cells and this approach is therefore reasonable. Other inputs, based on guessed or measured values, may be used resulting in the same qualitative information.

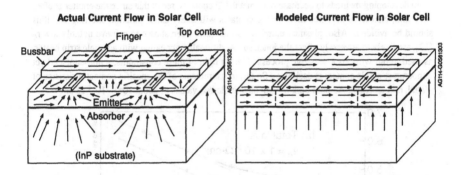

Actual Current Flow in Solar Cell **Modeled Current Flow in Solar Cell**

Fig. 1. Schematic of the simple rectangular grid used for the modeling study illustrating the actual and modeled current flow.

Table I
List of typical InP shallow homojunction parameters used in modeling.

R_E	=	600 Ω/sq (100 Ω/sq for concentrator)
r_c	=	10^{-2}, 10^{-3} or 10^{-4} Ω-cm^2
J_0	=	9×10^{-13} mA cm^{-2}
J_{mpp} (1 sun)	=	30 mA cm^{-2}
V_{mpp} (1 sun)	=	800 mV
Cell area	=	4 cm^2 (0.5 cm^2 for concentrator)
Minimum grid line width	=	10 µm
Grid line resistivity	=	2×10^{-6} (Pure Au) - 2×10^{-5} Ω-cm
Concentration ratios	=	1-1000 suns

RESULTS AND DISCUSSION

The requirements of contact resistance for solar cells depend on two (related) process aspects: 1) The minimum grid line width and 2) whether the cell will be used under concentration. Figure 2 illustrates how two different values of r_c will affect the total power loss of the model one-sun solar cell. As can be seen, although it is often believed that an r_c of 1×10^{-2} Ω-cm^2 is sufficient for one-sun grids, for the thinner finger widths used here, this value of r_c will cause the contact resistance to be the dominant loss mechanism and substantial loss reduction would be achieved if r_c can be reduced to 1×10^{-4} Ω-cm^2. Also shown in this figure is the effect that current crowding has on the calculation of power loss and why, for narrow lines, it is important to include this in the mathematical formalism. In Figure 3 is shown similar calculations, but for the model concentrator cell operated at 1000 suns. Note here that, unlike the case of the one-sun design, an r_c of even 1×10^{-4} Ω-cm^2 will still have a significant (although, not a dominant) effect on the power loss. For this case it is seen that contact resistance values typical of those usually required in VLSI would be advantageous. Finally, in Figure 4 is shown what effect increasing r_c from 1×10^{-4} Ω-cm^2 to 1×10^{-3} Ω-cm^2 would have on the cell as a function of increasing concentration. Here it is seen that even a slight non-uniformity or instability in r_c (as might be caused by progressive adhesion loss) would have a considerable effect on not only the efficiency but also the maximum concentration ratio.

In developing methods to decrease r_c of metal-InP contacts, recall that since the emitter is often very thin, excessive heat treatment of the contacts will likely cause junction shorting and thus should be avoided. Also, plasma treatments, which have been shown to achieve notably low r_c [12], are likely to adversely affect the junction. Additionally, annealing with a simple strip heater, as used in our work, is a sensitive process [13]. In light of these constraints, a non-annealed yet temperature stable contact is desirable. So far, measurements of the specific contact resistance of plated, non-annealed Au onto bulk n^+-InP have indicated values of about 10^{-3} - 10^{-4} Ω-cm^2.

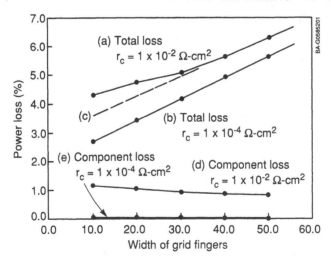

Fig. 2. Effect of contact resistance on 4 cm^2 model one-sun cell. Note that for thin grid lines (finger widths in μm) and $r_c \geq 1\times10^{-2}$ Ω-cm^2, contact resistance will dominate the total power loss. Shown in (c) is the effect of neglecting current crowding.

As indicated above, this range of r_c values is probably sufficient for one-sun devices but would likely cause measurable efficiency reduction for concentrator solar cells. However, recent work involving a Cr/Pd/Ag contact, as will be discussed below, indicates that specific contact resistance values of less than 10^{-5} Ω-cm^2 may be routinely achievable. Currently, we are investigating a device with a thicker emitter which would have the advantage of separating the junction from the top contact thereby allowing operation at higher temperatures. Finally, although it has not yet been attempted, rapid thermal annealing processing may be of considerable benefit. [14]

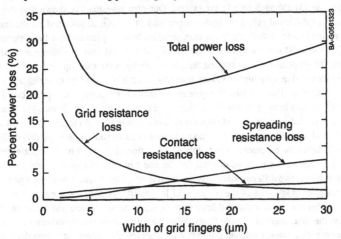

Fig. 3. Comparision of r_c loss to other resistance losses for the model concentrator cell operated at 1000 suns as a function of finger width where finger width/hight = 2. Note that spreading resistance loss is due to R_E and that shadowing losses are not shown.

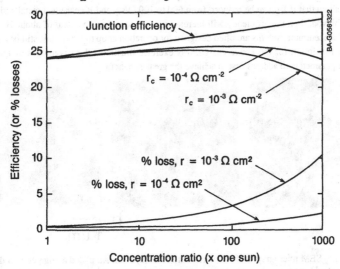

Fig. 4. Comparison of $r_c = 10^{-3}$ vs $r_c = 10^{-4}$ Ω-cm^2 as a function of concentration for the model concentrator cell. Note, efficiency gains would be expected if r_c were reduced below 10^{-4} Ω-cm^2.

The requirements and analysis of metal sheet resistance of the grid lines is probably one of the most critical, yet often overlooked aspects of solar cell processing technology. Even for a well established metallization processes, there are many mechanisms which can greatly affect the difference between the actual sheet resistance and that which was assumed for the design process. Consider, for example, the process of high purity plated Au contacts. In this case, it takes less than 1% of Fe contamination to increase the *ideal* resistivity (i.e., the pure bulk value) by a factor of 10 [15]. Increases in resistivity also occur for contamination of Au by other metals (such as Co, Sn, Ni, Sn and In) although to a lesser extent. Our own research has shown that the plating current has a substantial effect on both the porosity of the Au deposit and on the measured resistivity (See Figures 5 & 6). In particular, standard bus-bar patterns have been electroplated onto ITO films (100 nm thick) to a thickness of 1-4 μm and current-voltage measurements performed. These measurements suggest that the resistivity of the electroplated Au is ~2-8 times that of the ideal value - the higher values associated with higher current densities and a darker deposit appearance. Furthermore, the greater porosity of the deposit in Fig. 5a could enable oxidizing substances to enter the metal, causing the grid resistivity (and perhaps r_C) to be unstable. On actual cells, measurements indicate that the ratio of actual to ideal Au resistivity ratio is about 2-3. As will be shown, this variation is sufficient to necessitate consideration in the grid design.

Figure 7 shows the percentage power loss as a function of the ratio of the resistivities of the grid metal to that of the pure bulk metal for the one sun cell. For each of the three deposit thicknesses shown, two different situations are considered: The upper curve of each pair assumes that the designer is ignorant of the higher grid line resistivity and, the lower curves, that the designer had been aware that the grid lines were of higher resistivity and had accounted for this. Obviously, the first situation is more likely. With the information that resistivities may be up to eight times the ideal value, modeling shows that if higher resistivities are not assumed, then even at one-sun, these losses will certainly dominate. However, as also shown, a prudent design incorporating thicker and/or wider grid lines can be used to insure against these variations. The same effects of increasing the grid line resistivity is modeled for the concentrator cell in Figure 8. Here it is show that if R_E can be reduced from 600 to 100 Ω/sq, and if near ideal metal resistivity can be achieved, then the efficiency will increase up to a concentration ratio of about 100 suns. Although concentrator systems are often designed for operation at up to 1000 suns, still beyond the efficiency maximum for the above design, this exercise illustrates the advantages of knowing which cell parameters to address first to achieve the greatest benefit.

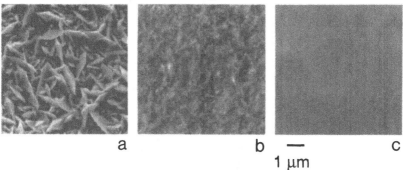

Fig. 5. (a&b) SEM micrographs of Au plated onto ITO. (a) 5 μm Au plated at high current density (~0.5 μm min^{-1}), (b) 1 μm Au plated at low current (~0.1 μm min^{-1}), (c) 120 nm evaporated Au. Deposit (a) had a resistivity ratio ~6-8 times that of pure gold, (b) had a resistivity ratio of ~ 2-3.

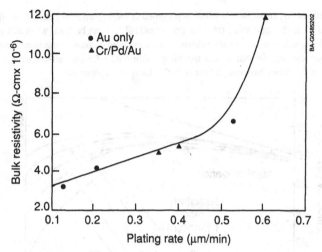

Fig. 6. Plot of plated Au resistivity as a function of plating current. As indicated, these results were obtained for grid lines plated both directly on an ITO film and onto Cr/Pd adhesion layers. Since the sheet resistance of the Au is low compared with that of the ITO and the Cr/Pd layers, a reasonable estimate of the resistivity may be made.

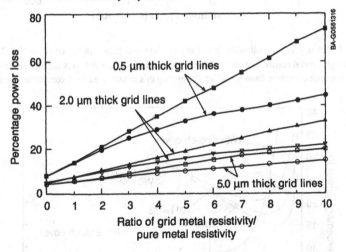

Fig. 7. Percentage power loss as a function of grid line resistance for three metal thickness.

Figure 9 shows that, although initially one may think that a thinner line width is always advantageous, this is not the case for concentrator cells. In this diagram it is seen that when the aspect ratio (width divided by height, W/H), is treated parametrically and the efficiency is calculated as a function of the grid line width, there is an optimum line width for a given aspect ratio. In practice, one would use as thick a grid line as readily achievable using techniques such as solder dipping. Also shown in this figure is the modeled effect of a technique developed by Entech Inc. in which the power loss due to shadowing is reduced. This technique is based on the use of a silicone plastic cover, the surface of which is patterned to form an array of corrugations (prisms) which act to focus the light away from the grid lines.[16] As can be seen, when this cover is incorporated into cell construction, the influence on the modeled performance is quite spectacular. However, one should note that in this modeling, additional reflection effects have not been considered and the covers have not yet been qualified for space applications.

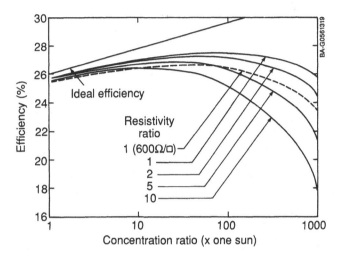

Fig. 8. Efficiency vs concentration ratio for a concentrator cell with R_E equal to 100 Ω/sq in which the grid metal resistivity ratio is treated parametrically. Also shown is the modeled result if R_E had not been decreased from 600 Ω/sq, illustrating this requirement for a concentrator design.

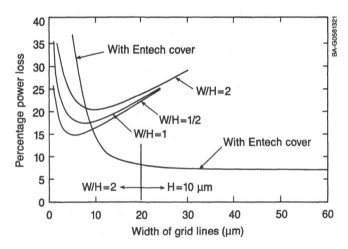

Fig. 9. Percentage power loss as a function of the grid line width and aspect ratio at 1000 suns. Note that there is a well defined minimum position for each aspect ratio. The grid line resistivity ratio is taken as unity. Also shown is the effect of an Entech cover.

In the above discussions, it was assumed that the adhesion of the metal grid lines onto the solar cell was such that it caused no problems in the contact resistance or reliability. However, for InP based solar cells, adhesion is found to be a significant problem, especially if Au plated contacts are used. The adhesion of plated Au is generally poor on many materials and this is certainly the case for InP. Indeed, we have observed that, even though the adhesion of plated Au on GaAs is usually sufficient for laboratory quality solar cells, if even a small amount of P is incorporated to make the GaAsP ternary, the adhesion of the plated Au begins to suffer. Many different techniques have been investigated in the hope of promoting the adhesion of plated Au including pre-plating surface treatments, acid Au striking deposits and different plating rates. Although these investigations have often led to observable variations in deposit adhesion, the resultant adhesion is usually not sufficient to survive a standard Scotch tape test. Recently however, we have observed substantial adhesion promotion through the use of Cr adhesion layers. This work closely followed earlier work in which a Cr/Pd/Ag metallization stack was used to provide the necessary contact to an ITO/InP cell [13]. In the case of the metal/InP contact, it has been found that the Cr/Pd/Ag (80 nm/40 nm/400 nm) will survive a Scotch tape test and provide sufficient conduction for cells up to 0.1 cm^2 in area. For larger area cells, it is necessary to *overplate* these layers so that a sufficiently thick conduction layer can be established.

CONCLUSION

The above modeling study has revealed several aspects of front contacts to solar cells which are not intuitively apparent. For one-sun applications incorporating relatively narrow grid lines, unless it is possible to establish a contact resistance to less than 10^{-2} Ω-cm^2, this term will dominate the resistance losses. For concentration applications however, a minimum value of 10^{-4} Ω-cm^2 must be achieved, while even lower values, comparable with those of VLSI, would be preferred. Although in practice there may not be complete freedom to lower the contact resistance without junction damage or adversely affecting cell performance, substantial reduction in r_c may be achieved by using techniques such as rapid thermal annealing in which the heating will occur only near the semiconductor/metal interface.

The resistivity of the grid line metal, its thickness and aspect ratio are important parameters in the grid design process. It has been indicated that variations in resistivity of up to 8 times the ideal value are more usual than may be expected and, if not accounted for, can dominate the solar cell power loss and seriously affect its stability. Also, although it is normally desired to have the narrowest grid line possible, it is shown that at a concentration of 1000 suns, there is an optimum grid line width which is dependent on the metal resistivity and the aspect ratio.

The adhesion of the grid lines is a critical aspect of cell power loss and stability from several viewpoints including that, if the adhesion is weak, the actual contact area may be much less than the area of the metallization. Recent results indicate that a Cr/Pd/Ag contact onto InP provides a highly adherent and low resistance grid contact.

Finally, although the basic procedures for studying grid losses were established by previous researchers, the practical application of these techniques is often neglected resulting in an unnecessarily pessimistic view of the potential many types of photovoltaic solar cells.

ACKNOWLEDGMENTS

The authors wish to thank X. Li for sample preparation and A. Mason for SEM photographs. This work was supported by the U.S. Department of Energy under Contract No. DE-ACH02-83CH10093 and by NASA Lewis Research Center under Interagency Order No. C-3000-K.

REFERENCES

1. A.W. Blakers, J. Zhao, A. Wang, A.M. Milne, X. Dai and M.A. Green, Proc. of the Ninth E.C. Photovoltaic Solar Energy Conference, Freiburg, FRG, Sept. 89 (Kluwer Academic Pub., Boston, 1989), p. 328.

2. L. Frass, J. Avery, J. Martin, V. Sundaram, G. Girard, V. Dinh, N, Mamsoori and J.W. Yerkes, Proc. of the 1989 DOE/Sandia Crystalline Photovoltaic Technology Project Review Meeting (Sandia National Laboratories, Albuquerque, NM, 1989), p. 173.

3. M. W. Wanlass, K. A. Emery, T. A. Gessert, G. S. Horner, C. R. Osterwald and T. J. Coutts, Solar Cells, 27, 191 (1989).

4. T. J. Coutts and M. Yamaguchi, "Indium Phosphide-Based Solar Cells: A Critical Review of Their Fabrication Performance and Operation", in Current Topics in Photovoltaics, edited by. T. J. Coutts and J. D. Meakin, (Academic Press, New York, 1988), p. 79.

5. W. Wanlass, T. A. Gessert, K. A. Emery and T. J. Coutts, Proc. NASA Conf. Space Photovoltaic Research Technology, April , 1988, Cleveland, OH (NASA Conf.Pub #3030, 1988), p. 41.

6. T. A. Gessert, M. W. Wanlass, T. J. Coutts, X. Li and G. S. Horner, Solar Cells, 27, 299 (1989).

7. "Standard Test Methods for Electrical Performance of Non-Concentrator Photovoltaic Cells Using Reference Cells", ASTM Standard E948.

8. M. A. Green, in Solar Cells (Prentice-Hall, Englewood Cliffs, NJ, 1982), pp. 153-161.

9. D. L. Meier and D. K. Schroeder, IEEE Trans. Electron. Dev. ED-31, 647 (1984).

10. H.B. Serreze, Proc. 13th IEEE Photovoltaic Specialists Conf. (IEEE, New York, 1978), p. 609.

11. H.H. Berger, J. Electrochem. Soc., Solid State Sci. Technol. 119 (4), 507 (1972).

12. W.C. Dautremont-Smith, P.A. Barnes and J.W. Staylt, Jr., J. Vac. Sci. Technol. B, 2 (4), 620 (1984).

13. T.A. Gessert, X. Li, T.J. Coutts, M.W. Wanlass and A.B. Franz, Proc. of the First International Conf. on Indium Phosphide and Related Materials for Adv. Electronic and Optical Devices, SPIE Proceedings Vol. 1144 (SPIE, Bellingham, WA, 1989) p. 476.

14. A. Katz, B.E. Weir, D.M. Maher, P.M. Thomas, M. Soler, W.C. Dautermont-Smith, R.F. Karlicek, Jr., J.D. Wynn and L.C. Kimerling, Appl. Phys. Lett. 55 (21), 2220 (1989).

15. F.H. Reid and W. Goldie, Eds., Gold Plating Technology, (Electrochemical Publications Ltd., Ayr, Scotland, 1974), p. 14.

16 M.J. O'Neill and M.G. Piszczor, Proc. 20th IEEE Photovoltaics Specialists Conf. (IEEE, New York, 1988), p. 1007.

Pd-Ge-Au OHMIC CONTACTS TO GaAs: RELIABILITY AND FAILURE ANALYSIS.

T. E. KAZIOR, H. HIESLMAIR, and R. C. BROOKS*
Raytheon Company, Research Division, 131 Spring St, Lexington MA., 02173
*Present address: Westinghouse Electric Corporation, Electronic Systems Group, P.O.Box 1521, Baltimore, MD 21203

ABSTRACT

We report on experiments that were performed to evaluate the temperature stability and long term reliability of non-alloyed Pd-Ge-Au ohmic contacts on N-type GaAs. Low resistance contacts ($\approx 1 \times 10^{-6} \Omega cm^2$) were obtained for samples that were sintered in a conventional furnace or flash sintered in a graphite susceptor. Elevated temperature storage (≈ 4000 hours at 280°C) showed improved contact stability when compared to Ni-AuGe-Ni-Au control samples. Gateless MESFETs subjected to bias temperature stress measurements ($I_{ds} \approx 300$-350mA/mm, 2000-4000 hours at 200°C) showed no significant change in device current. This result is in contrast to devices with Ni-AuGe-Ni-Au contacts which exhibited a 6-27% decrease in current under the same test conditions. Failure analysis reveals significant electromigration and Au diffusion in the drain fingers of devices with Ni-AuGe-Ni-Au contacts. In contrast, devices with Pd-Ge-Au contacts show no electromigration or Au diffusion in the GaAs.

INTRODUCTION

At present, alloyed Ni-AuGe metallurgy is routinely used to form low resistance ohmic contacts to N-type GaAs. However, chemical analysis (for example, Auger and Transmission Electron Microscopy (TEM)) has revealed a complex and inhomogeneous grain structure consisting of numerous intermediate compounds. This structure is not thermodynamically stable [1] and, as a result, changes in compound structure and continuing interdiffusion under normal device operation will lead to device failure (e.g., increases in contact and source resistance).

To address contact stability problems with the Ni-AuGe system numerous novel contact structures are under development [2-6]. These new approaches rely on solid phase reactions for contact formation. Recently much work has focused on Pd-based contacts. In particular, low resistance ($\approx 10^{-6} \Omega cm^2$) sintered Pd-Ge-based contacts to N+ ($\approx 10^{18}$ carriers/cm^3) GaAs have been demonstrated using two different methods [2,3]. These metal systems have the beneficial feature of being readily inserted into existing FET fabrication sequences. While these approaches show promise, very limited reliability or temperature stability data has been presented [2,3,7,8].

In a previous work, we presented data on process optimization and reliability evaluation of Pd-Ge-Au ohmic contacts to N-type GaAs [9]. The objective of this work has been to further evaluate the stability of the sintered Pd-Ge-Au contact metallurgy and begin to determine the failure mechanisms of these contacts. To accelerate chemical or physical reactions that would lead to changes in the contact structure and electrical resistance, samples were subjected to elevated temperature storage as well as conventional bias-temperature stress reliability measurements. The results were compared to similar measurements performed on samples with Ni-AuGe-Ni-Au contacts. Failure analysis was performed on selected samples.

EXPERIMENTAL

Ohmic metal structures were prepared on both N ($\approx 2 \times 10^{17}$ carriers/cm^3) and N+ ($\approx 10^{18}$ carriers/cm^3) ion implanted layers formed on semi-insulating GaAs substrates. All metal structures were prepared by standard e-beam evaporation and photoresist lift-off techniques. The base pressure just prior to evaporation was $\approx 5 \times 10^{-7}$ torr. Prior to metallization, all samples received an in-situ Argon ion beam sputter clean to remove any photoresist, hydrocarbon residues, and surface oxide. Films of 30nm Pd, 40nm Ge, and 400nm of Au were then sequentially deposited. One set of samples was furnace sintered for 30s at various temperatures. Another set of samples was flash sintered at various temperatures in an A. G. Associates Heatpulse 2101. The flash sintering was performed either by laying the samples on a 3" GaAs wafer or in a SiC coated graphite susceptor using thermocouple control. Control samples consisting of Ni(5nm)-AuGe(90nm)-Ni(15nm)-Au(400nm) flash alloyed at 450°C for 10s were also prepared.

Mat. Res. Soc. Symp. Proc. Vol. 181. ©1990 Materials Research Society

Contact resistance was measured using the transmission line method [10]. Van der Pauw patterns were used to determine the ohmic metal sheet resistance and 100µm wide test FETs were used to monitor device saturated current and low field resistance. Each sample contained a minimum of 100 test patterns to provide good statistics.

To accelerate thermally driven chemical or physical reactions that would lead to contact degradation, samples were stored on a 280°C hot plate in room air and remeasured every several hundred hours. Companion samples were fully processed, including Si_3N_4 passivation, and then sawn into chips. These devices were subjected to conventional DC bias-temperature stress tests where parts of the population were biased at different currents. The devices biased under constant drain voltage were stored at a 200°C base plate temperature in a N_2 ambient for extended periods of time and the drain current was monitored at two hour intervals. In order to simplify the interpretation of the reliability data, the electrical measurements were performed on gateless FETs, eliminating the possibility of device drift due to Schottky barrier degradation. To investigate the role of electromigration in the degradation of the ohmic contacts, some of the devices received a Ti-Pt-Au thickening layer to reduce the current density in the drain fingers.

Upon completion of the elevated temperature storage and bias-temperature stress measurements selected samples were subjected to failure analysis - optical and scanning electron microscope (SEM) inspection, energy dispersive x-ray analysis (EDX, in the SEM using a 5keV electron beam to minimize penetration depth), and Auger Electron Spectroscopy.

RESULTS and DISCUSSION

Figure 1 shows results of the experiment to determine optimal furnace and flash sintering temperatures for formation of low resistance contacts to N^+ GaAs. For furnace sintering the data indicates that there exists a broad range of temperatures for low resistance ohmic contact formation. Median values of specific contact resistance $\leq 1\mu\Omega cm^2$ have been achieved for sinter temperatures as low as 380°C and as high as 480°C. Sintering below 380°C resulted in poor ohmic contacts. Above 480°C the metal morphology and edge definition became poor. For flash sintering, low resistance contacts could only consistently be achieved using the graphite susceptor. The data presented in Figure 1 reveals a slightly smaller process window for the flash sintered contacts, flash sinter temperatures above 430°C resulted in poor metal morphology and edge definition. Samples flash sintered either free standing (wafer sitting on the quartz pins of the anneal tray), or laying either face up or in proximity on a 3" GaAs wafer yielded either poor or non-reproducible results. For comparison, the flash alloy temperature dependence for Ni-AuGe-Ni-Au contacts to N^+ GaAs is also presented. While low resistance contacts can be achieved over a broad range of temperatures (430-550°C, data for alloy temperatures above 460°C is not shown), the metal morphology and edge definition for Ni-AuGe-Ni-Au samples alloyed above 460°C begins to degrade. This is consistent with results presented in ref. 1. Therefore, we have found that the process window for minimizing contact resistance and maintaining good metal morphology and edge definition is very broad for furnace (≈100°C) or flash (≈40°C) sintered Pd-Ge-Au contacts.

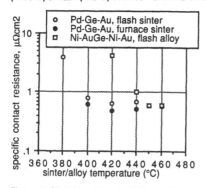

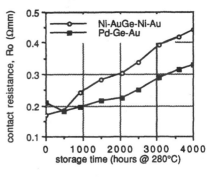

Figure 1: Plot of median values of specific contact resistance versus sinter temperature for furnace and flash sintered Pd-Ge-Au contacts and versus alloy temperature for Ni-AuGe-Ni-Au contacts.

Figure 2: Plot of median value of contact resistance versus storage time at 280°C for furnace sintered Pd-Ge-Au contacts and alloyed Ni-AuGe-Ni-Au contacts.

The data for the median value of contact resistance as a function of storage time are summarized in Figure 2. While the Pd-Ge-Au samples initially had a slightly higher value of R_0, no significant change in R_0 was observed up to 2000 hours. After 2000 hours R_0 began to increase slowly, increasing by ≈50% after 4000 hours. In contrast R_0 for he Ni-AuGe-Ni-Au control samples began to increase almost immediately and continued to increase with storage time, increasing by ≈200% after 4000 hours. The data in Figure 2 are for contacts formed on N layers only. Similar results were observed for contacts to N+ layers.

During the high temperature storage, ungated saturated current for the Pd-Ge-Au samples remained virtually unchanged for 4000 hours. For Ni-AuGe-Ni-Au samples, the ungated saturated current decreased at a fairly constant rate, reaching ≈84% of the initial current in 4000 hours. Therefore, based on the observed changes in the contact resistance and ungated saturated current the Pd-Ge-Au contacts appear to be more stable at elevated temperature than the Ni-AuGe-Ni-Au contacts.

Figures 3 shows typical drain current versus time for one set samples undergoing bias-temperature stress measurements. For these samples, I_{ds} ≈350mA/mm (200μm total gate periphery) and V_{ds} =1.25V. The Ti-Pt-Au thickening layer was not used. The measurements were performed initially at a base plate temperature of 150°C for 500 hours and then stepped to 200°C for an additional 3500 hours. (In Figure 3 are the data for 200°C only.) For the samples with Ni-AuGe-Ni-Au contacts, the current remained constant until ≈1500 hours and then decreased by >20% after 4000 hours. In contrast, current in the Pd-Ge-Au samples decreased only slightly (<1%) in 4000 hours.

Figures 4 shows drain current versus time data for typical samples with the Ti-Pt-Au thickening layer. For these samples, I_{ds} ≈300mA/mm (600μm total gate periphery), V_{ds}=2.5V. The results were similar to the results presented in Figure 3. The devices with Ni-AuGe-Ni-Au contacts and the thickening metal showed large decreases in drain current (≈6%). Companion samples without the thickening metal layer showed even larger decreases in drain current (8-10%). The devices with the Pd-Ge-Au contacts with or without the thickening metal showed current changes of <1%. Therefore, based on the reliability data, the Pd-Ge-Au contacts were more stable under bias-temperature stress tests. In addition, the samples with Ni-AuGe-Ni-Au contacts showed significant decreases in current independent of the current density in the drain fingers.

Optical and SEM micrographs revealed significant electromigration damage on the drain fingers of the Ni-AuGe-Ni-Au samples (see Figure 5 - left). Note that the samples in Figure 5 did not have the Ti-Pt-Au thickening layer and, therefore, were more susceptible to electromigration along the drain fingers. For the Ni-AuGe-Ni-Au samples subjected to 4000 hours of test (samples of Figure 3), the Au overlayer and a Ni-containing layer on the drain fingers are almost entirely eroded away exposing Ga and As in a large portion of the finger. Also Au nodules are growing on the source fingers and drain feed. For the Ni-AuGe-Ni-Au samples with thickening metal and subjected to 2000 hours of test (samples of Figure 4) significant electromigration damage was observed on the edges of the drain fingers not covered by thickening metal. For the Ni-AuGe-Ni-Au samples without thickening

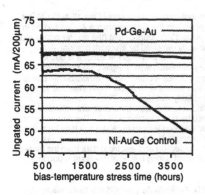

Figure 3: Plot of ungated drain current versus time for typical samples undergoing bias-temperature measurements.

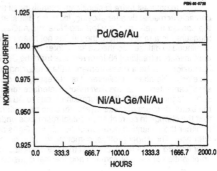

Figure 4: Plot of normalized ungated drain current versus time for typical samples with the Ti-Pt-Au thickening layer.

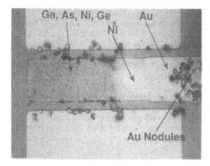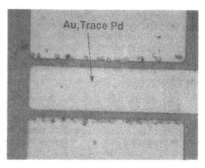

Figure 5: Optical micrographs showing electromigration of drain fingers of Ni-AuGe-Ni-Au contacts (left) and no degradation of drain fingers of Pd-Ge-Au contacts (right) after 4000 hours of reliability testing. EDAX was used to identify the primary components in each region.

metal and subjected to 2000 hours of test, approximately 50% of the Au overlayer on the drain fingers is entirely eroded away exposing underlying Ni and Ga. This is in stark contrast to the drain fingers of the Pd-Ge-Au samples which showed no signs of electromigration after 2000 hours of test and only slight signs of degradation (the beginning of Au nodule formation) after 4000 hours (see Figure 5 - right).

Auger analysis was performed on the drain fingers of selected devices. Figures 6 & 7 presents profiles of samples before (top) and after (bottom) bias-temperature reliability measurements. Immediately after sintering the Pd-Ge-Au contacts show no diffusion of Au, Ga or As across the PdGe contact layer. After 2000 hours of test the data suggests that the PdGe and Au layers are interdiffusing, whereas no significant change was observed at the PdGe-GaAs interface. Similar PdGe-Au intermixing was observed on samples subjected to 280°C storage only (no current). Since during aging there exists the possibility of either grain growth or the beginning of electromigration, the apparent interdiffusion could be due to nonuniform sputtering through a rough Au surface during Auger depth profiling. However, one would also expect to see significant broadening or intermixing at the PdGe-GaAs interface. Therefore, we believe that the intermixing is not a measurement artifact, but rather an indication of interdiffusion or grain growth. The exact structure of these layers needs to be confirmed by cross sectional TEM. We believe that under continued elevated temperature operation or storage the Au will eventually diffuse through the PdGe contact layer along grain boundaries and lead to contact degradation due to Au diffusing into the GaAs. Whether or not electromigration contributes to the degradation of these contacts is yet to be determined, although the data suggests that the ultimate failure mechanism may be thermal diffusion. Further reliability studies and chemical analysis are needed for confirmation.

In contrast there is significant intermixing of the individual components of the Ni-AuGe-Ni-Au contacts immediately after alloying, including diffusion of Au into the GaAs (from the AuGe) and As in the contact metals (possibly forming NiAs and AuGa). After 2000 hours of test the sputter profile was measured in a region where the Au has not been completely removed by electromigration, Au continued to diffuse into the GaAs. (Note: The slow rise in the Au profile at the surface is due to contamination (residual N) from removal from the Si_3N_4 passivation layer.) Again, the continued Au indiffusion may be a measurement artifact resulting from surface roughening or nonuniformity due to Au electromigration. However, it is interesting to note that while the Ni-AuGe/GaAs interface shows significant intermixing, the Au/Ni-AuGe interface remains fairly well defined. This observation supports the claim that the intermixing is due to Au indiffusion, and not a measurement artifact. Again, the exact structure of these layers needs to be confirmed by cross sectional TEM. The data suggest that the change in current and contact resistance under elevated temperature storage may be related to either the continued diffusion of Au into the GaAs or changes in the structure at the metal GaAs interface. The Auger profiles, combined with the optical inspection and electrical data from the bias-temperature stress tests, suggest that the decrease in current during the reliability tests is due to a combination of electromigration and thermal diffusion.

At this time it is not clearly understood why the Ni-AuGe-Ni-Au contacts are more susceptible to electromigration than the Pd-Ge-Au contacts. One possible explanation is the difference in ohmic metal sheet resistance. The sheet resistance of the Pd-Ge-Au contacts was approximately three times lower than the sheet resistance of the Ni-AuGe-Ni-Au contacts even though the Au overlayer

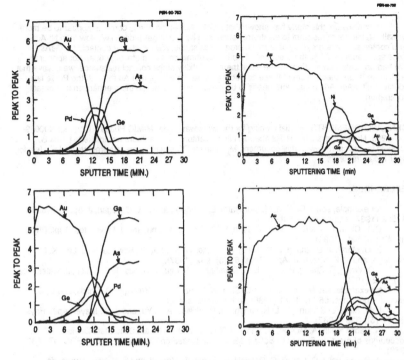

Figure 6: Auger profile of Pd-Ge-Au drain finger before (top) and after 2000 hours of reliability testing (bottom).

Figure 7: Auger profile of Ni-AuGe-Ni-Au drain finger before (top) and after 2000 hours of reliability testing (bottom).

was 400nm in both cases. This may result in a higher probability of Au electromigration for the same current density for the Ni-AuGe-Ni-Au contacts. Another possible explanation is that the Au overlayer in the Ni-AuGe-Ni-Au contacts may not be as strongly bound to the Ni-containing layer and, as a result, the Au is less resistant to electromigration. In comparison, under thermal stress the Au layer in the Pd-Ge-Au contacts appears to be diffusing into the PdGe contact layer (along grain boundaries) and, as a result, the Au layer may become more resistant to electromigration. Further reliability studies and chemical analysis are needed for confirmation.

SUMMARY

The results of this work can be summarized as follows:
1) Low resistance Pd-Ge-Au contacts can be achieved for a broad range of furnace (≈370-470°C) and flash sinter (≈390-430°C in graphite susceptor) temperatures.
2) In samples subjected to elevated temperature storage (4000 hours at 280°C) the contact resistance for the Pd-Ge-Au contacts increases at a much slower rate than Ni-AuGe-Ni-Au samples
3) The current in gateless FETs with the Pd-Ge-Au contacts did not decrease under the bias-temperature stress measurements conditions used in this work. In contrast, the current in devices with Ni-AuGe-Ni-Au contacts decreased more than 20% in 4000 hours.
4) Failure analysis has revealed significant electromigration along the drain fingers of devices with Ni-AuGe-Ni-Au contacts. In contrast, the devices with Pd-Ge-Au contacts show no appreciable electromigration.
5) Auger analysis has revealed significant Au diffusion into the GaAs for the Ni-AuGe-Ni-Au contacts immediately after alloying and after reliability test. In contrast, there is no apparent Au diffusion through the PdGe layer after sintering, and Au is just beginning to diffuse through the PdGe layer and into the GaAs after reliability test.

318

In conclusion, this study has shown that Pd-Ge-Au N-type ohmic contacts appear to be more thermally stable, more resistant to electromigration, and have a larger process window than Ni-AuGe-Ni-Au contacts. Therefore, they offer much promise as stable, low resistance contacts for GaAs device applications. While the presence of Au in this ohmic contact metallurgy does not appear to have had any detrimental effect on the initial thermal stability of the contact resistance, we believe that the ultimate failure mechanism of these contacts may be the diffusion of Au through the PdGe layer and into the GaAs. Additional work needs to be performed to establish the mechanism for contact degradation.

ACKNIOWLEDGEMENTS: Partial funding for this work was under MIMIC Phase 1 Contract N00019-88-C-0218, Navy Contract to the Raytheon-TI Joint Venture. The authors would like to thank S. Shanfield for critical review of this document, M. Bush for DC electrical measurements, and S. Hein for Auger measurements.

REFERENCES:

1: for example, see Y-C. Shih, M. Murakami, E. L. Wilkie, and A. C Callegari, J. Appl. Phys. **62** (2), 582 (1987) and references therein.
2: C. L. Chen, L. J. Mahoney, M. C. Finn, R. C. Brooks, A. Chu, and J. Mavroides, Appl. Phys. Lett. **48 (8)**, 535 (1986).
3: E. D. Marshall, B. Zhang, L. C. Wang, P. F. Jiao, W. X. Chen, T. Sawada, S. S. Lau, K. L. Kavanagh, and T. F. Kuech, J. Appl. Phys. **62 (3)**, 942 (1987).
4: L.C. Wang, B. Zhang, F. Fang, E. D. Marshall, S. S. Lau, T. Sands, T. F. Kuech, J. Mater. Res. **3 (5)**, 922 (1988).
5: for example, see M. Murakami, Y-C. Shih, W. H. Price, E. L. Wilkie, K. D. Childs, and C. C. Parks, J. Appl Phys. **65 (4)**, 1974 (1988) and references therein.
6: L. H. Allen, L. S. Hung, K. L. Kavanagh, J. R. Phillips, A. J. Yu, and J. W. Mayer, Appl. Phys. Lett. **51 (5)**, 326 (1987).
7: A. Paccagnella, A. Migliori, M. Vanzi, B. Zhang , and S. S. Lau, Solid State Devices, Proccedings of the 17th European Solid State Device Research Conference, ESSDERC '87, **839** (1988).
8: A. Paccagnella, C. Canali, G. Donzelli, E. Zanoni, R. Zanetti, and S. S. Lau, Journal de Physique, **49 (9)**, C4-441 (1988).
9: T. E. Kazior and R. C. Brooks, presented at 175th Meeting of the Electrochemical Society, Los Angelas, CA, May, 7-12 1989; submitted to J. Electrochem. Soc.
10: H. H. Berger, Solid State Electron. **15**, 145 (1972).

THERMAL BEHAVIOUR OF Au/Pd/GaAs CONTACTS

B. Pécz, R. Veresegyházy, I. Mojzes, G. Radnóczi, A. Sulyok
Research Institute for Technical Physics of the Hungarian
Academy of Sciences, H-1325 Budapest, P.O.Box 76, Hungary

V. Malina
Institute of Radio Engineering and Electronics of the
Czechoslovak Academy of Sciences, Lumumbova 1, 18251 Prague 8,
Czehoslovakia

ABSTRACT

Au(85nm)/Pd(55nm)/GaAs(100) samples were heat treated in
the 325-425°C temperature range. The annealed samples have been
investigated using Rutherford Backscattering Spectrometry,
Auger Electron Spectroscopy and Transmission Electron
Microscopy. The gold layer remained largely unreacted up to
300°C. Significant Pd diffusion into GaAs consuming a 50-60 nm
thick layer of GaAs is evident in the case of sample annealed
at 325°C and a slight Au diffusion is also noticeable. In the
sample annealed at 350°C the spreading of palladium was very
quick. A strong reaction took place between the GaAs and the
metallization in the case of sample heat treated at 375°C. At
this temperature we have identified the PdGa phase using
electron diffraction.

INTRODUCTION

The reliability and thermal stability of contacts to
semiconductor devices are very important. Gold based alloys are
the most commonly used materials for contacts to GaAs.
Preferential reactions between the metallization and the
gallium take place very often in these systems. These reactions
result in the loss of arsenic due to its high volatility. The
interfaces of these contacts are laterally non-uniform.

The Au/GaAs system is a good example for the laterally
inhomogenous interface [1, 2]. In this system rectangular based
pyramidal pits were grown into the GaAs during the heat
treatment. The cross section of sample heat treated at 400°C
for 10 min. is shown in the Fig. 1. These pits show in the
(110) section triangular form (Fig.1.), while in the
perpendicular (1$\bar{1}$0) section they show elongated form. Gold
reacts with GaAs forming first Au-Ga solid solution and at
higher temperatures the appearance of Au_7Ga_2 and Au_2Ga phase
take place. The excess arsenic evaporates from the sample [3].

Recently the Pd/GaAs system was also investigated in
several laboratories [4,5]. In this system the formation of
ternary compounds were reported.

Lamouche et. al. reported [5] that they obtained low
resistance ohmic contact applying a Au/Pd layered metallization
to GaAs.

In the course of the present work we have investigated the
Au/Pd/GaAs system using RBS, AES and a preliminary TEM
investigation was carried out as well.

Palladium and gold were evaporated onto GaAs substrate by
resistance heating to produce Au(85nm)/Pd(55nm)/GaAs(100)
samples. During the evaporation the vacuum was about 10^{-4} Pa
and the slices were held at 150°C temperature.

Mat. Res. Soc. Symp. Proc. Vol. 181. ©1990 Materials Research Society

Each sample was annealed in flowing forming gas
(5%H₂+95%N₂) for 10 minutes.

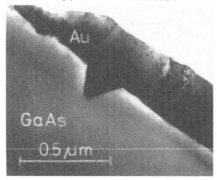

Fig.1. The cross section of Au/GaAs sample heat treated at 400°C for 10 min.

We prepared samples for TEM investigation [6]. Two small chips from each sample were bonded face to face, embedded into a small aluminium ring, which has a diameter of 3 mm. The samples were thinned by mechanical grinding to approximately 50 μm and finally they were thinned by ion milling. We have used Ar⁺ ions accelerated by a potential of 9 kV. The samples were milled at a beam angle of 3-5° to the specimen surface.

The thinned samples were examined using a JEOL 100U transmission electron microscope.

The RBS analysis was performed by a 3 MeV ⁴He⁺ ion beam. During the measurements the vacuum was about 2∗10⁻⁴ Pa.

Auger electron spectra were taken by conventional spectrometer detecting the Ga, As, Pd, Au and O Auger peaks. To produce depth profiles of the components continuous ion sputtering was applied with 2 keV Ne⁺ ions at 30° angle of incidence. Concentration of components was calculated from its peak to peak intensity considering the electron backscattering factors as correction. Preferential sputtering effect was neglected at the evaluation.

RESULTS

The RBS spectra of the as-deposited and heat treated samples can be seen in Fig.2. Our RBS investigations show, that the gold layer, which is the top layer of the sample, remained largely unreacted up to 300°C.

Significant Pd diffusion into GaAs is evident already in the case of sample annealed at 325°C and a slight Au diffusion is also noticeable. A strong reaction took place between the GaAs and the metallization in the case of sample heat treated at 375°C.

Even in the case of as-deposited sample some reaction products were observed on the cross section (Fig.3.a.). We observed, that the palladium reacted with GaAs forming intermetallic grains, which are about 30 nm thick. It is supposed that these reaction products are the phase I (Pd₅(GaAs)₂ as it was identified by Sands et. al. [4]). A very thin layer of native oxide was also observable (white stripe on the cross section), what dissappeared during the annealing.

In the case of sample annealed at 325°C palladium penetrated into the GaAs consuming a 50-60 nm depth layer of

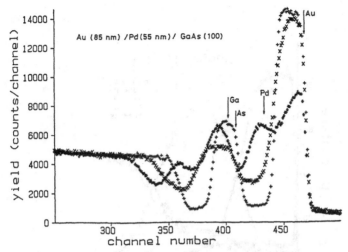

Fig.2. 3 MeV ⁴He⁺ RBS spectra of the as-deposited and heat treated samples: + as-deposited, x sample heat treated at 325°C, * sample heat treated at 375°C.

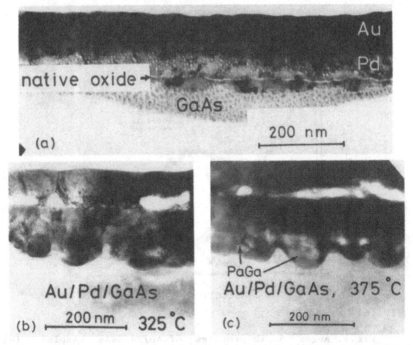

Fig.3. Cross sectional TEM images of Au/Pd/GaAs samples: (a) as deposited, (b) annealed at 325°C, (c) annealed at 375°C

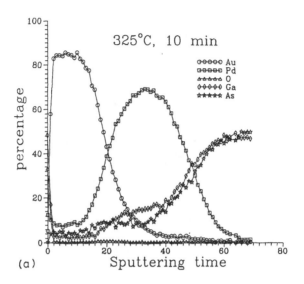

(a)

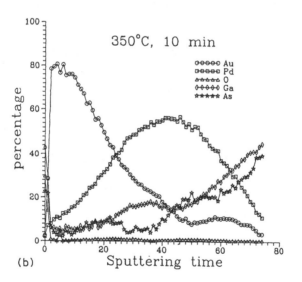

(b)

Fig.4. Auger depth profiles of Au/Pd/GaAs samples: (a) annealed at 325°C, (b) annealed at 350°C

GaAs (Fig.3.b). The deep penetration is supposed to consist of the phase II (Pd₄GaAs). On the Auger depth profile of the sample (Fig.4.a) we can see that the palladium diffused both into the GaAs and gold. The diffusion of gold was relatively lower. There is no arsenic on the surface of the sample. We have to mention that the detection limit of our AES equipment for As and Ga is about 4-5%.

In the case of sample heat treated at 350°C the gold content decreases monotonously with the depth (Fig.4.b). The spreading of palladium was very quick. Even from this system the loss of a small amount of arsenic is evident comparing the Ga and As content of the metallization.

The grains grown into the GaAs can be seen on the cross section of Au/Pd/GaAs sample heat treated at 375°C (Fig.3.c). Applying electron diffraction method and dark field technique we identified the PdGa phase in this part of the sample.

Sands et al. assumed that at about 400°C a certain amount of arsenic evaporates from the sample due to the

Pd₄GaAs + 3GaAs = 4PdGa + 4As↑ reaction.

Applying the EGA (Evolved Gas Analysis) method we have shown that at about this temperature arsenic really evaporates from the sample. Evolved Gas Analysis is a mass spectrometric method. The sample is situated in high vacuum, heated applying a constant heating rate and the volatile component loss is monitored using a quadrupole mass spectrometer. In this case

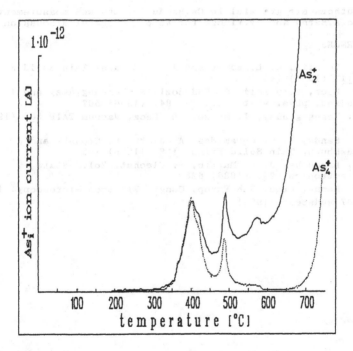

Fig.5. Volatile component loss vs. temperature curve of the Au(85 nm)/Pd(55 nm)/GaAs(100) sample. The applied heating rate was 30°C/min.

the heating rate was 30°C/min. A large volatile peak can be seen at about 400°C on the volatile component loss vs temperature curve (Fig.5). The cause of the second arsenic peak may be that finally the gold reacts with gallium and the excess arsenic evaporates. The temperature of this evaporation peak is 50-60 °C lower in the case of Au/GaAs system, but here the diffusion of gold and gallium species through the palladium layer needs higher temperature.

Lamouche et. al. found that the specific contact resistance of this system tends toward ohmic above 360°C. They reached the optimum of specific contact resistance at about 500°C. The correlation between their results and the temperatures of the arsenic evaporation peaks is obvious.

SUMMARY

The reaction in Au/Pd/GaAs system takes place in three consecutive steps. Our results are consistent with the formation of ternary Pd-Ga-As compounds. The further reaction of Pd_4GaAs with GaAs results in considerable loss of arsenic. PdGa phase was identified by electron diffraction. Finally the reaction of gold with GaAs causes further arsenic loss and the appearance of Au-Ga alloys with low melting points.

ACKNOWLEDGMENTS

The authors are grateful to Cs.Hajdu for the RBS measurements and to A.Barna and Gy.Vincze for help and useful discussion.

REFERENCES

1. T. Yoshiie, C. L. Bauer and A. G. Milnes, Thin Solid Films, 111, (1984) 149

2. B. Pécz, E. Jároli, G. Radnóczi, R. Veresegyházy and I. Mojzes, phys. stat. sol. (a) 94, (1986) 507

3. R. Veresegyházy, I. Mojzes, B. Pécz, Vacuum TAIP 36, (1986) 547

4. T. Sands, W. G. Keramidas, A. J. Yu, R. Gronsky and J. Washburn, Thin Solid Films, 136, (1986) 105

5. D. Lamouche, J. R. Martin, P. Clechet, Solid-State Electronics, 29, (1986) 625

6. A. Barna, Proc. 8th Europ. Congr. Electron Microscopy, 1, 107 Budapest (1984)

THERMOELECTRIC POWER IN QUANTUM CONFINED OPTOELECTRONIC MATERIALS UNDER CLASSICALLY LARGE MAGNETIC FIELD

KAMAKHYA P. GHATAK*, B. DE**, M. MONDAL* AND S. N. BISWAS*
*Department of Electronics and Telecommunication Engineering, Faculty of Engineering and Technology, University of Jadavpur, Calcutta-700032, INDIA.
**13 Little Brook Road, CT- 06820 , DARIEN, U.S.A.
***Department of Physics, Y. S. Palpara College, P. O. Box 721458, Midnapore, West Bengal INDIA.
****Department of Electronics and Telecommunication Engineering, B. E. College, Shibpur, Howrah-711103, West Bengal, INDIA.

ABSTRACT

We shall study the thermoelectric power under classically large magnetic field (TPM) in optoelectronic materials of quantum wells (QWs), quantum well wires (QWW's), quantum dots (QDs) and compare the same with the bulk specimens of optoelectronic materials by formulating the respective electron dispersion law. The TPM increases with decreasing electron concentration in an oscillatory manner in all the cases, taking n-$Hg_{1-x}Cd_xTe$ as an example. The TPM in QD is greatest and the least for quantum wells respectively. The theoretical results are in agreement with the experimental observations as reported elsewhere.

With the advent of MBE and MOCVD techniques, low-dimensional microstructures having quantum confinement in 1, 2 and 3 dimensions such as quantum wells (QWs), quantum well wires (QWW's) and quantum dots (QDs) find extensive applications in QW lasers /1/, FETs, high-speed digital networks /2/, optical modulators and other devices /3/. Microstructures based on various optoelectronic materials are currently being studied because of the enhancement of carrier mobility. It appears from the literature that the thermoelectric power under classically large magnetic field (TPM) in quantum confined optoelectronic microstructures has yet to be investigated. The TPM gives information about the band structure, the density-of-states function and can be related to the Einstein relation /4,5/. In what follows, we shall investigate the TPM in quantum confined optoelectronic materials taking n-$Hg_{1-x}Cd_xTe$ as an example. We wish to note that the compound $Hg_{1-x}Cd_xTe$ is the classic very narrow-gap optoelectronic materials and its band gap can be varical to cover the entire spectral range from 0.8 to 30 μm by adjusting the alloy composition /6/. Its use as an infrared detector material has spurred a $Hg_{1-x}Cd_xTe$ technology and it is ideally suited for quantum confined physics because the relevant physical parameters are within easy experimental reach /7/.

The energy spectrum of the conduction electrons in bulk specimens of optoelectronic materials can be expressed /8/ as

$$\hbar^2 k^2 / 2m_0^* = \gamma(E) \qquad (1)$$

where the notations are defined in /8/. The modified dispersion relation in the presence of a classically large magnetic field in quantum wells of optoelectronic materials assume the form

$$
(\hbar k_y + eBd_x) \left[2m_0^* \gamma(E) - \hbar^2 k_z^2 - (\hbar k_y + eBd_x)^2 \right]^{1/2}
$$

$$
- (\hbar k_y - eBd_x) \left[2m_0^* \gamma(E) - \hbar^2 k_z^2 - (\hbar k_y - eBd_x)^2 \right]^{1/2}
$$

$$
+ \left\{ 2m_0^* \gamma(E) - \hbar^2 k_z^2 \right\} \left[\sin^{-1} \left[(\hbar k_y + eBd_x) \left\{ 2m_0^* \gamma(E) - \hbar^2 k_z^2 \right\}^{-1/2} \right] - \sin^{-1} \left[(\hbar k_y - eBd_x) \left\{ 2m_0^* \gamma(E) - \hbar^2 k_z^2 \right\}^{-1/2} \right] \right]
$$

$$
= h n_x eB \qquad (2)
$$

where $2d_x$ is the width of the quantum well along x-direction, n_x is the size quantum number along x-direction and B is the large magnetic field. Putting $k_y = \pi n_y / 2d_y$, where n_y and $2d_y$ are the quantum number and the width along y direction respectively, in (2) we get the modified electron-dispersion law for quantum well wires for the present case. Putting $k_y = n_y \pi / 2d_y$ and $k_z = n_z \pi / 2d_z$, where n_z and $2d_z$ are the quantum number and the width along z-direction respectively, in (3) we get the corresponding law for quantum dots in the present case. It is worth remarking that the quantization of the energy in transverse plane of the direction of application of B, as valid for three-dimensional electron gases is not at all valid for quantum wells, quantum wires and quantum dots. In the absence of magnetic field we get

$$2m_0^* \gamma(E) = (\hbar n_x \pi / 2d_x)^2 + p_y^2 + p_z^2 \qquad (3)$$

$$2m_0^* \gamma(E) = (\hbar n_x \pi / 2d_x)^2 + (\hbar n_y \pi / 2d_y)^2 + p_z^2 \qquad (4)$$

$$2m_0^* \gamma(E) = (\hbar n_x \pi / 2d_x)^2 + (\hbar n_y \pi / 2d_y)^2 + (\hbar n_z \pi / 2d_z)^2 \qquad (5)$$

The TPM for the present case is given by [4-5]

$$G = S/en_0 \qquad (6)$$

where S and n_0 are the entropy and electron statistic respectively. Using the appropriate equations and the numerical integrations we can find the TPM in all the cases.

Using the appropriate appropriate equations together with the parameters /8/

$$E_g(x) = \lfloor -0.303 + 1.73 + 5.6 \times 10^{-4} (1-2x) \ T + 0.25 \ x^4 \rfloor eV,$$

$$m_o^*(x) = \lfloor 3\hbar^2 E_g(x)/4P^2(x) \rfloor, \quad P^2(x) = \lfloor (\hbar^2/2m_o) \ (18+3x) \rfloor$$

$\Delta = 0.9 eV$, B = 1 Tesla and T = 4.2 K we have plotted the normalized TPM in all the cited cases where the circular plots exhibit the experimental data as given elsewhere /9/.

It appears from Figs. 1, 2 and 3 the TPM increases with increasing electron concentration in an oscillatory manner for QW, QWW and QD of n. $Hg_{1-x} Cd_x Te$ where the simplified results in accordance with two band Kane model (i.e. $E(1+E \ E_g^{-1}) = \hbar^2 k^2/2m_o^*$) and that of parabolic energy bands (i.e. $E = \hbar^2 k^2/2m_o^*$) have also been shown for the purpose of numerical computations. The TPM is greatest in QDs and least in QWs in all the cases respectively. With varying magnetic field a change is reflected in the TPM through the redistribution of the electrons among the quantized levels. In quantum well the two directions are free and in QDs three directions are quantized. The three dimensional quantization leads to discrete energy levels, somewhat like atomic energy levels, which produces sharp changes. Consequently, the crossing of the Fermi energy by the size quantized levels in QDs would have much greater impact on the redistribution of the electrons among the allowed levels as compared to that for QWs and QWWs respectively. It appears from Fig. 4 that the TPM increases in an oscillatory was with decreasing film thickness in all the cases. Though the TPM varies non-linearly in all the limiting cases in all the figures, the rates of variations are totally band structure dependent.

We wish to note that the three-band Kane model is valid for III-V compound semiconductors but must be used as such for studying the electronic properties of n-InAs where the spin-orbit splitting parameter (Δ) is of the order of bandgap (E_g). However for many important semiconductor $\Delta \gg E_g$ (e.g. n-InSb) or $\Delta \ll E_g$ (e.g. InP). Under these limiting conditions the full expression of $\gamma(E)$ as given in (1) reduces to the form $E(1 + E. Eg^{-1})$ as stated above. Furthermore for $E_g \to \infty$, as for parabolic energy bands $\gamma(E)$ gets simplified as E. Thus the dispersion relation (1) covers various semiconductors having different band structures. We must note that the study of transport phenomena and the formulation of the electronic properties of semiconductors are based on the dispersion relations in such materials. Besides, by knowing TPM we can also determine the Einstein relation for the diffusivity-mobility ratio (a very important device

$\llcorner n_0 = 10^{20}$ m^{-3} for QD, $n_0 = 10^{14}$ m^{-2} for QW and $n_0 = 10^{11}$ m^{-1} for QWW respectively$\lrcorner$.

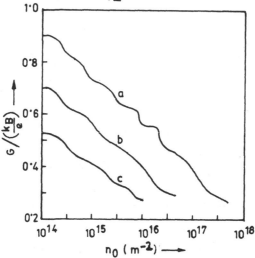

Fig. 1. Plot of normalized TPM in quantum well of n-$Hg_{1-x}Cd_x$ Te versus electron concentration in accordance with (a) three band Kane model, (b) Two-band Kane model, (c) parabolic energy bands ($2d_x$ = 40 nm).

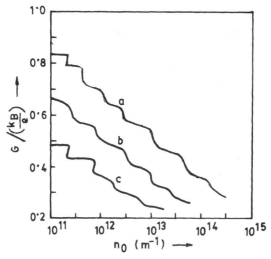

Fig. 2. Plot of normalized TPM in quantum wires of n-$Hg_{1-x}Cd_x$Te versus electron concentration in accordance with (a) three band Kane model (b) Two-band Kane model, (c) parabolic energy bands ($2d_x = 2d_y$ = 40 nm).

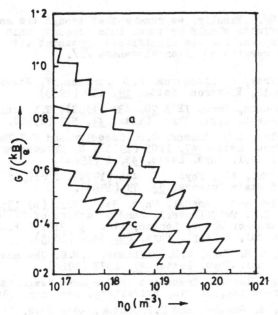

Fig. 3. Plot of normalized TPM in quantum dots of n-HG$_{1-x}$Cd$_x$
Te versus electron concentration in accordance with
(a) three band Kane model, (b) Two-band Kane model,
(c) parabolic energy bands (2d$_x$=2d$_y$=2d$_z$=40nm).

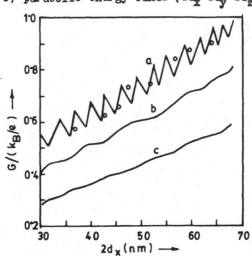

Fig. 4. Plot of normalized TPM versus film thickness in
Hg$_{1-x}$Cd$_x$Te in accordance with three band Kane model
in QD, QWW and QW as shown by a, b and c respectively. The circular plots exhibits the experimental
results

parameter ΔQ. Finally, we remark that though the many body and others effects should be taken into account, this simplified analysis exhibits the significant agreement with the experimental results as given elsewhere $\angle Q$.

1. D.R. Scifres, C. Lindstrom, R.D. Burnham, W. Streifer and T.L. Paoli, Electron. Letts. 19, 170 (1983)

2. P.M. Salomon, Proc. IEEE 70, 439 (1982); T.L. Schlesinger and T. Kuech, Appl. Phys. Letts. 49, 519 (1986).

3. H. Hieblum, D.C. Thomas, C.M. Knoedler and M.I. Nathan, Appl. Phys. Letts. 47, 1105 (1985); S. Torucha and H. Okamoto, Appl. Phys. Letts. 45, 1 (1984).

4. W. Zawadzki, Adv. Phys. 23, 435 (1974); Springer series in Solid State Sciences 53, 79 (1984).

5. K.P. Ghatak and M. Mondal, Phys. Stat. Sol. (b) 135, 819 (1986); I.M. Tsidilkovski, Band Structure of Semiconductors (Pergamon Press, London, 1982) p. 313; K.P. Ghatak and M. Mondal, J. Appl. Phys. 65, 3480 (1989).

6. P.Y. Lu, C.H. Wang, C.M. Williams, S.N.G. Chu and C.M. Stiles, Appl. Phys. Letts. 49, 1372 (1986).

7. B.R. Nag, Electron Transport in Compound Semiconductors, Springer-Verlag, Berlin, Heidelberg, New York, 1980.

8. B. Mitra, A. Ghoshal and K.P. Ghatak, Acta Phys. Slov. 39, 45 (1989).

9. B.I. Gelmont, Jour. Theor. and Exp. Phys. 98, 101 (1989).

10. K.P. Ghatak and M. Mondal, J. Thin. Solid Films, 148, 219 (1987).

PART VII

Novel and Au-Ge-Ni Ohmic Contacts to GaAs

THERMODYNAMIC STABILITY OF PtAl THIN FILMS ON GaAs

DAE-HONG KO and ROBERT SINCLAIR
Department of Materials Science and Engineering, Stanford University, Stanford, CA 94305

ABSTRACT

The thermal stability of PtAl thin films on GaAs substrates has been studied using transmission electron microscopy and Auger electron spectroscopy. The PtAl thin films were formed by sequential deposition of discrete Pt and Al layers on GaAs by e-beam evaporation followed by subsequent annealing processes. Interfacial reactions in the Al/Pt/GaAs system proceed in two stages. Upon low temperature annealing Pt and GaAs react to form PtGa and $PtAs_2$. Further high temperature annealing causes PtGa, $PtAs_2$ and Al to react together producing the desired PtAl on GaAs. We observed solid-phase epitaxial regrowth of GaAs during the second stage of reaction. The PtAl/GaAs interface is determined to be thermally stable during an 800°C/30 min. anneal, while remaining morphologically uniform on GaAs.

INTRODUCTION

There has been an increasing number of studies on the thermal stability of intermetallic compounds on GaAs in order to find stable contact materials which can be compatible with high temperature processing, such as that required for a self-aligned gate GaAs MESFET. Several investigators have suggested metal-gallides[Pt-Ga,Ni-Ga,Co-Ga,Rh-Ga][1,2,3,4] and metal-arsenides[Er-As][5] as stable materials by considering M-Ga-As phase diagrams. However, despite some successful results, predicting experimentally stable systems, the compounds proposed are not easily formed by conventional deposition techniques. Such difficulties have limited the application of metal-gallide and metal-arsenides in GaAs integration circuit processing. Metal aluminides can be considered as altenative candidate materials, as they simply replace the gallium by another group III element. They often have a large negative free energy of formation(and so a high melting point) and are well known alloys in the metallurgical field. There is a reasonable chance that they are stable with GaAs, as was demonstrated exprimentally by Sands for NiAl[6]. However, because four elements are present at the interface, quaternary phase equilibria are necessary to understand their behavior.

In this study, the formation of PtAl thin films and their thermal stability on GaAs after annealing at various temperatures were investigated using transmission electron microscopy(TEM), Auger electron spectroscopy(AES), and energy dispersive X-ray spectroscopy(EDAX). In addition, we found solid-phase epitaxial regrowth of GaAs in an Al/Pt/GaAs system, similar to the results of Sands et al. in a Si/Ni/GaAs system[7]. Rather than producing the compound directly, it was formed by a sequential reaction. This effectively shows that PtAl is thermodynamically stable with GaAs. The pertinent Pt-Al-Ga-As phase diagram is also discussed.

EXPERIMENTAL PROCEDURES

Semi-insulating (100) GaAs wafers were degreased with trichloroethylene, and etched in $H_2O_2:H_2O:H_2SO_4(1:1:5)$ and $H_2O_2:NH_4OH:H_2O(1:1:10)$ solutions. Pt(300Å) and then Al(400Å) were sequentially deposited by e-beam evaporation with a base pressure of 10^{-8} torr. Samples were annealed at 400°C for 10 min, followed by additional annealing treatments at 550°C, 650°C, and 800°C for 30 min. in a flowing Ar gas atmosphere. Before annealing, samples were capped with WSi_2 by sputtering to prevent the outdiffusion of As. This cap was removed by a plasma-etching process after the annealing treatments. Plan-view and cross-sectional samples were prepared following the procedure described by Bravman and Sinclair[8], where samples were glued, dimpled, and ion-milled. Unannealed and annealed

Mat. Res. Soc. Symp. Proc. Vol. 181. ©1990 Materials Research Society

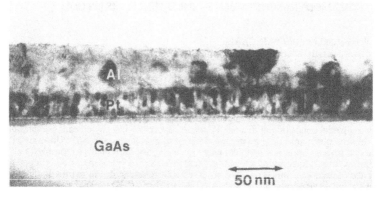

Fig. 1 Bright-field TEM micrograph of a (110) cross-sectional view of an as-deposited Al/Pt/GaAs sample

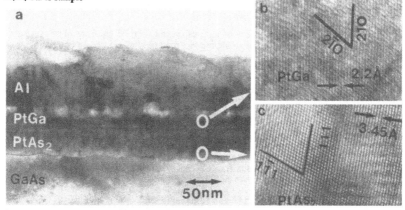

Fig. 2 Cross-sectional micrographs ((a)(110) bright-field and (b),(c) high-resolution) of an Al/Pt/GaAs sample annealed at 400°C for 10 min., showing Al/PtGa/PtAs$_2$/GaAs structure after annealing. Micrographs in (b) and (c) correspond to the circled area in (a) showing the (210) PtGa and (111) PtAs$_2$ planes, respectively.

samples were characterized *in-situ* as well as *ex-situ* in a Philips 430ST microscope operating at 300KV. In addition, Auger electron spectroscopy was performed on.

RESULTS AND DISCUSSION

The as-deposited Al/Pt/GaAs films appear as shown in Fig. 1. After annealing at 400°C for 10 min., the Pt layer reacts with the GaAs substrate, resulting in a layered structure, as is often the case in metal-GaAs reactions (see Fig. 2(a)). Cross-fringes in the high-resolution electron micrograph in Fig. 2(b), (c), and EDAX data demonstrate that these layers are Al/PtGa/PtAs$_2$ on GaAs, which is in agreement with previous studies on Pt/GaAs reactions.[9] The white band between PtGa and PtAs$_2$ layers is the same amorphous layer(presumed to be native "oxide") which can be seen at the Pt/GaAs interface in the as-deposited sample in Fig. 1. Even though the reaction between Pt and Al has been reported to occur at as low as 250°C[10], it was not observed at 400°C due to the consumption of most of Pt in the reaction with GaAs.

Voids are seen to be present at the PtGa/Al interface in Fig. 2(a) , due to diffusion of Pt toward the GaAs substrate during the Pt/GaAs reaction(i.e. the Kirkendall effect).

Although the PtGa and $PtAs_2$ phases are thermally stable with GaAs, they are not stable with the top Al layer. Consequently, the higher temperature annealing treatment drives the Pt-gallide and Pt-arsenide to react with Al. This process involves the decomposition of PtGa and $PtAs_2$, resulting in PtAl and reformed GaAs by the reaction $2PtGa + PtAs_2 + 3Al = 3PtAl + 2GaAs$. The sequence of this reaction process is shown in Fig. 3 and Fig. 4. Each sample was annealed at 400°C for 10 min. Subsequently, each underwent a second anneal, 30 min at 550°C for the sample shown in Fig. 3 and 30 min at 650°C for the one in Fig. 4. In Fig. 3 it is shown that the thickness of the $PtAs_2$ layer below the white oxide band has decreased substantially compared to its initial thickness in Fig. 2(a), by changing $PtAs_2$ phase into reformed GaAs phase. From this figure it is noticed that decomposed Pt in $PtAs_2$ diffused into Al forming PtAl at the top. Concurrently, it is presumed that Ga in PtGa diffuses to the $PtAs_2$/GaAs interface to react with decomposed As in $PtAs_2$, forming GaAs which grows epitaxially with the unreacted GaAs. The high resolution micrograph in Fig. 3(b) shows that this regrown GaAs layer is fully epitaxial with the unreacted GaAs. The regrown GaAs phase

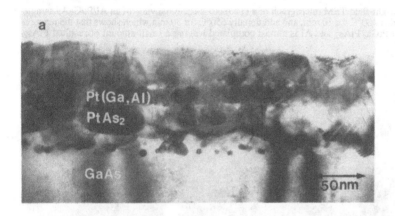

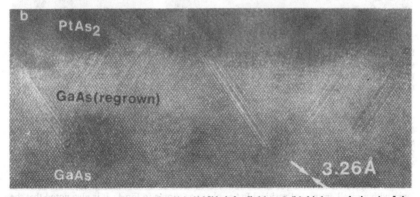

Fig. 3 Cross-sectional micrographs ((a) (110)bright field and (b) high resolution) of the Al/Pt/GaAs sample annealed at 400°C for 10 min, and additionally at 550°C for 30 min. Both micrographs show that regrown GaAs contains a high density of defects such as $PtAs_2$ precipitates and stacking faults generated from these precipitates.

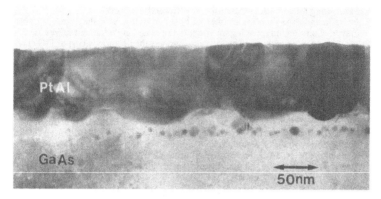

Fig. 4 Bright-field TEM micrograph of a (110) cross-sectional view of an Al/Pt/GaAs sample annealed at 400°C for 10 min, and additionally 650°C for 30 min.which shows that the reaction between PtGa, PtAs$_2$ and Al is almost completed leaving a small amount of residual PtAs$_2$ phase.

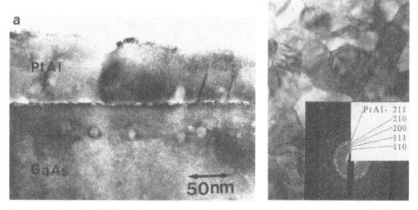

Fig. 5(a)Bright-field TEM micrograph ao a (110) cross-sectional view of an Al/Pt/GaAs sample annealed at 400°C for 10 min., and additionally at 800°C for 30 min , which shows uniform interface of PtAl layer on the regrown GaAs.(b)Plan-view, bright-field TEM micrograph and corresponding selected area diffraction pattern of the sample in (a)

contains a high density of defects such as an array of precipitates and associated stacking faults, as shown in Fig. 3(a) and (b). The precipitates are believed to be the PtAs$_2$ phase. *In-situ* TEM experiments demonstrated that the regrowing GaAs phase trapped some of the PtAs$_2$ grains, that did not react with Al. Equivalent precipitates were observed in the Si/Ni/GaAs system by Sands et al.[7]

High-temperature annealing at 650°C almost completes the above reaction, resulting in a PtAl layer with a small amount of residual PtAs$_2$ phase as shown in Fig 4. It is also noted that the density of stacking faults in the regrown GaAs is significantly decreased after the 650°C annealing treatment.

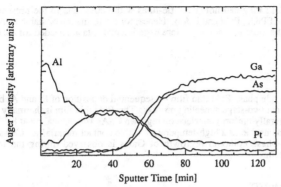

Fig.6 Auger electron spectroscopy(AES) sputter profile of an Al/Pt/GaAs sample annealed at 400°C for 10 min and 800°C for 30 min.

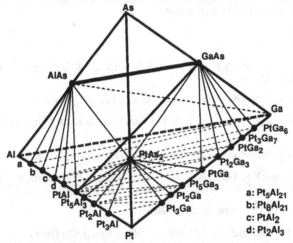

Fig. 7 Tentative Pt-Al-Ga-As quaternary phase diagram based on experimental results for the Pt-Ga-As ternary phase diagram and calculated heat-of-formation values by Miedema, showing the internal tie line between the PtAl and GaAs phase.

The PtAl layer is thermally stable after 800°C annealing showing a laterally uniform interface with regrown GaAs phase in Fig. 5(a). The PtAl phase after the 800°C annealing was identified by a selected area diffraction pattern in Fig. 5(b) obtained in the plan-view sample and Auger electron spectroscopy depth profile in Fig. 6.

Fig. 7 is the tentative Pt-Al-Ga-As quaternary phase diagram where internal tie lines apart from PtAl-GaAs are not yet determined. This phase diagram is drawn based on experimental data[11] for the Pt-Ga-As ternary phase diagram and calculated data by Miedema[12,13] for Pt-Al-As and Pt-Ga-Al ternary phase diagrams. From the result of this study an internal tie-line can be drawn between PtAl and GaAs phases, which shows the existence of the tie-tetrahedron consisting of PtAl, PtGa, $PtAs_2$, and GaAs. Even though this is not a complete phase diagram, we can predict that Pt rich phases such as Pt_5Al_3, Pt_2Al, and Pt_3Al are not stable on

GaAs. This is because they can not have tie-lines with GaAs, due to the presence of the competing tie-plane of PtGa, PtAl and PtAs$_2$. Hence, for the formation of stable Pt-Al phases, Al-rich phases may be more appropriate. More experimental data are needed for the complete phase diagram.

CONCLUSIONS

We obtained a single phase PtAl thin film by sequential deposition of Pt and Al bilayer on GaAs and a subsequent two-stage annealing process. This PtAl thin film is thermally stable on GaAs, showing a laterally uniform interface on regrown GaAs after annealing at 800°C for 30 min. Thus, it may be usable as a high-temperature stable contact material for GaAs devices. Furthermore, solid-phase epitaxial regrowth of GaAs is observed during the interfacial reactions in this system.

ACKNOWLEDGEMENTS

We would like to thank M. Kniffin and H.S. Park for their assistance on AES analysis and W.S.Lee and B.G.Park for providing samples. This work was sponsored by the Center for Integrated Systems, Stanford University, CA.

REFERENCES

1. Y.K. Kim, D.K. Shuh, R.S. Williams, S.P. Sadwick and K.L. Wang, Mat. Res. Soc. Proc. 148, 15 (1989)

2. A. Guivarc'h, R. Guerin and M.Secoue, Electron. Lett. 23, 1004 (1987)

3. C.J. Palmstrøm, K.C. Garrison, B.-O. Fimland, T. Sands and R Bartynski, J. Appl. Phys. 65, 4753 (1989)

4. A. Guivarc'h, M. Secoue, and B. Guenais, Appl. Phys. Lett. 52, 948 (1988)

5. C.J. Palmstrøm, K.C. Garrison, S. Mounier, T. Sands, C.L. Schwartz, N. Tabatabaie, S.J. Allen,Jr., J.L. Gilchrist, and P.F. Miceli, J. Vac. Sci. Technol. B7, 747, (1989)

6. T. Sands, Appl. Phys. Lett. 52, 197 (1988)

7. T. Sands, E.D. Marshall and L.C. Wang, J. Mater. Res. 3, 914 (1988)

8. J.C. Bravman and R. Sinclair, J. Electron. Microsc. Technol. 1, 53 (1984)

9. C. Fontaine, T. Okumura, and K.N. Tu, J. Appl. Phys. 54, 1404 (1983)

10. S.P. Murarka, I.A. Blech, and H.J. Levinstein, J. Appl. Phys. 47, 5175 (1976)

11. R. Beyers, K.B. Kim, and R. Sinclair, J. Appl. Phys. 61, 2195 (1987)

12. A.R. Miedema, R.Boom, and F.R. de Boer, J. Less Comm. Met. 41, 283 (1975)

13. A. K. Niessen, F.R. de Boer, R. Boom, P.F. de Chatel, W.C. Mattens and A.R. Miedema, CALPHAD 7, 51 (1983)

THE EFFECT OF MICROSTRUCTURE AND PROCESSING PROCEDURES ON
THE RESISTIVITY OF CO-SPUTTERED W-Si LAYERS ON GaAs SUBSTRATES.

S. Carter*, A.E. Staton-Bevan*, D.A. Allan** and J. Herniman**
*Dept. of Materials, Imperial College, London SW7 2AZ, U.K;
**British Telecom Research Laboratories, Martlesham Heath, Ipswich,
IP5 7RE, U.K.

ABSTRACT

The relationship between microstructure, composition, resistivity and
processing procedures of W-Si layers on (100) GaAs was examined for both
"as deposited" specimens and specimens annealed at temperatures between 100°C
and 1000°C. TEM, EDAX, SIMS, AUGER and four point probe resistivity
measurements were employed.
The layers, exhibiting a columnar growth structure typical of sputter
deposition, are amorphous below ≈ 800°C. At 700°C, the formation of pits,
attributed to the outdiffusion of Ga and As into the W-Si layer, is observed
at the W-Si/GaAs interface. The Ga and As outdiffusion was confirmed for
temperatures above 700°C. The layers annealed between 800°C and 1000°C
consist of a polycrystalline mixture of αW, βW and W5Si3 with coarse
particles, thought to be W5Si3 precursors, formed along the W-Si/GaAs
interface and protruding into the substrate. As the frequency of these
protrusions increases with increasing temperature, the resistivity of the W-
Si layers decreases.
Both the composition and the resistivity of the W-Si thin films are
affected by the processing procedure. The Si/W ratio of the W-Si thin films
decreases whilst their resistivity significantly increases as a result of
etching away the Si3N4 capping layer using HF. It is thought that this is
due to the removal of Si-oxides formed within the layer during the W and Si
sputtering. The decrease in the Si/W ratio and the increase in resistivity
are not observed if an AlN capping layer is used.

INTRODUCTION

Tungsten silicide Schottky contacts on GaAs exhibit excellent thermal
stability [1]. This together with their compatibility with already
established IC technology has created much interest in their use as gate
metallization for self-aligned GaAs MESFETs [2,3].
Allan et al. [4,5] have shown that co-sputtered WSix/GaAs contacts are
also potentially suitable as resistor materials for GaAs MMICs. This paper
reports results of TEM investigation, SIMS and Auger analyses, resistivity
measurements, and discusses the effect of the processing procedures on the
resistivity of such layers. A single nominal composition of both the "as
deposited" specimens and specimens annealed between 100° - 1000°C have been
investigated.

EXPERIMENTAL PROCEDURE

The layers were co-sputtered on (001) GaAs wafers from a single W-Si
target [4,5]. They were furnace annealed at temperatures and for times given
in Table 1. During annealing the W-Si layers were either uncapped or
protected with Si3N4 or AlN capping layer. In most cases, the capping layers
were subsequently removed. Si3N4 was etched away using HF, whilst AlN
capping was removed in a hot orthophosphoric acid or a diluted solution of an
alkaline resist developer (5% sol. of tetramethyl ammonium hydroxide). The

Mat. Res. Soc. Symp. Proc. Vol. 181. ©1990 Materials Research Society

uncapped specimens, and those capped with Si₃N₄ or AlN but with the capping layer removed after annealing were given codes starting with U, SR and AR respectively. If the capping layer was maintained during the subsequent investigation, the codes were SM and AM (see Table 1). The remainder of the code refers to the annealing temperature (e.g. AM700 is a specimen capped with AlN and annealed at 700°C, where the capping layer was not removed for further work).

The TEM examination was performed on plan-views and cross-sections of the specimens using a Jeol JEM-2000FX microscope equipped with an EDX LINK analysing system. TEM specimens were prepared by standard techniques using 5.5 keV Ar⁺ ion beam thinning. Plan view specimens were thinned from the substrate side only.

The SIMS analyses were performed by Cascade Scientific Ltd. using a 14.5keV Cs⁺ primary ion beam with a 20nA beam current and a 350x350μm² raster. Profile depths were measured on a Tencor Alpha-step. No ion implantation standards were available for WSiₓ and hence this study can only serve as a qualitative comparison between the samples.

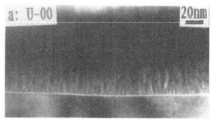

The Auger depth profiles and the resistivity measurement were performed by BTRL, Martlesham Heath. Auger analyses were performed on a JEOL JAMP 10S spectrometer operating at 10 keV and using a 5 keV Ar⁺ sputter beam. Resistivity was measured using a method described in [9]. An Alessi Industries four point probe was used.

RESULTS AND DISCUSSION

Microstructures (TEM)

Fig. 1 shows that in cross-section the W-Si layers exhibit a columnar growth structure typical of sputter deposition [6,10]. In plan view (Fig. 3), this columnar structure is seen as W-rich accumulations (dark) in a Si-rich matrix (light). The diffraction data (Fig. 2a-d) shows that for annealing temperatures below ≈ 800°C the W-Si layers are amorphous. At 700°C, the formation of pits, described also by [7] and thought to be associated with the outdiffusion of Ga and As into the W-Si layer [8], is observed at the W-Si/GaAs interface (Fig. 1c). The W-Si layers annealed between 800°C and 1000°C consist of a polycrystalline mixture of αW, βW and W₅Si₃ (Figs. 2e,f) with coarse particles, thought to be W₅Si₃

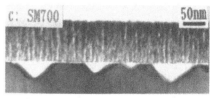

Fig. 1: Microstructures of the W-Si layers in cross-sections

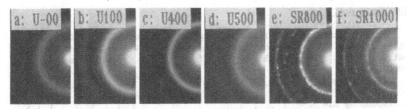

Fig. 2: Diffraction patterns of the W-Si layers

precursors, formed along the W-Si/GaAs interface and protruding into the substrate (Figs. 1d,e). The number and size of these protrusions increase with increasing annealing temperature.

EDAX analyses (TEM)

Table 1 gives the results of the TEM EDAX analyses of the W-Si layers performed on the cross-sectional specimens. They fall into two distinct groups. The SR-samples capped with Si_3N_4 and with capping layer removed show consistently lower Si/W ratios (≈ 0.4-0.6) than the remaining specimens (Si/W ≈ 0.8-1.1).

The above results indicate that the

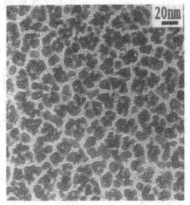

Fig. 3: Typical microstructure of a W-Si layer in plan view (U400)

Table 1: Composition of the W-Si layers

Specimen anneal code / time [min.]	Si/W ratio (x in WSi_x) Layer	Interface particles	Ga [at%]	As [at%]
U-00/-	0.8±0.1	-	9±1	8±1
U100/60	0.8±0.1	-	5±3	5±3
U200/60	0.9±0.1	-	12±1	9±1
U300/60	0.9±0.1	-	3±1	4±1
U400/60	1.0±0.1	-	6±1	4±1
U500/60	1.1±0.1	-	7±1	6±1
*SR600/30	0.4±0.1	-	-	-
SR700/30	0.6±0.1	-	11±1	10±1
SR800/15	0.6±0.2	0.9±0.3	25±7	23±8
SR950/10	0.5±0.1	1.2±0.2	23±3	18±3
SR1000/10	0.5±0.3	1.0±0.6	26±6	22±6
SR-00/-	0.5±0.1	-	-	-
SM-00/-	1.0±0.1	-	-	-
SM700/30	1.0±0.1	-	-	-
SR800/30	0.5±0.1	-	-	-
*AR800/30	0.8±0.1	-	-	-
SR850/15	0.4±0.2	-	-	-
AR850/15	0.8±0.2	-	-	-

*Specimens analyzed by EDAX-SEM

HF etch used for the removal of the Si_3N_4 capping layer also removes Si from the W-Si thin films. The original composition of the layers seems to be preserved in the composition of the coarse interface particles that were formed prior to the etching and that do not appear to be soluble in HF (see specimens SR800, SR950 and SR1000). As elemental Si and the W-silicides are insoluble in HF, a Si compound soluble in HF must be present within the layers. The possible candidates are SiO, SiO_2 and Si_3N_4. Both oxygen and nitrogen are present in the chamber during the sputtering process.

Increase in As and Ga content within the W-Si layer can be detected from $\approx 700°C$. This confirms the suggestion that the appearance of the pits at 700°C is accompanied by Ga and As outdiffusion.

SIMS depth profiles

Further confirmation of the above results was provided by the SIMS analyses (Figs. 4). The count rate intensities of the W-profiles for the sputtered layer show a slight decrease from the "as deposited" sample to 400°C and then they stabilize between 400°C-1000°C. The corresponding Si-count rate intensities (Fig. 4 (Si)) remain at the same level up to 500°C, they show a sharp drop at 700°C and then remain at this new lower level to 1000°C. These changes are reflected in the EDAX analyses as a slight increase in the Si/W ratios from 0.8 for the "as deposited" specimen to 1.1 for the 500°C specimen (U-specimens), followed by a sharp drop to Si/W = ≈ 0.5 for temperatures between 600°C-1000°C (SR-specimens)(Table 1).

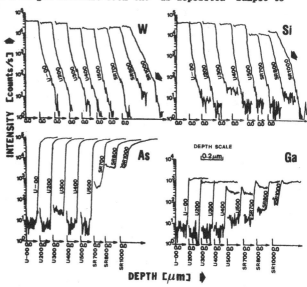

Fig. 4: SIMS analyses for W, Si, As and Ga

The SIMS depth profiles also confirm the outdiffusion of Ga and As into the W-Si layers at temperatures above 700°C as indicated by EDAX. The count intensities for both elements within the W-Si layer are higher at 700°C than at 0°C-400°C, and there is a further increase in counts between 700°C-1000°C. There is a possibility that the outdiffusion of Ga starts already at 500°C (Figs. 4(Ga)).

The depth profiles for Si and W for 1000°C, when compared to the lower temperature ones, show a change in the slope of the curves at the W-Si/GaAs interface, forming a shoulder (indicated by an arrow) that off-sets and increases the final depth of penetration of these two elements by ≈ 0.02-0.03μm (i.e. ≈ 20-30nm). Similar results were also obtained by [8] who suggested that indiffusion of W and Si into GaAs takes place. This penetration of W and Si to a depth of 20-30nm below the W-Si/GaAs interface is consistent with the formation of the coarse interface protrusions observed in the specimens annealed at temperatures between 800°C-1000°C and described above (compare Figs 1d,e with Figs. 4(W),(Si)).

Auger analyses

Auger depth profiles were obtained for SR-00, SM700 and AR-00 (Fig. 5). All the three specimens were found to contain oxygen (10at% for the as deposited samples, and 8at% for SM700). SM700 was also analysed for nitrogen, showing that there is no nitrogen present within the W-Si layer. Therefore the most likely HF-soluble compounds that can have formed during the W-Si deposition are Si oxides. During the etching procedure, these can be removed t'gether with the Si₃N₄ capping layer thus decreasing the Si/W ratio. This would also leave some voids between the W-rich columns of the W-Si layer which would, in effect, increase the resistivity of the thin film.

Resistivity measurements

The results of the resistivity measurements are given in Fig. 6. Again, two distinct groups of values emerged. The SR-samples had significantly higher resistivities with values nearly twice as large as those of the remaining specimens.

The above results indicate that the etching procedure used for the removal of the Si₃N₄ capping layers also increases resistivity the W-Si thin films. This observation was further confirmed by measuring the resistivity of an uncapped "as deposited" specimen before and after 30 sec. etch in HF. The resistivity roughly doubled after the etch. The increase in resistivity is probably associated with formation of voids within the W-Si layer due to the removal of the Si-oxides. No

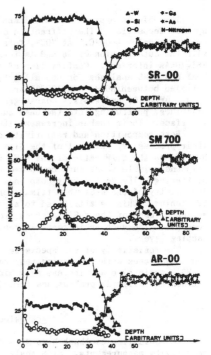

Fig. 5: Auger analyses

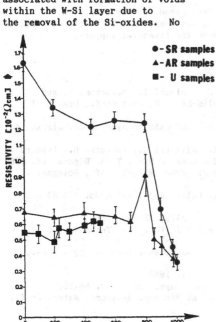

such an increase in resistivity was observed when AlN capping layers were used (Fig. 6).

Within each group, the resistivities of specimens remained, on average, constant for temperatures up to 800°C, but decreased significantly between 800°C to 1000°C. This decrease in resistivity is attributed to the formation of the silicide protrusions at the W-Si/GaAs interface. As the number and of the protrusion increase with increasing temperature, an interconnected silicide mesh is formed. It is likely that the resistivity decreases with increasing frequency of interconnecting points.

Fig. 6: Resistivity of the specimens versus annealing temperature

CONCLUSIONS

The WSi$_x$ layers exhibit a columnar growth microstructure consistent with sputter deposition. The diffraction evidence indicates that the layers are amorphous below ≈800°C. At 700°, formation of pits, thought to be associated with the outdiffusion of Ga and As into the W-Si layer, is observed at the WSi$_x$/GaAs interface. Confirmation of this outdiffusion was obtained from EDAX and SIMS analyses for annealing temperatures above 700°C. The samples annealed between 800°C and 1000°C are polycrystalline consisting of a mixture of αW, βW(=W$_3$O) and W$_5$Si$_3$. Coarse grained particles, thought to be tungsten silicides, protruding into the GaAs substrate can be seen along the WSi$_x$/GaAs interface. Their number increases with increasing annealing temperature.

Both composition and resistivity of the W-Si thin films are affected by their processing. Removal of the Si$_3$N$_4$ capping layer after annealing using HF reduces the Si/W ratio of the W-Si thin film. This indicates that a Si-compound soluble in HF must be formed within the layer during W and Si co-sputtering. The most likely compounds are Si-oxides (SiO or SiO$_2$). Resistivity of the W-Si thin films increased significantly as a result of the HF etching. This is attributed to a possible formation of voids between the W-rich columns of the W-Si layer due to the removal of the HF soluble Si-compounds. No such an increase in resistivity was oberved when using AlN capping layers.

The resistivity of the specimens annealed at temperatures higher than 700°C decreases with increasing temperature. This is related to an increase in WSi$_x$ grain size and increasingly higher frequency of interconnections between the silicide protrusions formed along the W-Si/GaAs interface.

<u>Acknowledgments</u>

We would like to thank Miss M. Madaninezhad for her help with resistivity measurements. SIMS analyses were performed by Cascade Scientific Ltd., U.K. S. Carter is grateful to SERC for financial support.

REFERENCES

[1] Yokoyama N., Ohnishi T., Onodera H., Shinoki T., Shibatomi T. and Ishikawa H: IEEE International Solid-State Circuit Conf., Digest of Tech. Papers, vol.XXVI, 44, 1983
[2] Kanamori M., Nagai K. and Nozaki T: IEEE GaAs IC Symp. Tech. Digest, 49, 1985
[3] Kotera N., Shigeta J., Ueyanagi K., Miyazaki M., Yanazawa H., Imamura Y., Tanaka H. and Hashimoto N: IEEE GaAs IC Symp. Tech. Digest, 41, 1985
[4] Allan D.A., Ng T.K. and Gilbert M.J: Proc. ESSDER Conf., Bologna, Italy, 801, 1987
[5] Allan D.A. and Sullivan P.J: U.K. Patent Application No. A23584 and A23741
[6] Thornton J.A: J. Vac. Sci. Technol., A4(6), 3059, 1986
[7] Lahav A.G., Wu C.S. and Baiocchi F.A: J. Vac. Sci. Technol., B6(6), 1785, 1988
[8] Takatani S., Matsuoka N., Shigeta J. and Hashimoto N: J.Appl. Phys., 61(1), 220, 1987
[9] Smits F.M: Bell Syst. Tech. J., 37, 711, 1958
[10] Carter S. and Staton-Bevan A.E: Inst. Phys. Conf. Ser. No.100: Section 8, 683-688, Paper presented at Microsc. Semicond. Mater. Conf., Oxford, 10-13 April, 1989

ON THE OHMIC CONTACT FORMATION MECHANISM IN THE Au/Te/n–GaAs SYSTEM

K. WUYTS*, G. LANGOUCHE*, H. VANDERSTRAETEN*, R.E. SILVERANS*,
M. VAN HOVE**, M. VAN ROSSUM**, H. MÜNDER***, AND H. LÜTH***
* Physics Department, K.U.Leuven, Celestijnenlaan 200D, B–3030 Leuven, Belgium
** Imec, Kapeldreef 75, B–3030 Leuven, Belgium
*** ISI, KFA Jülich, D–5170 Jülich, Germany

ABSTRACT

Alloyed Au/Te/n–GaAs ohmic contacts, with contact resistivities comparable to those of the AuGe device standard, have been developed and studied by Mössbauer spectroscopy, Raman scattering and X–Ray Diffraction. The formation of Au–doped Ga_2Te_3 crystallites, grown epitaxially on a defectively GaAs surface was observed. No evidence for the formation of an n^{++}–GaAs surface layer could be derived. The interpretation of all experimental results leads to a description of the ohmic conduction mechanism based on a resonant tunneling process assisted by defect/impurity related deep levels through low barrier metal/Te/(Au)Ga_2Te_3/GaAs interfaces.

INTRODUCTION

The ohmic contact formation to n–GaAs represents a long standing problem in III–V semiconductor research. Although extensively used over the last two decades [1], the nature of the Au–Ge(–Ni) alloyed ohmic contact still remains debated [2]. Also for the more recently developed Ge/Pd/GaAs system, no definite statement concerning the actual mechanism causing the ohmic type conductivity could be made [3]. Till now the discussion mainly centered upon the question whether the low resistance conduction across the metal/semiconductor interfaces can be explained by doping of the GaAs substrate, or by the formation of a graded heterojunction. The doping model requires the formation of a highly Ge doped (5×10^{19} donors/cm³) thin ($\simeq$ 50–100 Å) GaAs surface layer. The heterojunction model postulates the existence of a graded $Ge_x(GaAs)_{1-x}$ intermediate layer, the required grading of which, however, being strongly dependent on the doping level of the Ge top layer. The possibility of a defective regrown GaAs surface layer, causing ohmic conduction, has also been proposed [2, 4–6].

To extend the study of the nature of n–GaAs ohmic contacts, a Te based metallization scheme, with a specific contact resistivity (r_c) comparable to that of the standard AuGe contact metallization [7], was developed by our research group. In this paper, the experimental data obtained by Mössbauer spectroscopy, X–Ray Diffraction and Raman scattering on furnace alloyed Te contacts are presented ; I–V measurements, reported previously, will be summarized. The discussion is conducted in the frame of the above mentioned models.

EXPERIMENTAL PROCEDURE

Metallization of the GaAs substrate, uniformly Si–doped to a dose of 10^{17} at/cm³, was performed by successive deposition of 50 Å Au, 500 Å Te and 1200 Å Au on (100) oriented GaAs by resistive heating evaporation in a vacuum of the order of 10^{-6} Torr. For the I–V measurements, wafers with pads 100x100 μm in size and gaps of 100 μm were prepared by photolithography. Short time furnace heat treatments were performed in a graphite strip heater for 15 s under a forming gas ambient (90% N_2 + 10% H_2).

To prepare samples containing Mössbauer probes, ^{129m}Te was implanted (40 keV) in the as evaporated Te layer, after which the Au overlayer was deposited. The Mössbauer data were recorded at 4 K, using a CuI absorber. The X–Ray Diffraction experiments were done on a computer controlled Rigaku Dmax II rotating anode. Cu K_α radiation ($\lambda = 1.542$ Å) was selected with a flat pyrolitic graphite monochromator. The Raman experiments were performed at room temperature using a triple spectrometer with multichannel detection (Dilor XY). A focus diameter of 1–2 μm was used. The spectra were taken with the 514 nm line of an Ar ion laser at an input power of 0.4 mW.

RESULTS

I–V measurements

Specific contact resistivity (r_c) measurements, performed on AuGe–and Au/Te ohmic contacts using identically prepared GaAs wafers, yielded comparable values for the two systems, the lowest values for both being $\simeq$ 5x10^{-5} Ωcm² [7]. By adding a Ni layer on top of the metallizations, a similar improvement to the contact quality of both systems was found : the spreading on the measurements strongly reduced, the r_c–values decreased on the average by one order of magnitude ($r_c \simeq$ 5x10^{-6} Ωcm²) [7]. The AuGe–and Au/Te–metallization differ by the temperature at which ohmic conduction is observed. This onset, which for Ge contacts usually is reported at 350–400° C was found at $\simeq$ 500° C for Te contacts with a 1200 Å thick Au layer, and was even shifted to 550° C for contact structures with 700 Å of Au. The 350° C heat treatment yielded for both (700 Å and 1200 Å Au) Te metallizations only a strong degradation of the Schottky characteristics compared to anneals at lower (300° C) temperature [7].

Mössbauer measurements

As a first technique allowing to probe the microscopic configuration of the Te atoms in the alloyed contact structures, 129I Mössbauer spectroscopy was applied. The 129I Mössbauer spectra recorded from as deposited and alloyed samples are presented in fig. 1 ; as stated in the previous paragraph, only the 550° C spectrum corresponds to an ohmic contact structure. One component fits to the data are also shown, the parameters are listed in table 1. For the as deposited and 300° C annealed samples, clearly the hyperfine interaction parameters of 129I in resp. Te and AuTe₂ [8] were obtained. The parameters fitted to the high temperature (400° and 550° C) annealed spectra, correspond to those of Au–doped Ga₂Te₃ [9].

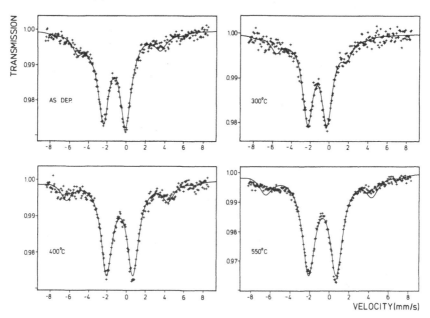

Fig. 1   129I Mössbauer spectra of alloyed Au/(129mTe)Te/Au/GaAs structures, using a CuI absorber, fitted with the parameter sets listed in table 1.

Table 1 ^{129}I Mössbauer parameters of one component fits to the spectra of alloyed Au/(^{129m}Te)Te/Au/GaAs structures.   QS = quadrupole splitting, η = asymmetry parameter, IS = absorber isomer shift given with respect to CuI, Γ = linewidth.

	QS (Mhz)	η	IS (mm/s)	Γ (mm/s)
as dep.	−402(10)	.76(5)	+1.19(5)	1.4(1)
300° C	+355(10)	.60(5)	+1.40(5)	1.3(1)
400° C	+479(10)	.87(5)	+0.77(5)	1.4(1)
500° C	+489(10)	.83(5)	+0.75(5)	1.6(1)

However a pronounced deviation between the 550° C one component fit and the experimental data can be observed : the rather big asymmetry (8%) between the spectral main lines is not reproduced, and the smaller side peaks are missing. Also the increase of the fitted linewidth (from 1.4(1) mm/s for the 400° C to 1.6(1) mm/s for the 550° C data) indicates the ensemble of ^{129}I probes in the 550° C contact structure is rather submitted to a distribution, or at least to a sum of different hyperfine fields, whereas the I nuclei in the 400° C and 300° C annealed samples are subjected to a single quadrupole interaction. Similar phenomena were observed in the spectra of Au–doped (.1–22 at%) and pure Ga$_2$Te$_3$ reference sources : an increase of the asymmetry between the main spectral lines and of the misfit at the smaller side peaks, accompanied by a line broadening, was found as a function of decreasing Au content [9]. Based upon this analogy, the spectrum was fitted with a distribution of hyperfine parameters, including those associated with pure and Au alloyed Ga$_2$Te$_3$; the result is shown in fig. 2a. However, a further significant improvement to the fit quality, accompanied by a reduction of the linewidth of the individual components in the distribution from 1.5(1) mm/s to 1.3(1) mm/s, was obtained by adding a component with an IS = +.85(5) mm/s, QS = +350(10) Mhz, η = 0 (see fig. 2b). This quadrupole interaction was also observed in high dose (10^{16} at/cm^2) Te–implanted GaAs [10], and could in a previous report on pulsed laser beam mixed Au/Te/n–GaAs contacts explicitly be associated with the transition to ohmic type conductivity [11]. As the parameters of this quadrupole interaction are closely related to those of the (assumed) Te DX–center configuration in GaAs [12], the defect was thought to correspond to a DX–related deep level in a possibly highly disordered GaAs matrix [11]. Notable finally is also the fact that fits to the 550° C data, including the parameter set associated with Te substitution on As sites (IS = +.70(5) mm/s, QS = 0) [12], always yielded as result a zero occupation of this lattice site.

XRD measurements

XRD measurements were performed on as deposited, 400 and 500° C annealed samples. As shown in fig. 3, XRD clearly confirms the presence of Ga$_2$Te$_3$ in both alloyed contact structures : the diffraction peaks labeled by b in the figure correspond to the (200) lattice planes of Ga$_2$Te$_3$, this compound having the deficit zincblende lattice structure with a lattice constant d = 5.899 Å. However, a reversal in the intensity ratio between the first and second order Ga$_2$Te$_3$ (200) diffraction is observed in the 500° compared to the 400° C

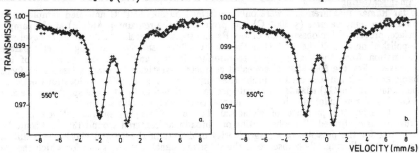

Fig. 2 ^{129}I Mössbauer spectra of 550° C alloyed Au/(^{129m}Te)/Au/GaAs, using a CuI absorber.   a. fitted with a distribution of (Au)$_x$(Ga$_2$Te$_3$)y parameters b. fitted with the DX–related defect parameters added.

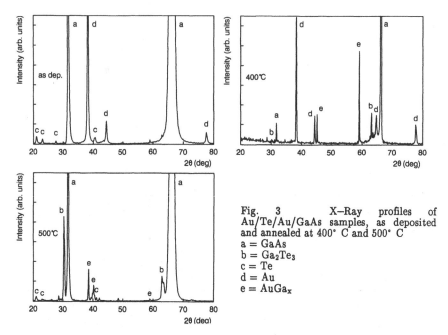

Fig. 3 X—Ray profiles of Au/Te/Au/GaAs samples, as deposited and annealed at 400° C and 500° C
a = GaAs
b = Ga₂Te₃
c = Te
d = Au
e = AuGaₓ

spectrum. This changing intensity indicates a rearrangement of the atoms constituting the Ga_2Te_3 lattice structure, which most probably can be explained as a difference in the Au doping level of these crystallites. Indeed, the same feature, though less pronounced, was also observed for highly Au—doped ($\simeq$ 20 at%) Ga_2Te_3 reference powders. One can also note that of the full X—Ray profile of Ga_2Te_3 only the signals arising from the (200) and (400) planes are apparent. An ω—scan around these angles indeed revealed the Ga_2Te_3 crystallites had grown epitaxially on the substrate, after the 400° C as well as the 500° C annealing : the width of the rocking curve was measured to be .8(2)°.

For the 400° C alloyed contacts, apart from the formation of (Au—doped) Ga_2Te_3, the X—Ray profile also reveals the presence of unreacted Au (peaks labeled d) and of AuGa—phases (peaks labeled e). The most intense peaks in the 500° C spectrum (besides those of GaAs and Ga_2Te_3) can most probably be associated with the formation of AuGa. No sign of unreacted Au is still observed ; rather unexpected however is the indication of some elemental Te (peaks labeled c).

Raman scattering
So far Raman measurements were only performed on 500–550° C annealed samples. At lower alloy temperatures ($\leq$ 400° C), the presence of the remaining Au(Ga) at the sample surface, prevented the observation of any Raman signal. Therefore, a small angle bevel will be polished on as well the 400° C as the 500° C alloyed samples, enabling to get separate information from the GaAs substrate—interface layers, and from the structure of the overlayer. For the 500–550° C annealed contacts, investigated so far, a laser light focus diameter of 1–2 μm was used to probe the sample structure at different locations between the remaining Au(Ga) islands. Typical spectra are shown in fig. 4. The spectra of Te and of Ga_2Te_3 (taken under identical experimental conditions) are also shown.

Clearly, the presence of elemental Te and of Ga_2Te_3 in the alloyed contact structures is confirmed. The observation of a broadband centered around 130 cm^{-1} could be indicative for the presence of a large density of defects in the Te and Ga_2Te_3 crystallites formed after alloying. The absence of the substrate signal at those spots for which the Te signal dominates the spectrum can be understood if one takes into account the optical thickness of Te as compared to that of Ga_2Te_3, yielding most probably a more reduced light penetration depth.

The line appearing at 293 cm⁻¹ is the GaAs LO phonon mode, the peak at 269 cm⁻¹, the intensity of which strongly varied with respect to the LO line intensity as a function of the location probed on the sample surface (their ratio changing from 1:5 to 1:2), could correspond to the GaAs TO lattice vibration, or to the L⁻ mode of the coupled LO phonon plasmon modes (PLP), which can be observed in highly (≥ 5x10¹⁷ donors/cm³) doped GaAs [13]. In the latter case, the fluctuations in the free carrier density interact with the electrical field of the LO phonon, giving rise to two coupled modes L⁻ and L⁺. Hence, in case the 269 cm⁻¹ line arises from the presence of a high density of free carriers in the GaAs surface layers, the corresponding high frequency L⁺ mode should be observed with a frequency depending on the free carrier concentration. However in the spectral range 500–1000 cm⁻¹, no PLP mode was observed, leading us to the conclusion the 269 cm⁻¹ line should be attributed to the GaAs TO lattice vibration. Also, the lineshape and linewidth of a plasmon mode should be different from those of the LO phonon mode. As can be seen in figure 4, both the LO and the 269 cm⁻¹ lines are identically shaped, which gives a second indication for this conclusion. However, according to the selection rules, the only phonon mode which should be observed from a GaAs (100) surface in backscattering geometry is the LO phonon frequency. Hence, the presence of this strong TO mode in the spectra indicates the presence of disorder or of defects in the GaAs surface layers, which relax the selection rules allowing for the observation of a TO phonon.

DISCUSSION AND CONCLUSIONS

The main models developed to explain the ohmic contact formation to n–GaAs for Ge based metallizations (see introduction and [6]), can also be applied to the Au/Te/n–GaAs system. A highly doped (10¹⁹–10²⁰ donors/cm³) GaAs surface layer can be formed by the substitution of Te atoms on As sites, forming a shallow donor level, in the regrown GaAs. The role of Ge in the formation of a graded $(Ge)_x(GaAs)_{1-x}$ heterojunction can be taken over by the semiconducting compounds Ga_2Te_3, $GaTe$, As_2Te_3 (band gaps $E_g \simeq 1.0$ eV), and/or by elemental Te ($E_g = .33$ eV). The observation that Te atoms can also form a deep level in GaAs [12], on the other hand, is compatible with the requirements of the so called amorphous heterojunction model, in which conduction is supposed to occur by

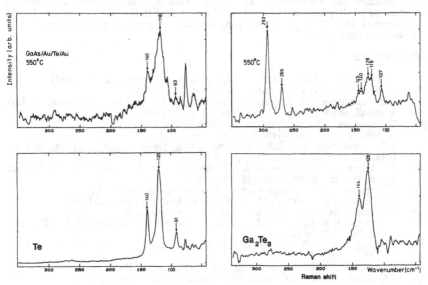

Fig. 4   Raman spectra of Au/Te/Au/GaAs samples anealed at 550° C, probed at different locations on the sample surface ; and reference spectra of Te and Ga_2Te_3.

resonant tunneling assisted by localized deep states in the forbidden energy gap of the semiconductor [2,4,6].

Clearly, no evidence in favor of the doping model can be deduced from the experimental data, although both Mössbauer and Raman spectroscopy were proven to be very sensitive in probing the substitution of Te dopants on electrically active sites in GaAs, the former by the observation of Te atoms on a regular (substitutional) lattice site [12], the latter by the direct evaluation of carrier concentrations $\geq 5 \times 10^{17}$ donors/cm^3 by the coupled LO phonon plasmon modes [13].

From the experimental results, on the contrary, strong evidence in favor of the heterojunction–like models can be deduced. XRD revealed the epitaxial growth of (Au–doped) Ga$_2$Te$_3$ on the GaAs substrate. This result was confirmed by Mössbauer spectroscopy and Raman scattering, the latter giving also some indication for the presence of defects in the Ga$_2$Te$_3$. Also the formation of elemental Te was indicated by both XRD and Raman spectroscopy for the 500–550° C alloyed (ohmic) contacts. Evidence in favor of a defective regrown GaAs surface layer is given by the observation of the TO phonon mode in the Raman spectra, and by the possible indication of a DX–related defect in the Mössbauer data. Thus, support for as well the defective as the graded crystalline heterojunction model is found. An intriguing point remains the change in the Au doping level of the Ga$_2$Te$_3$ compound, observed as well in the XRD as in the Mössbauer spectra, by increasing the contact alloy temperature from 400 to 500° C. As Cu–doped (< 1 at%) Ga$_2$Te$_3$ was reported to have a band gap $\simeq$ 0 eV [9], this feature could, if confirmed for Au–doped Ga$_2$Te$_3$, fit into a model explaining the ohmic conduction by the presence of a graded ((Ga$_2$Te$_3$)$_{1-x}$Au$_x$)GaAs transition layer. However, band gap measurements performed on a (Ga$_2$Te$_3$)$_{95}$Au$_{05}$ compound did not reveal any difference with the Ga$_2$Te$_3$ band gap [14].

Hence, a combination of the defective and graded heterojunction model can be proposed to explain the ohmic conduction in the furnace alloyed Au/Te/n–GaAs system. In this view, the high metal/GaAs barrier is divided in different smaller barriers at the metal/Te/(Au)Ga$_2$Te$_3$/GaAs interfaces, the conduction across these interfaces occurring by a resonant tunneling process assisted by defect/impurity related deep levels in the space charge region [15,16]. So far, the experimental data do not allow a conclusive statement about the relative importance of both effects : for a contact alloy temperature of 500° C, XRD indicates the formation of Te, not observed at 400° C, whereas Mössbauer spectroscopy possibly indicates the creation of a deep level in the GaAs substrate. Raman measurements on 400° C alloyed samples, currently in progress, probably can help to elucidate this discord.

REFERENCES

[1]   N. Braslau, J.B. Gunn, and J.C. Staples, Sol. Stat. electron. 10, 381 (1967)
[2]   D. Kirillov, and Y. Chung, Appl. Phys. Lett. 51(11), 846 (1987)
[3]   C.J. Palmstrom, S.A. Schwarz, E. Yablonovitch, J.P. Harbison, C.L. Schwartz, E.D. Marshall, and S.S. Lau, J. Appl. Phys. 67(1), 334 (1990)
[4]   T. Sebestyen, Sol. Stat. Electron. 25, 543, (1982)
[5]   A. Illiadis, J. Vac. Sci. Technol. B5(5), 1340 (1987)
[6]   E.H. Rhoderick, and R.H. Williams, Metal–Semiconductor contacts, 2nd ed. (Clarendon Press, Oxford, 1988), p. 204
[7]   K. Wuyts, A. Vantomme, R.E. Silverans, M. Van Hove, and M. Van Rossum, Mater. Res. Soc. Symp. Proc. 144, 545 (1989)
[8]   G. Langouche, aggregaatsproefschrift (Leuven, unpublished) (1986)
[9]   K. Wuyts, G. Langouche, and R.E. Silverans, Hyp. int. (in press) (1990)
[10]   H. Bemelmans, and G. Langouche (private communication)
[11]   K. Wuyts, R.E. Silverans, M. Van Hove, and M. Van Rossum, Mater. Res. Soc. Symp. Proc. 157 (in press) (1990)
[12]   G. Langouche, H. Bemelmans, and M. Van Rossum, Mater. Res. Soc. Symp. Proc. 104, 527 (1988)
[13]   G. Abstreiter, E. Bauser, A. Fisher, and K. Ploog, Appl. Phys. 16, 345 (1978)
[14]   J.C. Jumas, Université de Montpellier, (private communication)
[15]   A.M. Andrews, H.W. Korb, N. Holonyak, C.B. Duke, and G.G. Kleiman, Phys. Rev. B5(6), 2273 (1972)
[16]   G.H. Parker, and C.A. Mead, Phys. Rev. 184(3), 780 (1969)

THE EFFECTS OF ION BEAM MIXING ON RAPID THERMAL ANNEALED
OHMIC CONTACTS TO n-GaAs

Seemi Kazmi, Roman V. Kruzelecky, David A. Thompson, Centre for Electrophotonic Materials and Devices, McMaster University, Hamilton, Ontario, Canada L8S 4M1.

ABSTRACT

Ni/Ge/Au and Ni/Ge/Pd contacts have been made on 10^{18} cm^{-3} n-type GaAs. The contacts were subjected to ion beam mixing through the metallization using 70-130 keV Se$^+$ ions and subsequently subjected to rapid thermal annealing (RTA). These are compared with unimplanted contacts produced by RTA techniques on the same substrate. The specific contact resistance ,ρ_c, has been measured for the two systems. In addition, the contacts have been studied using Auger depth profiling and SEM studies have been used to determine surface morphology. Values of $\rho_c \sim 10^{-6}$ -10^{-7} ohm-cm^2 have been measured. It is observed that ion beam mixing or the addition of a Ti overlayer (to the Ni/Ge/Au) improves the contact morphology.

INTRODUCTION

Despite a considerable amount of research, problems still remain with a very important part of all GaAs integrated circuit fabrication: i.e., the formation of reliable ohmic contacts having a very low resistivity and a smooth morphology for fine pattern definition. These contacts rely on the formation of a thin heavily-doped layer at the semiconductor-contact interface. They are usually formed by appropriate heat treatment of metal multilayer systems containing a suitable dopant species deposited on the semiconductor surface. A commonly used combination for forming contacts on n-GaAs is based on the Au-Ge eutectic [1], usually in conjunction with a layer of Ni. Experimental work indicates that Ni helps in achieving a more homogeneous contact and prevents the contact surface from balling-up [2,3]. Alloying by rapid thermal annealing is reported to give better results by preventing irregular interdiffusion of metallic phases [4,5]. Although fairly low contact resistivities, in the 10^{-6} - 10^{-7} Ω-cm^2 range, have been reported [2,3], problems with contact morphology and stability upon ageing still remain inherent in the Au-Ge system since alloying involves a liquid phase. Ito et al [6] have observed that the addition of Ti/Au overlayers to conventional AuGe/Ni contacts improves alloying behaviour at lower temperatures (350-400°C), yielding $\rho_c < 10^{-6}$ Ω-cm^2.

Pd/Ge contacts have also been found to show ohmic behaviour after sintering between 400 - 600°C [7,8]. This system has an advantage over Au-Ge based contacts because only solid-phase reactions occur during sintering. This increases the reproducibility of contact formation and results in improved contact morphology. Ohmic behaviour is attributed to the preferential incorporation of Ge on Ga sites caused by interactions between Pd and GaAs. However, the resistivities reported are significantly higher than for the Au-Ge contacts and sintering required relatively long periods of time (30 to 120 min.)

More recently, ion beam mixing, followed by heat treatment, has been applied to Au-Ge based contacts to get a smoother surface morphology and improve reliability. Both inert ions, like Ar$^+$ and Kr$^+$[9], and dopant ions, like Si$^+$ and Se$^+$ [10,11], have been used for mixing. The results indicate an improvement in the surface morphology compared to traditional unmixed contacts. However the contact resistivities reported are much higher, 10^{-3} to 10^{-4} Ω-cm^2, probably due to incomplete annealing of the damage produced during ion bombardment.

Mat. Res. Soc. Symp. Proc. Vol. 181. ©1990 Materials Research Society

This study demonstrates improvements to both the Au/Ge and Pd/Ge systems.Studies of Ni/Ge/Au contacts formed by rapid thermal annealing with a Ti overlayer to act as a diffusion barrier are reported. Since the presence of Ni is believed to enhance Ge diffusivity into GaAs when it is used in the Au/Ge contact system [12], its potential role in Pd/Ge contacts has been investigated. The effects of Se$^+$ ion beam mixing on both the Ni/Ge/Au and Ni/Ge/Pd contacts have also been studied.

EXPERIMENTAL PROCEDURE

Surface preparation and removal of native oxides is crucial to obtaining adherent contacts and reproducible alloying. The n-type GaAs substrates (2.6-3.6 x 10^{18} Si cm^{-3}) were first degreased using organic solvents and rinsed in deionised H$_2$O. Contact patterns were defined by photolithography.For samples to be processed above 150°C, a 200 nm thick layer of SiO$_2$ was spun on for masking. Just prior to installation of the samples in the vacuum chamber for contact deposition, we employed UV ozone cleaning followed by oxide removal using dilute HF (1 HF: 10 D.I. H$_2$O) and a weak etch (1 HF: 1 HCl: 4 H$_2$O with 1 drop of H$_2$O$_2$ per 30 ml of solution).

Multilayer contact structures were prepared sequentially by e-beam evaporation performed in a vacuum system with a base pressure of about 4 x 10^{-7} Torr. Samples were outgassed at 300°C for a few minutes at the base pressure and then maintained at 100° C during evaporation. The metal layer thicknesses were 15 nm Ni/ 15 nm Ge/ 30 nm Au for the Au-based contacts and 10 nm Ni/ 40 nm Ge/ 40 nm Pd for the Pd-based contacts. After metallization, several of the samples were implanted with Se$^+$ ions at a dose of 10^{15} cm^{-2}, for implant energies between 70 and 130 keV [see Fig. 1(a) and 1(b)]. The implantations were carried out for sample temperatures in the range 30-300°C; the higher temperatures were to encourage annealing of defects created during bombardment. An additional layer of 30 nm Ti/ 30 nm Au was also deposited on some samples. Subsequently, the samples were rapid thermally annealed at various temperatures between 400 and 600°C in a covered carbon boat.

The morphology of the resulting contacts was characterized using scanning electron microscopy (SEM) in conjunction with Auger depth profiling. Contact resistance has been evaluated using the transmission line method (TLM) [13]. For TLM, the contact geometry consisted of rectangular coplanar pads, 100 μm long by 150 μm wide, spaced 10, 30, 50, 70, and 90 μm apart, respectively.

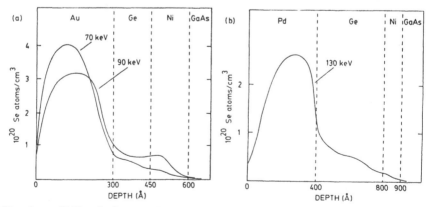

Fig. 1: TRIM calculations [14] of the distribution of Se in (a) GaAs-Ni/Ge/Au and (b) GaAs-Ni/Ge/Pd after implantation at a dose of 10^{15} Se$^+$ ions/cm^2.

RESULTS AND DISCUSSIONS

(1) Ni/Ge/Au

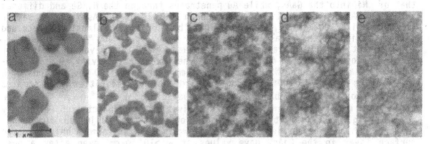

Fig. 2:   SEM micrographs of Ni/Ge/Au contacts after alloying at 450°C
for 30s: (a) unimplanted; (b) implanted with 100 keV Se$^+$; (c)
unimplanted with Ti overlayer; (d) implanted with 120 keV Se$^+$ + Ti/Au
overlayer; (e) implanted with 70 keV Se$^+$ + Ti/Au overlayer.

SEM analysis of as-deposited Ni/Ge/Au pads indicates a smooth, featureless
morphology. Figure 2(a) shows the effect of alloying at 450°C. The bright areas
are Au-rich regions that form a relatively interconnected network. The darker
regions mainly contain Ni and Ge. Increasing the alloying temperature to 500°C
results in the formation of isolated Au islands. SEM micrographs of the
Ni/Ge/Au contacts subjected to ion beam mixing shown in Fig. 2(b) exhibit a
finer-grained surface with grain size about 0.3 μm. The presence of a Ti
overlayer suppresses Au island formation during alloying, resulting in a
blotchy but well-interconnected, fine-grained structure as shown in Fig. 2(c).
However, there is still some segregation of Au and Ni/Ge, as indicated by
energy dispersive X-ray analysis, with a somewhat higher Au content and lower
Ge/Ni content in the brighter regions. Contacts which were coated with Ti/Au
after mixing with 120 keV Se$^+$ and prior to alloying, exhibit a more homogeneous
surface [Fig. 2(d)]. Relative to the unmixed sample, bright Au-rich regions
occupy a larger area fraction, forming an interconnected network. Decreasing
the implant energy to 70 keV resulted in a less pronounced grain structure
[Fig. 2(e)].

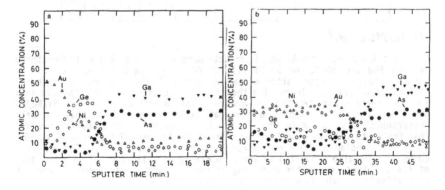

Fig. 3:   Auger depth profiles of GaAs-Ni/Ge/Au contacts after alloying at
450°C for 30 s: (a) unimplanted, (b) implanted with 100 keV Se$^+$
(T_i = 300°C).

Auger depth profiling [Fig. 3(a)] indicates that alloying the conventional GaAs-Ni/Ge/Au contacts at 450°C results in an Au-rich surface layer followed by an intermixed Ni/Ge layer in contact with the GaAs. The diffusion of Ge follows that of Ni into the GaAs, while Au penetrates through the Ni/Ge and diffuses into the GaAs. In contrast, the Auger depth profile of a GaAs-Ni/Ge/Au sample implanted with 100 keV Se$^+$, shown in Fig. 3(b), indicates almost complete and uniform mixing of the metallic layers after alloying at 450 °C.

Contact resistivity values, as determined by TLM, are given in Table I for samples alloyed by RTA at 450°C for 30 s. The lowest resistivity, 9×10^{-7} ohm-cm^2, was obtained for the Ni/Ge/Au contact implanted with 100 Kev Se$^+$ ions with the samples kept at 300°C during bombardment. Resistivities for samples subjected to a 90 keV bombardment were comparable at 1×10^{-6} Ω-cm^2. These are an order of magnitude lower than the resistivities for contacts subjected only to alloying. Resistivity measurements on contacts subjected to deeper implants, which were carried out in an attempt to achieve a more heavily doped surface layer in the GaAs, gave values of $\rho_c > 10^{-4}$ Ω-cm^2 even after a post implant anneal at 900°C for 2 seconds. Evidently, the annealing does not sufficiently remove the damage created in the GaAs during implantation. The addition of the Ti/Au layer resulted in lower contact resistivity in the unimplanted case but does not seem to have a significant effect in the implanted samples.

TABLE I: Summary of contact resistivities. Some samples were implanted with 10^{15} Se$^+$ ions cm^{-2} at temperatures T_I and energies E_I and subsequently alloyed at temperatures T_A for a time t_A.

Contact structure	E_I (keV)	T_I (°C)	T_A (°C)	t_A (s)	ρ_c (Ω - cm^2)
Ni/Ge/Au	-	-	450	30	$5 \pm 2 \times 10^{-5}$
	90	100	450	30	$1 \pm 0.3 \times 10^{-6}$
	90	300	450	30	$1 \pm 0.3 \times 10^{-6}$
	100	300	450	30	$9 \pm 3 \times 10^{-7}$
Ni/Ge/Au/Ti/Au	-	-	450	30	$1 \pm 0.5 \times 10^{-5}$
	70	300	450	30	$3 \pm 1 \times 10^{-5}$
	120	300	450	30	$2 \pm 0.6 \times 10^{-5}$
Ni/Ge/Pd	-	-	500	120	$6 \pm 2 \times 10^{-6}$
	-	-	550	60	$1.5 \pm 0.5 \times 10^{-7}$
	130	300	550	60	$8 \pm 3 \times 10^{-5}$
	-	-	600	30	$1.5 \pm 0.5 \times 10^{-6}$
	-	-	600	60	$7 \pm 2 \times 10^{-5}$
	120	300	600	60	$1.5 \pm 0.5 \times 10^{-5}$
	-	-	600	120	$9 \pm 2 \times 10^{-6}$

(2) Ge/Pd; Ni/Ge/Pd

GaAs-Ni/Ge/Pd and GaAs-Ge/Pd contacts were annealed by RTA at temperatures between 500 and 600°C for 30-300 s. Ge/Pd [Fig. 4(a)] contacts exhibit a pitted, grainy structure after annealing at 500° C with randomly distributed dark Pd-rich regions. Increasing the anneal temperature to 550°C results in a more prominent grain structure and growth of the Pd-rich regions. GaAs/Ni/Ge/Pd annealed at 500° C exhibits a somewhat textured morphology that is well interconnected as shown in Fig. 4(b). The surface is considerably smoother than that obtained using Ge/Pd alone. Increasing the anneal temperature to 550° C has no appreciable effect on the resulting contact morphology. The Ni/Ge/Pd contacts were also subjected to ion beam mixing with 120-130 keV Se$^+$ at a dose of 1×10^{15} ions cm^{-2}. After annealing at 550°C for 60 s, the contacts exhibit a relatively smooth, fine-grained morphology [Fig. 4(c)].

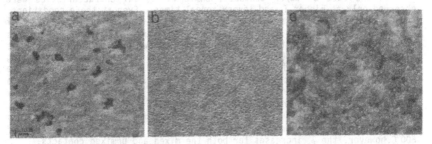

Fig. 4:   SEM Micrographs of Ge/Pd-based contacts after annealing at 550°C for
60 s: (a) GaAs-Ge/Pd, (b) GaAs-Ni/Ge/Pd, and (c) GaAs-Ni/Ge/Pd
implanted with 130 keV Se⁺ at T_I = 300°C.

Figure 5 shows a comparison of the AES depth profiles of the Pd-based
contacts after sintering at 550°C for 60 s. A profile of the Ge/Pd contact, as
shown in Fig. 5(a), indicates a fairly uniform interdiffusion of Ge and Pd with
a well-defined interface between the contact metals and the substrate.   By
contrast, a profile of sintered GaAs/Ni/Ge/Pd, shown in Fig. 5(b), indicates
a more complex interdiffusion process.   Significant amounts of As have outdiffused into the Ni layer, while Ga has diffused through the metal layers and
accumulated at the Pd surface.   The Ge and Pd intermix, with Ge diffusing
through the Ni into the GaAs. However, little Pd has diffused through the Ni.
For the implanted contacts, AES profiling indicates greater uniformity in the
intermixing of Pd and Ge and in the formation of a NiGe$_x$As$_y$ phase at the GaAs
interface [Fig. 5(c)].

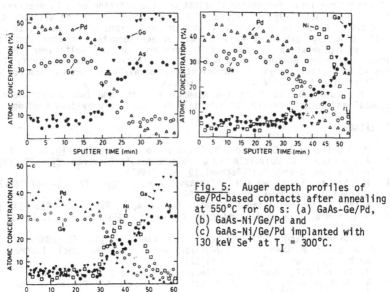

Fig. 5: Auger depth profiles of
Ge/Pd-based contacts after annealing
at 550°C for 60 s: (a) GaAs-Ge/Pd,
(b) GaAs-Ni/Ge/Pd and
(c) GaAs-Ni/Ge/Pd implanted with
130 keV Se⁺ at T_I = 300°C.

In the case of Ge/Pd contacts, ohmic electrical characteristics were observed only at higher T_A and longer anneal times. Even at these conditions, contact resistivities were generally in the 10^{-3} Ω-cm^2 range. In the case of Ni/Ge/Pd, ohmic contacts were obtained over the range of anneal temperatures studied even with anneal times as short as 30 s. However, contact adhesion was a problem for lower T_A (500°C) and shorter anneal times (30 s), preventing accurate determination of ρ_c. Resistivities for n-GaAs-Ni/Ge/Pd contacts are listed in Table I. Contacts which were rapid thermally annealed at 550°C had a specific contact resistance of 1.5×10^{-6} Ω-cm^2 which is which is significantly lower than ρ_c measured for the Ge/Pd contacts (10^{-3} - 10^{-4} Ω-cm^2). A lower $\rho_c = 9 \times 10^{-7}$ ohm-cm^2 is measured for contacts which were subjected to 10^{15} Se$^+$ ions cm^{-2} bombardment prior to annealing. At a higher annealing temperature of 600°C, however, the ρ_c increases for both the mixed and unmixed contacts.

The low ρ_c observed for Ni/Ge/Pd contacts can be explained in terms of the AES profiling results. Selective trapping of As in the Ni layer to form a NiGe$_x$As$_y$ phase, together with out-diffusion of Ga through the Ni and Pd, reduces the out-diffusion of As relative to Ga from the GaAs. The resulting excess As in the GaAs encourages Ge, diffusing in from the NiGe$_x$As$_y$ phase, to occupy Ga sites where it acts as a donor.

CONCLUSIONS

In summary, our experimental results show that the addition of a thin layer of Ni in contact with the substrate in the Ge/Pd contacts greatly improves the alloying chemistry, allowing low resistivity contacts ($\rho_c < 10^{-6}$ ohm-cm^2) to be obtained by rapid thermal annealing at an optimum temperature of 550°C for 60 seconds. The resulting homogeneous contact morphology is superior to that observed for conventional Ge/Pd contacts and would facilitate fine patterning at the micron level for device fabrication. The addition of a Ti overlayer to Ni/Ge/Au contacts helps to suppress Au island formation during alloying at $T_A < 500$°C, resulting in a uniform, fine-grained morphology. However, there is still some segregation of Au which may affect long-term contact stability. Shallow implantation at elevated temperatures induces a more uniform, fine-grained morphology with accompanying lower resistivity after alloying. Further studies of Ni/Ge/Pd contacts and ion beam modification of metallic layers are underway.

We would like to thank the Natural Sciences and Engineering Research Council and Ontario Centre for Materials Research for providing funding.

REFERENCES

1. N. Braslau, J.B. Gunn and J.N. Staples, Sol.St.Electron. 10, 381 (1967).
2. A.Piotrowska, A. Guivarc'h and G. Pelous, Sol.St.Electron. 26, 179 (1983).
3. N. Braslau, J.Vac. Sci. Technol.19, 803 (1981).
4. N.Yokoyama, S. Ohkawa and H. Ishikawa, Jap. J. Appl. Phys.14 1071 (1975).
5. M.Ogawa,J. Appl.Phys. 51 (1980) 406.
6. H.Ito, T. Ishibashi and T. Sugeta, Jap. J. Appl. Phys. 23, L635 (1984).
7. A.K. Sinha, T. E. Smith and H. J. Levinstein, IEEE Trans. El. Dev.ED-22, 218 (1975).
8. H.R.Grinolds and G. Y. Robinson, Sol.St.Electron. 23, 973 (1980).
9. A. J. Barcz, M. Domansky, J. Jagielski and E. Kaminska, Nucl. Instrum. and Meth. Phys. Res. B19/20, 773 (1987).
10. Z. Jie and D.A. Thompson, J. Electronic Materials 17, 249 (1988).
11. R.S Bhattacharya, A. K. Rai, A. Ezis, M. Rashid and P. Pronko, J. Vac. Sci. Technol. A3, 2316 (1985).
12. M.N.Yoder, Sol.St.Electron. 23, 117 (1979).
13. H.H. Berger, Sol.St.Electron. 15, 145 (1972).
14. J.F. Ziegler, J.P. Biersack and U. Littmark, TRIM-89 code, IBM (1989).

Ni-Ge INTERMIXING ON GaAs PRODUCED BY TEMPERATURE STANDARDIZED RAPID THERMAL ANNEALING

Michael B. Brooks and Thomas W. Sigmon, Stanford Electronics Labs, Stanford University, Stanford, CA 94305-4055

ABSTRACT

A new Rapid Thermal Annealing technique has been developed which uses very small samples of low melting elements and eutectic alloys to standardize temperature measurements. Temperatures of thin films during anneal are shown to lie within -3°C to +13°C of nominal for measured short anneal times using a special sample geometry. Results of annealing Ge/Ni/GaAs and Au/Ge/Ni/GaAs at 200°C and 250°C are presented. Rutherford backscattering is used to analyze the films and the resulting channeling spectra show that Ge consumption follows parabolic kinetics, with an activation energy estimate of ~ 0.7eV. The data is consistent with a controlling process of fast Ni diffusion through grain boundaries at these low temperatures. We expect that a Ni_2Ge phase is being formed preferentially, along with a Ni_xGaAs phase at the GaAs/Ni interface. In this contact metal system Au interacts only minimally for the times and temperatures studied. Our results agree with those of previous workers who studied Ni diffusion during Ni_xGaAs phase formation at comparable temperatures.

INTRODUCTION

Rapid thermal annealing is finding greater use in producing ohmic contacts of lower resistivity and improved morphology for III-V materials [1-3]. Accurate temperature control during annealing is a key issue affecting contact reproducibility, as the contact metallization may reach temperatures that differ substantially from nominal (for example, that measured by the monitor thermocouple). For this reason low temperature RTA studies of the Au/Ge/Ni ohmic metal system for GaAs are few, despite the fact that important interdiffusion is known to occur at 200°C [4,8,18]. This paper describes a new method of temperature standardized RTA, using a specialized annealing geometry and time-temperature cycles, thus permitting the low temperature -short anneal time behavior of Ni and Ge to be quantitatively analyzed for the first time. Samples are examined using Rutherford backscattering and a kinetics study is presented.

EXPERIMENTAL PROCEDURE

The substrates used in this work are 50 mm diameter, semi-insulating (100) GaAs (n type, 1.5 x10^8 - 3.1 x10^7ohm-cm), purchased from DOWA Mining Co. (Tokyo). These are chemically cleaned in trichloroethylene, acetone and methanol. After a deionized water rinse, the wafers are etched in 1:1:16 NH_4OH : H_2O_2 : DI H_2O, rinsed and blown dried, dipped in 20:1 buffered oxide etch (BOE) for 60 sec., rinsed again in DI H_2O and blown dry in N_2. Prior to e-beam evaporation of the thin metal films, the 20:1 BOE dip, rinse, and dry is repeated. Wafers were prepared with layers ordered Au/Ge/Ni/GaAs (1580 Å, 550 Å, 500 Å respectively) and Ge/Ni/GaAs (420 Å, 350 Å). These thicknesses have been verified via Rutherford backscattering and RUMP [19] simulations (See Results and Discussion, below and Figs. 1 and 2).

Temperature standards used for the Heatpulse 210T Rapid Thermal Annealer (A.G. Associates) were obtained from Alfa Products (Danvers, Mass.). Table I shows these low melting eutectics and pure elements. Two eutectic alloys were made by conventional furnace melting of the elements. Melting ranges of the standards are verified in a melting point apparatus, with all ranges within ± 3°C of the manufacturer's reported values (Table I). The standards employed in the RTA are cut to sizes less than 200μm in the largest dimension and placed on silicon nitride capped GaAs samples 4-5mm on a side (these are rectangular, cap thickness equaled 1050Å). Ohmic metal samples are the same size, without caps.

The annealing geometry is conserved for all samples, and consists of sandwiches of RTA thermocouple support wafer, sample (with or without standard), and Si cover piece (bottom to top ordering). The latter is a polished (111) Si section about 2.5cm on a side, placed over each standard or sample. Thus all samples or standards are annealed metal side up, using the top bank of lights only, while covered by the (111) Si piece. The sample plus cover piece is displaced (no shadowing) from the immediate vicinity of the embedded thermocouple, which contacts the support wafer from the backside. We expect that the use of Si cover pieces, combined with this annealing geometry, yield approximately equal temperatures at the thermocouple junction and coverpiece-sample interface.

Multistep cycles of heating are programmed into the Heatpulse 210T using the open loop intensity mode (the closed loop mode can not be used below 300˚C). Reproducible time temperature cycles have been developed for anneal times under 30 seconds, and are pre-run at least 5 times before sample insertion. A "time-course-trial" method of repetitive annealing is used to frequently cycle the RTA, maintaining a residual heat within the chamber. Thus all samples are inserted at temperatures of 50-54˚C, which appears to be the key for reproducible time-temperature profiles. Allowing the chamber to cool to ambient temperature gives inconsistent peak or plateau temperatures and profile widths. Forming gas ($10\%H_2 + N_2$) is flowed through the chamber during the procedure.

Nominal (programmed) anneal times and temperatures are not usable as reliable figures for kinetics analysis, and so more accurate values are measured from a profile obtained by a strip chart recorder. The Ge/Ni and Au/Ge/Ni samples are annealed in parallel, at nominal times from 1-30 seconds, with short time cycles for the temperature standards generally being used (see Table I).

RESULTS AND DISCUSSION

RTA Temperature Standardization

The temperature standards of Table I are used to bracket the nominal temperatures of 200˚C and 250˚C, thus establishing absolute limits on their ranges of variation. Standards which show melting after rapid thermal anneal must have reached at least the requisite melting temperatures--- these are labeled "M" in Table I. The 200˚C and 250˚C nominal anneal temperatures have standards that are close in melting temperature, and we note that in each case a lower bound of about -3˚C may be fixed (see the Sn/Zn and Pb/Sb standards). This is because annealed samples (or standards) must have reached at least these nominal temperatures, based on the given RTA measured thermocouple readings. For a higher bound at 200˚C, the 30 second anneal of the Sn/Ag sample is examined; the lack of melting has a special interpretation. This standard shows that the high end plateau temperature of 208˚C, as measured by the thermocouple, is still below 221˚C in an absolute sense. Thus there is at most a -3˚ to +13˚C error in the thermocouple temperature measurement, as checked against the temperature standards, for 200˚C. The non-melting behavior of the Cd/Zn standard similarly allows us to fix an upper bound on measured temperature of +13˚C for a nominal 250˚C anneal. Therefore the variation of thermocouple measured temperature during RTA cycles is limited to -3 to +13˚C at 200˚C, and -4 to +13˚C at 250˚C.

Table I also gives additional data on the time at these temperatures. The "Measured anneal time" column describes the concave down curve at the peak for anneal profiles. The listed peak temperature is the maximum of this curve, and the time spent above the nominal temperature but at or below the peak is given by data in the last column. We note that only through the use of a particular annealing geometry (and special annealing cycles) are reproducible time-temperature cycles accurately obtained, and these require temperature standardization.

As-Deposited and Ge/Ni/GaAs Samples

Figures 1 and 2 show RBS spectra for the as deposited samples, including RUMP [19] simulations, using the 170˚ backscattering geometry. The simulation for the Ge/Ni/GaAs samples (Fig.1) represents a best fit based on including initial intermixing of the Ni/Ge and Ni/GaAs layers. The implication is that some limited intermixing must occur on deposition, however without further study we are unable to specify the nature of this in detail. In the Au/Ge/Ni/GaAs spectrum (Fig. 2), a small peak just to the left of the low energy edge of the Au signal can be seen. Simulations indicate that this is a surface peak of either Ge, Ga, or As.

Table I. RTA Temperature Standardization Results, from rapid thermal annealing of the indicated materials, giving measured melting temperature ranges and anneal times.

Composition elements, % by weight	Melting temperature range °C specified[a]	measured[b]	RTA measured[c] peak temp °C & results[d]	Measured anneal time (seconds)
In, pure	157	156-158	202; M	2
Sn/Cd; 67.8/32.2	177	176-179	200; M	1.5
Sn/Pb; 61.9/38.1	183	182-183	200; P	3
Sn/Zn; 91.9/9.0	199	199-202	202; M	4
Sn/Ag; 96.5/3.5	221	221-224	197-208; N	30
Sn/Ag; (repeat)	221	221-224	256; M	4.5
Sn; pure	232	232-236	253; M	3
Pb/Sb; 87.0/13.0	252	249-255	253; M	3
Cd/Zn; 82.6/17.4	266	266	251; N	3

[a]specified by the manufacturer
[b]determined in this study by using a melting point apparatus
[c]measured by the Heatpulse 210T Thermocouple (chromel-alumel) in a passive or open loop mode
[d]Results Key: M= sample melted and reformed; P= partial or very limited melting; N= no signs of melting

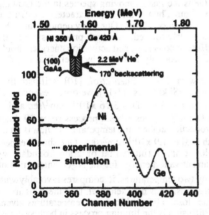

FIG. 1. 2.2 MeV ^{4}He$^+$ RBS spectrum for as deposited Ge/Ni/GaAs sample with corresponding RUMP simulation (random orientation for both). Inset details scattering geometry.

FIG. 2. 2.2 MeV ^{4}He$^+$ RBS spectrum for as deposited Au/Ge/Ni/GaAs sample with corresponding RUMP simulation (random orientation for both). Inset details scattering geometry. Asterisk * corresponds to Ge surface signal.

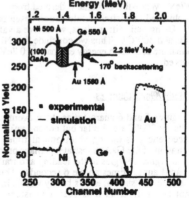
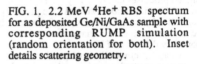

However, we believe that this peak is due to low temperature movement of Ge through Au, as has been documented by Ingrey and Maclaurin [17].

The rapid thermal annealed Ge/Ni/GaAs samples have been examined with Rutherford backscattering using a 2.2 MeV ^{4}He$^+$ beam. We simultaneously employ detectors at both the backscattering (170°) and grazing angles (5° from the plane or 85° from the sample normal). Use of channeling methods permit suppression of the substrate GaAs edge signal and facilitate resolution of Ni peak displacements. Figure 3 shows results taken from Ge/Ni/GaAs samples that were rapid thermal annealed at a nominal temperature of 250°C (results from samples annealed at 200°C are similar though not shown). Enhanced resolution permits measurements of the Ge signal full width half maximum (FWHM). Using a stopping power ε_{Ge} (calculated for this geometry) [16], and assuming a density of 4.42 x10^{22} atoms/cm^3 for Ge, we convert the energy loss ΔE to a Ge layer thickness. The decrease of Ge width vs time$^{1/2}$ is plotted in Fig.4.

Inspection of Fig. 3 shows that for increasing anneal times the growth of a peak at about channel 320 occurs. RUMP simulations indicate that this peak should correspond to the expected position of a Ni$_x$Ge phase. Recent work argues that x = 2 for the growth of a Ni$_2$Ge phase at low temperatures [5-7]. Slope changes of the higher energy edge of the Ni peak, together with its shift to the right, the dissappearance of the valley between the Ni and Ge peaks, and the consumption of Ge, all imply fast growth of a Ni$_2$Ge phase at 250°C.

By examining Fig. 4 we see that that diffusion limited kinetics are obeyed. Linear kinetics cannot be fitted to the observed data, although for very short times some fall-off occurs. Consumption of Ge is most likely interface dominated at first, perhaps affected by surface exides, then changing to diffusion controlled at later times. Since Ni reacts on deposition with GaAs [9,14,15] growth of the Ni$_x$GaAs [8-9] probably occurs in competition with the reaction of Ge. Ge is evaporated via e-beam onto the Ni layer in our samples, possibly heating the Ni and causing further reaction with GaAs. Unfortunately, without an appropriate marker experiment, we cannot be sure whether Ni or Ge is the fastest moving species in the growing Ni$_x$Ge phase. Lateral Ni$_x$GaAs growth studies from Chen et al. [10], characterize diffusion limited processes controlled by Ni movement at low temperatures. Our study demonstrates that Ni and Ge interactions on GaAs are probably similarly limited. As both processes are likely controlled by Ni diffusion, rates of reaction, diffusion constants, and activation energies should be similar. We expect that consumption of Ge is probably mediated by interstitial or grain boundary diffusion of Ni.

A careful comparison of our kinetics data with that of Chen et al. [10] is illustrative. Borrowing methods used by workers in analyzing Ni$_2$Si kinetics [11-13], we obtain estimates of the diffusivities of Ni during Ge consumption: for 200°C; D = 3.76 x10^{-14} cm^2/sec and at 250°C; D = 1.93 x 10^{-13} cm^2/sec. Althought D$_{Ni}$(200°C) describes a process some 20 times slower that that indicated for lateral Ni$_2$GaAs production at the same temperature [10], it is much greater that the value observed for Ni$_2$Si growth [13] of 1.9 x10^{-16} cm^2/sec. One should expect this as different diffusion mechanisms no doubt apply (interstitial or grain boundary vs substitutional). Given the above, we estimate an activation energy for the Ni-Ge interaction of $\approx$ 0.7eV. While this estimate is derived from only two temperatures, it compares favorably with one obtained by a similar calculation based on diffusivities of Chen et al. (see their Figure 8) [10]. Using their diffusivities, we derive a value close to 0.7eV (from two temperatures given, 200°C and 300°C). Thus we conclude that Ni diffusion is the limiting process in both systems, and that the activation energy estimate of 0.7eV is consistent with the very rapid intermixing rates observed.

Au/Ge/Ni/GaAs Samples

Figures 5 and 6 characterize the n ovements of Ni and Ge with and without Au (at 250°C, results for 200°C are similar but not shown). The early shift of the low energy Ni edge toward reduced energies during short anneal times is apparent. This is consistent with rapid Ni penetration of the GaAs substrate, perhaps producing Ni$_2$GaAs [8-10]. Much more unusual is a shift to higher energies in the leading and trailing edges of the Ni peak, seen in Figure 5, with increased anneal times above 6 seconds. This feature is absent at 200°C for the same anneal times (data not shown) and is barely visible in the Au/Ge/Ni/GaAs samples (Fig. 6). This shift is not an artifact, as it is seen consistently in Fig. 3 for RTA at 250°C (and also in a

FIG. 3. <100> Aligned RBS spectra for Ge/Ni/GaAs samples rapid thermal annealed at 250°C. Thick solid line (as deposited), plus signs (+) (4.5s anneal), thin solid line (12s), squares (28.5s). Curves for annealed samples merge below channel ~275.

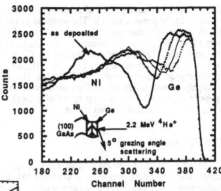

FIG. 4. Parabolic kinetics of Ge layer consumption for rapid thermal annealed Ge/Ni/GaAs samples. (time measurement error ± 0.5s).

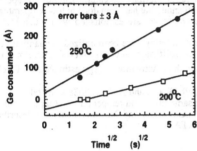

FIG. 5. RBS spectra (random) for Ge/Ni/GaAs samples annealed at 250°C. Same geometry as Fig. 1. Letter "x" sumbols (as deposited), plus signs (+) (6s), solid squares (28.5s).

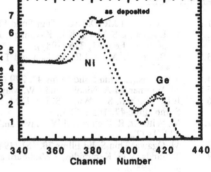

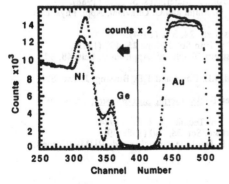

FIG. 6. RBS spectra for Au/Ge/Ni/GaAs samples (random, same geometry as Fig. 2), that were rapid thermal annealed at 250°C. Letter "x" symbols (as deposited), plus signs (+) (6s), solid squares (28.5s). Counts below channel 400 are multiplied by two.

This work was supported by DARPA (contract number DAAL01-88-K-0828)

362

complementary grazing angle channeling plot for samples annealed at 200°C, though this is not included in the present study). The amounts of Ge shifting to lower energies are similar in both metallizations for anneals of the same times and temperatures. This is also true for Ni. These data (Figs. 5, 6) show Ni movement into GaAs and Ge simultaneously, with Ge penetrating to deeper levels. The changes in slope of the Ge and Ni signal edges (both leading and trailing) with increased anneal times, plus the "filling in of valleys" between Ge and Ni, imply interdiffusion and growth of a Ni_xGe phase. The spectra of the Au/Ge/Ni samples show that Au is largely passive while Ni and Ge interdiffuse. At 250°C some Au is shifting to lower energies (greater depths) for the longest times of anneal. Gold may also serve as a medium through which Ge moves, as described above.

CONCLUSIONS

Small samples of low melting temperature elements and eutectic alloys have been used for temperature standardization of a commercial rapid thermal annealer. Employing special techniques, a kinetics analysis using rapid thermal annealing has been done at low temperatures and short anneal times. Anneal times and temperatures are established with much higher accuracy than previously possible.

Rapid thermal annealing and Rutherford backscattering have been utilized to quantitatively characterize the kinetics controlling Ni and Ge interdiffusion on GaAs. Ni diffusion limits Ge consumption, which follows parabolic kinetics with an activation energy estimate of ≈ 0.7eV, similar to Ni behavior in other studies.

Ge-Ni interdiffusion is similar in both the Ge/Ni/GaAs and Au/Ge/Ni/GaAs metallizations, with like amounts of Ge being consumed at the same times and temperatures. Au plays only a limited role under these initial circumstances. Gold, Ge, and Ni movements are described without ambiguity, and may be controlled using RTA.

REFERENCES

1. S.S. Gill, and J.R. Dawsey, Thin Solid Films, **167**, 161 (1988).
2. L.H. Allen, L.S. Hung, K.L. Kavanagh, J.R. Phillips, A.J. Yu, and J.W. Mayer, Appl. Phys. Lett. **51**, 326 (1987).
3. G. Bahir, J.L. Merz, J.R. Abelson and T.W. Sigmon, J. Electron. Mat., **16**, 257 (1987).
4. T.G. Finstad, Thin Solid Films **47**, 279 (1977).
5. M. Wittmer, M-A. Nicolet, and J.W. Mayer, Thin Solid Films **42**, 51 (1977).
6. E.D. Marshall, S.S. Wu, C.S. Pai, D.M. Scott, and S.S. Lau, Mater. Res. Soc. Symp. Proc.**47**, 161 (1985).
7. M. Wittmer, R. Pretorius, J.W. Mayer and M-A. Nicolet, Solid-State Electron.**20**, 433 (1977).
8. M. Ogawa, Thin Solid Films **70**, 181 (1980).
9. T. Sands, V.G. Keramidas, J.Washburn and R.Gronsky, Appl. Phys. Lett. **48**, 4042.(1986).
10. S.H. Chen, C.B. Carter, C.J. Palmstrøm, J. Mater. Res. **3**, 1385 (1988).
11. L.R. Zheng, L.S. Hura, J.W. Mayer, G. Majni and G. Ottaviani, Appl. Phys. Lett. **41**, 646 (1972).
12. J.M. Poate, T.C. Tisone, Appl. Phys. Lett. **24**, 391 (1974).
13. K.N. Tu, W.K. Chu, and J.W. Mayer, Thin Solid Films **25**, 403 (1975),
14. S.H. Chen, C.B. Carter, C.J., Palmstrøm, T. Ohashi, Appl. Phys. Lett. **48**, 803 (1986).
15. W.F. Egelhoff, jr., D.A. Stiegerwald, J.E. Rowe and T.D. Bussing, J. Vac. Sci. Technol. A **6**, 1495 (1988).
16. W.K. Chu, J.W. Mayer, M-A. Nicolet, Backscattering Spectrometry, (Academic Press, Orlando, 1978), p. 145.
17. S. Ingrey and B. MacLaurin, J. Vac. Sci. Technol. A **2**, 258 (1984).
18. E. Relling and A.P. Botha, App. Surface. Sci. **35**, 380 (1988-89).
19. M.O. Thompson, R.C. Cochran, and L. Doolittle, computer code RUMP (Cornell, 1985).

Au-Ge-Ni-Ti OHMIC CONTACTS ON GALLIUM ARSENIDE

K. B. ALEXANDER* AND W. M. STOBBS**
*Metals and Ceramics Division, Oak Ridge National Laboratory, P. O. Box 2008, Oak Ridge, TN 37831-6376
**Department of Materials Science and Metallurgy, Cambridge University, Cambridge, England CB2 3QZ

ABSTRACT

The influence of titanium additions on the microstructure of Au-Ge-Ni ohmic contacts on gallium arsenide has been evaluated. Sequentially deposited layers of Ge, Au and Ni were topped with a titanium layer. The titanium formed a native titanium oxide on the upper surface of the contact which helped to maintain a continuous film during the anneal. Both the as-deposited and the annealed microstructures were studied with the use of electron microscopy techniques. In order to examine thoroughly the various phases which form in the annealed contact, unique specimen preparation procedures were used to fabricate a single plan-view specimen in which it was possible to examine the microstructure at various depths.

INTRODUCTION

As GaAs device technology becomes more sophisticated, the performance of metallic contacts to GaAs becomes much more critical. Understanding the relationship between the microstructural development, the interfacial properties and the electrical performance is therefore crucial to the development of these contacts. A metallic contact to a semiconductor must result in very low contact resistivity and be as microstructurally uniform as possible. As device dimensions become smaller and smaller, the extent of uniformity becomes especially critical as the contact pad size approaches the scale of the inhomogeneity.

Ohmic contacts on GaAs are often formed by depositing a thin film on the GaAs, then annealing the film above the eutectic temperature. As the film melts, co-mixing "cleans" the surface of any oxides and hydrocarbons, gallium and arsenic are incorporated into the melt and rapid intermixing occurs. Au-Ge-Ni contacts on GaAs have been extensively used for many years.[1] The germanium forms a low melting eutectic (360°C) with the gold, enhances the reactivity of the gold with the GaAs[2] and serves as a dopant replacing gallium in the GaAs structure. Nickel enhances the diffusion of germanium into the GaAs[3,4], catalyzes the reaction of gold with the GaAs[2], changes the wetting characteristics of the melt[5], and enhances gallium and arsenic solubility into the melt.[4] A common problem observed in Au-Ge-Ni metal contacts on GaAs is that during the high temperature anneal used to make the contact ohmic, the film often agglomerates, resulting in a very nonuniform surface morphology. It has been proposed that this could be prevented by a capping layer such as silicon nitride or silicon oxide.[6] Ito and co-workers[7] used a titanium overlayer in an attempt to prevent a metallic Au-based film from agglomerating. In the present study, the effect of titanium on the microstructure of a Au-Ge-Ni-Ti metallic layer on GaAs was evaluated.

EXPERIMENTAL PROCEDURE

The GaAs substrate used had a 0.55 μm undoped buffer layer and a 0.2 μm Si-doped layer with N = 1.9 x 10^{18} cm^{-3}; both layers were grown by an MBE process. All samples were etched with 1:1 HCl/H$_2$O for one minute and subsequently rinsed in water. The germanium was deposited by a boat evaporation technique, whereas the Au, Ni, and Ti were all deposited by an electron gun technique. The anneal used was 40 sec at 430 C in a N$_2$/H$_2$ atmosphere. This annealing condition yielded the lowest contact resistivity, which was similar to that obtained with Au-Ge-Ni ohmic contacts. Unlike the specimens fabricated for the TEM studies, those fabricated for the electrical property measurements were subjected to an argon plasma treatment (250V, 3 mTorr) prior to contact deposition. Specimens were examined with a Philips EM400T electron microscope equipped with an energy dispersive spectrometer (EDS) and a Philips CM30 equipped with a Gatan parallel collection electron energy loss spectrometer (PEELS). For the EDS work, probe sizes between 5 and 20 nm were used, depending on the scale of the phase being examined. For the PEELS analysis, nanoprobe mode was used to obtain a 7 nm probe. The PEELS spectra were collected in diffraction mode (image-coupled). All specimens were studied in both edge-on and plan-view sections. The specimen preparation technique used was similar to that outlined in Newcomb et al.[8] Care was taken while preparing the plan-view specimens to ensure that the thin area traversed through the substrate and all subsequent layers. The geometry of the thinned plan-view specimen thus produced is shown in Figure 1.

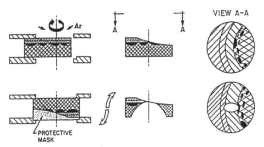

Figure 1: Geometry for fabricating plan-view specimens.

RESULTS AND DISCUSSION

The thicknesses of the as-deposited layers were measured from a cross-sectioned TEM specimen to be 17 nm, 55 nm, 45 nm and 17 nm for the Ge, Au, Ni and Ti layers respectively. A cross section of the annealed specimen is shown in Fig. 2. The microanalysis results performed on the cross-sectioned and plan-view specimens will be presented beginning from the GaAs substrate and working upwards through the overlayers. A schematic of Fig. 2 will be shown with each spectrum with the analysis location marked. The spectrum from the GaAs is shown in Figure 3a. The Ga/As ratio is larger than one, probably due to arsenic outgassing at the annealing temperature as well as to gallium being fluoresced by the arsenic-K radiation. Figure 3b shows the spectrum obtained from the darkly-imaging pit protruding into the GaAs substrate. These pits which form in the GaAs contain primarily gold and gallium. A wide range of gold/gallium ratios was observed suggesting that the phases formed are non-equilibrium phases. Some of pits are polycrystalline and contain some porosity, as shown in the dark field image in Fig. 4a. The diffraction patterns from several pitted regions were indexed consistent with the Au_7Ga_2 phase.

The pits are formed as a result of liquid-phase etching and wetting. The gallium and gold react to form a liquid phase which etches into the GaAs surface, forming a pit in the GaAs bounded by the low energy {111} planes. Since GaAs is a non-centrosymmetric structure, the {111} surfaces terminate in either gallium or arsenic depending on the specific {111} orientation.[9] The pits are rectangular when viewed along the [001] direction. When the {111}-bounded pit is viewed in a [011] orientation, it will be bounded by the {001} and {111}-type planes as is seen in Fig. 2. Plan-view sections of the pit distribution at various levels in the GaAs are shown in Figs. 4b and 4c. The deepest regions show rectangular pits (Fig. 4b); as one gets closer to the uppermost GaAs surface, the cross-section through the deeper pits becomes larger and the small, recently nucleated pits are also observed (Fig. 4c).

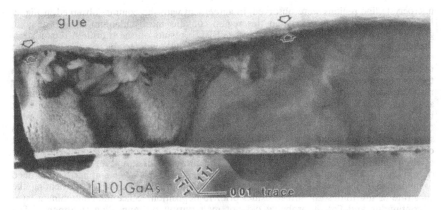

Figure 2: Bright field TEM image of the annealed contact on GaAs in cross-section.

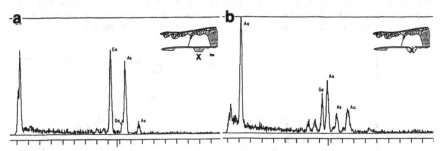

Figure 3: (a) EDS spectrum from the GaAs substrate (b) EDS spectrum from the dark pits observed in Fig. 2.

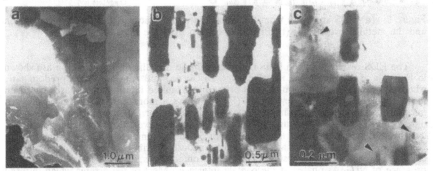

Figure 4: (a) Dark-field TEM image of the pits in plan-view. (b) Plan-view bright-field TEM image of the pits deep in the Ga-As substrate. (c) Image of the smallest pits near to the GaAs/metallic overlayer interface.

The transition region between the pits and the metallic phases appears as a layer about 10 nm thick. Small probe (5 nm) convergent beam electron diffraction patterns from this region are primarily consistent with amorphous material, however small microcrystalline regions were observed, in agreement with the contrast observed in Fig. 2. Microanalysis shows (Fig. 5a) that this region contains mostly titanium as well as gold, gallium and arsenic. Electron energy loss (PEELS) results on the transition region show that the layer is rich in titanium and oxygen (Fig. 5b). Any detectable carbon levels from this region were the same as those obtained from adjacent metallic phases and are probably due to TEM specimen surface contamination. A similar layer was observed in the as-deposited specimen; therefore, the layer must initially be a contaminant layer resulting from the fabrication process. The layer is thicker in the annealed specimens, suggesting that it is stabilized during the anneal by the addition of titanium. Recent work by Sands et al.[10] showed that an annealed Ni/GaAs couple retains a stable oxide/hydrocarbon layer; while annealed Pt and Pd layers on GaAs do not. It was suggested that the stronger affinity of nickel for oxygen may be the cause. In the present case, titanium has an even stronger affinity for oxygen, which explains the apparent stability of the contaminant layer upon annealing. The presence of this layer would clearly be detrimental to the contact resistance. The persistence of this layer emphasizes the need to minimize native oxide layers by either ion-beam sputter-cleaning or by heating the substrate in situ to desorb surface oxides prior to deposition. Since the contaminant layer is continuous and flat, it probably did not melt during the annealing treatment. The obvious question is then: How does the Au diffuse through this layer to form the pits in the GaAs? Studies[11] have shown that any holes or discontinuities in an amorphous film will serve as a conduit for diffusive flow and a pit will form directly below the hole in the film. It is suspected that a similar process has occurred in these specimens.

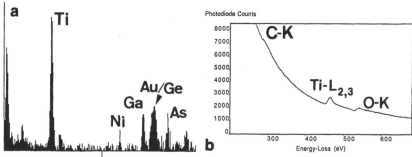

Figure 5: (a) EDS spectrum from the transition region between the GaAs/pitted region and the metallic overlayers (b) PEELS spectrum from the transition region.

The EDS results obtained from the phases in the metallic overlayer region are shown in Figs. 6 and 7. Within the metallic layer, Ni-Ge(As) and Au-rich phases predominate, whereas the uppermost region consists primarily of faulted intermetallic phases based on nickel-titanium. The plan-view image (Fig. 8) shows a more representative view of the microstructure since more grains are typically viewed. For example, in the cross-section image (Fig. 2) Ni-Ge(As) phases are about 200 nm; however, from the plan-view image it is clear that the phase varies in size from 50 to 500 nm. It is evident that the Ni-Ge(As) and Au-based phases are fairly homogeneously distributed throughout the metallic contact. It has been suggested[12,13] that low contact resistance is associated with the presence of Ni_xGeAs phases at the contact interface. Most of the arsenic which diffused to the metallic layers is found in the Ni-Ge phase, which is consistent with the lack of solubility of arsenic in Au.

An amorphous layer, indicated in Fig. 2, forms on the uppermost surface. Chemical analysis of this layer (Fig. 9a), taken from a region of the film adjacent to a gold-containing grain, demonstrates that the film contains titanium and oxygen. The upper surface of the ohmic contact is continuous (Fig. 9b) and undulated (Fig. 2). Obviously, the titanium oxide forms an effective capping layer, preventing the film from agglomerating while molten.

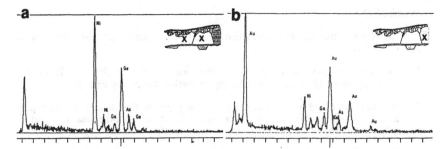

Figure 6: (a) Typical EDS spectrum from areas indicated in the schematic. (b) EDS spectrum from the area indicated in the schematic.

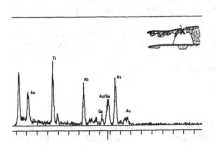

Figure 7: Typical EDS spectrum of the faulted grains in Fig. 2.

Figure 8: Dark-field plan-view TEM image of the metallic overlayers shown shown in Fig. 2. Several of the phases present are indicated.

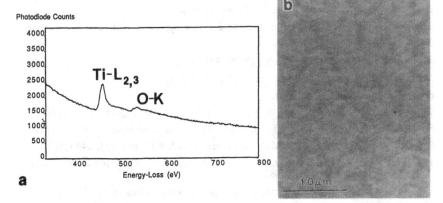

Figure 9: (a) PEELS spectrum of the uppermost capping layer. (b) SEM of the surface of the annealed ohmic contact.

CONCLUSIONS

The effect of titanium on the microstructure of these particular Au-Ge-Ni films on GaAs is threefold:

1. The titanium forms a titanium oxide on the surface of the metallic film. This forms a capping layer which prevents film agglomeration during the anneal.

2. Intermetallics based on Ni-Ti form primarily adjacent to the titanium oxide capping layer. This effectively ties up some nickel which would otherwise be used to enhance the ability of the film to form a low resistance contact.

3. The titanium diffuses down to the contaminant layer and stabilizes the contaminant layer.

The microstructure of these films after annealing consists of Ni-Ge(As), Ni-Ti and Au-based phases, with Au-Ga phases filling the etched-in pits in the GaAs. The Au-Ga phases vary somewhat in composition, but several of the etched-in pits have been identified as containing Au_7Ga_2. Adding less titanium to the ohmic contact would allow the titanium oxide capping layer to form while reducing the Ni-Ti phase formation and the thickness of the contaminant layer.

The use of a single multiply-sectioned plan-view specimen gives a much more representative view of the microstructure than cross-sectioned specimens. Cross-sectioned specimens, however, must be used in conjunction with the plan-view specimen in order to clearly map out the geometry of the microstructure being examined.

ACKNOWLEDGEMENTS

The authors thank Dr. R. Broom from IBM Zurich for support and for providing the specimens. Research sponsored in part by the Division of Materials Sciences, U.S. Department of Energy under contract DE-AC05-84OR21400 with Martin Marietta Energy Systems, Inc.

REFERENCES

1. N. Braslau, J. B. Gunn and J. L. Staples, Solid State Electron., 10, 381 (1967).

2. M. Ogawa, J. Appld. Phys., 51 (1), 406 (1980).

3. W. T. Anderson, Jr., A. Christou and J. E. Davey, IEEE J. of Solid State Circuits, SC-13 (4), 430 (1978).

4. N. Braslau, J. Vac. Sci. Tech., 19 (3), 803 (1981).

5. G. Y. Robinson, Solid State Elect., 18, 331 (1975).

6. F. Vidimari, Electron. Letters, 15, 675 (1979).

7. H. Ito, T. Ishibashi and T. Sugeta, Japan. J. Appl. Phys., 23 (8), L635 (1984).

8. S. B. Newcomb, C. B. Boothroyd and W. M. Stobbs, J. Microsc. 140 (2), 195 (1985).

9. T. Yoshiie, C. L. Bauer and A. G. Milnes, Thin Solid Films, 111, 149 (1984).

10. T. Sands, V. G. Keramidas, A. J. Yu, K.-M. Yu, R. Gronsky and J. Washburn, J. Mater. Res., 2 (2), 262 (1987).

11. C. B. Boothroyd, PhD Thesis, Cambridge University, Cambridge, U. K. (1986).

12. T. S. Kuan, P. E. Batson, T. N. Jackson, H. Rupprecht and E. L. Wilkie, J. Appl. Phys., 54 (12) 6952 (1983).

13. M. Grimshaw and A. Staton-Bevan, MRS Symp. Proc., 144, 589 (1989).

A Two-Step Process for the Formation of Au-Ge
Ohmic Contacts to n-GaAs

M.A.Dornath-Mohr[*], M.W.Cole[*], H.S.Lee[**], C.S.Wrenn[***], D.W.Eckart[*],
D.C.Fox[*], L.Yerke[*], W.H.Chang, R.T.Lareau, K.A.Jones[*], and F.Cosandy[****]
[*]Electronics Technology and Devices Laboratory, Fort Monmouth, NJ 07703.
[**]GEO-CENTERS, INC., NJ Operations, Lake Hopatcong, NJ 07849. Work
performed at U.S. Army ETDL, Fort Monmouth, NJ 07703.
[***]Vitronics, Inc., 15 Meridian Road, Eatontown, NJ 07724. Work performed
at U.S. Army ETDL, Fort Monmouth, NJ 07703.
[****]Department of Mechanics and Materials Science, Rutgers, The State
University of New Jersey, Piscataway, NJ 08855.

ABSTRACT

The formation of low temperature Au-Ge contacts to n-GaAs is a two-
step process. In the first step, the metals segregate into Au and Ge rich
regions and the intermixing of the Au and Ge with the Ga and As causes a
reduction in the barrier height. The second step occurs after extended
annealing, during which time Au and Ge continue to diffuse into the
substrate. An orthorhombic Au-Ga phase is formed and it is likely that
other Au-Ga or Ge-As phases are formed. The length of the extended anneal
is dependent upon the atomic percent of Ge in the film, with the 10 at. %
Ge taking 6 hr., the 27 at. % Ge taking 3 hr. and the 50 at. % Ge taking 9
hr. to become ohmic. The 75 at. % Ge sample doesn't show ohmic behavior
even after 33 hr. of annealing. The metal-semiconductor interface
configuration appears abrupt, showing no protrusions into the GaAs
substrate.

INTRODUCTION

A substantial effort has been made in the development of ohmic contact
fabrication processes due to their associatoin with GaAs devices. Although
a standard process for low resistance ohmic contacts to n-type GaAs by the
annealing of a Au-Ge film is available, the mechanisms for these contacts
are not well understood. Most studies center around the electrical
behavior [1-3] rather than the mechanisms or microstructure of ohmic
contacts. Reactions between the metals and semiconductor and the resulting
microstructure are especially important because the accompanying electrical
behavior may be affected significantly.

In order to obtain a current through the metal-semiconductor
interface, the barrier must be either removed or modified in the contact
formation process. Since low values of specific contact resistance, r_c,
are routinely obtained with alloyed contacts, a modification is taking
place, but there is no clear picture to what is happening
microstructurally. In this paper we investigate a possible mechanism for
ohmic contact formation, which contends that heavy Ge doping [4,5] of the
GaAs allows tunneling through the barrier layer in the semiconductor near
the metal semiconductor interface.

EXPERIMENTAL

The substrates used for transmission line method (TLM) and Schottky barrier height measurements were (100) grown GaAs wafers doped with Si to 10^{17} at/cm^3. The substrates used for the material characterization techniques were (100) semi-insulating GaAs wafers. Following an organic degrease step, the standard lithographic techniques were used to define the TLM and Schottky barrier height test patterns. Mesas were defined by an etch of $H_2SO_4{:}H_2O_2{:}DI$ (1:8:100). Prior to metallization, the wafers were etched with a solution of $NH_4OH{:}DI$ (1:100) to remove the native oxide layer. Au-Ge films with compositions of 10, 27 (eutectic composition), 50, and 75 atomic percent Ge were formed by sequential electron beam evaporation of the specified amount of Au and Ge with the total thickness held constant at 75 nm. To study the effects of an encapsulation layer, a 100 nm Si_3N_4 layer was deposited by PECVD on some of the samples before annealing. Anneals < 10 minutes were carried out in a nitrogen ambient using a commercial RTA system (HEATPULSE - 410) and longer anneals were performed in an argon ambient using a conventional furnace. The temperature was held constant at 320|C while the time was varied up to 33 hr. The 320|C temperature was chosen to insure that the annealing temperature stayed below the Au-Ge (356|C) and Au-Ga (341|C) eutectics.

TLM and Schottky barrier height measurements, scanning electron microscopy/energy dispersive spectroscopy (SEM/EDS), transmission electron microscopy (TEM), Rutherford backscattering spectroscopy (RBS), Auger electron spectroscopy (AES), and secondary ion mass spectroscopy (SIMS) were used in this study as described in our previous work [6].

RESULTS

Figure 1 [6] shows the electrical history of a low temperature ohmic contact (uncapped, 27 at. % Ge). A significant increase in current flow can be seen within the first few minutes of annealing. The TLM curve remains essentially stable after 5 min., until at some point following a 3 hr. anneal when the devices show ohmic behavior with $r_c = 5 \times 10^{-6}$ ohm-cm^2. The other compositions studied exhibit similar behavior, stabilizing after a few minutes but taking longer than 3 hr. to become ohmic.

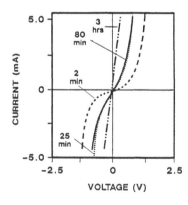

Figure 1. IV curves as determined by TLM measurements for samples annealed for 2, 25, 80, and 180 min.

The 10 at. % Ge samples become ohmic after a 6 hr. anneal with r_c = 8 x 10^{-6} ohm-cm^2. The 50 at. % Ge samples require a 9 hr. anneal before exibiting ohmic behavior with r_c = 1 x 10^{-2} ohm-cm^2. The 75 at. % Ge samples are not ohmic even after a 33 hr. anneal. As seen in Figure 2, a minimum in r_c occurs in the vicinity of 27 at. % Ge, indicating that there is an optimum composition for low temperature, low resistance contacts. Samples with a Si_3N_4 cap, (27 at. % Ge) are ohmic after 3 hr. of annealing with r_c = 1 x 10^{-4} ohm-cm^2. Most of the decrease in the barrier height, from 0.75 to 0.40 eV, occurs during the first 5 min. of annealing as shown in Figure 3, corresponding to the initial rapid change in current flow.

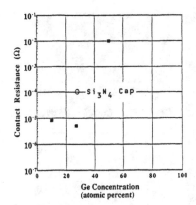

Figure 2. Specific contact resistance determined by TLM, plotted as a function of composition.

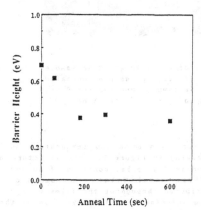

Figure 3. Barrier heights determined electrically, plotted as a function of annealing time.

SEM plan view analysis shows segregation of the metals into Au and Ge rich regions almost immediately upon annealing [6]. This segregation of the metals occurs for all the compositions examined. However, as seen in the SEM micrographs in Figure 4, the morphology of the ohmic contact varies greatly with composition. Figure 4a shows a 10 at. % Ge sample after 6 hr. of annealing. Figure 4b shows a 27 at. % Ge sample after a 3 hr. anneal and Figure 4c shows a 50 at. % Ge sample following a 9 hr. anneal. Figure 4d shows a 75 at. % Ge sample after 33 hr. of annealing. A backscatter electron image of the 75 at. % Ge samples is shown in Figure 4e.

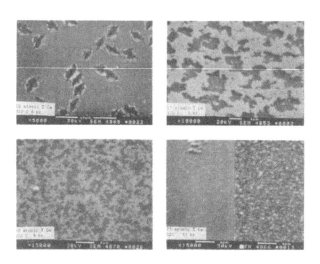

Figure 4. SEM micrographs of a.) a 10 at. % Ge sample
annealed for 6 hr., b.) a 27 at. % Ge samples annealed
for 3 hr., c.) a 50 at. % Ge sample annealed for 9 hr.,
and d.) a 75 at. % Ge sample annealed for 33 hr.

The sequence of events in the formation of a low temperature ohmic
contact is shown in the schematic drawing in Figure 5. The structure of
the contact before annealing, as shown in Figure 5a, consists of Au and Ge
films evaporated sequentially. These metal films are composed of
polyscrystals that are randomly distributed throughout the films with some
overlapping occuring at the Ge and Au interface. TEM analysis shows that
an oxide layer, approximately 2 nm thick, is present at the metal-
semiconductor interface in all the samples. During a brief, 5 min.
anneal, the metals segregate, with the Au and Ge grains extending the
entire vertical length of the metal film, creating Au and Ge rich regions.
There is some diffusion of the As and Ga out of the substrate and diffusion
of the Au and Ge into the substrate as seen in Figure 5b. During a 3 hr.
anneal the As and Ga continue to diffuse into the metal layer. RBS shows
the rate of diffusion decreases with increasing anneal time. AES and SIMS
show indications of Ga and As at the metal surface and estimate the
concentration of these species to be in the range of 0.0001 to 0.1 % Ga and
As at the top surface layer. SIMS ion images show that the As
predominately occupies the same regions as the Ge and the Ga is spread
randomly throughout the area. Some As has evaporated from the surface of

the metal layer and some Ga has formed a compound with Au as shown in Figure 5c. Preliminary TEM results reveal this compound to be an orthorhombic Au_2Ga phase [7]. It is likely that other Au_xGa_y and Ge_xAs_y phases exist. High Resolution TEM reveals a smooth, abrupt interface between the metal and substrate deliniated by a 2 nm GaO_x layer. When a Si_3N_4 cap is deposited on the sample before annealing, the As cannot evaporate from the surface and gathers at the interface between the Si_3N_4 and metal layer as seen in Figure 5d.

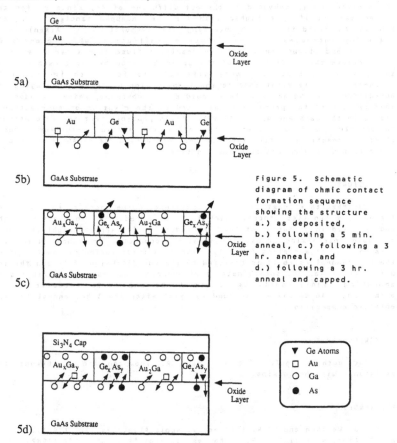

Figure 5. Schematic diagram of ohmic contact formation sequence showing the structure a.) as deposited, b.) following a 5 min. anneal, c.) following a 3 hr. anneal, and d.) following a 3 hr. anneal and capped.

DISCUSSION

We believe that the process to form these low temperature, shallow ohmic contacts has at least two steps. The first step occurs within the first few minutes of annealing and is characterized by dramatic changes in the electrical properties and substantial atomic movement across the metal-semiconductor interface. The second step occurs after several hours of annealing, the exact length of which depends on the composition. Preliminary studies show that during this extended annealing time, solid phase transformation occurs, causing an orthorhombic Au_2Ga phase to appear in the eutectic composition. Other Au-Ga and Ge-As phases have not been ruled out. Since the 75 at. % Ge sample does not have the large, distinct Au and Ge rich regions typical of the other compositions and it did not show ohmic behavior, these two phenomena might be related.

The metal segregation and barrier height reduction that occur in step 1 agrees with Sinha and Poate's [8] belief that interface mixing seems to account for the reduction in the barrier height. Our results suggest that Ge in-diffusion is enhanced by the out-diffusion of As, since r_c for the capped samples is much higher than for the uncapped samples. The cap hinders As out-diffusion by blocking its evaporation. Arsenic out-diffusion increases the rate of Ga out-diffusion, which creates Ga vacancies and allows the Ge to more readily diffuse into these sites. We also believe the in-diffusion of the Ge is assisted by the presence of Au since ohmic contacts are more difficult to form with increased Ge concentration. Au might enhance the in-diffusion of Ge by encouraging the out-diffusion of Ga and As. This rapid atomic movement, which includes the in-diffusion of Au, probably is assisted by the creation of Au electronic states in the GaAs energy gap [9,10]. It is probable that the Au continues to promote the in-diffusion of Ge after the initial interface induced chemical reactions since the samples containing less Au form poorer contacts and require longer annealing times.

CONCLUSIONS

Our results suggest that low temperature, Au-Ge contacts to n-GaAs are formed by a two step process which includes a reduction in barrier heigh. and assisted in-diffusion of Ge. The reduction in barrier height is caused by rapid intermixing across the metal-semiconductor interface during the first 5 min. of annealing (step 1). The slow in-diffusion of the Ge during the long anneal (step 2) is assisted by the out-diffusion of As and the in-diffusion of Au. The smooth metal-semiconductor interface is attributed to annealing significantly below the Au-Ge and Au-Ga eutectics. An orthorhombic Au_2Ga phase was found to exist after the 3 hr. anneal for the eutectic composition.

ACKNOWLEDGMENTS

The authors would like to thank R. Thompson and A. DeAnni for assistance with processing.

REFERENCES

1. J. G. Werthen and D. R. Scifres, J. Appl. Phys. 52, 1127 (1981).
2. O. Aina, W. Katz, and B. J. Baliga, J. Appl. Phys. 53, 777 (1982).
3. A. Iliadis, J. Vac. Sci. & Technol. B5, 1340 (1987).
4. G. Y. Robinson, Solid-State Electron. 18, 331 (1975).
5. W. J. Devlin, R. A. Stall , C. E. C. Wood, and L. F. Eastman, Solid-State Electron. 23, 823 (1980).
6. M. A. Dornath-Mohr, M.W. Cole, H.S. Lee, D.C. Fox, D.W. Eckart, L. Yerke, C.S. Wrenn, R.T. Lareau, W.H. Chang, K.A. Jones, F. Cosandey, to be published in J. Electon. Mater., Special Edition.
7. M. W. Cole, et al, to be published.
8. A. K. Sinha and J. M. Poate, Appl. Phys. Lett. 23, 666 (1973).
9. J. Gyulai, J. W. Mayer, V. Rodriguez, A. Y. C. Yu, and H. J. Gopen, J. Appl. Phys. 42, 3578 (1971).
10. M. Jaros and H. L. Hartnagel, Solid-State Electron. 18, 1029 (1975).

THE RELATIONSHIP BETWEEN MICROSTRUCTURE AND CONTACT RESISTANCE IN NiAuGe/ZrB$_2$/Au OHMIC CONTACTS TO GaAs.

M.P. GRIMSHAW*†, A.E. STATON-BEVAN*, J. HERNIMAN** and D.A. ALLAN**.
*Imperial College of Science, Technology and Medicine, Department of Materials, Prince Consort Road, London SW7 2BP, U.K.
** British Telecom Research Laboratories, Martlesham Heath, Ipswich IP5 7RE, U.K.
† now at The Cavendish Laboratory, Cambridge University, Cambridge CB3 OHE, U.K.

ABSTRACT.

The microstructure and contact resistance of NiAuGe contacts to n-type GaAs were determined as a function of initial contact composition. The contact microstructures were found to contain varying amounts of of α, α' and β (or Au$_7$Ga$_2$) Au-Ga, epitaxial Ge, NiGe and NiGeAs phases. A previously unidentified NiAs$_x$(Zr,B) phase was also observed. The contact resistance was found to vary between 0.22-0.38±0.03Ωmm. Comparison of the microstructural and contact resistance data revealed that the ohmic formation models based on (i) the formation of a recrystallised n+ GaAs layer and (ii) the presence of a graded Ge/GaAs heterojunction were not applicable to this contact system.

I. INTRODUCTION.

The NiAuGe ternary metallization is widely used as an ohmic contact to n-type GaAs. A variety of contact compositions and annealing treatments are employed. Generally the Au:Ge ratio is near the eutectic composition (12.4wt.% Ge) and the Ni concentration varies from approximately 4 to 25 wt.%, refer to Fig.1, [1-19]. The Au-rich deviations from the Au/Ge eutectic composition are the result of an additional Au overlayer necessary for electrical contact. The additional Au has been shown to result in increased metal penetration into the GaAs [20]. The contacts studied herein contain a ZrB$_2$ diffusion barrier which separates the additional Au layers from the NiAuGe ohmic contact, Fig.2. In this study the effects of varying (i) the Au layer thickness (specimens A-F) and (ii) the Ni layer thickness (specimens D, G & H) in the NiAuGe ohmic contact are investigated. A discussion of the relationship between the microstructure and the contact resistance is also presented.

II. EXPERIMENTAL.

The experimental details of the contact fabrication process and transmission electron microscopy (T.E.M.) have been presented in previous publications [21,22] therefore only a brief outline will be made here. The e-beam deposited metals and diffusion barrier layer thicknesses are shown in Table I. The weight percentage values quoted in the text are calculated from bulk density data and are therefore only approximate. Both blanket layer and patterned specimens were prepared. All specimens were given identical optical annealing treatments

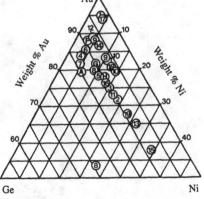

Fig.1. Pseudo-ternary phase diagram of as-deposited NiAuGe contact compositions [1-19].

Table.I. As-deposited layer thicknesses of
specimens A-H.

	Layer Thicknesses (nm)				
Specimen	Ni	Au	Ge	ZrB_2	Au
A	5	30	20	50	20
B	5	35	20	50	20
C	5	40	20	50	20
D	5	45	20	50	20
E	5	50	20	50	20
F	5	55	20	50	20
G	10	45	20	50	20
H	15	45	20	50	20

Au
ZrB_2
Ge
Au
Ni
GaAs

Fig.2. Typical as-deposited layer structure.

(440°C peak temperature reached in 25 s). The contact resistance, R_c (Ωmm), was measured using the extrapolation method on a T.L.M. pattern with a $100\times150\mu m^2$ pad size [22,23]. The contact resistance values represent means of at least 40 measurements. A Jeol 120CX transmission electron microscope with E.D.X. analysis was used for specimen characterisation.

III. RESULTS.

Variation in the Au Layer Thickness (Specimens A-F).

The microstructure of the least Au-rich specimen, A (Ni:Au:Ge, 6.1:79.4:14.5 wt.%), is shown in Fig.3(a). The metallization has balled-up upon melting forming cavities (light contrast) between the substrate and diffusion barrier. The remaining metal islands consist of an α Au(Ga) matrix phase containing plates of hexagonal α' Au-Ga, epitaxial Ge and orthorhombic NiGe phases [21]. A high proportion of the ternary NiGeAs phase is found to nucleate at the interfaces

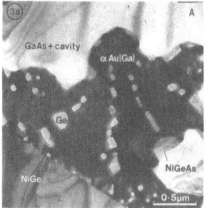

Fig.3(a)&(b). Plan-view T.E.M. images of (a) specimen A and (b) specimen F.

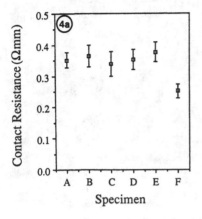 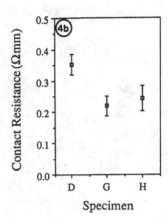

Fig.4(a)&(b). Contact resistance as a function of (a) Au layer thickness and (b) Ni layer thickness.

between the metal islands and cavities. It is proposed that the formation of this phase is enhanced by the presence of the adjacent free GaAs surface.

The increased Au layer thickness in specimen B (Ni:Au:Ge, 5.4:81.8:12.8 wt.%) effectively prevented cavity formation. The microstructure of specimen B, not shown, was found to be the same as that observed in the non-balled-up regions of specimen A, except for a dramatic decrease in the proportion of the NiGeAs phase. A previously unidentified tetragonal phase with lattice parameters of $a_o=0.39\pm0.02$ and $c_o=0.94\pm0.05$nm was observed. This phase occurred as rods with average dimensions of 180×30nm, refer to Figs.3(b)&5(b) (light contrast). The composition of this phase was tentatively estimated to be $NiAs_x(Zr,B)$, where x=2.7±0.7. The small grain size and close proximity of the diffusion barrier made accurate E.D.X. analysis difficult, therefore it is not known if Zr or B (which cannot be detected) is actually associated with this phase.

Further increase in the Au layer thickness (specimens C-F) resulted in increased dissociation of the GaAs substrate. This effect was illustrated by an increase in the proportions of the NiGeAs and $NiAs_x(Zr,B)$ phases. The β Au-Ga (or Au_7Ga_2) phase was observed in the Au-rich matrix of specimen F (Ni:Au:Ge, 3.7:87.6:8.7 wt.%), Fig.3(b). Formation of the β Au-Ga phase resulted in local penetration of the GaAs. The variation in the contact resistance as a function of Au concentration, Fig.4(a), is relatively uniform with the exception of specimen F which shows a significant decrease in contact resistance.

Variation in the Ni Layer Thickness (Specimens D, G & H)

Increase in the Ni layer thickness resulted in an increase in the proportions of the β Au-Ga, NiGeAs and $NiAs_x(Zr,B)$ phases and a decrease in the Ge and NiGe phases, Figs.5(a)&(b). The cross-sectional T.E.M. images of specimens D (Ni:Au:Ge, 4.4:85.2:10.4 wt.%) and H (Ni:Au:Ge, 12.0:78.4:9.6 wt.%), Fig.6(a)&(b), show that the increase in the Ni layer thickness caused increased metal penetration into the substrate resulting in the formation of a bilayered morphology. The increase in the Ni layer thickness is thought to result in an increase in the proportion of Ni-Ga-As formed during the initial stages of the annealing process. The Ni-Ga-As phase is subsequently transformed into the NiGeAs phase, refer to the discussion below. The microstructure of such bilayered high Ni concentration contacts has been reported previously [14,10,11].

The contact resistance was found to decrease as a function of increasing Ni layer thickness between specimens D and G (Ni:Au:Ge,8.4:81.7:9.9 wt.%), Fig.4(b), with a possible slight increase between specimens G and H.

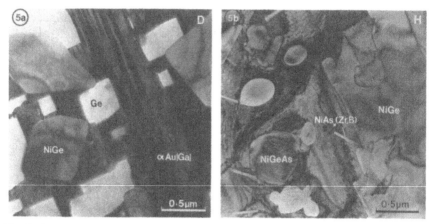

Fig.5(a)&(b). Plan-view T.E.M. images of (a) specimen D and (b) specimen H.

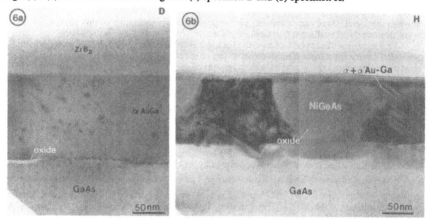

Fig.6(a)&(b). Cross-sectional T.E.M. images of (a) specimen D and (b) specimen H.

IV. DISCUSSION.

In this section the relationship between the microstructure and contact resistance is discussed. An attempt is made to reconcile the results with the various models for ohmic formation.

The most popular explanation for ohmic conduction is the tunneling of electrons through a narrow Schottky barrier formed by the presence of a thin heavily doped n+ layer in the GaAs. The formation of such a layer may occur by indiffusion of a dopant species or by recrystallisation of doped GaAs. The Ge_{Ga} substitutional defect is the only defect known to form a suitable donor level in the GaAs, however, formation of a high concentration of such a defect is inhibited by the amphoteric nature of Ge and the effect of dopant compensation.

The diffusion based model [24,25] relies on the formation of an excess of Ga vacancies, V_{Ga}, as a result of the strong Au-Ga reaction. The indiffusing Ge may then occupy a majority of V_{Ga} sites. If the formation of an ohmic contact does depend on this proposed interdiffusion process then the increase in the proportion of the more Ga-rich β Au-Ga phase associated with the fall in contact resistance observed for specimen F could be interpreted as being due to a corresponding increase in the concentration of V_{Ga} sites. The Au-Ga reaction may not necessarily result in a stoichiometric imbalance in the GaAs substrate since As has been shown both to evaporate from [26] and react with [14] the metallization.

Recrystallisation of GaAs has been shown to occur through the decomposition of Ni-Ga-As phases [27] as in reaction (1) below. In order that the recrystallised GaAs has n+ doping Ge must

$$Ni\text{-}Ga\text{-}As + Ge \rightarrow Ni\text{-}Ge + GaAs \qquad (1)$$

diffuse into the Ni-Ga-As phase preferentially on the Ga sublattice before or during decomposition. It could be proposed the Ni-Ga-As decomposition reaction occurs simultaneously with the formation of the NiGeAs phase, reaction (2) below, depending on the local Ge concentration. The volume of recrystallised GaAs would be expected to increase with increase in the proportion of the Ni-Ga-As phase which would in turn be dependent on the Ni layer thickness, however no evidence of GaAs recrystallisation was observed even in the most Ni-rich specimen, H.

The formation reaction of the NiGeAs phase, reaction (2) below, has also been proposed as

$$Ni\text{-}Ga\text{-}As + Au + Ge \rightarrow NiGeAs + Au(Ga) \qquad (2)$$

the means of introduction of the Ge dopant [5,9,14]. Here it must be assumed that the Ge diffusion on the Ga sublattice of the Ni-Ga-As phase continues into the Ga sublattice of the GaAs substrate below. The slight increase (within the experimental error) in the contact resistance with significant increase in the proportion of the NiGeAs phase observed between specimens G and H suggests that this model is not responsible for the ohmic properties, however these results may also be explained by the effects of dopant compensation

The possibility of low Schottky barrier formation has recently been demonstrated for thin (9Å) As doped Ge layers under intermetallic phases such as NiAs [28]. It was proposed that such a structure could exist under a NiGeAs grain and thus be the cause of ohmic conduction. No such Ge layer was observed under the NiGeAs grains, however the resolution limit of the microscope (5Å) does not enable us to completely discount this model.

The presence of the epitaxial Ge phase suggests the possibility of ohmic conduction via a graded Ge/GaAs heterojunction [29]. The microstructural transition between specimens D and G in which there is a significant reduction in the proportion of the epitaxial Ge phase for a corresponding decrease in the contact resistance suggests that the epitaxial Ge phase is not essential for ohmic contact formation.

Evidence has been presented suggesting that the doping under a Au/Ge contact resembles a continuum of defect states throughout the GaAs band gap and that the hypothesized n+ layer is not present [30]. It is proposed that these defect states are the result of the disordering effect of the interdiffusion process on the GaAs band structure. Conduction in this environment may be due to tunneling [30] and phonon-assisted hopping [31] of electrons between defect states and recombination. The question arises as to whether such a combination of conduction mechanisms is capable of producing the reported temperature dependence of the contact resistance [6]. This model is consistent with the experimental results if the amount of disorder is considered to be related to the degree of dissociation observed in the contact metallization.

The ohmic properties may also be explained by the defect induced movement of the Fermi-level pinning position or the effects of strain on the GaAs band structure [32].

V. SUMMARY AND CONCLUSIONS.

The lowest contact resistance results occurred in specimens in which a significant amount of substrate dissociation was observed, as indicated by increase in the proportions of β Au-Ga, NiGeAs and NiAs$_x$(Zr,B) phases. These results were found to be consistent with ohmic contact formation models based on (i) Ga depletion of the substrate via the Au-Ga reaction, (ii) the formation of the NiGeAs phase and (iii) the introduction of a high density of defect states into the substrate via a diffusion based disordering process.

The models based on (i) the formation of a thin recrystallised n^+ GaAs layer and (ii) the presence of a graded Ge/GaAs heterojunction were found not to be responsible for the ohmic properties of this contact system.

Further information on point defect behaviour is required in order to elucidate the ohmic formation mechanism in these contacts.

ACKNOWLEDGEMENTS.

The authors wish to thank B.T.R.L. for funding and use of equipment, Professor D.W. Pashley for research facilities at Imperial College and S.E.R.C. for financial support.

REFERENCES.

[1] H. Goronkin, S. Tehrani, T. Remmel, P.L. Fejes and K.J. Johnson, I.E.E.E. Trans. Electron. Dev. 36(2), 281 (1989)

[2] W.O. Barnard, H.J. Strydom, M.M. Kruger and C. Schildhauer and B.M. Lacquet, Nucl. Inst. & Meth. in Phys. Res. B35, 238 (1988).

[3] R.K. Ball, Thin Solid Films 176, 55 (1989).

[4] F. Vidimari, Electron. Letts. 15, 674 (1979).

[5] R.A. Bruce and G.R. Piercy, Sol.State Electron. 30, 729 (1987).

[6] M. Heiblum, M.I. Nathan, C.A. Chang, Sol. State Electron. 25, 185 (1982).

[7] M.I. Nathan and M. Heiblum, Sol. State Electron. 25, 1063 (1982).

[8] M. Wittmer, R. Pretorius, J.W. Mayer and M.A. Nicolet, Sol. State Electron. 20, 433 (1977).

[9] M. Procop, B. Sandow, H. Raidt and L. Do Son, Phys. Stat. Sol. (a) 104, 903 (1987).

[10] Y-C. Shih, M. Murakami, E.L. Wilkie and A.C. Callegari, J.Appl.Phys. 62(2), 582 (1987).

[11] X. Zhang and A.E. Staton-Bevan, Inst. Phys. Conf. Ser. 87(4), 303 (1987).

[12] T.K. Higman, M.A. Emanuel, J.J. Coleman, S.J. Jeng and C.M. Wayman, J. Appl. Phys. 60(2), 677 (1986).

[13] A. Iliadis and K.E. Singer, Solid State Comm. 49, 99 (1984).

[14] T.S. Kuan, P.E. Batson, T.N. Jackson, H. Rupprecht and E.L. Wilkie, J. Appl. Phys. 54(12), 6952 (1983).

[15] M. Ogawa, J. Appl. Phys. 51(1), 406 (1980).

[16] A. Christou, Sol. State Electron. 22, 141 (1979).

[17] A.K. Rai, A. Ezis, A.W. McCormick, A.K. Petford-Long and D.W. Langer, J. Appl. Phys.61(9),4682 (1987).

[18] M. Murakami, K.D. Childs, J.M. Baker and A. Callegari, J. Vac. Sci. Technol. B4(4), 903 (1986).

[19] K. Heime, U. König, E. Kohn and A. Wortmann, Sol. State Electron. 17, 835 (1974).

[20] J.R. Shappirio, R.T. Lareau, R.A. Lux, J.J. Finnegan, D.D. Smith, L.S. Heath and M. Taysing-Lara, J. Vac. Sci. Technol. A5(4), 1503 (1987).

[21] M.P. Grimshaw and A.E. Staton-Bevan in Advances in Materials, Processing and Devices in III-V Compound Semiconductors, edited by D.K. Sadana, L.E. Eastman and R. Dupuis (Mater. Res. Soc. Proc. 144, Pittsburgh, PA 1988) pp. 589-594.

[22] J. Herniman, D.A. Allan and P.J. O'Sullivan, I.E.E. Proc. 135(1), 67 (1988).

[23] G.K. Reeves and H.B. Harrison, I.E.E.E. Electron. Dev. Letts. EDL-3(5), 111 (1982).

[24] R.P. Gupta and W.S. Khokle, Sol. State Electron. 28(8), 823 (1985).

[25] W. Dingfen and K. Heime, Electron. Letts. 18(22), 940 (1982).

[26] I. Mojzes, T. Sebestyen and D. Szigethy, Sol. State Electron. 25(6), 449, (1982).

[27] T. Sands, E.D. Marshall and L.C. Wang, J. Mater. Res. 3(5), 914 (1988).

[28] J.R. Waldrop and R.W. Grant, Appl. Phys. Letts. 50(5), 250 (1987)

[29] T. Sebestyen, Sol. State Electron. 25, 543 (1982).

[30] D. Kirillov and Y. Chung, Appl. Phys. Lett. 51(11), 846 (1987).

[31] N.F. Mott, Phil. Mag. 19, 835 (1969).

[32] M. Jaros and H.L. Hartnagel, Sol. State Electron. 18, 1029 (1975)

CORRELATION BETWEEN THE INTERFACIAL NONUNIFORMITY AND THE SPECIFIC CONTACT RESISTANCE OF OHMIC CONTACTS TO GAAS

T.Q. TUY, I. MOJZES, V.V. TUYEN AND I. CSEH

Research Institute for Technical Physics of The Hungarian Academy of Sciences, Budapest, P.O.Box 76. H–1325, Hungary

ABSTRACT

Considering the effect of the simultaneous presence and interaction of the different phases at the contact, a modification of the model presented by Wu and coworkers [Solid–St. Electron. $\underline{29}$ (1986) 489] for explanation of ohmic contact resistance of n–GaAs was developed. The modified model combines the existence of the mixed phase structure of AuGeNi/n–GaAs contact with assumptions proposed by Wu et al. that the specific contact resistance R_c contains two parts R_{c1} and R_{c2}, where R_{c1} is the specific contact resistance of the alloyed and underlaying doped contact region, and R_{c2} is that of the high–low junction between the heavily doped contact region and the bulk semiconductor. The R_{c1} depends strongly on the apparent barrier height and the effective impurity concentration formed by doping from the contact alloys during annealing. In the present paper a new theoretical model for R_{c1} is proposed and compared with the experimental results.

1. INTRODUCTION

Wu et al. have proposed an improved model to explain ohmic contact resistance of AuGeNi/n–GaAs contact assuming that the specific contact resistance R_c is containing two parts R_{c1} and R_{c2} [1]. R_{c1} is due to the contact between the alloy and the high conductivity region formed by doping from the contact materials. The contact resistivity R_{c2} is caused by the barrier of the high–low junction between the heavily doped contact region and the bulk. The results indicated that if the bulk material is lightly doped, i. e. $N_D < N_C$, where N_D is the donor concentration for n–type material and N_C is the effective density of states in conduction band, then the barrier height ϕ_{h1} coincides with the barrier height of the high–low junction and it increases with decreasing of N_D. When the bulk material is degenerated, i. e. $N_D > N_C$ then the barrier height ϕ_{h1} vanishes and R_C is mainly determined by R_{c1} depending solely on the effective donor concentration N_{De} of the heavily doped region. If N_{De} is high enough, the field emission is the main mechanism determining the carrier transport. In this case R_c is predominantly determined by R_{c2} and inverse proportionality between R_c and N_D can be found. In this case the barrier height ϕ_{bn} of contact between the alloy and underlaying heavily doped region is fixed and the effect of the nonuniformity of metal–semiconductor contact (MSC) should not be taken into account.

However, in recent years many authors [2–7] indicated that the interface of MSC is not planar but spatially inhomogeneous containing some different phases. In the AuGeNi/n–GaAs contact Ni_2GeAs, NiAs and Au(GaAs) phases exist at the interface, and R_{c1} depends on the relative contact areas occupied by these phases [2]. The low R_{c1} can be achieved when the Ni_2GeAs phases are dominating while a higher R_{c1} may be occurred when Au−rich phases are predominating. The barrier height of this MSC was found in the range of (0.25−0.9)eV [3].

For AuGe/n−GaAs contact Braslau [4,5] indicated that the interface consists of two metallic phases corresponding to the ohmic regions where the Ge accumulates, and the other areas of the interface are barriers. The current flows through the matrix of the Ge−rich and Au−rich phases which are connected through the overlaying metal. The contact resistance will be determined by combination of the true contact resistance in these submicron region in series with the spreading resistance in the semiconductor.

Murakami [7] presented that the dominating value of the resistance of AuGeNi/n−GaAs contact were found to be strongly influenced by the compounds which are directly in contact with GaAs. Low contact resistance was obtained when NiAs(Ge) compound were formed in the vicinity of the interface.

The nonuniformity at the AuGeNi/n−GaAs interface may be occurred by the appearance of the protrusions, too [8].

Many authors have been concluded that the minimum R_{c1} for AuGeNi and AuGe/n−GaAs contacts is obtained when the interface area of phases associated with a low barrier is dominating [2−7].

On the basis of the above mentioned results a modified model was developed for investigation more perfectly the specific contact resistance R_{c1}.

2. THE MODIFIED MODEL OF THE AuGeNi/n−GaAs OHMIC CONTACT.

2.1. The basis of the modified model.

Waldrop's assumptions indicated that the current transport for tunnel or ohmic contacts depends on both the effective barrier height ϕ_{bn} and the effective donor concentration N_{De} [3].

a) The interface of AuGeNi/n−GaAs contact is inhomogeneous. The barrier heights of contacts can be found in the range of (0.25−0.9)eV for the Ni_2GeAs, NiAs and Au(rich) phases [3] and in the range of (0.3−0.9)eV for Ge−rich and Au−rich phases [4,5]. In our analysis only the current flowing perpendicular to the interface will be taken into account. Practically, for AuGeNi/n−GaAs contact in many cases there is a coexistence of the two phases at the interface. The effective barrier height will be changed according to the barrier heights and the distribution of the components. The variation of the effective barrier height is calculated on the basis of the extended theory of the mixed phase parallel ohmic MSC.

The R_{c1} can be determined by thermionic field emission (TFE) and field emission (FE) with corresponding of the effective barrier heights.

b) About the role of the Ge in the contact we assume that Ge may contribute to reducing both barrier height and the width of the effective barrier. The variation of the barrier height of the effective contact by the appearance of Ge in the different metallic phases and their distribution leads to the correlation between the nonuniformity of the interface with the specific contact resistance.

2.2. Calculation.

Usually the specific contact resistance R_{c1} is calculated from the J (current density)$-$V (voltage) equations [9]:

$$R_{c1} = \left[\frac{\delta J}{\delta V} \bigg|_{V=0} \right]^{-1}$$

a) For the TFE model at the condition $E_{00} \approx kT$, R_{c1} will be [1]

$$R_{c1} = \frac{kE_{00} \; [\coth(E_{00}/kT)]^{1/2} \; \cosh(E_{00}/kT)}{A^* \; T \; [\; \pi E_{00}(\phi_{bn1} + V_n)]^{1/2}} \cdot \exp[((\phi_{bn1} + V_n)/E_0 - (V_n/kT)] \tag{1}$$

where ϕ_{bn1} is the effective barrier height of the mixed phase ohmic contact. For metallic phase$-$nGaAs when the effective donor concentration N_{De} is in the range of $(5 \times 10^{17} - 3 \times 10^{18})$ cm^{-3} the TFE mechanism of the current transport is dominating. In this case the effective barrier height can be calculated from the corresponding effective current density:

$$J_{TFE} = \Sigma \; P_i \; J_{TFEi} \tag{2}$$

where J_{TFEi} is the current density flowing through the i$-$th component contact with corresponding barrier height ϕ_i, and $P_i = S_i/S$ is the area ratio. The effective barrier height of the mixed phase contact in the case of the applied voltage being equal to zero was calculated numerically from equation:

$$(\phi_{bn1} + V_n)^{1/2} \; \exp(-\phi_{bn1}/E_0) = \Sigma \; P_i(\phi_i + V_n)^{1/2} \; \exp(-\phi_i/E_0) \tag{3}$$

b) For FE model in the case when $E_{00} >> kT$ the specific contact resistance R_{c1} will be [1]

$$R_{c1} = [C_1 k E_{00} \; \sin \; (\pi C_1 kT)/\pi A^* T \;] \; \exp \; (\phi_{bn2}/E_{00}) \tag{4}$$

Here similarly, the effective barrier height ϕ_{bn2} can be calculated numerically from the effective current density for FE model by the following equation:

$$\exp(-\phi_{bn2}/E_{00})/[\ln(2\phi_{bn2}/V_n)\sin\{(\pi kT/2E_{00})\ln(2\phi_{bn2}/V_n)\}] =$$

$$\Sigma \; P_i \exp(-\phi_i/E_{00})/ \; \Sigma[\ln(2\phi_i/V_n)\sin\{(\pi kT/2E_{00})\ln(2\phi_i/V_n)\}] \tag{5}$$

where V_n is the energy of the Fermi level measured from the bottom of the conduction band. In the nondegenerated case:

$$V_n = E_c - E_v = (E_g/2) - \ln(N_{De}/N_i) \tag{6}$$

In the case of the totally degenerated semiconductor layer ($N_{De} \gtrsim 3 \times 10^{18}$ cm^{-3}) the Fermi level is above bottom of the conduction band and V_n was calculated numerically.

In Eqs.(1$-$5) :

$$E_0 = E_{00} \; \coth(E_{00}/kT) \quad \text{or} \quad E_0 = 1,85 \times 10^{-12}(N_{De}/m^* \epsilon_s) \tag{7}$$

In these equations m^* is the effective mass of electron, ϵ_s is the permittivity of n$-$GaAs and equals $\epsilon_{sr}\epsilon_0$, ϵ_{sr} is the dielectric constant , ϵ_0 is the free space permittivity, A^* is the Richardson constant and

$$A^* = 2\pi m^* qk/h^3;$$

$$C_1 = (1/2E_{00})\ln(4\phi_{bn2}) \qquad \text{for the FE model [1].}$$

On the basis of our calculations (Eqs.(2–7)) the variation of the R_{c1} with the parameters of the mixed phase contact will be analyzed.

3. ANALYSIS AND COMPARISON TO EXPERIMENTAL RESULTS

3.1) If $N_D > 4\times7\ 10^{17}$ cm^{-3}, the specific contact resistance is $R_c \approx R_{c1}$ [1]. The variation of the R_c is mainly determinated by the parameters of the component contacts. At the interface of the AuGeNi or AuGe/n–GaAs contact often exist two metallic phases (see the experimental results [2–7]). One phase usually is the Au–rich one with high barrier heights of (0.7–0.9)eV and the other one is the phase containing Ge with barrier height in the range of (0.25–0.4)eV [3].

The figure shows the variation of the barrier height of contact formed by a pair of metallic phases (0.3, 0.9)eV from J–V measurement of FE mechanism with different effective donor concentrations. The effective barrier height depends simultaneously on the barrier heights, the area ratios of the component contacts and the effective donor concentration. In the range of $N_{De} < 3\times10^{18}$ cm^{-3} the solid phases having low barrier height play dominant role, although these phases occupied a relatively small area at the interface.

The calculated variation of the specific contact resistance R_{c1} can be seen on the same figure according to the variation of the area ratio occupied by component contacts at the interface. If Ge forms a heavily doped layer at the MSC and this contact is planar and uniform with the barrier height of 0.8eV, in order to obtain specific contact resistance $R_c \approx 10^{-6}$ Ωcm^2 then the effective donor concentration must be $N_{De} > 5\times10^{19}$ cm^{-3}[4]. In the mixed phase contact, as it is demonstrated in Fig. 2, $R_{c1} \approx 10^{-6}$ Ωcm^2 when the Au–rich phase of the pairs (0.3, 0.9)eV, (0.4, 0.9)eV, (0.6, 0.9)eV phases occupied 85%, 70%, and 0% of the total contact area, respectively. For the pair (Ni$_2$GeAs, Au) (0.3, 0.9)eV, R_{c1} gradually decreases with the increment of the area ratio occupied by Ni$_2$GeAs phase at the interface. If this phase occupied the whole contact area then R_{c1} can reach the value of 10^{-7} Ωcm^2. When $N_{De} \approx 10^{19}$cm^{-3} and at least 70% of the contact area occupied by Ni$_2$GeAs phase then R_{c1} can reach the value of 10^{-6} Ωcm^2. Qualitatively a similar circumstance can be obtained for the variation of R_{c1} at the range of $(4.7\times10^{17}–3\times10^{18})$ cm^{-3} of N_{De} according to the Eqs.(1,2) for TFE mechanism.

3.2) When $N_D < 4.7\times10^{17}$ cm^{-3} and N_{De} is very high then $R_{c2} >> R_{c1}$. Our results will coincide with that of presented by Wu et al.[1]. In our calculation the ratio of N_{De} and N_D plays an important role in determining R_{c2}.

It is interesting to indicate that for explanation of the dependence of contact resistance of AuGeNi/n–GaAs and AuGe/n–GaAs contacts on the donor concentration of the substrate Braslau presented a model of the spreading resistance under Ge–rich island that neglected the influence of high–low barrier. Otherwise Wu et al. regarded that it is the important part in total contact resistance and neglected the spreading resistance in their calculation. However, the two models successfully explained the inverse proportionality between specific contact resistance and N$_D$.

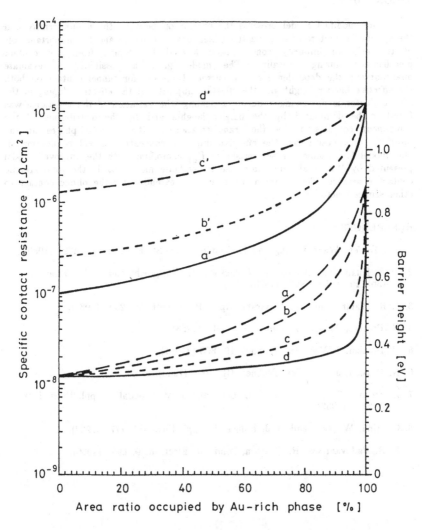

Barrier height and specific contact resistance of AuGeNi/n–GaAs contact from J–V measurement of FE mechanism for a pair of metallic phases with corresponding barrier heights (0.3, 0.9)eV and different N_{De} effective donor concentrations: a'– 10^{20}; b'– 5×10^{19}; c'– 10^{19} and d'– 3×10^{18} cm^{-3}.

Specific contact resistance R_{cl} is indicated for of AuGeNi/n–GaAs contact ($N_{De} = 5 \times 10^{19}$ cm^{-3}) for FE mechanism with corresponding component barrier heights: a/ (0.3, 0.9)eV; b/ (0.4, 0.9)eV; c/ (0.6, 0.9)eV; d/ (0.9, 0.9)eV.

4. CONCLUSIONS

Our modified model presents the correlation between the nonuniformity and the specific contact resistance via the mixed phase structure and the important role of the effective impurity concentration formed by doping from the contact metallization during annealing. The model gives a possibility to evaluate quantitatively the dependence of the current transport for tunnel contact on both the effective barrier height and the effective impurity. If the effective doping in the semiconductor is not changed during annealing, the variation of the resistance was found to be dominated by the barrier heights and by the distribution of the component contacts or by the transformation of the metallic phases at the interface. In opposite case the effective impurity concentration will be dominated. Our model for specific contact resistance R_{c1} combining with the improved model presented by Wu et al. will contribute to evaluate more exactly the total specific contact resistance. The common model can be extended for the ohmic contact to other semiconductors,too.

REFERENCES

1. Wu Dinfen, Wang Dening and K. Heime, Solid–St. Electron. 29, 489 (1986).

2. T. S. Kuan, P.E. Batson, T. N. Jackson, H. Rupprecht and E. L. Wikie, J. Appl. Phys. 54, 6952 (1983).

3. J. R. Waldrop and R. W. Grant, Appl. Phys. Lett. 50, 250 (1987).

4. N. Braslau, Thin Solid Films, 104, 391 (1983).

5. N. Braslau, J. Vac. Sci. Technol. 19, 803 (1981).

6. N. Braslau, J. Vac. Sci. Technol. B1, 700 (1983).

7. L. Freeouf, T. N. Jackson, S. E. Laux and J. M. Woodall, Appl. Phys. Lett. 40, 634 (1982).

8. O. Aina, W. Katz and B. J. Baliga, J. Appl. Phys. 53, 777 (1982).

9. F. R. Padovani and R. Stratton, Solid–St. Electron. 9, 695 (1966).

PART VIII

Contacts to InP and
Related Materials

INTERFACIAL MICROSTRUCTURE AND CARRIER CONDUCTION PROCESS IN Pt/Ti OHMIC CONTACT TO p–In$_{0.53}$Ga$_{0.47}$As FORMED BY RAPID THERMAL PROCESSING

S. N. G. CHU, A. KATZ, T. BOONE, P. M. THOMAS, V. G. RIGGS,
W. C. DAUTREMONT-SMITH, AND W. D. JOHNSTON, Jr.
AT&T Bell Laboratories, Murray Hill, New Jersey 07974

ABSTRACT

The strong dependence of electrical properties of Pt/Ti ohmic contact to p–In$_{0.53}$Ga$_{0.47}$As (Zn: 5×10^{18} cm^{-3}) on the interfacial microstructure formed by rapid thermal processing (RTP) were intensively studied by transmission electron microscopy, Auger Spectroscopy, and transmission line model (TLM) measurements. The rapid decrease of the specific contact resistance with an increase in RTP temperature was correlated with the development of an interfacial reaction zone. Significant interdiffusion of Ti, In and As across the interface occurred at temperature above, 350°C for a 30 second of RTP. A minimum specific contact resistance (3.4×10^{-6} Ω–cm^2) was achieved at RTP temperature of 450°C. The corresponding interfacial microstructure revealed a complicated solid state reaction zone with InAs as one of the major interfacial compounds. The low contact resistance is attributed to the carrier conduction through the InAs regions. This is also consistent with the results of Pt/Ti contact experiments to p-type InAs, InP and GaAs binary surfaces, where the lowest contact resistance was achieved on InAs (3.0×10^{-7} Ω–cm^2 at Zn: 5×10^{18} cm^{-3}). The temperature dependence of specific contact resistance of as-deposited Pt/Ti contact to InGaAs agrees very well with the thermionic emission dominated carrier transport mechanism with an effective barrier height, ϕ_b, of 0.13V. The rapid decrease in the contact resistance as well as its reduced temperature dependence after RTP treatment at elevated temperatures suggesting a partial conversion of thermionic emission dominated contact area to field emission dominated regions. A phenomenological theory of multiple parallel carrier conduction processes was proposed to analyse the temperature dependence of specific contact resistance for contacts with complicated interfacial microstructure. It was found that, for low resistance contacts, majority of the carriers conducted through only a fraction of the contact area via a tunneling mechanism.

INTRODUCTION

The developing a reliable low resistance ohmic contact to p-type InGaAs(P) is essential to all InP-based optoelectronic devices, especially for the long wavelength laser diodes and LEDs.[1–7] Recently, Katz et al, demonstrated that by using a rapid thermal processing (RTP), specific contact resistance as low as 3.4×10^{-8} cm^2 can be achieved in Au/Pt/Ti ohmic contact to Zn-doped p–InGaAs at a doping level of 1.5×10^{19} cm^{-3}.[1,2] From the device reliability point of view, the Au/Pt/Ti is a preferrable contact because the Ti layer acts as a barrier for Au diffusion into the semiconductor, which is believed to cause device degradation.[6] Due to a strong solid state reaction between the contact metals and semiconductor surfaces [6] , the contact resistance in InP based compound semiconductors is determined by the interfacial microstructure. For metal contacts to GaAs, the interfacial solid state reaction has been extensively studied and reported by various authors.[8–21] In this paper, we study the interfacial microstructure of Pt/Ti ohmic contacts on p-InGaAs formed by rapid thermal processing (RTP). In order to obtain insight on the carrier conduction process, a phenomenological theory, based on a simultaneous parallel carrier conduction processes across the interface is formulated to explain the temperature dependence of the specific contact resistance developed under different RTP conditions. A microstructural parameter is also introduced to account for the effect of complicated interfacial microstructure on carrier transport processes.

EXPERIMENTAL

Thin films of Pt (75 nm thick)/Ti (75 nm thick) were e-beam deposited successively onto Zn-doped (5×10^{18} cm^{-3}) In$_{0.53}$Ga$_{0.47}$As layer (~1.2 µm thick) lattice-matched to InP prepared by hydride VPE, and heat treated by means of RTA using a model 410T A. G. Associates, Heatpulse annealer under a controlled forming gas ambient (15% H$_2$) at temperatures between 300°C and 600°C for 30 sec. The interfacial microstructure were studied by TEM using a Philips TM420 electron microscope operated at 120 Kev. The TEM cross-sectional samples were prepared by Argon ion-

Mat. Res. Soc. Symp. Proc. Vol. 181. ©1990 Materials Research Society

milling at liquid nitrogen temperature, under the condition of 5 KV and 15 mA, in a Gatan Duomill system. Bulk films of Ti, Pt and TiPt (15 nm each) were used as reference samples for electron diffraction studies of the interfacial compounds. These films were prepared by depositing onto SiO_2/InP and InP substrate followed by chemical etching from the backside of the substrate. Auger depth profiles of the elements were obtained to study the extend of interdiffusion.

The electrical measurements involved both current-voltage (I-V) and specific contact resistance analysis using a standard transmission line model (TLM). The later was carried out over six different temperatures ranged from 20° to 150°C. The temperature dependence measurements enable the understanding of the nature of carrier transport mechanism across the contact.

RESULTS AND DISCUSSIONS

Interfacial Microstructure of the Pt/Ti Contact to p-InGaAs

Figure 1 shows a XTEM micrograph of the interface of as-deposited Pt/Ti film onto InGaAs at room temperature. Macroscopically, the interface seems abrupt. Since the as-deposited contact is ohmic (with a specific contact resistance of 1.7×10^{-4} Ωcm^2), interfacial reaction on the atomic level already occurs at room temperature. The strong solid state reactions of the contact metal with low melting point group III elements and the volatile group V elements are further promoted by RTP at elevated temperature.

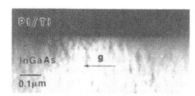

Figure 1.   XTEM micrograph of as-deposited Pt/Ti film on p-InGaAs showing distinct interface.

The Auger depth profiles of all the elements of as-deposited contact, and contacts subjected to RTP at temperature range from 400°C to 500°C are shown in Figure 2. The major outdiffusion elements at RTP temperature above 450°C are In and As, while the main indiffused element is Ti.

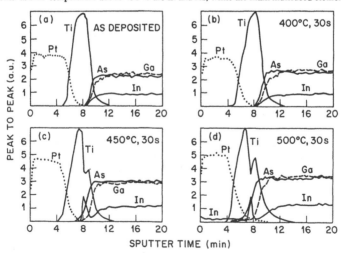

Figure 2.   Auger depth profiles of (a) as-deposited Pt/Ti film on p-InGaAs, (b) RTP at 400°C for 30 sec., (c) RTP at 450°C for 30 sec., and (d) RTP at 500°C for 30 sec.

Significant changes in the Auger depth profiles of In, As and Ti at the interface occurred at 450°C. The corresponding interfacial microstructure of the 450°C RTP sample is shown in Figures 3(a) and 3(b). A complicated reaction zone about 800 Å thick developed at the interface as a result of the heat treatment. The microstructure of the reacted zone consists of a defective interfacial layer and solid state regrown second phase regions as indicated by the arrows. The microstructure of the metal film is shown in Figure 3(b). The average grain size of the polycrystalline Pt/Ti film is slightly larger than the thickness of the film.

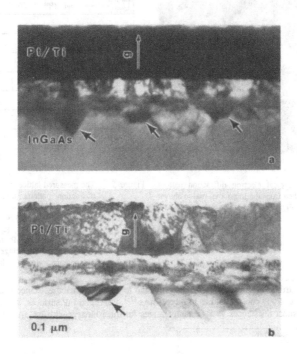

Figure 3.   XTEM micrographs of Pt/Ti film on p-InGaAs RTP at 450°C for 30 sec showing (a) complicated interfacial microstructure developed due to interfacial solid-state reaction (bright field), and (b) a thinner region revealing the microstructure of the bulk Pt/Ti film (dark field).

The electron diffraction patterns taken from the interfacial region were compared with those taken from the bulk Ti/Pt film and InGaAs to identify the dominating interfacial compound developed during RTP at 450°C. Figure 4 shows a typical diffraction pattern taken from the reacted zone of this sample. The lattice spacings measured from the diffraction patterns of 450°C and 350°C samples are listed in Figure 5 along with those obtained from the diffraction pattern of pure Ti, Pt and Ti/Pt films to unambiguously single out the diffractions from the interfacial compounds. The lattice spacings from the interfacial compound are then compared with the available diffraction data[22] of all the possible binary compounds of In, Ga, Ti, Pt, As and P. The best agreement was found to be with InAs, where there is a match of six lattice spacings. This also agrees qualitatively with the Auger profiles, see Figure 2, which shows that both In and As outdiffused faster than the Ga profile. A few lattice spacings also match In_2Pt, $GaPt_3$, and $GaPt_2$. Stable phase such as $PtAs_2$ was not observed, probably due to the short RTP time.

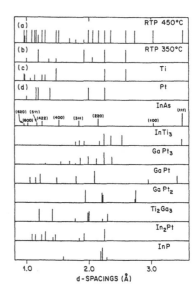

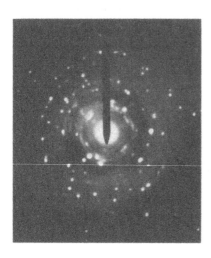

Figure 4. A typical electron diffraction pattern taken from the interfacial region shown in Figure 3.

Figure 5. The measured lattice spacings from the diffraction patterns taken (a) in the interfacial regions of a sample RTP at 450°C for 30 sec, (b) of a sample RTP at 350°C for 30 sec, (c) from the Ti films, (d) from the Pt films, and the known lattice spacings of various compounds from the "Powder Diffraction File".

The contribution of InAs interfacial compound to the low contact resistance is also consistent with the result of Pt/Ti contact studies to p-type InAs [23], GaAs and InP surfaces. Figure 6 compares the specific contact resistances vs RTP temperatures for three binarys with InGaAs at the same Zn-

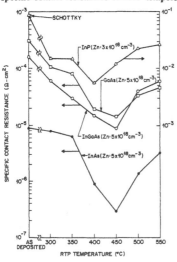

Figure 6. RTP (30 sec) temperature dependence of specific contact resistances for Pt/Ti on p-type InGaAs, InAs, GaAs and InP.

dopant level except for InP where the maximum carrier concentration can be achieved is 3.0×10^{18} cm^{-3}, even though the total Zn concentration can be much higher. The lowest contact resistance is achieved on InAs (3.0×10^{-7} Ω-cm^{-2}). The stability of the bulk Pt/Ti film microstructure under different RTP condition is also examined in Figure 7. The as-deposited film has an average grain size of 11 nm. The grains grew in size with increasing RTP temperature and reached a maximum size of 56 nm at 450°C. Further increase in RTP temperature showed no effect on the grain size. This suggested a stabilization of the bulk Pt/Ti film microstructure at temperature of 450°C.

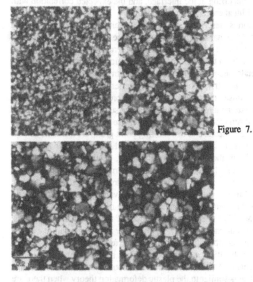

Figure 7. Plan-view TEM micrographs of Pt/Ti films showing the change the grain size as a function of RTP temperatures: (a) as-deposited, (b) 300°C, (c) 450°C, and (d) 500°C.

Change of Specific Contact Resistance with RTP Temperature

The rapid decrease in the specific contact resistances with increasing RTP temperature for a 30 sec heating treatment, as seen in Figure 6, indicates the development of low R_c microstructures at the interfaces in all these systems. The extent of these low R_c microstructural areas increase with increasing RTP temperature, and reach a maximum value in all cases at 450°C for InGaAs, InAs and GaAs, where the respective R_c is minimum (9×10^{-6} Ω-cm^2 for InGaAs). Since the dominate interfacial compound on p-InGaAs formed at 450°C was found to be InAs, this low R_c interfacial microstructure can naturally be referred to InAs based compounds. Further increase in RTP temperature resulted in an increasing in R_c, indicating the initiation of a different, high R_c, microstructure at the interface.

Phenomenological Analysis of the Specific Contact Resistance Temperature Dependence

Although the formation of the rectifying barriers, such as Schottky barrier and tunneling barrier, etc., in metal-semiconductor contacts is well understood,[24] the mechanism for the ohmic contact formation is still not clear, especially with metal contacts to InP-based material systems. Theoretically, if the barrier height is relatively low, or the barrier is extremely narrow, the metal/semiconductor interface becomes transparent to the carriers, and an ohmic contact can be expected simply because the carriers experience only the bulk resistance of the semiconductor.[24] In the Au/Pt/Ti contact to n$^+$-InGaAs with different InAs mole fraction, Nittono et al[25] found that specific contact resistance decreases with increasing InAs mole fraction x until $x = 0.65$. They concluded that the decrease in the contact resistance was due to a decreasing in the barrier height with an increasing in InAs mole fraction reported by Kajiyama et al[26] for a Au contact on n-InGaAs. The same argument, however, would not apply to metal contact to p-In$_{0.53}$Ga$_{0.47}$As. The calculated barrier height ϕ_b for the Au contact to p-In$_{0.53}$Ga$_{0.47}$As was 0.6V,[26] which is about three times larger than the barrier height on $n-In_{0.53}Ga_{0.47}As$. In practice, despite a much larger calculated barrier height for $Pt/Ti/p-In_{0.53}Ga_{0.47}As$ contact as compared to $Pt/Ti/n-In_{0.53}Ga_{0.48}As$ contact, specific contact resistance as low as 3.4×10^{-8} Ω cm^2 can still be achieved at a similar doping level. Furthermore, we have demonstrated that a three order of magnitude reduction of the specific contact resistance can be achieved simply by changing the RTP condition, and thus the interfacial microstructure. Therefore, the dominant factor which controls the contact resistance is, in fact, the interfacial microstructure. The interfacial microstructure is also recognized as a main parameter to control the Schottky-barrier height in InP.[8] Robinson, in his extensive review,[8] concluded that the Schottky-barrier height in InP was strongly affected by the chemical reaction occurring at the interface, and the expected correlation with the work function of the contact metals for ideal contacts did not exist.

The present carrier transport theories describing the metal to semiconductor contacts are formulated based on the assumption that the metal-semiconductor interface is chemically abrupt and structurally uniform. The carrier transport nature of the interface can, therefore, be described by an interfacial energy barrier derivable from the band theory of metals and semiconductors. Such an ideal interface does exist for most of the metals deposited on atomically clean semiconductor surfaces. However, achieving an ideal interface is a non-trivial process. The carrier transport nature is often found to deviate from the ideal property considerably depending on the processing conditions. The implication is again that it is very sensitive to the interfacial microstructure such as oxides.

For metals on InP based compound semiconductor, an ideal interface is difficult to achieve due to strong solid state reactions between contact metals and low melting point group III elements as well as with the volatile group V elements, especially for metals on InGaAs and InGaAsP after RTP. The carrier transport nature across the interface in these systems is, therefore, determined solely by the interfacial microstructure. Furthermore, the complicated interfacial microstructure, as seen in Figure 3 containing regions of different interfacial compounds as well as massive interfacial crystalline defects, renders the application of the simple band theory impossible. It is conceivable that the simple band theory is still valid locally in each interfacial segment containing a uniform microstructure. However, it is not until these interfacial segments are fully characterized in terms of the chemical compounds, their geometrical configuration and distribution, and individual carrier transport nature, that a prediction of the contact resistance for such a complicated interface may be feasible.

In view of the above mentioned difficulties, a phenomenological theory is proposed in the following to describe the effect of the interfacial microstructure on the contact resistance of $Pt/Ti/p$-InGaAs ohmic contacts. The basic approach is similar to the plastic deformation theory when there are several deformation processes operating simultaneously.[27] In this case, due to the complicated interfacial microstructure, the carrier transport mechanism is no longer describable by a single process. Instead a parallel carrier conduction mechanism is considered. We also introduce a microstructure parameter f to account for the effect of microstructure on the carrier conduction.

Let us consider an interface with a single microstructure, characterized by a structure parameter f, the current density of J across the interface is then given by:

$$J = J(f,T,V), \tag{1}$$

where T is the temperature, and V is the applied voltage.

To simplify the formulation based on the existing carrier transport processes, i.e. the thermionic emission, the field emission and the thermionic-field emission processes, we further assume that the variables T and V and the structural parameter f are separable. Therefore,

$$J = j(f) j(T) j(V). \tag{2}$$

For a thermionic emission process, $j(T) = T^2 \exp(-e\phi_b/kT)$, where ϕ_b is the barrier height, and $j(V) = \exp(eV/kT) - 1$. Furthermore, if the interface is ideal, $j(f) = A^*$, where A^* is the Richardson constant. For a field emission process, $j(T) = 1$, and $j(V) = C_1 \exp[-C_2(\phi_b - V)/\sqrt{N}]$, where C_1 and C_2 are constants and N is the carrier concentration.

For a general case of an ohmic metal, the interface contains microstructure of M different compounds distributed uniformly across the interface, each occupying a fractional area A_i, where the total contact area $A = \sum_{i=1}^{M} A_i$. The average current density across the interface is then a weighted sum of the current densities across each microstructural regions. Thus, J is a summation of the M current components,

$$J = \sum_{i=1}^{M} x_i \, j_i \, (f_i) \, j_i \, (T) \, j_i \, (V) \tag{3}$$

where $x_i = A_i/A$ is the fraction of the interfacial area occupied by the "*ith*" compound.

The specific contact resistance for an ohmic contact based on the multiple parallel carrier transport mechanisms is then given by

$$R_c = \left[\left(\frac{\delta J}{\delta V} \right)_{V=0} \right]^{-1}$$

$$= \left\{ \sum_{i=1}^{M} x_i \, j_i \, (f_i) \, j_i \, (T) \left[\frac{\delta j_i(v)}{\delta V} \right]_{V=0} \right\}^{-1} \tag{4}$$

Again, for a nearly ideal interface where the carrier transport mechanism is dominated by a single thermionic emission process, the R_c reduces to a single term, i.e. $x_1 = 1$ and $x_i = 0$ for $i \neq 1$. Hence

$$R_c = \frac{k}{j(f)eT} \exp \frac{e\phi_b}{kT}, \tag{5}$$

or

$$\ln R_c T = \ln \frac{k}{j(f)e} + \frac{e\phi_b}{kT}. \tag{6}$$

Thus, for a thermionic emission dominated process, a linear relationship between $\ln R_c T$ versus $1/T$ should be expected. Figure 8 plots the measured $\ln R_c T$ versus $1/T$ for as-deposited sample as well as for samples after RTP at different temperatures. A linear relationship exists only in the as-deposited sample, indicating a thermionic emission controlled carrier transport process. The

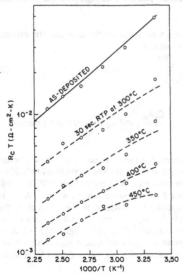

Figure 8.   A $\ln R_c T$ versus $1/T$ plots of the temperature dependence of specific contact resistance for samples RTP at different temperatures showing a progressive deviation from linearity with increasing RTP temperature.

calculated barrier height from Figure 8 is 0.13 V, and the corresponding microstructure factor $j(f)$ is $0.29 \, AK^{-2} \, cm^{-2}$. The low barrier height and the structure parameter characterize the interfacial microstructure of the as-deposited Pt/Ti/p-InGaAs ohmic contact. It will be seen later that this thermionic emission dominated interfacial microstructure remains in the major portion of the contact area even after RTP at 450°C. Since RTP enhances the solid state reaction, new interfacial microstructures are expected to be developed in the expanse of the existing interfacial microstructure. At the same time, changes may also be expected in the existing microstructure, which is reflected in a change of the structure parameter. The sharp decrease in the specific contact resistance with increasing RTP temperature, as shown in Figure 7, indicates the development of a new, low contact resistance microstructure. The interface now contains multi-segment regions, dominated by different carrier transport properties. Thererfore, the overall carrier transport mechanism can no longer be described by a single process, and a deviation from the linearity in $\ln R_c T$ vs $1/T$ is expected, which is indeed observed in Figure 8. Empirically we found that only a second, temperature independent component is necessary to be included in equation (4) to fit the data. The implication is that the carrier transport mechanism across the newly developed interfacial microstructure is by field emission process. The corresponding interfacial microstructure is schematically shown in Figure 9, where the current is conducting through regions of two different types of compounds. For simplicity we have avoided the introduction of the possible thermionic-field emission process.

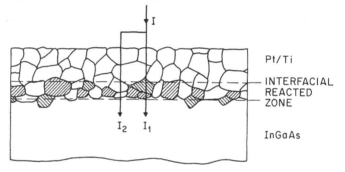

Figure 9.   Schematic diagram of a two carrier-conduction processes through interface containing two different compounds.

Let x and $1-x$ be the fraction of interfacial areas occupied by the new and the original microstructures respectively. Thus, equation (4) can be written as:

$$R_c = \left[\frac{x}{R_{tu}} + \frac{1-x}{R_{th}} \right]^{-1} \qquad (7)$$

where $R_{th} = \dfrac{k}{j_1(f_1)eT} \exp \dfrac{e\phi_b}{kT}$, $R_{tu} = \dfrac{\sqrt{N}}{j_2(f_2)C_1C_2} \exp \dfrac{C_2\phi_{tu}}{\sqrt{N}}$, and ϕ_{tu} is the tunneling barrier.

As shown in Figure 10, Equation (7) fits the measured R_c verses T values of all samples treated by RTP at different temperatures very well. The parameters used for the best fit are listed in Table I. As expected, the fractional area x of the low R_c, field emission regions increases from 0.07 in sample after RTP temperature at 300°C to 0.3 in sample after RTP at temperature of 450°C. A slight changing in the original microstructure with RTP temperature is reflected in the increase of the structure parameter f. The same effect is likely to cause the slight decreasing in the values of R_{tu} (from $6 \times 10^{-6} \, \Omega cm^2$ to $4 \times 10^{-6} \, \Omega cm^2$) in fitting the measured data. The higher R_{tu} value as compared to $3 \times 10^{-7} \, \Omega\text{-}cm^2$ for Pt/Ti on p-InAs is attributed to the complicated microstructure and dopant incorperation in the interfacial compound. It is conceivable that the p-type dopant level in the interfacial compound is lower or even compensated.

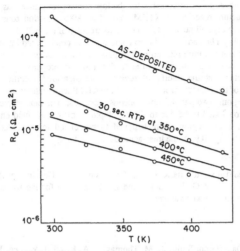

Figure 10. R_c versus T plots showing good agreement between the predicted values (solid lines) from the theory and the measured data at different RTP temperatures.

TABLE I. Summary of The Parameters Using To Fit The Measured Rc vs T Dependences

T_{RTP} (°C)	1-x	ϕ_b (V)	$j(f_1)(AK^{-2}\,cm^{-2})$	x	R_{tu} ($10^{-6}\Omega\,cm^2$)
NO RTP	1.00	0.13	0.28	0.00	
300	0.93	0.13	1.23	0.07	6.0
400	0.80	0.13	1.84	0.20	5.0
450	0.70	0.13	2.45	0.30	4.0

This phenumenological analysis thus arrives at the following important conclusions.

(1)   The majority of the carriers at the Pt/Ti/p–$In_{0.53}Ga_{0.47}As$ (Zn: 5×10^{18} cm^{-3}) contact after RTP at 450°C for 30 sec. are transported through the interfacial area with low contact resistance, which occupies only 30% of the total contact interface. The remaining carriers are conducted across the higher contact resistance areas via a thermionic emission process.

(2)   The low contact resistance microstructure involves InAs based compounds. In view of the large barrier height 0.5V for p-InAs [28], as compared to 0.13V for the remaining contact area, the transport mechanism through the low contact resistance area has to be by tunneling. This is consistent with a second temperature independent current component required to fit the measured data.

(3)   The phenomenological theory can be applied to a variety of metal contacts to InP-based materials.

SUMMARY

We have reported on the formation of Pt/Ti ohmic contact to Zn doped, 5×10^{18} cm^{-3}, p-type InGaAs by means of rapid thermal processing. The as-deposited contact is ohmic and has a specific contact resistance of 1.6×10^{-4} Ω cm^2. The specific contact resistance decreases rapidly with increasing RTP temperature down to a minimum value of 9.0×10^{-6} Ωcm^2 at 450°C for 30 sec. The interfacial compound believed to be responsible for the low contact resistance was found to be InAs. The interfacial microstructure revealed by XTEM showed a ~800Å reaction zone with complicated microstructure and massive crystalline defects. The microstructure of the bulk Pt/Ti film was found to be stabilized at 450°C RTP. The carrier conduction across a such complicated interface was analyzed by a phenomenological theory of multiple carrier transport processes. A structure parameter was also introduced to take into account the change in the carrier transport property due to a change in the micro- and defect structures at the interfacial region. The measured specific contact resistance temperature dependence of the contacts processed at different RTP temperatures, agreed very well with the proposed theory using only two parallel simultaneous processes. The rapid decrease in the specific contact resistance was concluded to be due to the development of low R_c field emission dominated regions at the interface. These regions were believed to be related to the InAs based interfacial compounds, and occupied only 30 % of the interfacial area.

ACKNOWLEDGEMENTS

The authors would like to thank R. H. Saul and J. E. Geusic for their support and encouragement in this work and G. J. Fisanick and L. C. Feldman for the use of TEM. Valuable discussion with M.Murakami is acknowledged.

REFERENCES

[1] A. Katz, W. C. Dautremont-Smith, P. M. Thomas, L. A. Koszi, J. W. Lee, V. G. Riggs, R. L. Brown, J. L. Zilko and A. Lahav, J. Appl. Phys., 65, 4319 (1989).

[2] A. Katz, W. C. Dautremont-Smith, S. N. G. Chu, P. M. Thomas, L. A. Koszi, J. W. Lee, V. G. Riggs, R. L. Brown, S. G. Napholtz, J. L. Zilko and A. Lahav, Appl. Phys. Lett., 54, 2306 (1989).

[3] A. Katz, P. M. Thomas, S. N. G. Chu, W. C. Dautremont-Smith, R. G. Sobers, and S. G. Napholtz, J. Appl. Phys., 66, 2056 (1989).

[4] R. Kaumanns, N. Grote, H-G. Bach, F. Fidorra, Inst. Phys. Conf. Ser. No. 91, 501 (1987).

[5] M. Fukuda, O. Fujita, and S. Uehara, J. Lightwave Technol. 6, 1808 (1988).

[6] W. C. Dautremont-Smith, P. A. Barnes, and J. W. Stayt, Jr., J. Vac. Sci. Technol. B2, 620, (1984).

[7] A. K. Chin, C. L. Zipfel, M. Geva, I. Camlibel, P. Skeath, and B. H. Chin, Appl. Phys. Lett., 45,37 (1984).

[8] G. Y. Robinson, "Physics and Chemistry of III-V Compound Semiconductor Interfaces", Edited by Carl W. Wilmsen, p. 73, Plenum Press, New York (1985).

[9] T. Sands, V. G. Keramidas, A. J. Yu, K-M Yu, R. Gronsky, and J. Washburn, J. Mater. Res. 2, 262 (1987).

[10] T. S. Kuan, J. L. Freeouf, P. E. Batson, and E. L. Wilkie, J. Appl. Phys. 58, 1519 (1985).

[11] T. Sand, V. G. Keramidas, R. Gronsky and J. Washburn, Thin Solid Films, 136, 105 (1986).

[12] T. Sand, E. D. Marshall, and L. C. Wang, J. Mater. Res. 3, 914 (1988).

[13] L. H. Allen, L. S. Hung, K. L. Kavanagh, J. R. Phillips, A. J. Yu, and J. W. Mayer, Appl. Phys. Lett. 51, 326 (1987).

[14] M. Murakami, Y-C. Shih, W. H. Price, E. L. Wilkie, K. D. Childs and C. C. Parks, J. Appl. Phys. 64, 1984 (1988).

[15] M. Murakami, W. H. Price, Y-C. Shih, N. Braslau, K. D. Childs and C. C. Parks, J. Appl. Phys. 62, 3295 (1987).

[16] M. Murakami, W. H. Price, Y-C. Shih, K. D. Childs, B. K. Furman, and S. Tiwari, J. Appl. Phys. 62, 3288 (1987).

[17] M. Murakami, Y-C. Shih, W. H. Price, N. Braslau, K. D. Childs, and C. C. Parks, Inst. Phys. Conf. Ser. No. 91, 55 (1988).

[18] M. Murakami, Y. C. Shih, H. J. Kim, and W. H. Price, Proc. of the 20th Int. Conf. Sol. Stat. Dev. Mat., D-2-3, 283, Jap. Soc. of Appl Phys. (1988).

[19] J. M. Vandenberg, H. Temkin, R. A. Hamm, and M. A. DiGiuseppe, J. Appl. Phys. 53, 7385 (1982).

[20] J. M. Vandenberg, and H. Temkin, J. Appl. Phys. 55, 3676 (1984).

[21] V. G. Keramidas, H. Temkin, and S. Mahajan, Inst. Phys. Conf. Ser. No. 56, 293 (1981).

[22] Power Diffraction File, Joint Committee on Powder Diffraction Standards, International Center for Diffraction Data, Swarthmore, PA, 1980.

[23] A.Katz, S.N.G.Chu, B.E.Weir, W.C.Dautremont-Smith, R.A.Logan, T.Tanbun-Ek, W.Savin, and D.W.Harris, to be published.

[24] S. M. Sze, Semiconductor Devices: Physics and Technology, John Wiley and Sons, New York (1985).

[25] T. Nittono, H. Ito, O. Nakajima, and T. Ishibashi, Jap. J. Appl. Phys., 26, 10, L865 (1986).

[26] K. Kajiyama, Y. Mizushima, and S. Sakata, Appl. Phys. Lett., 23 458 (1973).

[27] J. C. M. Li, Rate Processes in Plastic Deformation of Materials, Edited by J. C. M. Li and A. K. Mukherjee, pp. 479, ASM Materials/Metalworking Technol. Series #4, ASM, Cleveland, Ohio (1975).

[28] M.Murakami, P.-E.Hallali, W.H.Price, M.Norcott, N.Lustig, H.-J.Kim, S.L.Wright, and D.LaTulipe, Mat. Res. Soc. Symp. Proc. this volume, (1990).

PROCESS DESIGN FOR NON-ALLOYED CONTACTS TO InP-BASED LASER DEVICES

A. Katz, W. C. Dautremont-Smith, S. N. G. Chu, S. J. Pearton, M. Geva, B. E. Weir, P. M. Thomas, and L. C. Kimerling
AT&T Bell Laboratories, Murray Hill, NJ 07974

ABSTRACT

Pt/Ti and W thin films on n- and p- type InP and related materials have been investigated for potential use as a refractory ohmic contacts for conventional, single-side coplanar contacted and self-aligned barrier heterostructure laser devices. Pt and Ti films were deposited sequentially by electron gun evaporation, while the W layer was rf sputtered, both onto p^+ -$In_{0.53}Ga_{0.47}As$ (Zn doped $5x10^{18} cm^{-3}$) and n^- - InP (S doped, $5x10^{18} cm^{-3}$). The deposition parameters of the two metal systems were optimized to produce adherent films with the lowest possible induced stress. Almost all the studied systems performed as ohmic contacts already as-deposited and were heat treated by means of rapid thermal processing in the temperature range of 300-900°C. The final contact processing conditions were tuned to provide the lowest possible contact resistance values accompanied by low mechanical stress and stable microstructure.

1. INTRODUCTION

InP-based optoelectronic devices are strongly influenced, both in short and long term performance and reliability, by the quality of their ohmic contacts. The most sensitive contact is the one which is in the immediate vicinity of the active layer, and thus more attention has to be paid to its properties. Correct design of the ohmic contacts to these devices has to take into consideration the following issues: selection of the metallization scheme to form a pure ohmic contacts with the lowest possible resistance, creating a stable microstructure over a large processing temperature range and tuning the deposition and the subsequent thermal process in order to introduce the lowest possible stresses into the thin metallic film. Subsequently, and however as important as the previous consideration, one has to define the process sequence which will yield the optimum performance of as many properties from the above mentioned list and successfully incorporate it into the overall device fabrication process sequence. Any simplification in the overall device manufacturing sequence attributed to the utilization of correct metal scheme is certainly considered as a further advantage. Thus, for example, Pt/Ti metallization scheme, has recently been shown to provide ohmic nature contacts to both n-type InP[1] and p-type InP-based materials such as p^+ - InGaAs[2,3] and p-InGaAsP[4,5]. It can, therefore, be used as a common contact to both conventional and coplanar contacted devices, eliminating a few of the required process steps to produce two different contact metallization schemes. W shows also an ohmic behavior on both p^+ - InGaAs[6] and n-InP[7] and can be used, in an analog to GaAs technology[8-10], as a ohmic and self-aligned contact scheme for laser devices.

The purpose of this work is to demonstrate the unique design concept of ohmic contacts to InP-based laser devices by using entirely refractory metallization schemes.

2. EXPERIMENTAL PROCEDURE

Experiments were performed using <100> semi-insulating (SI) Fe-doped InP substrates of resistivities greater than $10^6 \Omega$cm. p^+–$In_{0.53}Ga_{0.47}As$ (Zn doped $5x10^{18} cm^{-3}$) layers (0.5 μm thick) were grown on these substrates by liquid phase epitaxy (LPE)[11]. This Zn-doped ternary layer was grown at 630°C with a lattice mismatch ($\Delta a/a$) less than $5x10^{-4}$, which ensures a misfit-free layer. On other substrates, n^- - InP (S doped $5x10^{18} cm^{-3}$) layers (0.5 μm thick) were grown by metalorganic chemical vapor deposition (MOCVD)[12-13]. The carrier concentrations were measured in all cases by Hall effect and polaron profiling.

The refractory metallization schemes were designed in the following fashion: Pt(60 nm)/Ti(50nm) metallization was selected as the vehicle for the evaluation of a common and similar contacts for both n-InP and P-InGaAs layers in coplanar and two sides contacted devices. W(100nm) film was selected as the ohmic and self-aligned contact masking metallization scheme for n-InP and P-InGaAs as well. The Ti(50nm thick) and Pt(60nm thick) layers were deposited by electron gun (e-gun) evaporation with a background pressure better than 1×10^{-7} Torr. The W films (100nm thick) were rf sputtered from a planar 8" diameter round target at an argon pressure in the range of 5 to 28 mTorr and power of 240W.

For the electrical measurements, using the transmission line method (TLM)[14], the metal films were deposited onto square openings ($200 \times 200 \mu m^2$) linearly spaced (with intervals of 10 to 50μm) that were wet etched through a SiO_2 (300nm thick) plasma-deposited layer. Subsequently, semiconductor mesas were etched to give the required one-dimensional current flow. In addition broad area monitor n-InP and p-InGaAs layers were metallized for metallurgical studies.

The contacts were rapid thermally processed (RTP) using an A. G. Associates 410T Heatpulse™ annealer under controlled forming gas ambient (15% H_2) at temperatures between 300 and 600°C for the Pt/Ti metallization and up to 900°C for the W metallization. The proximity heating approach was used to prevent substrate decomposition by placing the InP-based samples with the metalized side up on a silicon wafer, all in the furnace chamber.

The analytical examination involved a variety of techniques such as high resolution field emission scanning electron microscopy (SEM), transmission electron microscopy (TEM) both in cross-sectional (XTEM) and top-view (flat-on) modes, Auger electron spectroscopy (AES) with sputter depth profiling, and 1.8 MeV $^4He^+$ Rutherford backscattering spectrometry (RBS). Stresses were measured in-situ through heating cycles at temperatures up to 500°C by the 2-300s FleXus system. The electrical characterization involved measurements of current- voltage-temperature (I-V-T) and specific contact resistance by TLM. The argon (Ar) concentration in the W films was analyzed by secondary ion mass spectrometry (SIMS), using a CAMECA IMS-3F system, with Cs^+ primary beam and detecting positive secondary ions.

InP wafers with full area tungsten metallization were evaluated for dry etching, as a preliminary stage for forming W ohmic self-aligned contacted devices. These wafers were lithographically patterned with AZ1350J photoresist to give a test structure with a variety of different size (1-50μm) and shape openings. This enabled us to determine the effect of the geometry of the openings on the tungsten etch rate. The samples were exposed to a 50W 0_2 plasma in a standard barrel reactor to ensure the unmasked areas were free of residual photoresist scum. Reactive ion etching was performed in a conventional stainless shell parallel plate reactor (Materials Research Corporation Model 51). The wafers were laid on the 6 inch diameter, water-cooled lower electrode (cathode) to which 13.56 κH_z power was applied through an impedance matching network. The electrode spacing was 2.75 inches, and the cathode was covered with a quartz cover plate. The chamber pressure was lowered to 7×10^{-7} Torr by mechanical and diffusion pumps before the introduction of CF_4 and 0_2 through electronic mass flow controllers. All etching was performed at 5 mTorr with flow rates of 10 sccm Cf_4 and 10 sccm 0_2, and a self-bias of 300V on the cathode. After the RIE treatments the photoresist was removed in acetone and the etch depth measured by Dektak stylus profilometry.

3. RESULTS AND DISCUSSION

a. Pt/Ti Contacts to n⁻–InP and p⁺–InGaAs

The interfacial microstructure and phase sequence evaluation of the Pt(60nm)/Ti(50nm) contacts to both n⁻-InP (S doped $5 \times 10^{18} cm^{-3}$) and p⁺-InGaAs (Zn doped $5 \times 10^{18} cm^{-3}$) were widely reported elsewhere[15]. In both contacts relatively limited reactions occurred due to heat treatments up to temperature of 450°C. RTP at this temperature led to significant interdiffusion reactions in the Ti-semiconductor interface as well as solid state reaction between the Ti and Pt to form mostly the Pt_3Ti intermetallic. Intensive RBS analysis revealed the existence of this layer (about 12nm thick) in both contacts. AES of the Pt/Ti/p-InGaAs contacts provided evidence for the formation of two separated layers on top of the ternary substrate, containing Ti and As layers (about 10 nm thick) adjacent to the substrate and In and As layer in between the latter and the Ti layer.

Figure 1 shows the XTEM micrographs of the corresponding interfacial microstructure of the Pt/Ti/p-InGaAs (Fig. 1a) and Pt/Ti/n-InP (Fig. 1b) samples, after RTP at 450°C. For the former, a complicated reaction zone of about 80nm thick was developed at the interface due to the heat treatment. The microstructure of the reacted zone consists of a defective interfacial layer and solid state regrown second phase region as indicated by the arrows. The electron diffraction patterns taken from the interfacial region were compared with those taken from the bulk Ti/Pt film and InGaAs in order to identify the dominating interfacial compound developed through RTP at 450°C[5]. From this analysis, a relatively high concentration of the InAs phase was detected at the Ti/InGaAs original interface. Compilation of the AES and XTEM observations suggests that the outdiffusion of In and As from the substrate resulted in a significant In depletion near the InGaAs surface and built an In high concentration region in between the InGaAs and Ti. Thus, the accompanying decrease in the lattice constant in the InGaAs layer induced misfit stresses which may account for the deformed zone observed near the interface. For the Pt/Ti/n-InP (Fig. 1b), an almost abrupt metal semiconductor interface is observed, which agrees with the limited nature interfacial reaction that was observed by means of AES and RBS. It is clear that all the intermetallics formation took place at the original metal layer volume as a result of the RTP and did not cause creation of any damaged zone in the InP substrate.

In summary it has to be emphasized that even as a result of heat treatment at high temperature of 450°C, the Pt/Ti contact to both InP and InGaAs layers contained relatively thin interfacial reacted layers (maximum of 80nm thick), which is about 4 times smaller than a typical interfacial reacted layer in the standard Au-based contacts to both InP and InGaAs.

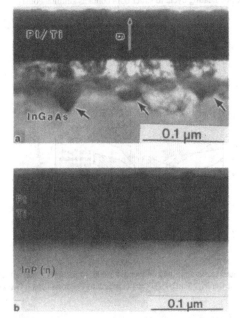

Figure 1. XTEM micrographs of (a) Pt(75nm)/Ti(50nm)/p-InGaAs and (b) Pt(75nm)/Ti (50nm)/n-InP samples, after RTP at 450°C for 30 sec.

The mechanical and thermal stresses in these Pt/Ti/InP-based contacts were extensively studied and reported elsewhere[16], as well as their electrical properties[2-5]. Summary of those is given in Figure 2 which shows both the specific contact resistance characteristics and the thin layer biaxial stress values of the common Pt/Ti contact to p-InGaAs and n-InP, as a function of the RTP temperatures. In addition it presents the induced stress in the Pt/Ti/SiO$_2$ structure on InP substrate. The specific resistance values of the Pt/Ti contacts to both p-InGaAs an n-InP are given as a function of the semiconductor doping level, over the range of 5x10^{18} to 1.5x10^{19}cm^{-3} Zn doped, for the former, and 5x10^{17} to 5x10^{18}cm^{-3} S doped for the latter. The vertical line at temperature of 470°C shows the highest RTP temperature at which degradation of the contact microstructure has not yet occur, as was revealed by the TEM, RBS and AES analysis[1,3]. The compilation of the above mentioned three properties enables optimization of the contact formation process conditions, which results in the lowest specific contact resistance, lowest induced biaxial stress both in the metallization films and in the bonding pad structure, and a stable microstructure. For the Pt/Ti contacts to both n-InP and p-InGaAs one can see that the best sintering temperature range in order to satisfy these three major requirements is between 400° and 450°C. The high end of this range provides a contact with an extremely low specific contact resistance, achieved due to the limited interfacial structure evaluation. Reducing the sintering temperature, however, to 400°C buys a considerable improvement in the biaxial induced stress of both the metal/SiO$_2$ and metal/semiconductor systems. This wide available window of processing temperature reveals·one of the most attractive advantages of the Pt/Ti metallization contact scheme, namely, its flexibility in processing,which enables the user to optimize the fabrication conditions due to his specific needs and constraints. Fortunately, the best electrical performance of both Pt/Ti/n-InP and Pt/Ti/p-InGaAs are achieved as a result of sintering at the same temperature of 450°C. This allows not only using one common metallization scheme to form the front and back contacts, as is shown in the schematic attached to Figure 2, or the two adjacent contacts in the case of coplanarized devices, but also enables application of only one heating cycle to sinter both contacts simultaneously.

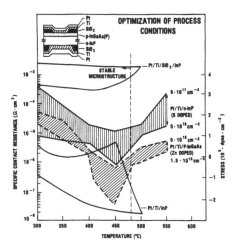

Figure   2.   Specific contact resistance and induced biaxial stress as a function of the processing temperature of the Pt(75nm)/Ti(50nm) contacts to Zn doped In$_{0.53}$Ga$_{0.47}$As and S doped InP.

b.   W Contacts to n-InP and p-InGaAs

W and W-alloys are considered the metals of choice for the Schottky gate metallization in self-aligned GaAs metal-semiconductor field effect transistor (MESFET) and heterostructure field effect transistor (HFET) devices[10,17]. In these GaAs device technology the W gates perform as a self-aligned contacts for ion implantation into the source and drain regions, which involves high temperature (800-850°C) activation annealing, and thus sets the major requirement for this self-aligned metal scheme, which is to form a stable and reproducible Schottky contact through the ion implantation and the subsequent severe heat-treatment[9]. In the InP-based laser device technology, however, the terminology of self-aligned devices refer to Etched Mesa Buried Heterostructure (EMBH) lasers that are processed through all the required manufacturing steps of mesa etching, regrowth of the blocking layers and final processing, with the existence of the metal ohmic contact on top of the semiconductor base-structure. This metal pattern, therefore, has to serve as a selective mask for both the InP-based material dry or wet etching and the regrowth of the blocking material, surrounding the mesa. From the metallurgical point of view the metal of choice has to be etchable in order to enable the geometrical definition, which serves both as the contact and as the mesa etching mask, to be inert to the semiconductor etching procedure, to prohibit any semiconductor growth on top of it during the regrowth of the blocking layers, to perform as an inert layer toward the InP-based material under it and to be stable through the regrowth cycle (~650°C), to have good adhesion to the InP-based material and thus to produce low stress through the process, and should also be compatible with the existing patterning and intercontacting materials and techniques. Above all, the metal of choice has to perform as a good ohmic contact to the semiconductors layer under it. Only by fulfilling these requirements one can benefit from using self aligned device technology for manufacturing InP-based laser devices.

Figure 3 shows a backscattered cross-section micrograph of the $W/In_{0.53}Ga_{0.47}As/InP$ sample after RTP at 500°C (Fig. 3a) and 700°C (Fig. 3b) for 30 sec taken by high resolution SEM. The latter represents the conditions which take place in the regrowth process. These micrographs show an almost abrupt metal-semiconductor interface, which agrees with the AES and RBS observations of these samples[6]. The tungsten layer morphology, as is revealed by means of the SEM, looks polycrystalline and columnar, with a grain size of about 100nm, following RTP at 700°C. One can see that the W contact had an almost abrupt interface with both InP and InGaAs layers.

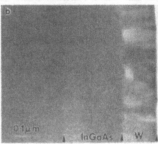

Figure 3.   SEM high-resolution field emission cross section of the W(100nm)/p-InGaAs sample after RTP at (a) 500°C and (b) 700°C for 30 sec.

Figure 4 presents the TLM-derived resistance measurements for W contacts to $5x10^{18}$ and $1x10^{19} cm^{-3}$ Zn-doped InGaAs and to $5x10^{18}$ and $1x10^{19} cm^{-3}$ S-doped InP layers as-deposited and after RTP at different temperatures. The plotted values are averages over 20 measured points and are associated with standard deviations <30%. One can see that by doubling the Zn doping level, the minimum specific contact resistance values were dropped from $8.5x10^{-3}$ to $7.5x10^{-6} \Omega cm^2$ in the contacts to p-InGaAs and from $2.2x10^{-5}$ to $3.5 -10^{-6}$ in the contacts to n-InP. The contact to the lower doped InGaAs layer was rectifying as-deposited as well as after heating at temperatures up to 450°C, while converting to an ohmic contact as a result of RTP at higher temperatures. In the former, doubling the doping level to $1x10^{19} cm^{-3}$ yielded an ohmic contact already as-deposited for the W/n-InP contacts, and after RTP at 300°C for both cases. The as-deposited sample had relatively high contact resistivity ($R_c = 0.6 \Omega mm$) and specific contact resistance ($p_c = 1.3x10^{-3} \Omega cm^2$). Following the heating of the contacts to the $5x10^{18} cm^{-3}$ Zn-doped InGaAs, and due to the transformation from rectifying to ohmic contact, its resistivity and specific resistance dropped, and yielded the minimum values after RTP at 600°C for 30 s ($R_c = 0.15 \Omega mm$ and $p_c = 8.5x10^{-5} \Omega cm^2$). This heating condition yielded the best electrical performance of all the W/n-InP contacts, as well. Heating the samples to higher temperatures resulted in the increase of all the measured electrical property values, namely, the contact resistance, the resistivity, and the heterolayer sheet resistance.

Figure 5 shows the in-situ biaxial stress measurements during heating and cooling cycles of the rf diode sputtered W films as a function of the Ar pressure through the deposition. Since no phase evaluations were observed in these films through heating cycles up to 500°C, the changes in the film stress during the thermal cycles are attributed to the combined effect of other factors discussed elsewhere[16]. The stress of the W films, however, was found to be unchanged in all cases during heating cycles up to 280°C. Since the coefficient of thermal expansion of the rf sputtered tungsten film

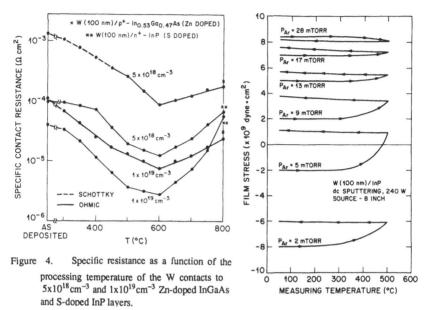

Figure 4. Specific resistance as a function of the processing temperature of the W contacts to $5 \times 10^{18} \text{cm}^{-3}$ and $1 \times 10^{19} \text{cm}^{-3}$ Zn-doped InGaAs and S-doped InP layers.

Figure 5. Biaxial stresses in the rf sputtered W film as a function of the in-situ measuring temperatures.

was found to be almost similar to that of InP substrate (see Fig. 6), this kind of flat dependence is expected. Above 280°C the stress curve of all the studied films showed the existence of plastic deformation, indicating densification or stress relaxation effects. In these particular cases, with the exception of the film deposited under Ar pressure of 28mTorr, the stress change is in the direction of higher tensile stress, suggesting that the densification component of the resulting plastic deformation is the dominant one. During cooling the stresses showed only an elastic component, indicating a pure thermal stress change. Another interesting result is the correlation between the Ar sputter pressure and the level of the stress hysteresis, decreasing from about an overall value of $3 \times 10^{19} \text{dyne·cm}^{-2}$, in the film that was deposited under Ar pressure of 5mTorr, to about $0.1 \times 10^{9} \text{dyne·cm}^{-2}$, when deposited under Ar pressure of 28mTorr. From the above mentioned in-situ results one can conclude that a zero stress condition in W films may be achieved while deposited onto InP substrate under an Ar pressure of about 8mTorr. Minimizing the mechanical and thermal stresses in the film is one of the most required properties of the W thin films, and thus we have adopted these deposition conditions for all our far future experiments.

Figure 6 shows the in-situ stress measurements of W/InP and W/GaAs samples (W deposited under Ar pressure of 8mTorr), as a function of the sintering temperature. Measuring stresses of identical films deposited parallel on two or more different substrates, enables the extraction of the biaxial elastic modulus and coefficient of thermal expansion of the film[18]. These elastic characteristics of this W film were calculated, using this technique, to be $0.97 \times 10^{12} \text{Pa}$ and $5.843 \times 10^{-6} °\text{C}^{-1}$, respectively. It is important to emphasize that the thermal expansion coefficient of the W film at the regrowth temperature (~650°C) was calculated to be very similar to that of the InP, namely, 4.83 and $4.75 °\text{C}^{-1}$, respectively, which suggest minimum stress mismatch in the interface through the thermal cycle.

In order to complete the metallurgical study of the W metallizations, we have investigated the Ar concentration in the W film, induced by different Ar pressure deposition conditions. Figure 7 shows the SIMS depth profile of the sample that was deposited under Ar pressure of 10mTorr, and which was found to be typical for all the observed profiles. The profiles show that the Ar is incorporated into the W layer, with a constant concentration across it, but did not diffuse into the InP substrate. Figure 8 summarizes the stress measurements and the Ar to W concentration ratio in the as deposited samples, as a function of the Ar pressure during the rf sputter deposition. In addition it presents the RIE etching rates of these different films. One can see that by increasing the Ar pressure the induced biaxial stress in the W film is increased from a compressive stress of about 7×10^9 dyne·cm^{-2}, in film that was sputtered under Ar pressure of 2mTorr, to a tensile stress of about 8×10^9 dyne·cm^{-2}, when deposited under pressure of 28mTorr. Zero stress situation was indeed achieved in the film that was deposited under Ar pressure of about 8mTorr. Usually, an increase in the Ar pressure corresponds to a lower density of energetic Ar reflected from the metal target. The decrease in ion irradiation during the film depositions leads to increase in the tensile stress, as a result of decreasing ion-to-vapor ratios[19], and thus agrees with the current measurements.

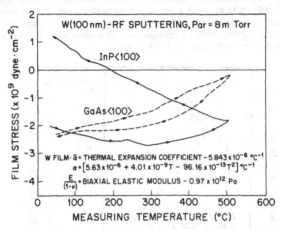

Figure 6.   Stress as a function of temperature in the W/InP and W/GaAs as-deposited samples.

SIMS examination indicated that the amount of Ar incorporated into the W films during sputtering was inversely proportional to the Ar pressure (see Fig. 8). The Ar/W ratio in the film sputtered at 5mTorr was 0.0357 whereas at 28mTorr it was reduced to 0.001, which is consistent with previous reports of Ar entrapment in $W_x Si_y$ films[9,20,21]. The entrapment mechanism is known to be the shallow implantation of Ar neutrals reflected from the target. At higher pressures the mean free path

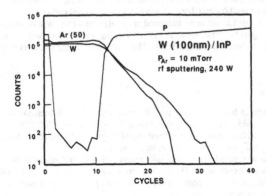

Figure 7.   SIMS depth profile of W/InP sample rf sputter deposited under Ar pressure of 8mTorr.

of Ar atoms substantially decreases and the enhanced scattering lowers its incorporation into the film.

The etch rate of the W films showed almost a linear dependence on the Ar pressure during the deposition (Fig. 8). Since these higher Ar pressures led to films with higher tensile stresses it might be expected that the more highly stressed films would display a higher etch rate due to the weaker bondings. The ion bombardment component of the RIE is therefore able to produce more efficient desorption of the W etch products. The samples were overetched for up to 8 minutes in order to determine the selectivity of the CF_4/O_2 discharge for W over InP. Under our conditions we observed InP sputter removal rates of ~20Å·min^{-1}, with selectivity for W over InP of ~10:1. This appears to be adequate for most self-aligned contact and regrowth applications. Since the original W layers were only 800-1000Å thick it is difficult to say much about the anisotropy of the etching, as is seen in the SEM micrographs at Figure 9. We did not see any significant differences in appearances between the four different types of etched tungsten. Further investigation is still required to verify the latter conclusion and in particular to look more carefully into the film morphology.

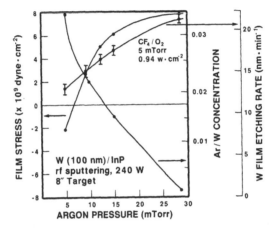

Figure 8.   Induced stress, Ar/W concentration ratio, and RIE etching rate of W(100nm) film on InP substrates, as a function of the sputter deposited Ar pressure.

By summarizing the electrical and metallurgical properties of the W contact to n$^-$-InP and p$^+$-InGaAs one can conclude that these contacts provide all the desired performance required for the ohmic and self-aligned contact masking metallization scheme, as was defined earlier.

The only issue that has yet to be addressed is the viability of the W metallization as a selective mask for the InP regrowth. This issue is currently being studied and will be reported later on, accompanied by further electrical characteristics of both contacts and the optoelectronic performance of the laser devices.

4. CONCLUSIONS

1. Ohmic contacts to both p- and n-type InP-based materials have been formed using refractory metal schemes, which are thermally stable up to 500°C (Pt/Ti) and 700°C(W).
2. The contact interfacial microstructure and morphology, due to the solid-phase reactions which took place, contain only a relatively thin intermixing layer.
3. Pt/Ti contacts to both n-InP and p-InGaAs reveal superior performance to those of the standard Au-based alloyed contacts. Pt/Ti metallization scheme can be used beneficially in the device manufacturing ambient as a common contact for both p- and n- InP-based structures, on planar and two sides contacted devices.

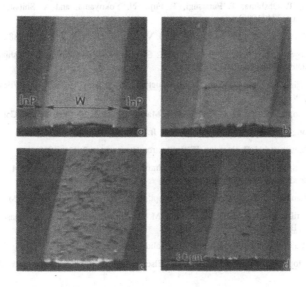

Figure 9.   SEM top view micrographs of the W stripes etched in the W films that had been sputter deposited under argon pressure of (a) 5, (b) 10, (c) 15 and (d) 28 mTorr.

REFERENCES

[1]   A Katz, B. E. Weir, S. N. G. Chu, P. M. Thomas, M. Soler, T. Boone, W. C. Dautremont-Smith, J. Appl. Phys., To be published, Issue of April 1, (1990).

[2]   A. Katz, W. C. Dautremont-Smith, S. N. G. Chu, P. M. Thomas, L. A. Loszi, J. W. Lee, V. G. Riggs, R. L. Brown, S. G. Napholtz and J. L. Zilko, Appl. Phys. Lett. 54, 2306 (1989).

[3]   S. N. G. Chu, A. Katz, T. Boone, P. M. Thomas, V. G. Riggs, W. C. Dautremont-Smith, and W. D. Johnston, Jr., J. Appl. Phys., To be published, issue of April 15, (1990).

[4]   A. Katz, P. M. Thomas, S. N. G. Chu, W. C. Dautremont-Smith, R. G. Sobers and S. G. Napholtz, J. Appl. Phys. 67. 884 (1990).

[5]   A. Katz, W. C. Dautremont-Smith, P. M. Thomas, L. A. Koszi, J. W. Lee, V. G. Riggs, R. L. Brown, J. L. Zilko, and A. Lahav, J. Appl. Phys. 65, 4319 (1989).

[6]   A. Katz, B. E. Weir, D. M. Maher, P. M. Thomas, M. Soler, W. C. Dautremont-Smith, R. F. Karlicek, Jr., J. D. Wynn and L. C. Kimerling, Appl. Phys. Lett. 55, 2220 (1989).

[7]   A. Katz, S. J. Pearton and M. Geva, Submitted to Appl. Phys. Lett..

[8]   K. Ishii, T. Ohshima, T. Futatsugi, T. Fujii, N. Yokoyama, and A. Shibatomi, IEDM 86 proceedings, 274 (1986).

[9]   A. G. Lahav, C. S. Wu and F. A. Baiocchi, J. Vac. Sci. Technol. B, $\underline{6}$, 1785 (1988).

[10]   N. Yokoyama, T. Ohnishi, H. Onodera, T. Shinoki, A. Shibatomi, and H. Ishikawa, IEEE J. Solid-State Circuits 18, 520 (1983).

[11]   K. Nakajima, in GaInAsP alloy semiconductors, edited by T. P. Pearsall (Wiley, New York, 1982), pp. 43-60.

[12]   J. L. Zilko, E. J. Flynn, D. C. T. Huo, A. T. Macrander and T. M. Shen (Private Communication).

[13]   J. A. Long, V. G. Riggs and W. D. Johnson, Jr., J. Cryst. Growth 69, 10 (1984).

[14]   G. K. Reeves and H. B. Harrison, IEEE Electron Device Lett. 5, 111 (1982).

[15]   A. Katz, S. N. G. Chu, W. C. Dautremont-Smith, M. Soler, B. E. Weir and P. M. Thomas, SPIE-89 Proceedings, To be published.

[16]   A. Katz and W. C. Dautremont-Smith, J. Appl. Phys., To be published, Issue of May 15, (1990).

[17]   N. C. Cirillo, Jr., H. K. Chung, P. J. Vold, M. K. Hibbs-Brenner and A. M. Fraasch, J. Vac. Sci. Technol. B3, 1680 (1985).

[18]   T. F. Ratajczyk, Jr. and A. K. Sinha, Thin Solid Films, 70, 241 (1980).

[19]   T. Itoh, Ion Beam Assisted Film Growth, (Elsevier Science Publisher, New York, 1989) pp. 121-123.

[20]   J. L. Vossen and W. Kern , Thin Film Processes, (Academic Press, New York, 1978), p.59.

[21]   R. A. Levi and P. K. Gallagher, J. Electrochem. Soc. 132, 1986 (1985).

SUBMICRON PSEUDOMORPHIC HEMT's USING NON-ALLOYED OHMIC CONTACTS WITH CONTRAST ENHANCEMENT.

Ph. Jansen, W. De Raedt, M. Van Hove, R. Jonckheere, R. Pereira, and M. Van Rossum
Interuniversity Micro-Electronics Center (IMEC) Leuven, Belgium.

Abstract

We report for the first time the realization of submicron pseudomorphic $Al_{.15}$ $Ga_{.85}As$-$In_{.20}Ga_{.80}As$ HEMT's with non-alloyed Pd/Ge ohmic contacts. Best results of contact resistance were obtained at a sintering temperature of 340°C with values as low as 0.057 Ωmm. Enhanced contrast, needed for accurate alignment of the gate by electron-beam lithography, was obtained by using Pd/Ge/Ti/Pd and Pd/Ge/Ti/Pt metal sequences. These contacts exhibited even lower contact resistances than the standard Pd/Ge contacts. Although Pd/Ge/Ti/Pd exhibits good morphology, reaction is witnessed at the edges, reducing the accuracy of alignment.

Processed enhancement mode devices exhibit maximum transconductances in excess of 520 mS/mm and currents of 300 mA/mm for 0.3 micron gatelength. This study shows that the contact resistance is no longer a restriction for obtaining very high transconductances in high performance devices.

1 Introduction

Since the first realization of High Electron Mobility Transistors (HEMT's) ohmic contacts were made using gold based metallizations. These metallizations exhibit low series and contact resistance and are well known from the Metal Semiconductor Field Effect Transistor (MESFET) technology. For HEMT's however, further improvement was needed due to deeper penetration to contact the two dimensional electron gas. [1, 2]

The drawback of these gold based metallizations is the complexity of the reactions occuring at the metal semiconductor interface. The contact resistance is a function of cleaning technique, surface quality and alloy parameters. A major improvement has been made by the capped alloy technique [3, 4]. An oxide or nitride layer prohibits As outdiffusion and so improves the quality of the interface and the contact resistance. However, a good oxide or nitride deposition process is required.

In the mid 80's, non-alloyed ohmic contacts based on Solid Phase Epitaxy (SPE) on GaAs gained much interest. As shown by S.S. Lau and coworkers [5], Pd/Ge ohmic contact formation allows for a better control resulting in repeatability, good thermal stability and small resistivity value dispersion. A. Paccagnella and coworkers [6] showed a lowest contact resistance of 0.16 Ωmm by evaporating 50 nm of Pd and 130 nm of Ge and by sintering at a temperature of 325°C for 30 minutes.

In this contribution a comparison between gold based alloys and Pd/Ge on pseudomorphic HEMT's is presented as well as improved Pd/Ge metallizations for electron beam detection for sub-half micron gatelengths.

2 Optimization of Pd/Ge ohmic contacts

Si-doped <100> GaAs substrates ($\approx 3.10^{17}$ cm^{-3}) were chemically cleaned and rinsed in a HCl:H_2O 1:1 solution prior to loading in a vacuum system. Different thicknesses of Pd

Material	300° C	320° C	340° C
Pd(50)/Ge(150)	3.15 ± 0.35 Ω	2.90 ± 0.45 Ω	2.27 ± 0.12 Ω
Pd(40)/Ge(120)	2.85 ± 0.30 Ω	2.65 ± 0.30 Ω	2.20 ± 0.10 Ω
Pd(30)/Ge(70)	2.77 ± 0.12 Ω	2.78 ± 0.13 Ω	2.80 ± 0.12 Ω
Pd(25)/Ge(75)	4.05 ± 0.15 Ω	3.35 ± 0.12 Ω	3.68 ± 0.20 Ω

Table 1: Resistance measured between two Pd/Ge contacts (100x100 μm^2) after sintering for 2 minutes, without correction for metal resistance nor parasitic resistance of the leads. Thicknesses are written in nm.

and Ge were evaporated in an electron beam evaporation system with a base vacuum of 10^{-8} torr. Pattern definition was realized by a standard lift-off technique. A scan of the sintering temperature from 300°C to 400°C was performed in forming gas (10% H_2, 90% N_2) ambient for 2 minutes in a rapid thermal anneal oven. This sintering time, shorter compared to the results of [5], was chosen to fit to our standard GaAs processing. The most significant results are listed in Table 1.

Best results are obtained using a 40/120 nm Pd/Ge layer with sintering for 2 minutes at 340°C in forming gas ambient. These results are comparable to the results of [6], except the much shorter time used in our experiments.

3 Pd/Ge versus AuGe/Ni/Au contacts

Comparison tests were performed on pseudomorphic HEMT layers. The layers were grown by Molecular Beam Epitaxy (MBE). For all samples the growth sequence was as follows: an undoped GaAs substrate, 1 μm GaAs buffer layer, 15 nm undoped $In_{.20}Ga_{.80}As$, 3 nm undoped $Al_{.15}Ga_{.85}As$ spacer layer, 40 nm $Al_{.15}Ga_{.85}As$ (2.10^{18} cm^{-3}) and 40 nm GaAs (4.10^{18} cm^{-3}) top layer.

Three metallization structures were compared, namely AuGe/Ni/Au with and without Si nitride cap layer and Pd/Ge. After a HCl:H_2O (1:1) cleaning the gold based contacts were made by evaporating a sequence of 120 nm of eutectic AuGe, 15 nm of Ni and 60 nm of Au. This multilayered contact was optimized for lowest contact resistance. The 200 nm thick Si nitride was deposited in a plasma enhanced chemical vapour deposition system. For Pd/Ge optimum conditions as determined previously were used.

Measurements were performed on a Transmission Line Model (TLM) structure [7] after depositing an extra 200 nm Pd/Au (C-gate) overlayer. Results are shown in Figure 1. These values are corrected for the internal resistances of the measurement system. Pd/Ge exhibits a lower contact resistance than uncapped AuGe/Ni/Au ohmic contacts. Even on a highly doped GaAs top layer the contact resistance of the capped AuGe/Ni/Au is comparable to the contact resistance of the Pd/Ge scheme.

When normalizing the contact resistance, values as low as 0.057 Ωmm are obtained for the Pd/Ge sample. However, the accuracy of fitting the TLM model has a very strong impact on the absolute value of the extrapolated intercept. Only fits with error values less than 10^{-1} are taken into consideration. Further analysis shows that the spread of the results for Pd/Ge is lower than for AuGe/Ni/Au, and has a better fit on the TLM line. This is believed to be due to the cleaning action of Pd, and makes the ohmic resistance less sensitive to precleaning and surface condition.

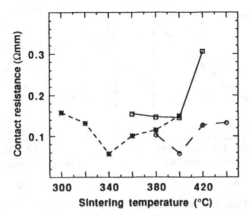

Figure 1: Plot of the contact resistance in function of the sintering temperature for different ohmic metals: AuGe/Ni/Au (□), capped AuGe/Ni/Au (o) and Pd/Ge (∗).

Similar tests were performed at cryogenic temperature. Due to the high mobility of the two dimensional electron gas at low temperatures the layer resistance is significantly decreased. See Figure 2a for a plot of the decreasing sheet resistance as a function of temperature.

The contact resistance is plotted versus temperature dependence in Figure 2b. From literature we expect a different temperature behaviour for contact resistance between AuGe/Ni/Au and Pd/Ge. AuGe/Ni/Au is directly contacting the InGaAs quantum well whereas Pd/Ge is not. However, the observed differences are within the accuracy of the measurements.

We believe that due to the very high doping level in the HEMT structure, the barriers through which tunneling occurs, are extremely thin. Field emission probably is the main effect since the tunneling current through the AlGaAs layer is not affected by the temperature change.

4   Contrast enhanced contacts

The realization of sub-half micron gate devices requires writing a very small gate between the ohmic source and drain. The electron beam lithography system needs to detect the edges of source and drain regions. If the contrast of the ohmic contacts is low, detection has to be improved by using a higher gain and beam current. This however, affects the minimum spot size of the system so that minimum gatelengths can increase drastically.

To solve this problem we tried high reflective materials as a cover for Pd/Ge. Au is a first candidate for enhanced contrast. From the literature results, we find that a diffusion barrier is needed between Pd/Ge and Au. Best barrier was obtained with Ti/Pd and Ti/Pt according to [6]. We performed tests on Ti/Pd/Au and Ti/Pt/Au and reaction was witnessed starting at 300°C and 350°C respectively. This explains the results of A. Paccagnella concerning the steep increase of the contact resistivity of Pd/Ge/Ti/Pd/Au after annealing for multiple hours at 300°C. This reaction also deteriorates the resistance when sintering at 340°C.

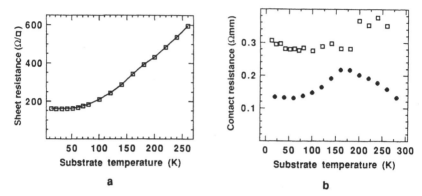

Figure 2: (a) Plot of the sheet resistance as a function of the substrate temperature. (b) Plot of the contact resistance as a function of the substrate temperature for AuGe/Ni/Au (□) and Pd/Ge (•).

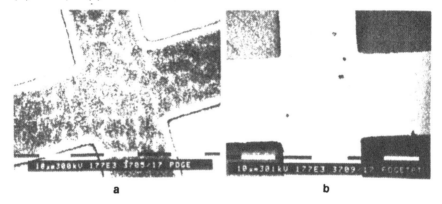

Figure 3: SEM photograph of a Pd/Ge (a) and Pd/Ge/Ti/Pt (b) alignment structure, sintered at 340°C for 2 minutes.

Therefore we decided to eliminate Au and we used Ti/Pd or Ti/Pt on top of Pd/Ge. The contrast of Pd is somewhat lower than Au but Pt has a higher contrast, due to the higher density. A comparison was also made with a Pd/Ge and Pd/Ge/Pt contact.

The contrast of the different metallization structures were compared by Scanning Electron Microscope (SEM) pictures. The contrast of a SEM picture is comparable to the detection sensitivity seen by the electron beam lithography system and therefore is a good tool for analyzing relative contrast enhancement. SEM photographs of Pd/Ge and Pd/Ge/Ti/Pt are shown in Figure 3a and b respectively. Pd/Ge has the worst contrast and on large surfaces grain boundaries, probably originated from the columnar growth, are witnessed.

Pd/Ge/Pt, Pd/Ge/Ti/Pd and Pd/Ge/Ti/Pt are very bright with the best contrast for the contacts with Pt on top. When analyzed under an optical microscope the Pd/Ge/Pt surface is rough and not as shiny as the Ti/Pd and Ti/Pt metal cap. This suggests the

Material	Ohmic	C-gate	Four point meas.
Pd/Ge	6.93 ± 0.29 Ω	2.67 ± 0.29 Ω	0.88 ± 0.13 Ω
Pd/Ge/Pt	5.27 ± 0.42 Ω	4.12 ± 0.30 Ω	
Pd/Ge/Ti/Pd	4.04 ± 0.35 Ω	2.40 ± 0.18 Ω	0.55 ± 0.34 Ω
Pd/Ge/Ti/Pt	4.00 ± 0.26 Ω	2.49 ± 0.15 Ω	0.57 ± 0.16 Ω

Table 2: Contact resistance measured on a TLM-structure with a width of 100 μm. By using a four point measurement setup parasitic resistance can be eliminated.

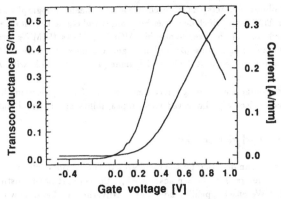

Figure 4: Current and transconductance as a function of gate-voltage for a 0.3x80 μm^2 high performance pseudomorphic HEMT.

occurrence of a reaction between Pt and Ge.

The Pd/Ge/Ti/Pd sample has rough edges, originating from reactions caused by slight imperfections in the covering of different metal layers. As a result Ti/Pt is the best cap with most contrast and highest repeatability.

The results of the contact resistance are displayed in Table 2. Best results, 0.055 and 0.057 Ωmm were obtained for Pd/Ge/Ti/Pd and Pd/Ge/Ti/Pt respectively, which have even lower contact resistances than the standard Pd/Ge ones.

5 Realization of high performance devices

A Pd/Ge metallization with Ti/Pt overlayer has been used for source and drain contacts in pseudomorphic HEMT's. The 0.3 μm gate was defined by electron beam lithography. The gate was recessed by a wet etchant to obtain the right threshold voltage and metallized with Pd/Au as gate contact. Best results were obtained for enhancement mode devices. These devices exhibit peak transconductances in excess of 520 mS/mm and currents of 300 mA/mm. Figure 4 shows a measured current and transconductance curve.

This realization proves that by using Pd/Ge/Ti/Pt, contact resistance is no longer a restriction for obtaining very high transconductances in high performance devices. Moreover an increased yield was witnessed compared to previously realized devices with AuGe/Ni/Au ohmic contacts.

6 Conclusion

Pd/Ge ohmic contacts on pseudomorphic HEMT's are shown to be more stable than AuGe/ Ni/Au which is inherent to the contacting scheme. Moreover the contact resistance is lower than the standard AuGe/Ni/Au and as low as the cap alloyed metallization structure. Comparison is also made at cryogenic temperatures and no major change in behaviour was noticed. From these results one can conclude that Pd/Ge is an excellent substitute for Au-based contacts on pseudomorphic HEMT's both at room and cryogenic temperatures. In addition, the absence of Au may favourably affect the reliability of these contacts. This point certainly deserves further investigation.

Non alloyed ohmic contacts can be used in electron beam lithography technology by using a Ti/Pd or Ti/Pt cap, which decreases both series and contact resistance. Realization of these enhanced contacts on pseudomorphic AlGaAs-InGaAs HEMT's resulted in very high transconductances exceeding 520 mS/mm and currents in excess of 300 mA/mm, which are comparable to the best devices with same gatelength made with AuGe/Ni/Au ohmic contacts.

Pd/Ge/Ti/Pt contacts are strongly recommended in a GaAs processing line to achieve high performance devices with improved yield, repeatability and thermal stability.

7 Acknowledgements

We wish to thank W. van de Graaf for the MBE growth and G. Borghs for useful discussions about growth parameters. Ph. Jansen acknowledges the support of the Instituut tot Aanmoediging van het Wetenschappelijk Onderzoek in Nijverheid en Landbouw (IWONL) and R. Pereira would like to thank CNPq (Conselho Nacional de Pesquisa e Desenvolvimento) and TELEBRAS-BRASIL for the support. This work was supported by the ESPRIT Basic Research Action 3042.

References

[1] A.Ketterson, F.Ponse, T.Henderson, J.Klem, H.Morkoç, J. Appl. Phys. 57, No.6, 1985

[2] A.Ketterson, F.Ponse, T.Henderson, J.Klem, C-K.Peng, and H.Morkoç, IEEE Trans. Elec. Dev., Vol.ED-32, pp.2257-2261, 1985

[3] A.J.Barcz, IEEE Elec. Dev. Lett., Vol.EDL-8, No.5, 1987

[4] A.Barcz, E.Kaminska, and A.Piotrowska, Thin solid films, 149, pp. 251-260, 1987.

[5] A.Paccagnella, A.Migliori, M.Vanzi, B.Zhang, and S.S.Lau, Proc. 17th ESSDERC pp.605, 1987

[6] A.Paccagnella, C.Canali, G.Donzelli, E.Zanoni, R.Zanetti, and S.S.Lau, Proc. 18th ESSDERC pp.441, September 1988

[7] H.H.Berger, Solid State Electron., Vol. 15, pp.145-158, 1972

LATERAL SPREADING OF Au CONTACTS ON InP

NAVID S. FATEMI* AND VICTOR G. WEIZER**
*Sverdrup Technology, Inc., Lewis Research Center Group, 21000
Brookpark Rd., Cleveland, Oh 44135
**NASA Lewis Research Center, 21000 Brookpark Rd., Cleveland,
OH 44135

ABSTRACT

We have investigated the contact spreading phenomenon
observed when small area Au contacts on InP are annealed at
temperatures above about 400 C. We have found that the rapid
lateral expansion of the contact metallization which consumes
large quantities of InP during growth is closely related to the
third stage in the series of solid state reactions that occur
between InP and Au, i.e., to the Au_3In-to-Au_9In_4 transition. We
present detailed descriptions of both the spreading process and
the Au_3In-to-Au_9In_4 transition along with arguments that the two
processes are manifestations of the same basic phenomenon.

INTRODUCTION

When small area Au-based contacts on InP are sintered at
temperatures above 400 C, a rapid lateral spreading of the
metallization along specific crystallographic directions on the
semiconductor surface can be observed.[1,2] The material in the
spread regions has been identified as the compound Au_9In_4,[1] and
the extent to which the spreading occurs has been shown to be
related to the amount of Au originally present.[2] The process is
apparently thermally activated with an activation energy somewhere
between 5.6 and 11.6 eV.[2] The purpose of this paper is to
provide a more detailed description of the spreading phenomenon in
the Au-InP system and to show that it is closely related to the
pink-to-silver (Au_3In-to-Au_9In_4) transition that is observed to
take place during contact sintering.

CONTACT SPREADING

The spreading phenomenon is illustrated in figure 1 where we
show the results of heating a 2000 A thick, 30 micrometer diameter
circular gold pad deposited on a polished, (100) oriented InP
surface for 2.5 hours at 435 C. As can be seen, the
metal-semiconductor contact area has been increased by more than a
factor of two. It can also be seen that a significant amount of
InP surrounding the circular pad has been replaced by the
silver-colored metallization. The metallization, which has been
identified as Au_9Ir_4 by Keramidas, et al.[1], can be seen to be
more or less flush with the unreacted InP along the reaction
front. Closer to the original disc, however, the surface of the
metallization drops considerably below the level of the
surrounding InP. Since the disc itself is not observed to
increase in size, the drop in the level of the metal indicates
that spreading is accompanied by a significant mass loss.

The amount of InP dissolved in the reaction can be
determined by chemically removing the metallization in the spread
regions. The resulting topography (for an identically processed
sample) is shown in figure 2. What we find when we remove the
metallization is that the InP has been eroded to an undulating,
but on the average constant, depth. The metallization in the

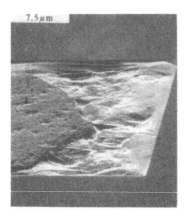
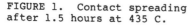

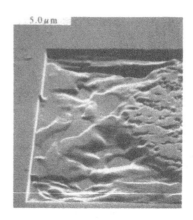

FIGURE 1.   Contact spreading
after 1.5 hours at 435 C.

FIGURE 2.   InP topography
after removal of metallization
from spread region.

spread region is therefore much thicker near the reaction front
than it is closer to the disc. From the orientation of pits that
were etched in the InP surface with a bromine-methanol etchant we
have determined that the (111) plane defining the reaction front
of the spread region is the In-faced plane. More importantly, the
spread regions were found to contain no trace of the Au_2P_3 layer
that has been observed to form at the metal-semiconductor
interface during contact sintering.[3,4]

With regard to the growth dynamics, we found that the
spreading process requires an incubation time at temperature
before growth begins. The incubation time appears to be material
sensitive, low dislocation density material being more susceptable
to spreading than lower quality material.

As we followed the progress of spreading we noted that the
spreading regions were always silver in color, and that when
spreading starts at a given point on the periphery of a disc, the
adjacent regions of the disc are also silver in color. As spreading proceeds, the entire disc changes from pink to silver.

It is evident that during the spreading process Au is being
transported from the Au-rich disc to the reaction front in the
spread region. If the only mass flow during spreading were the
diffusion of Au from the disc to the reaction front, however, then
at the front

$$9Au + 4InP \rightarrow Au_9In_4 + 4P.$$

This reaction would result in a 28% increase in the volume of the
spread region even if the P atoms dissipated. Since we actually
see a substantial decrease in the volume of the metal in the
spread region, we must conclude that In is being transported from
the reaction front to the disc.

Elias, et al., have measured the spreading rate constants at
various temperatures and have determined an activation energy for
the spreading process of 8.6 + 3 eV. If we assume that the
transport of either Au or In through the metallization is limiting
the rate of spreading, the diffusion coefficients determined by
Elias, et al., are not in conflict with those expected in an
intermetallic alloy with a solidus temperature T_m

in the 460 C range where, according to the empirical rule [5],
$$D = 0.3 \exp[-18T_m/T]$$
where D is the diffusivity at temperature T. However, the 8.6 ± 3
eV activation energy determined by Elias, et al.,[2] from
spreading rate measurements is much higher than would be expected
for a normal diffusion process (i.e., $Q = 18kT_m$[5]). It appears,
therefore, that a process other than diffusion is limiting the
rate of spreading, the actual diffusion of Au and In being faster
than Elias' calculations indicate.

When we sought to determine the effect of an SiO_2 cap layer
on the spreading process, we found that placing an SiO_2 on the
disc itself did not suppress spreading. However, while the cap
layer does not affect the ability of a disc to spread when it is
deposited on the disc, it is very effective in preventing
spreading when deposited on the InP surface adjacent to the disc.
Figure 3, for instance, shows the spreading that emanated from
three partially capped (600 A SiO_2) Au discs (2000 A thick) after
1 hour at 442 C in forming gas. As can be seen, spreading took
place only into uncapped InP. Figure 4 demonstrates how spreading
from an uncapped disc is terminated upon encountering an SiO_2 cap
on the InP surface.

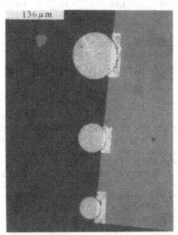 

FIGURE 3. Spreading from
partially capped contacts.

FIGURE 4. The attenuation of
spreading from an uncapped
disc upon encountering a cap
on the InP surface.

While a cap layer is effective in stopping the advance of
spreading, it does not do so abruptly. A close look at the
intersection of the spread region with capped InP reveals that the
spread region tunnels under the cap for about 1 micrometer before
stopping. An interesting feature of this tunnelling is that the
metallization under the cap does not shrink in volume as does the
uncapped metallization. Figure 5 illustrates the planar character
of the metallization under the cap (removed for clarity) as
compared to that along the adjacent uncapped reaction front.
While the details of the reaction suppressing mechanism are not
completely clear, we will show that it is apparently associated
with the cap's effect on the escape of phosphorus.

FIGURE 5. Topography of spread metallization under SiO$_2$ cap (cap removed).

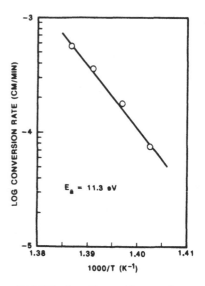

FIGURE 6. The effect of temperature on the Au$_3$In-to-Au$_9$In$_4$ conversion time.

THE Au-InP INTERACTION PROCESS

As stated in the introduction, there is evidence that the spreading phenomenon is closely related to the third stage in the Au-InP interaction. Let us first review what is known about the Au-InP reaction mechanisms and then draw some comparisons.

Recent investigations have shown that the solid state reaction of Au with InP takes place in several distinct steps or stages.[3,4] The first two stages have been shown to involve the entry of In and P into the contacting Au metallization. In stage I, In enters the Au lattice via a dissociative diffusion mechanism, the rate of which is controlled by the vacancy generation rate at the free surface of the metallization. In entry continues until the amount of In in the Au lattice reaches the solid solubility limit (about 10%). Stage I is quite rapid. Its results can be observed after aging at room temperature for several months. The P atoms that enter the metal with the In either take non-lattice sites in the metal or exit the system.

If stage I is allowed to go to completion a second In entry mechanism becomes active.[4] This stage proceeds rapidly in the temperature range used for contact sintering. In this stage In continues to enter the metallization, but this time via an interstitial exchange or kickout mechanism. In entry continues until the metallization is converted from the 10% Au(In) saturated solid solution to the compound Au$_3$In. The rate limiting process in stage II has been shown to be the release of In from the InP and its insertion interstitially into the Au lattice. Whereas in stage I the P atoms released with the In leave the system without reacting, in stage II they react to form the compound Au$_2$P$_3$ at the metal-semiconductor interface.

We report here on the results of an investigation into a

third stage in the Au-InP reaction that takes place upon further
sintering. In this stage In continues to enter the Au lattice
with the result that the pink compound Au_3In is converted to a
silver colored compound. Our first task was to identify the
silver colored product of stage III. To do this we used an XPS
system which we had calibrated by making up the various Au-In
intermetallic compounds and then determining their sensitivity
factors and binding energies. The resulting compositional and
binding energy analysis of the stage III metal indicated the
presence of the compound Au_9In_4. Vandenberg, et al.[6], also
found Au_9In_4 to form in this temperature range.

We then investigated the kinetics of the stage III
reaction. To do this we deposited Au_3In (the stage II product) on
a number of InP substrates and measured the rate at which Au_3In
was converted to Au_9In_4 at a given temperature. We found that the
conversion rate was proportional to the area of the interface
between the two phases. Thus the measured conversion rate per
unit interfacial area is invariant with sintering time. (In these
calculations the geometry of the converted regions was assumed to
be cylindrical.) A plot of the logarithm of this rate versus the
reciprocal temperature (figure 6) yields an activation energy of
11.3 eV.

There is a major difference in the behavior of the
phosphorus atoms during stage III as compared to stage II. In
samples where stage II has gone to completion, one observes the
presence of the compound Au_2P_3 at the InP-metal interface.
However, when stage II is bypassed by depositing and annealing
Au_3In, no sign of Au_2P_3 is found. Thus in stage III the P atoms
either take non-lattice sites or they leave the system.

THE EQUIVALENCE OF STAGE III AND CONTACT SPREADING

From the preceeding analysis we find that stage III and
contact spreading have a great deal in common. 1) In both cases
the reaction product is Au_9In_4. 2) The activation energies for
the two processes agree within published limits of uncertainty.
3) In neither case does the interfacial Au_2P_3 layer form. 4)
Stage III conversion in the deposited Au disc is observed to take
place concurrently with the growth of the spread regions in the
adjacent InP. On the basis of these similarities, it seems
reasonable to conclude that spreading and the stage III reaction
are manifestations of the same phase transformation process.

THE SPREADING/STAGE III MECHANISM

We have shown, in the preceeding analysis, that the
spreading mechanism involves the transport of Au and In in
opposite directions through the metallization in the spread
region. It is well known that In diffuses interstitially in all
the intermediate Au-In alloys.[7] In addition, since Au diffuses
interstitially in pure In [8] as well as in the intermediate
alloys in the closely related Au-Sn system [9,10,11], it is
reasonable to assume that the Au diffusion mode in Au-In alloys is
also interstitial.

Thus it appears that the diffusion of Au to the reaction
front and the diffusion of In from it are both interstitial in
nature. This is strong evidence that an interstitial exchange or
kickout mechanism [12] is involved during spreading/stage III.

The kickout process, already proposed to explain the stage
II phase transformation[4], starts with the release of In from the

InP and its entry into the metallization as an interstitial. The In interstitial then diffuses rapidly to the phase boundary where energy considerations favor an In-Au interchange. In this interchange, called the kickout step, the In interstitial displaces a Au substitutional atom and takes its place on the metal lattice. The resulting Au interstitial then diffuses away from the kickout site to react (in the case of stage II) with unbound P atoms to form Au_2P_3. The unique characteristic of this process is the fact that both species diffuse interstitially.

Indeed it is the interstitial diffusion of both Au and In during spreading that prompts us to suggest that the kickout process is active in the spreading/stage III phase transformation. We thus propose that the spreading process consists of the interstitial entry (at the spread region reaction front) and diffusion of In in the metal lattice, and its exchange with a substitutional Au atom (kickout) at the Au_3In-Au_9In_4 interphase boundary in the Au-rich disc where Au_3In is being transformed to Au_9In_4. The reaction at the interphase boundary is proposed to proceed as

$$13\ Au_3In + 3\ In_i \rightarrow 4\ Au_9In_4 + 3\ Au_i$$

where the subscript indicates interstitial siting. The resulting Au interstitials are transported back to the reaction front where they combine with interstitial In and vacant lattice sites to form Au_9In_4 there.

The proposal that a kickout mechanism is active during stage III provides an explanation for the observed dependence of the stage III reaction rate on the area of the interphase boundary. If the kickout rate is proportional to the number of potential kickout sites, and if these sites are located only at the boundary between the two phases, it follows that the kickout rate should indeed be proportional to the Au_3In-Au_9In_4 interfacial area.

As mentioned, P atoms released at the reaction front along with In do not react chemically during the spreading process and thus are free to exit the system. If these atoms (and/or the vacancies they leave behind at the InP-metal interface) are slow in diffusing out of the system, a slow (but sizeable) decrease in the volume of the metallization would be observed with time. This is, in fact, what is observed. As we have mentioned previously, the newly generated metallization near the reaction front in the spread regions is considerably thicker than the metal remote from the front where the P atoms and/or the vacancies they left behind have had time to diffuse out of the system. It is felt that the effect of InP capping on the geometry of the spreading metallization (figure 5) is related to its ability to affect the ease of phosphorus escape.

REFERENCES

1. V.G.Keramidas, H.Temkin, S.Mahajan, Inst.Phys.Conf.Ser. 56, 293 (1981)
2. K.R.Elias, S.Mahajan, C.L.Bauer, A.G.Milnes, and W.A.Bonner, J. Appl. Phys. 62, 1245 (1987)
3. N.S.Fatemi and V.G.Weizer, J. Appl. Phys. 65, 2111 (1989).
4. N.S.Fatemi and V.G.Weizer, J. Appl. Phys. 67, 1934 (1990).
5. H.Bakker in Diffusion in Crystalline Solids, G.E.Murch and A.S.Nowick, Eds. (Academic Press, Orlando, 1984) p. 193.
6. J.M.Vandenberg, H.Temkin, R.A.Hamm, and M.A.DiGiuseppi, J. Appl. Phys. 53, 7385 (1982).
7. G.W.Powell and J.D.Braun, Trans. AIME 230, 694 (1964).
8. T.R.Anthony and D.Turnbull, Phys. Rev. 151, 495 (1966).
9. M.Cretdt and R. Fichter, Metall (Berlin) 25, 1124 (1971).
10. L.Buene, Thin Solid Films 47, 159 (1977).
11. B.F.Dyson, J. Appl. Phys. 37, 2375 (1966).
12. W.Frank, U.Gosele, H.Mehrer, and A.Seeger, Diffusion in Crystalline Solids, G.E.Murch and A.S.Nowick, eds., (Academic, Orlando, 1984) p 63.

ROLE OF IN–SITU RAPID ISOTHERMAL PROCESSING (RIP) IN THE METALLIZATION AND PASSIVATION OF INDIUM PHOSPHIDE DEVICES

R. Singh, R. P. S. Thakur, A. Katz*, and A. J. Nelson**
University of Oklahoma, School of Electrical Engineering and Computer Science, Norman, Oklahoma 73019
*AT&T Bell Laboratories, Murray Hill, NJ 07974
**Solar Energy Research Institute, Golden, CO 80401

ABSTRACT

As compared to a stand alone rapid isothermal annealing unit, the integration of deposition system and rapid isothermal processing unit is very attractive for the next generation of micro, opto and cryoelectronics. We have used in–situ rapid isothermal processor for in–situ rapid isothermal chemical cleaning of InP, solid phase epitaxial growth of II–A fluorides, and in–situ metallization of InP capacitors. As compared to ex–situ annealed films, SrF_2 films deposited on InP by in–situ rapid isothermal processed films show less thermal stress and lower thermal hysteresis for the identical thermal budget. Similarly, as compared to ex–situ annealing, in–situ cleaning of InP before metallization followed by in–situ annealing results into improved high frequency capacitance–voltage (C–V) characteristics of $Al–SrF_2–InP$ capacitors.

INTRODUCTION

As a reduced thermal budget processing technique, rapid isothermal processing [1] technique based on incoherent sources of light as the energy source is emerging as a major processing technique. The short time feature of RIP is the key driver for the development of this new technology. In addition to the short time and high temperature processing feature, RIP at lower temperatures also has advantages over furnace processing due to the difference in the spectral contents of the two radiation sources of energy. At a given substrate temperature, due to the higher (compared to substrate temperature) filament temperature of the lamps, RIP provides photons in the visible and in ultraviolet region. Thus, certain physical and chemical phenomena can be promoted or initiated in the case of RIP [2,3]. Additionally, a reduced activation energy will be observed for the RIP case. Integration of the deposition and the rapid isothermal processing unit is attractive both for the cost reduction of the equipment as well as for improving the performance and reliability of the devices. In this paper we report the use of in–situ rapid isothermal processing for the solid phase epitaxial growth of II–A fluorides as well in–situ metallization of $Al–SrF_2–InP$ capacitors.

EXPERIMENTAL DETAILS

Undoped n–type (100) oriented InP wafers were used as substrates. After chemical cleaning, the InP samples were immediately loaded in the vacuum deposition system. The details of the integrated rapid isothermal processor are described elsewhere [4]. The substrate temperature was determined using a chromel alumel thermocouple.

In–situ cleaning of the InP substrates was carried out by using a 5% hydrogen and 95% argon mixture and heating the substrate at 475°C for 10 sec. at a base pressure of 3×10^{-8} Torr. The InP substrates were then allowed to cool to room temperature and SrF_2 films were deposited at room temperature followed by in–situ rapid isothermal annealing at 500°C for 5 sec. For comparison purposes, SrF_2 films prepared under

Mat. Res. Soc. Symp. Proc. Vol. 181. ©1990 Materials Research Society

conditions identical to those described earlier were subjected to ex–situ rapid isothermal annealing (identical annealing time and temperatures).

For electrical characterization of SrF$_2$ films, Al–SrF$_2$–InP capacitors were formed. In one case before depositing the back contact, InP back surface was in–situ rapid isothermal cleaned at 300°C for 3 seconds in the presence of Ar/H$_2$. Deposition of ohmic contact was followed by in–situ annealing at 450°C for 5 sec. In other case no in–situ cleaning of InP surface was employed before metallization. After metallization the ohmic contact was ex–situ postannealed at 450°C for 5 sec.

The stress measurements on SrF$_2$ films were performed using the flexus 2–300S thin film system [5]. The substrates were measured for their initial curvatures prior to any deposition or heat treatment. The initial curvatures were later substrated from all the following curvature measurements of the corresponding wafers after each processing step. The difference in curvature, caused by the processing steps was used to calculate the films stress. The film stress was assumed to be isotropic in the film plane and was calculated by using the equations described in Ref. 5. For the measurement of thermal stress as a function of temperature all the samples were heated and cooled in the rate of about 7.6°C/min.

In order to study the origin of the difference in the behavior of the thermal stress of ex–situ and in–situ annealed films, x–ray photoemission (XPS) measurements were used to characterize the surface of the SrF$_2$ films as well as at the SrF$_2$/InP interface for both the ex–situ and in–situ annealed films.

RESULTS AND DISCUSSION

Fig. 1(a) and 1(b) represent the high frequency C–V curves for the in–situ annealed and ex–situ annealed ohmic contact. The advantage of in–situ rapid isothermal cleaning is obvious from the improved C–V characteristics observed in Fig. 1(a) compared to Fig. 1(b).

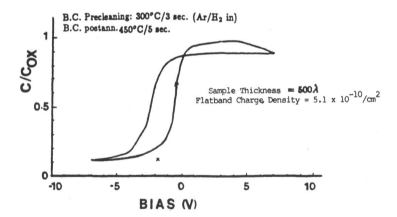

Fig. 1(a)

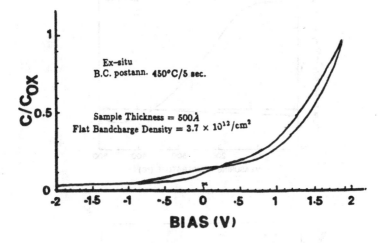

Fig. 1 (b)

Fig. 1. High frequency (1 MHZ) C–V characteristics of A*l*–SrF₂–InP capacitors. Case (a) corresponds to in–situ chemical cleaning of InP surface, and in–situ annealing of the ohmic contact. In case (b) the surface of InP was not cleaned before metallization and ex–situ annealing was carried out.

Figures 2(a) and 2(b) give the stress–temperature $(\sigma - \tau)$ curve of the ex–situ (Fig. 2(a)) and in–situ (Fig. 2(b)) rapid isothermal annealed films. The former was highly tensile stressed prior to the in–situ measurement cycle, and underwent a significant relaxation process, releasing the tensile stress of 2.6×10^{10} dyne cm^{-2} to zero stress conditions at 360°C. Further heating even introduced a slight compressive stress into the films, which was found to perform as a stressless film upon cooling back to room temperature. This $\sigma - \tau$ behavior and the value of the stress hysteresis suggest the occurrence of a complete stress relaxation of the initial induced stress, frequently accompanied by the formation of severe crystal defects. The in–situ annealed sample, on the other hand, was introduced to the measurement with a zero stress condition, and up to 200°C went only elastic deformation due to the evolution of thermal stress. A moderate amount of plastic deformation is reflected from the small change in the $\sigma - \tau$ slope. Cooling the sample back to room temperature led to the relaxation of the thermal elastic stress due to the difference in the coefficients of thermal expansion between the films and the InP substrate. A final tensile stress of about 1.5×10^9 dyne cm^{-2} was measured at the cooled sample which showed, however, only a moderate value of stress hysteresis.

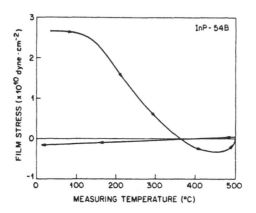

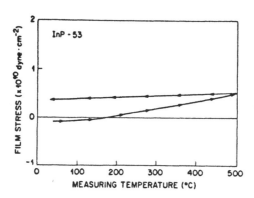

Fig. 2. Measured thermal stress of SrF₂ films on InP as a function of temperature for (a) ex–situ, and (b) in–situ annealed films.

Figures 3(a) and 3(b) show the high resolution XPS spectra of the Sr 3d core level for the ex–situ annealed films at the SrF_2 surface and at the SrF_2/InP interface, respectively. Similar results for the in–situ annealed films are presented in Fig. 4(a) and 4(b) respectively.

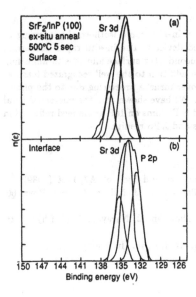

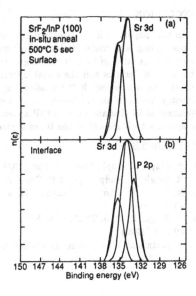

Fig. 3. XPS spectra of Sr 3d core level for (a) the SrF$_2$ surface, and (b) the SrF$_2$/InP interface of ex–situ rapid isothermal annealed films.

Fig. 4. XPS spectra of Sr 3d core level for (a) the SrF$_2$ surface, and (b) the SrF$_2$/InP interface of in–situ rapid isothermal annealed films.

It is clear from the XPS results presented here that there is a very small amount of oxygen is present in the in–situ rapid isothermal annealed films. The ex–situ annealed films show less barrier to the diffusion of oxygen. During the transfer os amples from the deposition system to the ex–situ annealer, exposure of the film results in the absorption of different gases, in particular oxygen. Such films when subjected to ex–situ rapid isothermal annealing are full of defects and are polycrystalline in nature [1]. The thermal stress behaviors of the ex–situ and in–situ rapid isothermal films is also different, and clearly the microstructure of the two films is different. The details of the microscopic nature of the phenomena is unknown and is currently under investigation, however it appears that the presence of oxygen in ex–situ annealed films resulting into Sr–O bonding hinders the solid phase epitaxial growth. The mixture of Sr–O and Sr–F bonds will result into an entirely different thermal stress as compared to almost Sr–F bonded material.

CONCLUSION

In this paper we have described the role of in–situ rapid isothermal processing in the passivation and metallization of InP based devices. The in–situ rapid isothermal processor is capable of in–situ cleaning of semiconductor surface and in–situ annealing of both the dielectrics and the metal layers. In addition to the well recognized features of short time processing, RIP has advantages over furnace processing due to the photo-chemistry driven by short wavelength photons. We have shown that the microstructural differences of in–situ and ex–situ RIP annealed SrF_2 films on InP are indeed reflected in the significant differences in the thermal stress and XPS results.

REFERENCES

1. R. Singh, J. Appl. Phys. <u>63</u>, R59 (1988).
2. R. Singh, F. Radpour, and P. Chou, J. Vac. Sci. and Technol. <u>A7</u>, 1456 (1989).
3. R. Singh, P. Chou, F. Radpour, H. S. Ullal, and A. J. Nelson, J. Appl. Phys. <u>66</u>, 2381 (1989).
4. R. Singh, R. P. S. Thakur, A. Kumar, P. Chou, and J. Narayan, Appl. Phys. Lett. <u>56</u> 247 (1990).
5. A. Katz and W. C. Dautremont–Smith, J. Appl. Phys. (In Press).

ANTI-REFLECTION COATINGS (ARC) FOR USE WITH ALUMINUM METALLIZATIONS ON GaAs ICs

MICHAEL F. BRADY AND AUBREY L. HELMS, JR.
AT&T Bell Laboratories, PO Box 900, Princeton, NJ 08540

ABSTRACT

The need for an Anti-Reflection Coating (ARC) on aluminum metallizations is well known. A new class of materials based on tungsten-silicide and tungsten-silicon-nitride has been developed for use as an ARC. It has been shown that the reflectivity (relative to aluminum = 100%) can be decreased to ˜55% for the tungsten-silicide material and to ˜6% for the tungsten-silicon-nitride materials. These materials are easily etched in fluorine containing plasmas and are not as sensitive to thickness uniformity issues as dyed resists or amorphous silicon ARC materials. The option of leaving these ARC materials on the aluminum lines may lead to an increase in the electromigration resistance. The dependence of the reflectivity on nitrogen content has been investigated. Additionally, the reflectivity reducing properties have been studied on a variety of substrates such as aluminum, gold, tungsten, tantalum, and silicon.

INTRODUCTION

The control of critical dimensions on fine line geometries with aluminum metallization requires a means of minimizing the amount of light reflected from the aluminum surface. This reflected light exposes areas of the photoresist which are protected by the mask leading to a loss of dimensional control. The reduction of the reflected light is particularly important for aluminum lines which must run over severe topography. Typical methods of reducing the reflected light include the use of dyed photoresists and the application of an amorphous silicon layer. Both of these methods have been used successfully, but suffer from processing limitations. The dyed resists are dependent upon thickness, shelf life, and formulation. The amorphous silicon layer has problems with photoresist adhesion and is very sensitive to thickness uniformity. Recently, interest in materials such as titanium nitride and titanium-tungsten as anti-reflection coatings (ARC) has grown [1,2]. These materials have the advantage that they can be left on the aluminum runners, which has been shown to enhance their electromigration resistance [1]. Additionally, ARC materials based on oxides, and nitrides of metals such as titanium and aluminum have been proposed [1-5].

EXPERIMENTAL

The films were deposited via the reactive sputtering of a WSi target in a mixture of argon and nitrogen in a MRC 943 sputtering machine. The maximum nitrogen content of the sputtering gas was 20% with the present experimental system. The deposition parameters are given in Table I. The films which were investigated include $WSi_{0.45}$ and three WSiN materials. The first WSiN material is a film deposited using the sputtering parameters for the standard gate material. This material has been given the designation, WSiN-G. The second WSiN material is also deposited using the deposition parameters of the standard gate material, but at high pressure (25 mtorr). This material has been given the designation, WSiN-HPG. A third WSiN material was deposited at low power and been given the designation WSiN-HR.

The reflectivity of the films at each thickness was measured by using a Nanospec connected to a strip chart recorder. The reflectivity results which are discussed in this paper were taken at a wavelength of 436 nm. This corresponds to the exposure wavelength of the Nikon G-line steppers used in the AT&T GaAs IC manufacturing line in Reading, PA. A sample of bare aluminum was used to set the reflectivity at 100% to serve as a reference. For the samples where the ARC was investigated on different metal films, a sample of the film without the ARC was used as a reference. For many of these films, the reflectivity as a function of wavelength was not constant as was the case with aluminum. For these materials, the maximum reflectivity was set to 100% regardless of wavelength. The reported reflectivity numbers for the ARC were then corrected to the actual reflectivity of the reference material observed at 436 nm.

TABLE I

ARC Deposition Parameters

Material	Target Composition	Power (KW)	Pressure (µm)	% Nitrogen Feed Gas	Rate Å/Pass	Sample Speed (cm/min)
$WSi_{0.45}$	1:1	1.0	9.0	0.0	125	100
WSiN-G	1:1	1.0	9.0	20	100	200
WSiN-HPG	1:1	1.0	25.0	20	100	200
WSiN-HR	1:1	0.3	25.0	20	100	100

RESULTS

The reflectivity of bare aluminum was constant as a function of wavelength from 390 to 800 nm as shown in Figure 1. The reflectivity of the $WSi_{0.45}$ and WSiN films generally changed as a function of wavelength as illustrated in Figure 2. The $WSi_{0.45}$ film did not show a large decrease in reflectivity at 436 nm as a function of thickness as shown in Figure 3. The reflectivity of this film is ¯55% relative to bare aluminum over the range in thickness from 125 Å to 750 Å. In contrast to this film, the WSiN films showed a strong dependence of the reflectivity at 436 nm as a function of thickness. This is illustrated in Figure 3. These films both showed a dramatic change in appearance as a function of thickness. The color of the films were a silver metallic color at ¯100 Å and changed to a dark maroon at the minimum reflectivity of ¯300-400 Å.

The influence of nitrogen content in the sputtering gas was investigated at four levels, 5, 10, 15, and 20%. In each case, the WSiN-HR film showed the lowest reflectivity followed by the WSiN-HPG and finally the WSiN-G material. This trend is illustrated in Figures 3. The $WSi_{0.45}$ had a reflectivity greater than any of the three WSiN materials as noted earlier.

The lowest reflectivity was found for materials sputtered in a 20% nitrogen mixture for each of the three WSiN ARC materials. This is illustrated in Figure 4 for the WSiN-HR material. It should be noted that the films sputtered in a 15% mixture exhibit a lower reflectivity at thicknesses greater than ¯300 Å for the WSiN-HR and WSiN-HPG materials.

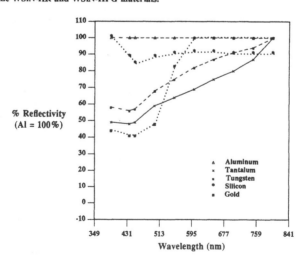

Figure 1. Plot of % Reflectivity (Al = 100%) versus wavelength from 390 to 800 nm for Al, Au, Ta, W, and Si.

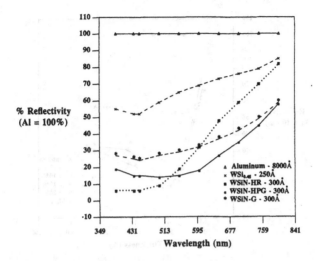

Figure 2. Plot of % Reflectivity (Al = 100%) versus wavelength from 390 to 800 nm for Al, WSi$_{0.45}$, WSiN-HR, WSiN-HPG, and WSiN-G.

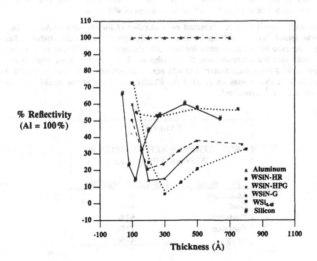

Figure 3. Plot of % Reflectivity (Al = 100%) versus thickness at 436 nm for Al, WSi$_{0.45}$, WSiN-HR, WSiN-HPG, and WSiN-G deposited in a 20% nitrogen sputtering ambient.

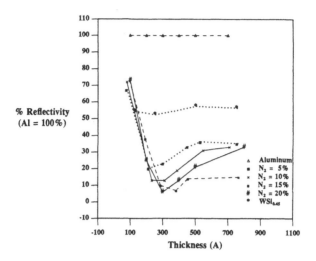

Figure 4. Plot of % Reflectivity (Al = 100%) versus thickness at 436 nm for WSiN-HR deposited in nitrogen sputtering ambients from 5 - 20%.

A number of experiments were performed to investigate the effect of changing target composition on the reflection properties of the ARC materials. It was found that equivalent films could be deposited using either a WSi or a $WSi_{0.7}$ target. It was found that the silicon in the films is a requirement to produce a good ARC. Films of tungsten nitride (WN) were deposited at thicknesses of 325 and 650 Å and produced a film with a reflectivity of 44%. It should be noted that the $WSi_{0.45}$ film has a reflectivity of 54% at thicknesses of 125 - 750 Å. Additionally, the reflectivity of tantalum nitride (TaN) has been measured to be 45% at a thickness of 400 Å.

The WSiN ARC materials were evaluated on a number of metals such as Al, Al, Ta, W, $WSi_{0.45}$, and Si. It was found that the reflectivity is reduced for all the materials studied. The reflectivity reduction was observed to be the greatest for the ARC deposited on Al. These results are summarized in Table II. Note that the reported reflectivity reduction has been corrected relative to the absolute reflectivity measured for the base material at 436 nm. The reflectivity for the WSiN-HR film appears to saturate at 24% at thicknesses up to 2354 Å. Finally, the stress of the WSiN ARC materials was measured to be ⁻0.8 x 10^9 dynes/cm^2.

TABLE II	
REDUCTION OF REFLECTIVITY BY WSiN-HR ARC (414 Å)	
<u>Base Metal</u>	<u>Reflectivity Reduction</u>
Aluminum	90%
Gold	49%
Tantalum	62%
Tungsten	73%
Silicon	49%

The results of an experiment which used a matrix of exposures on the Nikon G-line stepper indicated that the time to fully expose a typical interconnection pattern dropped from ⁻350 msec for bare aluminum to 170 msec for the WSiN-HPG ARC material. This will allow more process latitude resulting in a tighter control on the critical interconnect dimensions. Interconnect test patterns which were exposed over bare aluminum in the presence of severe topography did not yield acceptable photoresist patterns. This clearly indicated the need for an ARC. These same patterns were successfully patterned using the $WSI_{0.45}$ ARC material.

DISCUSSION

The use of an aluminum metallization system has a number of benefits in the manufacture of ICs. Since it can be patterned using reactive ion etching (RIE) processes, the process has a much higher yield. The typical method of patterning Ti/Pt/Au metallization systems with a lift-off technique often results in flaps and burrs which can lead to the shorting of parallel lines. Additionally, the yield of the lift-off process is very low for design rules which are below 1.5 μm. These small design rules are required for high levels of integration in GaAs ICs.

Aluminum is a common material used for interconnection in silicon IC technology. It is easily deposited via sputtering techniques with high rates, good step coverage, and good uniformity. Some of the disadvantages of using aluminum include its low melting point, high reflectivity, and its susceptibility to electromigration. This last point has made the use of Al + Cu and Al + Si alloys common.

The implementation of an ARC in the manufacture of GaAs ICs requires several conditions to be met. The ARC must be compatible with subsequent processing. The ARC must be deposited at low temperatures with good uniformity control. Finally, the ARC must be able to be patterned by RIE techniques. All of the materials investigated above meet these requirements.

The $WSI_{0.45}$ material is the standard film used as the gate material in the Self-Aligned Refractory Gate Integrated Circuit (SARGIC) process used in the manufacture of GaAs ICs at AT&T. Therefore, the deposition, definition, and etching of this material are well known. As shown in Figure 3, the reflectivity of this material is ~55% of bare aluminum. This has been shown to be an acceptable level for the etching of 1.5 μm lines of aluminum over severe topography.

The WSiN-G and WSiN-HPG materials are films which are deposited under the same conditions as the standard gate material cited above with the exception of the addition of 20% nitrogen to the gas stream and an increase in the pressure to 25 mtorr for the WSiN-HPG material. As shown in Figure 2, a 300 angstrom film of the WSiN-HPG material has a reflectivity which is only 15% that of bare aluminum. This would lead to a significant improvement in the ability to print very fine lines in aluminum. This material has been used on top of the gate material to define gate structures which range from ~0.6 μm to 1.5 μm. The addition of the nitrogen does not affect the etching characteristics of the film. It can be etched in any fluorine containing plasma such as NF_3, SF_6, or CF_4.

The WSiN-HR material is deposited at a lower power than the gate material. This results in a film which has a higher silicon content. The reflectivity of 300 Å of this film at the stepper wavelength of 436 nm is only 6% when compared to bare aluminum. Clearly, this is the best candidate for use as an ARC. As with the WSiN-(HP)G films above, it can also be easily patterned in any fluorine containing plasma.

The addition of nitrogen to the sputtering ambient reduces the reflectivity of WSi films deposited under a number of process conditions. For each of the four nitrogen levels investigated, the WSiN-G had the highest reflectivity, followed by the WSiN-HPG and finally the WSiN-HR material. This is illustrated in Figure 3. The WSiN-HR material has the highest silicon content. This may impact on the reflectivity of the film. The measured reflectivity of 44% for the WN films indicates that silicon is a very important part of the WSiN ARC material.

The minimum reflectivity decreases as the nitrogen content in the sputtering ambient increases over the range from 0 to 20%. This was true for all three of the WSiN materials investigated. The films with the lowest reflectivity were deposited at a nitrogen content of 20% in the sputtering ambient.

The results are very similar for the films deposited with both the WSi and the $WSi_{0.73}$ targets. The process parameters have been adjusted to yield films with the same W:Si ratio during the development of the gate process. These results indicate that the composition of the target is not critical to the deposition of the ARC material as long as the W:Si ratio can be preserved.

These WSiN ARC materials are effective at reducing the reflectivity of a number of high reflectivity materials. As indicated in Table II, there is not a discernible trend in the reduction of the reflectivity for different materials. As indicated in the table, the reflectivity can be reduced by 50 - 90% depending on the base metal. These results indicate that this class of materials may have a wide variety of other applications outside the present use of aluminum metallization systems on ICs.

The present ARC material used in the manufacture of silicon ICs is a layer of amorphous silicon. In Figure 3, the reflectivity properties of an amorphous silicon material is shown with the WSiN ARC materials. Note that the silicon ARC film has a very narrow range in thickness over which it has good ARC properties. All of the WSiN materials studied have relatively wide processing windows. In each case, the films with the lowest reflectivity had a brilliant gold color. This would serve as an easy process check for the operators on the manufacturing line. If the film is not the correct color, the thickness is probably wrong. Additionally, the measured stress for these films is very low. Similar materials made from TiN have reported stresses of 1×10^{10} dynes/cm^2. This is an order of magnitude higher than the WSiN films deposited in this study.

The advantages of these WSiN ARC materials over other systems such as dyed resists or amorphous silicon are their wide processing windows and the fact that they can be left on the aluminum lines. Since they do not have to be removed, they will enhance the electromigration resistance of the aluminum lines and also help prevent the growth of hillocks. This layer will also serve to protect the aluminum during subsequent processing. The dielectric layers are patterned using a CF$_4$ plasma. This will remove the WSiN ARC material in the bottom of the via hole, insuring a good contact to the underlying aluminum. In general, the implementation of these materials leads to a very high yield, robust, aluminum interconnection process and should have applications to many facets of IC manufacturing technology.

CONCLUSIONS

Four materials have been investigated as ARCs for use on GaAs ICs. The standard gate material has been shown to have a reflectivity which is 55% as compared to bare aluminum and has been chosen to be implemented in the first aluminum process. Three materials which are tungsten-silicon-nitrogen alloys have been shown to be far superior to this film. The lowest reflectivity films are deposited using deposition conditions which form a high resistance material in a nitrogen ambient of 20%. These materials have been shown to decrease the reflectivity of a number of materials such as aluminum, gold, tantalum, tungsten, and silicon. The measured stress of these films is an order of magnitude lower than similar films made from TiN. The films are easily patterned in fluorine plasmas and can be left on the aluminum runners to enhance the electromigration resistance.

REFERENCES

1. M. Rocke and M. Schneegans, J. Vac. Sci. Technol. B6(4), 1113 (1988).

2. Y-C. Lin, A.J. Purdes, S.A. Saller, and W.R. Hunter, International Electron Devices Meeting, Technical Digest p. 399, San Francisco, Dec. (1982).

3. T.R. Pampalone, M. Camacho, B. Lee, and E.C. Douglas, J. Electrochem. Soc. 136(4), 1181 (1989).

4. J.-S. Maa, D. Meyerhofer, J.J. O'Neill, Jr., L. White, and P.J. Zanzucchi, J. Vac. Sci. Tech. B7(2), 145 (1989).

5. K. Kamoshida, T. Makino, and H. Nakamura, J. Vac. Sci. Tech. B3(5), 1340 (1985).

PART IX

General Metallizations, Deposition and Applications

VAPOUR DEPOSITED NANOCOMPOSITE THIN FILMS

L. S. WEN
Institute of Metal Research, Academia Sinica,Shenyang 110015,China.

ABSTRACT

Fundamental principles of nanocomposites are discussed. The basic features of nanocomposites are low dimensionality of their composition components, widespreadness of the electronic interactions between the components and the great variety of microstructure of nanocomposites ranging from high level ordered 3-dimensional periodic structure to stochastically dispersed medium of superfine particle. All these offer much more potential for tailoring the property of materials than by formation of chemical compound or mixture including microcomposites. Vapour deposition enhanced by ion beam and electrical discharge plasma provides a class of versatile processes both for preparing low dimensional components of nocomposites and for synthesizing nanocomposite film itself. Elements of designing nanocomposite thin films and their preparing by ion beam and plasma enhanced vapour deposition are also discussed. Examples of nanocomposite films are given.

INTRODUCTION

Nanocomposites are a class of new materials with many unusal physical and mechanical properties. It is well known, that miniaturization becomes now an important trend in development of functional devices. They can be made recently with submicron size and are towards molecular size in the future. Two questions are arised in the work: how to obtain the necessary function by different microstructures of submicron including nanometer size? how to guarantee the stability in the work of such miniature devices?

On the other hand, refining grain size is an effective way to improve the mechanical property of materials known since long. The correlation of mechanical property of materials with their grain size is described by Hall-Petch equition:

$$\sigma_{1y} = \sigma_o + k \, d^{-1/2} \qquad (1)$$

where σ_{1y} is the lower yield strength of materials, σ_o and k are constants and d is the grain size. Recently, it was shown when grain size falls down to nanometer order, great improvement of mechanical property has been obtained and this kind of materials is called nanocrystalline materials.

Similar interesting results were obtained in area of composite materials. When the size of composition components of composite materials decreases to nanometer scale, their effect of modifying property of composite materials is also much more significant than usual. Therefore, composites with components of nanometer size are called nanocomposites and the ordinary types of composite materials with component of micrometer scale and of > 0.1 mm scale are called microcomposits and macrocomposites respectively.

Mat. Res. Soc. Symp. Proc. Vol. 181. ©1990 Materials Research Society

In one word, modification of materials property by artificial structure of nanometer scale shows great perspective in different areas of science and attracts more and more attention at present time. Resent paper deals with the fundamental side of nanocomposites in general.

LOW DIMENSIONALITY OF MATTER

Why the nanometer scale microstructure of materials plays such an important role in modifying property of materials? Nanometer scale is the macro-to-micro state transition region of matter. When the characteristic size of matter decreases to nanometer scale , its surface electron state prevails over bulk electron state and different quantum effects become significant, resulting in losing its macroscopic property. Matter in such a state is called low dimensional materials.

There are two types of effect of low dimensionality: surface effect and volume effect. The former appears when considerable part of atoms of materials are at the surface. For example, when the total number of atoms n_t in a superfine particle is of order 10^9, the percentage of surface atoms of the particle N_s is negligible (Fig 1). When n_t is about 10^6, N_s is already more than

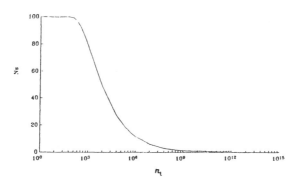

Fig 1   Surface effect of superfine particle

10%. When n_t decreases to 10^3, N_s becomess much more than that of bulk atoms.

Surface effect of low dimensionality leads to the unique physical and chemical characteristics of low dimensional materials. For example, as the particle size of powder falls, increases its specific surface energy resulting in growth of grain boundary diffusion coefficient, decrease of activation energy of sintering and falling of sintering temperature. All these are favorable for powder metallurgy and ceramic processes. On the other hand, as the particle size falls, aggregation of powder becomes ever more serious and fractal structure was formed (Fig 2), making much trouble for processing.

Volume effect appears when the total number of atoms in materials becomes much less than in the usual macroscopic bulk mate-

rials. For example, the gap between the neighboured sublevels of
electron in energy band is given by

$$\delta = \frac{1}{N(0)\Omega} \qquad (2)$$

where N (0) is the state density at the Fermi level in unit volume, Ω is the volume of particle.   When the total number of atoms
in the superfine particle is much less than that in bulk materials,
its energy sublevel gap δ   shall be considerably higher than that
of bulk.   Therfore, the electron structure of low dimensional materials differs considrably from that of bulk materials (Fig 3)[1].
As the result of volume effect of low dimensionality, for example,
increases electrical resistance of thin film and wire(Fig 4).

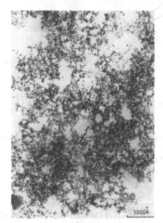

Fig 2   Fractal structure of
superfine particle

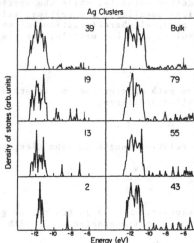

Fig 3   Electron structure of silver of different particle size[1]

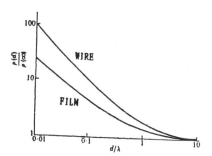

Fig 4   Electric resistance of low dimensional materials

To describe the spatial dimensionality of matter, characteristic length Λ was introduced. Its critical value Λ_c is the threshold of macro-to-low dimensional state transition of matter. When the characteristic length of materials satisfies

$$\Lambda \; < \; \Lambda_c \qquad\qquad (3)$$

they become low dimensional.   The spatial dimension of geometric graph is a pure geometric quantity, while the spatial dimension of materials is a physical quantity. Its critical value Λ_c depends , therefore, on the physical characteristics of materials and physical processes involved.   For example, for the motion of electron in solids

$$\Lambda_c \; = \; l_x \qquad\qquad (4)$$

where l_x is the free path of electron in the solid.
For superconducting processes,

$$\Lambda_c \; = \; \xi_o \qquad\qquad (5)$$

where ξ_o is the correlation length of Kubo electron pair given by

$$\xi_o \; = \; \frac{h \, v_r}{\pi \, \Delta \, (0)} \qquad\qquad (6)$$

where v_r is Fermi velocity, $\Delta \, (0)$ is BCS energy gap at 0 K.
For interaction of electromagnetic waves with materials

$$\Lambda_c \; = \; K \, \delta_x \qquad\qquad (7)$$

where K is constant, δ, is the skin depth of materials,

$$\delta_, = \frac{1}{\sqrt{\pi \nu \mu \sigma}} \qquad (8)$$

where ν is the frequancy of electromagnetic waves, μ is magnetic susceptbility of materials, σ is their electrical conductivity.

FUNDAMENTALS OF NANOCOMPOSITES

We consider now the development and fundamental problems of nanocomposites in detail. In late 70s, supermodulus effect was discovered in compositionally modulated multilayer films of Au-Ni and Cu-Pd systems by Hilliard et al.[2]. Bi-axial elastic modulus was shown to be the function of modulation wavelength Λ and remarkable increases of 2 to 4 times were observed when Λ is close to a critical value on the order of 2 nm depending on the system. Because the elastic moduli are usually structure insensitive, these reults initiate great interest in mechanical property of artificial modulated structure. In the subsequent experimental works, similar results were obtained in Cu-Ni and Ag-Pd systems, but not in Cu-Au system [3-5]. Anomalously high microhardness of factor 2 in TiN/VN multilayer film was obtained by Torodova et al. at Λ =5.2 nm [6]. Increase of microhardness of TiN-based film from 17 GPa to 45 GPa was obtained by Wen et al. through refining the grain size to 20-30 nm by addition of boron and low energy ion bombardment[7]. The best wear resistance was obtained in TiC/TiB, , TiN/TiB, and TiC/TiN at Λ = 20-50 nm by Holleck et al.[8,9]. Superplastisity of TiO, and CaF, ceramics was obtained by Gleiter et al., when they decrease the grain size down to nanocrystalline [10]. They refered the superplastisity so obtained to the diffusional flow of atoms along the intercrystalline interfaces. As a new way to obtain toughed ceramics, they suggested to recover partially the strength of ceramics through heat treatment. Modifying functional characteristics of materials was also obtained by both multilayer and particulate nanocomposites. For example, it is known that localized surface states play an important role in the enhancement of surface magnetism for 3d transition metals. Magnetic coupling between ferromagnetic films separated by nonmagnetic thin film, has been observed experimentally by using ferromagnetic resonance[11]. Enhancement of magnetism at the transition metal / noble metal (or simple metal) interfaces has been predicted theoretically[12]. X-ray reflection mirror has been made with multilayered structure consisting of elements of high and low scattering cross section [13]. The nanocomposite of superfine particle in insulating medium shown a magnitude of infrared emissivity considerably larger than predicted by classical theory[14].
Progress of different kinds of nanocomposites puts forward the need of studying the fundamentals of nanocomposites. The most prominant feature of nanocomposites is low dimensionality of their composition components which can be two-, one- or zero-dimensional. Homogeneous nanocomposites consist of one type of component. Heterogeneous nanocomposites consist of two or more types of component. They are multiphase materials. All kinds of composition components of nanocomposites are with electronic structure and energy state significantly different from that of bulk materials. That serves the physical ground in which nanocomposites differ from both chemi-

cal compound and mixture. The composition components of a chemical
compound are limited in range of the atoms of chemical elements.
In contrast, nanocomposite can be made of low dimensional components varying both chemical composition and characteristic length.
Another basic feature of nanocomposites is the widespreadness
of the electronic interactions between their composition components. Ordingnary microcomposites are a mixture of their components in nature. The chemical reactions at their internal interfaces does not change their nature. Therefore, they behave in ac-
cordence with the rule of mixture. Their property shall be obtained as the linear combination of the initial property of components with the volume percentage as the weighting factor. That determines the limit of the property improvement by microcomposites
which shall not be higher than that of the component of best property. On the contrary, for nanocomposites, low dimensionality
of their components leads to their much higher specific interface
area,greater interface energy and lower activation energy of interface chemical reaction than in bulk materials. That results usually in creation of new electronic structure or processes, in the
stability of such new electronic structure and in their widespreadness. All these offer new possibility for obtaining different nonlinear effect for tailoring the property of nanocomposites. The
property so obtained is, therefore, often several times or even several orders higher than that of every component. Another interesting result is that due to the very great interface area, the interface concentrition of detrimental trace elements shall be much
lower than in bulk materials, even its total quantity is the same
in both cases.
    The third feature of nanocomposites is the great variety of
their microstructure. As it was shown experimentally, the microstructure of nanocomposites ranges from high level ordered 3-dimen-
sional periodic structure to stochastically disordered particulate
dispersed nanocomposite of superfine particle in different medium
providing new possibility for modifying property of materials. The
orderliness of nanocomposites can be the periodic modulation of
crystallographic structure or other active factors including chemical composition and its spatial distribution, interface adsorption
and segregation, micromorphology, internal strain and its distribution, interface structure and defects, interface electron structure
and band structure(especially quantum well). Such periodic modulation can be at most three-dimensional, that is the modulation in
three each to other mutually perpendicular spatial directions. Such
high level orderliness is the limit case of nanocomposites. It has
been shown that such a periodic modulation has great effects on the
mechanical, optical, electrical, magnetic and functional transformation characteristics. Semiconductor superlattices are a typical
nanocomposite with periodically modulated structure. In the case
of one-dimensional periodic modulation, quantum well has been obtained. The motion of electron was limited in two-dimensional quantum well. Two-dimensional periodic modulation makes the electron
motion limited in one-dimensional quantum wire, while three-dimensional modulation limits the electron motion in zero-dimensional
quantum dot of nanometer scale. Variation of demensionality of electron motion creates a series of new characteristics of materials
and shows a broad perpective for development of microelectronic and
opto-electronic devices. Recently, strained layer superlattices
shows great potential of modifying optical property of materials.
Band gap is the basic characteristics of materials which has effect
on optical property of materials. It is shown now that band gap can

be modified continuously with strained layer superlattices at constant chemical composition.   Besides periodically modulated structure, there are also aperiodically modulated structures which are
widely used for sensor and function transformation devices.   Another limit case of microstructure of nanocomposites is stochastically diapersed composite of superfine particle in different media.
    The active factors for modifying the property of materials and
the types of nanocomposites are summarized in the following table.

Table   Effective factors modifying the property of materials and
        the types of nanocomposites

№	Effective factor	Type of nanocomposites
1	Periodic modulation of chemical composition	Semiconductor superlattices with compositional modulation and with doping modulation, metallic multilayer films with supermodulus effect,high strength multilayer nanocomposite metal films, superhard multilayer nanocomposite films of interstitial phases
2	Periodic modulation of crystallographic structure and chemical bond	Intercalated compound
3	Periodic modulation of crystallographic defects	Strained layer superlattices, semiconductor superlattices with interface dislocation structure
4	Periodic modulation of order and disorder	Superlattices of amorphous semiconductors
5	Periodic modulation of atomic scattering cross section	nanocomposite multilayer films for X-ray reflective mirror
6	Aperiodic modulation of different factors	Chemical conversion layer, nanocomposite films for sensors and function transformation devices
7	Homogeneous and heterogeneous nanocrystalline structure	High toughness nanocrystalline ceramics, high strength nanocrystalline metal, TiN based multiphase nanocrystalline superhard films
8	Stochastic dispersed media of superfine particle	Optical composite media of superfine particle, microwave absorbing coating of superfine powder, magnetofluid

    It is evident now that nanocomposites (or nanosolids) are a
new kind of materials different from both chemical compound and
mixture.   Their composition components are low dimensional.   The
interface interreaction between their components are strong and widespread.   In addition to this, they have a great variety of different microstructures.

DESIGNING AND PREPARATION OF NANOCOMPOSITES

    Analysis of fundamental principles of nanocomposites revealed
that the designing and preparation are of great importance for such
a new class of high performance materials.   The first step of materials designing is to outline an appropriate physical model.   The
detailed mechanism of the model with corresponding data obtained

both from theoretical anlysis and experimental measurement, should be able to provide the necessary property of nanocomposites. That means to make structure-property designing. The key to the settlement of the question consist in the soundness of designing physical model and selection of characteristic length and its critical value. Fractal dimension analysis must be an useful mathematical means for the later question. Therefore, the designing of nanocomposites can be theoretical, semiempirical and empirical. Theoretical representation of physical mechanism and its detailed mathematical analysis is important for sketching the first physical models and for making a new break through, when the old physical mechanism is already in serious disagreement with experimental data or is in principle not able to provide necessary property of materials. However, pure theoretical analysis is usually difficult due to the shortage of necessary physical and chemical data, especially for low dimensionl materials. On the other hand, pure empirical analysis is a profitable stage for accumulating the first experimental results and forming the idea of new physical mechanism and designing model. However, this process could be made more effective when the experimental results are correlated with theoretical representation.

Preparation is by no means less important than designing. In this respect, vapour deposition enhanced by ion beam and electic discharge plasma is a kind of versatile prcesses for preparing both low dimensional materials (thin films, wiskers and superfine powders) and their nanocomposites. Molecular beam epitaxy (MBE) and metalorganic chemical vapour deposition (MOCVD) have advantage over other methods for precisely layer by layer growing at the atomic scale. However, for more popular application, there is now a trend to use more inexpensive process. Recently, vacuum evaporation with resistance and induction heating, and chemical vapour growth with arc plasma heating are more and more applied for preparing superfine powder. For preparation of multilayer nanocomposite films, vacuum evaporation by electron beam heating, magnetron sputtering and other physical vapour deposition and chemical vapour deposition processes enhanced by ion beam and electric discharge plasma , are widely applied.

RESULTS OF NANOCOMPOSITE THIN FILMS

Now, we cite some our new results of nanocomposites. The first type is nanocomposite multilayer films for superhard coating,$Ti_,N/$ TiN and Ti/TiN. They were prepared by using hollow cathode discharge ion plating coater. Each kind of samples were made with a determined single layer thickness which ranges from 10 to 100 nm. Results of measurement with supermicrohardness tester shown that as the thickness of single layer falls, increases the hardness of nanocomposite films (Fig 5). X-ray diffractionanalysis revealed TiN, $Ti_,N$ and Ti phases in the film (Fig 6). Results of transmission electron microscopy observation shown, that the grain size of the film decreases to less than 20 nm, when the thickness of single layer falls down to $0.1\mu m$ (Fig 7).

Another example of nanocomposite film is microwave absorbing materials. Based on quantum percolation, free electron interaction and other physical effects of nanocomposition, a composite film was designed and prepared. Results of microwave absorptivity measure-

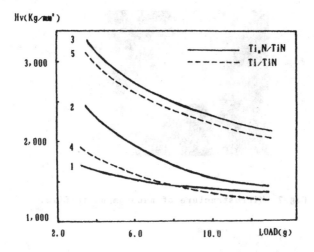

Fig 5 Microhardness of nanocomposite films,   thickness of the single layer: 1 - 100 nm, 2 - 20 nm, 3 - 10 nm, 4 - 40 nm, 5 - 20 nm

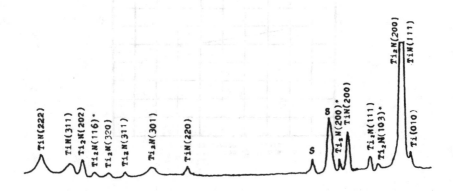

Fig 6 X-ray diffraction diagram of nanocomposite film

ment gave a absorption frequency band width of 3.6 GHz at 10 dB absoption level in 7-17 frequency band (Fig 8). That is of the same order as for the ordinary microwave absoption coating.   While the thickness and the surface density of the nanocomposite film together with the thickness and the weight of the deposition substrate are of one order lower than that of ordinary microwave absorbing coating.

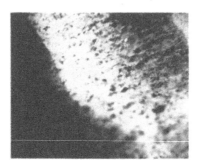

Fig 7 Microstructure of nanocomposite film.

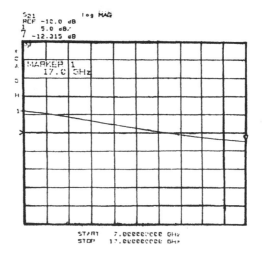

Fig 8 Microwave absorptivity of nanocomposite film

CONCLUSION

Artificial submicronstructures have both the important scientific meaning and broad perspective for application. They make the modifying characteristics of materials possible by means of utilization of different physical effects of nanocomposition. The most important points in studying nanocomposites seems to be the formation of a rational physical model and processing for their preparing. It is still difficult to evaluate their potential for improving the property of materials at the present time. However, it is already evident now that the research and development of nanocomposites initiate a new great progress in materials science which will strongly promote forward the science, technology and economy of mankind.

REFERENCES

1. R. C. Baetzold, Surface Science, 106, 243,(1981).
2. W. M. C. Yang, T. Tsakalakos and J. E. Hilliard, J. Appl. Phys. , 48, 876 (1977).
3. G. E. Henein and J. E. Hilliard, J.Appl.Phys., 54,728(1983).
4. T. Tsakalakos and J. E. Hilliard, J.Appl.Phys.,54,734(1983).
5. D. Baral, J. B. Ketterson and J. E. Hilliard, J. Appl. Phys., 57, 1076 (1985).
6. S. Todorova, U. Helmersson, S. A. Barnett, J. -E. Sundgren, J. E. Greene, IPAT 87, 6th Int. Conf. Ion and Plasma Asisted Techniques, Brighton, UK, May 1987, p248.
7. L. S. Wen, X. Z. Chen, Q. Q. Yang, Y. Q. Zheng, Y. Z.Chuang, 1st Int. Conf. Plasma Surf. Engineering, Sept. 23-27, 1988, Garmisch-Partenkirchen, FRG.
8. H. Holleck and H. Schulz, Thin Solid Films, 153, 11 (1987).
9. H. Holleck, Metall,43, 7, 614 (1989).
10. J. Karch, R. Birringer & H, Gleiter, Nature, 330, 556(1987).
11. M. Pomerantz, J. C. Slonczewski and E. Spiller, J. Magnetism & Magnetic Materials, 54-57,781,(1986).
12. C. L. Fu, A. J. Freeman, J. Magnetism & Magnetic Materils, 54-57, 777,(1986).
13. E. Spiller, in Low Energy X-Ray Diagnostics— 1981, eds. D. Attwood and B. L. Henke, AIP Conf. Proc. №75, AIP, New York, 1981, p.131.
14. C. G. Granqvist, R. A. Buhrman, J. Wyns and A. J. Sievers, Phys. Rev. Lett.,37,625(1976) .

Au OHMIC CONTACTS TO P-TYPE $Hg_{1-x}Cd_xTe$ UTILIZING THIN INTERFACIAL LAYERS

V. KRISHNAMURTHY *, A. SIMMONS **, AND C.R. HELMS *
* Department of Electrical Engineering, Stanford University, Stanford, California 94305
** Texas Instruments Inc., Dallas. Texas 75265

ABSTRACT

Ohmic behavior is observed in electroless Au contacts to p-type $Hg_{1-x}Cd_xTe$. Our investigation of the interfacial chemistry of such contacts suggest that this ohmic behavior may be due to the presence of a Te,O, and Cl layer. To verify this correlation with interfacial chemistry, thin plasma oxide layers were used in evaporated Au contacts. The annealed plasma oxidized contacts exhibited low contact resistances. This behavior was attributed to a low interface state density at the interfacial layer/$Hg_{1-x}Cd_xTe$ interface. In comparison, as-deposited and annealed Au contacts without a thin interfacial layer were rectifying with a large barrier height.

INTRODUCTION

P-type $Hg_{1-x}Cd_xTe$ is being used in next generation infrared systems which combine photon detection and signal processing into the same device [1,2]. One of the requirements for this upcoming technology is a low resistance contact to p-type $Hg_{1-x}Cd_xTe$. Unfortunately, most metal contacts to p-type $Hg_{1-x}Cd_xTe$ have been characterized by relatively large barrier heights. In fact, theoretical predictions as well as studies on cleaved surfaces [3-6] for metal contacts to p-type $Hg_{1-x}Cd_xTe$ give barrier heights larger than the bandgap for x<0.4 [7].With this in mind, we were surprised to discover that electroless Au contacts exhibited low contact resistances. Overlayer doping by the Au can possibly be excluded as a mechanism for barrier lowering resulting in low contact resistances because: (1) Au has a relatively low solid solubility in $Hg_{1-x}Cd_xTe$ [8] and (2) Au is an inefficient p-type dopant in $Hg_{1-x}Cd_xTe$ [9] .Material analysis performed with Auger sputter profiling revealed the presence of a thick interfacial layer composed primarily of Te, O, and Cl. These results will be discussed later in the text but nevertheless we should note that this interfacial layer may be responsible for the observed ohmic behavior. In fact, interfacial layers have been used to modify the barrier heights of various metal-semiconductor systems [10-15].To better understand the role of an interfacial layer, we performed electrical and material analysis on as-deposited and annealed Au contacts with and without an interfacial layer.

EXPERIMENTAL

The samples used in this study were single crystal p-type liquid phase epitaxy $Hg_{0.7}Cd_{0.3}Te$ and $Hg_{0.8}Cd_{0.2}Te$ substrates doped to approximately 10^{15} /cm^3. The samples were lightly polished with 1/8% Br in methanol solution to remove 5-10 μm. The thin interfacial layer was formed by bombarding the sample in an oxygen plasma for approximately 30 s. The oxygen plasma was produced at a pressure of 400 μm with a power of 50W at 13.5 MHZ. X-ray photoelectron spectroscopy (XPS) analysis of the oxidized surface indicated that Te and O were the primary constituents and relatively small amounts of Hg and Cd were observed. The thickness of the plasma oxide was estimated to be between 20 and 30Å. 200Å of Au was evaporated on samples with and without an interfacial layer. The electroless contacts were formed by immersing $Hg_{1-x}Cd_xTe$ in a 0.5% $AuCl_3$ solution for 30 s. The thickness of the Au layer as measured by a step profiliometer was approximately 600Å. Some of the samples were annealed in a quartz ampoule at 100°C for 216 h in a 400 Torr N_2 ambient. Samples on which electrical measurements were performed had dots with 2000Å of Au surrounded by ZnS and then encapsulated with a layered structure of Ni-In-Pb-In.

The material study was performed using Auger electron spectroscopy (AES) with sputter profiling performed at an approximate temperature of -130°C with a 1keV Ne^+ ion beam with a current density of approximately 7.8 µA/cm^2.The electron beam was operated at an energy of 3 KeV and a current density of 0.36 mA/cm^2. The Auger sputter profiles were performed at a low temperature in order to minimize the preferential sputtering of Hg. Electrical measurements were performed on sets of 28 and 70 Au dots with each dot having an area of 1.03×10^{-6} cm^2. For the other contact, 1300 interconnected Au dots were used. These measurements were performed at 77°K with a Tektronics 576 curve tracer.

MATERIAL ANALYSIS

In this section we will discuss the Auger sputter profiles of the as-deposited and annealed Au contacts with and without an interfacial layer. Figures 1(a) and 1(b) are Auger sputter profiles of as-deposited and annealed Au / Hg$_{0.7}$Cd$_{0.3}$Te films. Due to interference with the Au signal, the Hg signal could only be resolved approximately 200Å into the sputter profile. Analysis of the Au, Hg, and Cd lineshapes revealed that no significant amounts of Hg or Cd were dissolved in the film. For both contacts, Au reaction with the substrate is observed as seen by the broad Au profile. The 10%-90% widths of the Au, Cd,and Te signals are approximately 80Å for the as-deposited contact and 120Å for the annealed contact. The anneal accelerated the reaction between Au and the substrate but resulted in only a small increase in the reacted region and most of the Au remained unreacted. Regarding Au diffusion into the substrate, no detectable Au was observed with AES 100Å below the interface for the as-deposited contact (Fig. 1(a)) whereas some Au was detected for the annealed contact (Fig. 1(b)).

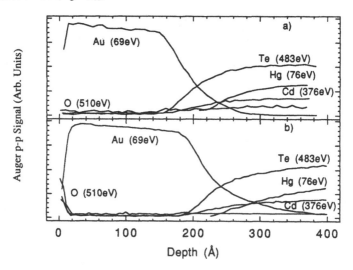

Figure 1. Auger sputter profiles of (a) as-deposited Au (200Å)/ Hg$_{0.7}$Cd$_{0.3}$Te and (b) Au (200Å)/Hg$_{0.7}$Cd$_{0.3}$Te annealed at 100°C for 216 h in a 400 Torr N_2 ambient.

Figure 2 shows an Auger sputter profile of an as-deposited electroless Au / Hg$_{0.8}$Cd$_{0.2}$Te. Also in this profile, no significant amounts of Hg or Cd were dissolved in either the as-deposited or annealed contact. In addition to the deposited Au, a layer (possibly non-uniform) composed of Te, O, and Cl a few hundred angstroms is present at the interface. There also appears to be some Au present in the interfacial layer but this may be an artifact of preferential sputtering or a non-uniform interfacial layer.

For the Au/ plasma oxidized Hg$_{0.7}$Cd$_{0.3}$Te, the Auger sputter profiles of the as-deposited and annealed contacts are shown in Figures 3(a) and 3(b). Similar to the Au / Hg$_{0.7}$Cd$_{0.3}$Te contacts, no significant amounts of Hg or Cd were present in the overlayer. The Au profiles for the plasma oxidized contacts have a much sharper slope compared to the unoxidized Au contacts (Figs. 1(a) and 1(b)). The 10%-90% widths of the Au,Cd, and Te signals for the as-deposited and annealed contacts are approximately 60Å. This can be attributed to the plasma oxide which has inhibited any reaction between Au and the substrate. Also, the anneal of the plasma oxidized contact, as shown in Figure 3(b), did not produce any reaction between the Au and the substrate. In addition, no significant Au diffusion was observed with AES 100Å below the interface for both plasma oxidized contacts.

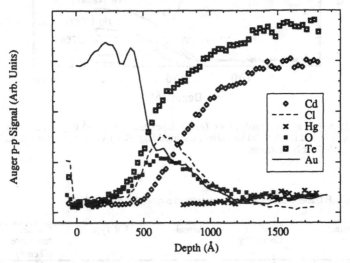

Figure 2. Auger sputter profile of as-deposited electroless Au (600Å)/ Hg$_{0.8}$Cd$_{0.2}$Te. Note the existence of an interfacial layer composed of Te, O, and Cl.

ELECTRICAL DATA

Electrical measurements performed on the as-deposited electroless contacts indicated ohmic behavior with contact resistances (shown in Table I) varying approximately from 2×10^{-1} to 6 Ω cm^2. The annealed electroless contacts were more reproducibly ohmic and had lower contact resistances ranging from 7×10^{-2} to 4×10^{-1} Ω cm^2. However, I-V data on as-deposited unoxidized contacts shown in Table I indicated a high contact resistance. The annealed unoxidized contacts shown in Figure 1(b) also exhibited this behavior. As can be seen from Table I, these contacts were rectifying with a high barrier height giving contact resistances greater than 70 Ω cm^2. Using the I-V data we obtained an approximate barrier height on the order of the bandgap (0.25eV at 77°K). For the plasma oxidized contacts, the as-deposited devices shown in Figure 3(a) were rectifying with an approximate barrier height of 0.1eV and contact resistances ranging from 8×10^{-1} to 6 Ω cm^2. Following the 100°C anneal, most of these contacts became ohmic and the contact resistances, as reported in Table I, varied for 3×10^{-1} to 1Ω cm^2.

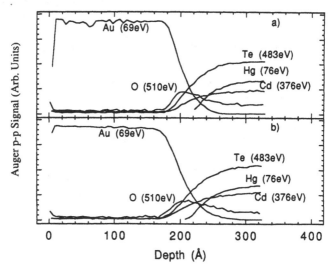

Figure 3. Auger sputter profiles of (a) as-deposited Au (200Å)/ plasma oxidized Hg$_{0.7}$Cd$_{0.3}$Te and (b) Au(200Å)/ plasma oxidized Hg$_{0.7}$Cd$_{0.3}$Te annealed at 100°C for 216 h in a 400 Torr N$_2$ ambient.

TABLE I. Electrical data for Au contacts to p-type Hg$_{0.7}$Cd$_{0.3}$Te

Contact type	Interface Chemistry	Anneal	I-V Type	Contact resistance $(\Omega \text{ cm}^2)$
Evaporated	------------	no	not measured	>70
Evaporated	------------	yes	Schottky $\phi_b \approx 0.25\text{eV}$	>70
Evaporated	Plasma Oxide	no	Schottky $\phi_b \approx 0.1\text{eV}$	0.8-6.0
Evaporated	Plasma Oxide	yes	mostly ohmic	0.3-1.0
Electroless	Te, O, and Cl layer	no	mostly ohmic	0.02-6.0
Electroless	Te, O, and Cl layer	yes	ohmic	0.07-0.4

DISCUSSION

If we correlate the electrical data with the Auger sputter profiles of the Au contacts, we observe that a reaction between Au and the p-type $Hg_{1-x}Cd_xTe$ substrate results in a rectifying contact with a large barrier height. This is consistent with the work of Davis et al.[16] on Au / cleaved $Hg_{1-x}Cd_xTe$. This observed behavior can possibly be the result of a large number of interface states produced by the Au reaction pinning the Fermi level near the conduction-band minimum (CBM). Spicer and co-workers have espoused that these interface states are a result of defects such as the Te vacancy and the Te_{Cd} antisite [7]. Regarding Au doping, Jones et al.[9] have reported that less than 3% of the Au introduced into the $Hg_{1-x}Cd_xTe$ act as acceptors. Thus Au is an inefficient p -type dopant in $Hg_{1-x}Cd_xTe$. Therefore overlayer doping does not play a large role in determining the electrical characteristics of reacted Au contacts.

The addition of an interfacial layer considerably alters the electrical characteristics of the Au contacts. For the as-deposited contacts, the thin interfacial oxide layer reduced the barrier height to approximately 0.1eV. This behavior is possibly attributed to a reduction in the interface state density which improves the barrier height dependence on the metal work function. Nevertheless, these contacts are rectifying and not ohmic. One possible explaination is that the interface state density is still high enough to pin the Fermi level. Following an anneal, contacts with a thin interfacial layer were mostly ohmic whereas contacts without an interfacial layer were rectifying. As seen in Figure 3(b), the interfacial layer is thermally stable so therefore the anneal seems to have further reduced the interface state density of the interfacial layer/$Hg_{0.7}Cd_{0.3}Te$ interface. For the electroless contacts, the interfacial layer may play a similar role. In addition, the non-uniformity and roughness of the interfacial layer may also contribute to the ohmic behavior.

In conclusion, electroless contacts on p -type $Hg_{1-x}Cd_xTe$ exhibit ohmic behavior due to the presence of an interfacial layer .It should also be noted that interfacial non-uniformities may also contrbute to the ohmic behavior.In addition, annealed Au contacts with a thin plasma oxide exhibit ohmic characteristics. This ohmic behavior is attributed to the low interface state density at the plasma oxide/ $Hg_{1-x}Cd_xTe$ interface. Au contacts without a thin plasma oxide were rectifying with a large barrier height. The electrical characteristics of these contacts are strongly dependent on the morphological properties of the thin interfacial layer. The existence of pinholes and variations in oxide thickness and composition across the $Hg_{1-x}Cd_xTe$ sample can result in problems with obtaining reproducible electrical characteristics. One or a combination of the above factors may have been responsible for the variations in contact resistance observed for Au contacts with interfacial layers. Utilizing interfacial layers to alter the Schottky barrier height to obtain ohmic contacts is a viable approach for p -type $Hg_{1-x}Cd_xTe$. Nevertheless, further work is necessary to fully understand the material and electrical characteristics of metal/thin interfacial layers/ $Hg_{1-x}Cd_xTe$ structures to produce reproducible,low resistance contacts.

ACKNOWLEDGEMENTS

We would like to acknowledge R. Strong for the XPS analysis of the plasma oxide and E. Weiss for useful discussions. This work was supported by DARPA through NRL with subcontract 7482253 to Stanford University from Texas Instruments.

REFERENCES

[1] M.B. Reine, A.K. Sood, and T.J. Tredwell, in Semiconductors and Semimetals, edited by R.K. Willardson and A.C. Beer (Academic, New York, 1981), Vol.18, p.304.

[2] M.A. Kinch, R.A. Chapman, A. Simmons, D.D. Buss, and S.R. Borello, Infrared Phys. 20, 1 (1980).

[3] A. Franciosi, P.Philip, and D.J. Peterman, Phys.Rev. B 32, 8100 (1985).

[4] D.J. Friedman, G.P. Carey, C.K. Shih, I. Lindau, W.E. Spicer, and J.A. Wilson, J. Vac. Sci. Technol. A 4, 1977 (1986).

[5] D.J. Friedman, G.P. Carey, I. Lindau, and W.E. Spicer, Phys. Rev. B 35, 1188 (1987).

[6] G.D. Davis, W.A. Beck, M.K. Kelly, M. Tache, and G. Margaritondo, J. Appl. Phys. 60, 3157 (1986).

[7] W.E. Spicer, D.J. Friedman, and G.P. Carey, J. Vac. Sci. Tecchnol. A 6, 2746 (1988).

[8] A.I. Andreivskii, A.S. Teodorovich, and A.D. Schneider, Sov. Phys. Semicond. 7, 1112 (1974).

[9] C.L. Jones, P. Capper, M.J.T. Quelch, and M. Brown, J. Crystal Growth 64, 417 (1983).

[10] M.A. Sobolewski and C.R. Helms, Appl. Phys. Lett. 54, 638 (1989).

[11] D.L. Pulfrey, IEEE Trans. Electron Devices 32, 33 (1984).

[12] H.C. Card, IEEE Trans. Electron Devices 23, 538 (1976).

[13] K. Rajkanan and W.A. Anderson, Appl. Phys. Lett. 35, 421 (1979).

[14] H. Tseng and C. Wu, J. Appl. Phys. 61, 299 (1986).

[15] M.A. Green, F.D. King, and J. Shewchun, Solid State Electronics 17, 551-572 (1974).

[16] G.D. Davis, W.A. Beck, N.E. Byer, R.R. Daniels, and G. Margaritondo, J. Vac. Sci. Technol. A 2, 546 (1984).

PREPARATION AND CHARACTERIZATION OF LPCVD TiB₂ THIN FILMS

CHANG CHOI[1], G.A. RUGGLES[1], C.M. OSBURN[1,2] , P. SHEA[1] AND G.C. XING[1]
[1]Department of Electrical and Computer Engineering, North Carolina State University, Raleigh, NC 27695
[2]MCNC, P.O.Box 12889, Research Triangle Park, NC 27709

ABSTRACT

Titanium diboride is an interesting new candidate for VLSI interconnection applications due to its high electrical conductivity and its excellent chemical inertness at high temperatures.

A thermodynamic analysis of the chemical vapor deposition of TiB₂ from gaseous mixtures of $TiCl_4$, B_2H_6 and H_2 onto silicon or SiO_2 substrates was performed using the SOLGASMIX computer program. Results indicate that at an input gas ratio corresponding to stoichiometry, TiB₂ should form in the solid phase. For non-stoichiometric input gas mixtures, other solid phases, including oxides and silicides, are expected to result from the reaction with the Si or SiO_2 substrate. Both the addition of hydrogen to the system and increased deposition temperature are expected to enhance the deposition efficiency of TiB₂.

Thin films of TiB₂, 50 - 120 nm thick, were experimentally deposited on thermally grown SiO_2 by low pressure CVD using various gas mixtures in the temperature range 400 - 700°C. Depth profiling of the deposited films, using XPS and RBS, indicated the successful growth of a uniform, stoichiometric TiB₂ film. After deposition, the films were rapid thermal annealed at various temperatures in an effort to reduce their resistivity. Resistivity was reduced by 80 % after an 1150°C, 10 sec RTA in Ar to a value as low as 40μΩ-cm, only 4 times larger than that reported for hot pressed polycrystalline TiB₂. X-ray diffraction of as-deposited samples showed no crystalline peaks; however, annealed films exhibited the expected TiB₂ peaks as well as those of Ti_3O_5 and TiO_2. The titanium oxide was most likely produced during annealing due to trace amounts of oxygen in the Ar atmosphere. The average grain size measured 0.1μm via TEM. More careful control of growth and annealing conditions is expected to result in even lower resistivity TiB₂ films.

INTRODUCTION

In the last 20 years titanium diboride has been investigated for several applications because of its interesting properties. It is an extremely hard, refractory material with a high thermal and electrical conductivity. It also has good stability with respect to oxidation and chemical attack. Commercial interests in TiB₂ have been focused on its use as abrasive, as a protective coating, as a high temperature electrode or in nuclear fusion applications.

Borides of refractory metals have been pointed out as interesting potential materials for microelectronic device fabrication. Nicolet [1] suggested the use of metal boride films as diffusion barriers and Shappirio et al. [2] showed that ZrB_2 is stable in contact to aluminum up to 600°C. Ryan et al. [3] studied TiB_x-Si thin film couples, finding that sputtered $TiB_{2.1}$ films did not react with the silicon substrate up to 1092°C, but TiB films reacted with silicon to form silicide. Feldman et al. used TiB₂ as an electrode in polycrystalline silicon thin film solar cells [4].

Titanium boride films have been fabricated by sputtering [2,3,5-8], by chemical vapor deposition [9-12], by reactive ion plating method [13], by laser induced vapor-phase synthesis

[14], and by reaction of Ti-B thin film couples [15]. In this paper, a thermodynamic study of TiB$_2$ deposition from TiCl$_4$, B$_2$H$_6$ and H$_2$ is presented and characterization of titanium boride films prepared by LPCVD is described.

THERMODYNAMIC CALCULATIONS

The calculation of equilibrium states in a chemical system is an important step towards the design of processes involving chemical reactions. A thermodynamic study is typically used as a guide to determine the effects of changing the controllable experimental variables such as temperature, total pressure, and the ratio of the chemical elements in the input gas ratio on the deposited film properties. For a given set of experimental conditions, the equilibrium calculations can be used to determine: (a) whether a deposition process should occur, (b) the possibility of multiple reactions, (c) the number of coexisting solid phases in equilibrium with the gases, (d) the efficiency of the deposition process in the absence of kinetic limitations, (e) which gaseous species are expected and their equilibrium partial pressures, and (f) the possibility of substrate reactions with either the gas or the deposited films.

Initial attempts to analyze CVD of TiB$_2$ were made by Peshev an Niemyski [16], and a more extensive thermodynamic analysis of deposition from gaseous TiCl$_4$, BCl$_3$, and H$_2$ was conducted by T.M. Besmann and K.E. Spear [11]. In the absence of a substrate reactions, TiB$_2$ and boron are the only solid phases which can be deposited. The theoretical efficiency of deposition appeared to generally increase with increasing temperature, decreasing Cl/Cl+H fraction in the reactant gas, and decreasing total pressure.

In this work, a thermodynamic analysis of the chemical vapor deposition of TiB$_2$ from gaseous TiCl$_4$, B$_2$H$_6$ and H$_2$ was performed using the SOLGASMIX computer program for the calculation of heterogeneous equilibrium compositions. The effect of the substrate on TiB$_2$ deposition from TiCl$_4$, B$_2$H$_6$ and H$_2$ at 400 - 800°C and 3 Torr is presented in Table I. In Figure 1, the expected solid phases deposited on Si and SiO$_2$ substrates at various temperatures are illustrated as a function of reactant gas ratio. Multiple phase deposits are observed on both substrates for low temperature deposition. In a diborane rich gas mixture, elemental boron is expected to accompany TiB$_2$ deposition at low temperature on both Si and SiO$_2$ substrates, while boron oxides are also expected on the SiO$_2$ substrates. In TiCl$_4$ rich gas mixtures, multiphase deposits of titanium oxide and titanium diboride are expected on an SiO$_2$ substrate while titanium silicides and titanium diboride are expected on a silicon substrate. However, at an input gas ratio corresponding to the stoichiometry of TiB$_2$, the amount of second phase deposition is considerably reduced compared to that of TiB$_2$. The permissible range of input gas ratios needed to deposit only TiB$_2$ increases with increasing temperature. This implies that very precise control of input gas ratio will be necessary to control the film composition. As can be seen in Table I, multiple phase deposition is predicted on most of the other substrates. The deposits consist of TiB$_2$, boron, titanium silicides, titanium oxides, and boron oxide, depending on the substrate. The substrate effect would be important at the initial stage of deposition, since the reaction between substrate and reactant would cease once a film significantly thicker than a diffusion length was deposited.

Figure 2 shows the effect of temperature and chlorine/chlorine + hydrogen ratio on the calculated deposition efficiency, defined as the number of moles of TiB$_2$ deposited divided by the number of moles of diboride which could deposit if the reactants formed the maximum amount of TiB$_2$, *ie.*, the amount which can form independent of any kinetic or thermodynamic limitations. It is evident from the figure that the addition of hydrogen to the input gas mixture increases the deposition efficiency. This is brought about by a

Table I. Phases deposited on different substrates determined using SOLGASMIX at 400 - 800°C and 3 Torr as a function of input gas ratio.

Substrate	B_2H_6 / $TiCl_4$ ratio			
	$TiCl_4$ only	$TiCl_4$ rich	1:1	B_2H_6 rich
None	None	TiB_2	TiB_2	TiB_2, B
Si	TiSi, $TiSi_2{}^+$	TiSi, $TiB_2{}^*$	TiB_2	B, $TiB_2{}^*$
SiO_2	Ti_2O_3, $Ti_5Si_3{}^{**}$	TiB_2, $Ti_2O_3{}^*$, $B_2O_3{}^{**}$	TiB_2, $B_2O_3{}^{*,+}$, $Ti_2O_3{}^*$	TiB_2, B, $B_2O_3{}^+$
TiB_2	None	TiB_2	TiB_2	TiB_2, B
Ti	$TiCl_3{}^+$	TiB_2	TiB_2	TiB_2
$TiSi_2$	TiSi	TiSi, TiB_2	TiB_2, TiSi, Si^* $TiSi_2{}^+$	Ti,TiB_2, Si
B	TiB_2	TiB_2	TiB_2, B	B, TiB_2

* one order of magnitude smaller than primary phase
** deposit only at high temperature
+ deposit only at low temperature

reduction in available chlorine, due to formation of HCl which, in turn, reduces the amount of volatile titanium compounds formed, thereby increasing the overall efficiency. The addition of H_2 to the input gas mixture, while increasing the efficiency, reduces the total amount of TiB_2 deposited at a fixed total input gas flow rate due to dilution. Increased deposition temperature also enhances the deposition efficiency.

EXPERIMENTAL SET-UP AND PROCEDURE

The overall reaction chamber system, depicted schematically in Figure 3, consists of a cylindrical high purity 316 stainless steel chamber penetrated by various ports, feedthroughs, and windows. The substrate is heated in the center of the chamber by means

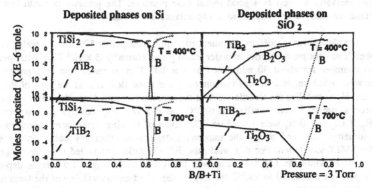

Fig. 1. Use of SOLGASMIX to calculate phases and phase width of films. It indicates that a very precise control of input gas ratio is needed to deposit only TiB_2. The allowable input window for single phase TiB_2 deposition increases with increasing temperature.

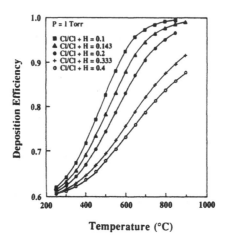

Fig. 2. Calculated deposition efficiency of TiB_2 as a function of temperature for various Cl/Cl+H ratios.

of a pyrolitic boron nitride coated graphite heating element, or by optional tungsten halogen lamps. The heater employs closed loop temperature control using a thermocouple mounted within a graphite susceptor on top of the element, and is capable of heating the substrate to approximately 1000°C. High purity gases are brought to the reaction chamber in stainless steel tubing using VCR fittings for ultimate vacuum integrity. The gases are introduced just above, and parallel to, the substrate to help promote uniform film growth. The source cylinder of $TiCl_4$ is maintained at ~60°C resulting in a $TiCl_4$ vapor pressure of 70 Torr. The flow of the vapor of $TiCl_4$ is controlled by a high temperature mass flow controller without using an additional carrier gas. A temperature gradient, maintained from the $TiCl_4$ source to the mass flow controller, prevents $TiCl_4$ vapor from condensing in the line. The pumping system consists of a corrosive service turbo pump backed by a corrosive service rotary vane pump. This combination provides sufficient pumping speed at the expected process pressures as well as a good initial base pressure. The pressure is controlled by an automatic throttle valve coupled to a capacitance manometer.

The deposition procedure commenced with loading of a 100 mm Si wafer into the chamber. The base pressure prior to wafer loading was normally 5×10^{-8} Torr. After pumping on the chamber for about 30 min down to 1×10^{-7} Torr to reduce the H_2O pressure, the heater was turned on, and Ar was introduced at the same flow rate as the total of the $TiCl_4$, B_2H_6 and H_2 to stabilize the throttle valve at the process pressure. Once the wafer temperature and process pressure were stabilized at 670°C and at 5 Torr respectively, Ar was terminated and H_2, $TiCl_4$ and B_2H_6 were turned on. The ratios of flowing gases were chosen to obtain stoichiometric TiB_2 and to get the maximum deposition efficiency based on the earlier SOLGASMIX simulation. For this work, the B/B+Ti ratio was varied over the range from 0.36 to 0.67, while the Cl/Cl+H ratio was fixed at 0.1. During a 1.5 minute deposition, the temperature decreased to 630°C due to film deposited on the window of the lamp heating port. After deposition, the heater and the reactant gases were turned off. The wafer was unloaded after the temperature of the susceptor decreased to 200°C. The thicknesses of the TiB_2 films were measured using a Dektak profilometer and verfied by RBS. The films

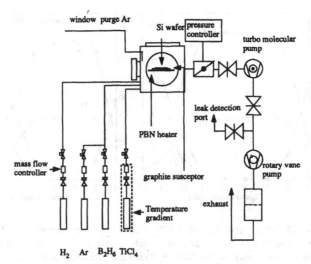

H₂ Ar B₂H₆ TiCl₄

Fig. 3. Experimental deposition chamber

deposited on the wafer were rapid thermal annealed at various temperatures in an Ar or a N_2 atmosphere. The as-deposited and annealed films were analyzed by four point resistance probe, x-ray diffraction, XPS and RBS. The grain size of annealed samples was measured via plan-view TEM.

RESULTS AND DISCUSSION

The thickness of films measured by Dektak ranged from 50 - 120 nm. Many of the growth conditions lead to severe thickness nonuniformity due to mass transport limited growth. In addition, some conditions were observed to result in films with a high Cl content. At a B/B+Ti ratio of 0.67, a temperature of 650°C, and a pressure of 5 Torr, gas depletion was observed, apparently due to an insufficient supply of $TiCl_4$ for reaction (flow rate = 5 sccm). At temperatures at or below 500°C, deposited films were observed to contain significant amounts of Cl, even though the gas input ratio corresponded to stoichiometric TiB_2 (B/B+Ti = 0.67). Higher deposition temperatures appear to favor single phase deposition of TiB_2, which is in agreement with predictions bases on SOLGASMIX calculations. By adding more $TiCl_4$ to the input gas mixture (decreasing the B/B+Ti ratio), the depletion effect was reduced and thickness uniformity improved. Thus, the data presented here focuses on a B/B+Ti ratio of 0.36 and a temperature of 650°C, the combination of which gave good thickness uniformity and pure, stoichiometric TiB_2 films. X-ray diffraction patterns of the films were obtained on the as-deposited samples, and on samples after each annealing procedure. The as-deposited sample showed only silicon substrate lines, indicating a very fine grained or amorphous deposit. The films RTA annealed above 1000°C gave patterns corresponding to the hexagonal TiB_2 structure. X-ray photoelectron spectroscopy was employed for depth profiling of film composition and chemical bonding. The XPS results in Figure 4 indicate that a stoichiometric TiB_2 film was formed uniformly throughout the bulk of the film. TiO_2 and B_xO_y are detected only at the surface of the films. No significant difference in composition and phases was noticed at the interface between the film and the underlying oxide layer, whether the sample was annealed or not, indicating very limited

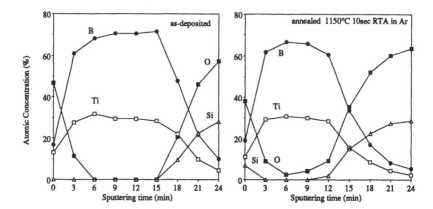

Fig. 4. XPS depth profiling of a TiB$_2$ film on SiO$_2$ as deposited and after 1050°C annealing in Ar. A uniform stoichiometric film of TiB$_2$ is formed, and TiB$_2$ is oxidized only at the surface. The films are deposited with a gas input of B/B+Ti = 0.36 at 650°C, 5 Torr. The thickness of the films are 70nm.

reactions of TiB$_2$ with the oxide during annealing. A small amount of oxygen was incorporated in the bulk of the film of a sample as a result of Ar annealing. The oxygen was incorporated during annealing. The films annealed in nitrogen exhibited a large amount of nitrogen (17 at.%) at the surface.

To characterize the films further, their resistivities were measured. The resistivity of TiB$_2$ as a function of annealing temperature and ambient is shown in Figure 5. The reduction of resistivity found in this work, and in that of Ryan et al. [3], are compared in this graph. The titanium diboride films of Ryan et al. were prepared by sputtering, and

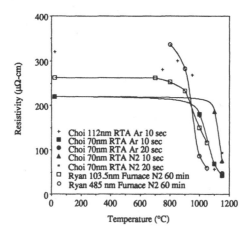

Fig. 5. Resistivity reduction of TiB$_2$ films after annealing. The resistivity of TiB$_2$ dreastically reduced around 1000°C.

were annealed in a conventional furnace for 60 min in nitrogen. In all cases, the resistivity of TiB_2 was drastically reduced around 1000°C, and the reduction was most pronounced for Ar annealing. The higher resistivity of the film annealed in nitrogen could be related to the surface nitrogen detected by XPS. The resistivity was reduced by 80% after an 1150°C, 10 sec RTA in Ar to a value as low as 40 $\mu\Omega$-cm, only 4 times larger than that reported for hot pressed polycrystalline TiB_2 and slightly lower than the reported value of Ryan et al for sputtered TiB_2 [3]. Annealing at 1050°C for 10 sec in Ar produced typical grain sizes of about 100 nm, with some grains as large as 150 nm, as measured by plan-view TEM.

CONCLUSIONS

According to equilibrium thermodynamic calculations the substrate effect on film deposition is very crucial, and multiple solid phases can be deposited by the reaction of $TiCl_4 + B_2H_6$ gases with a substrate. Precise control of input gas ratio, deposition temperature and pressure are important to deposit single phase TiB_2. Experimentally, stoichiometric TiB_2 films are obtained for B/B+Ti ratios varied over the range from 0.36 to 0.67 at a deposition temperature of 650 ± 20°C. The observed deposition rate is approximately 50nm per min. The resistivity of deposited TiB_2 is reduced after RTA annealing above 1000°C. Annealing in Ar produces the best results, with the lowest resistivity obtained to date of 40$\mu\Omega$-cm.

ACKNOWLEDGEMENT

This work has been supported by the Semiconductor Research Corporation (Contract #88-MP-132). The authors also gratefully acknowledge helpful discussions with J.J. Wortman and A. Reisman of the Nanometric Engineering Laboratory of NCSU/MCNC. They would like to thank C.U. Ro, P.L. Smith and Z.-G Xiao for XPS, TEM and XRD analysis.

REFERENCES

1. M.-A. Nicolet, Thin Solid Films, 52, 415 (1978).
2. J.R. Shappirio, J.J. Finnegan, R.A. Lux, J. Vac. Sci. Technol. A3, 2255 (1985).
3. J.G. Ryan, S. Roberts, G.J. Slusser and E.D. Adams, Thin Solid Films, 153, 329 (1982).
4. C. Feldman, F.G. Satkiewicz and N.A. Blum, J. Less-Common Met., 82, 183 (1981).
5. T. Shikama, Y. Sakai, M. Fukutomi and M. Okada, Thin Solid Films, 156, 287 (1988).
6. T. Shikama, Y. Sakai, M. Fujitsuka, Y. Tamauchi, H. Shinno and M. Okada, Thin, 164, 95 (1988).
7. T. Larsson, H.-O Blom , S. Berg, and M. Östling, Thin Solid Films, 172, 133 (1989).
8. H.-O Blom, T. Larsson, S. Berg, and M. Östling, J. Vac. Sci. Technol. A 6 (3), May/ Jun, 1693 (1988).
9. H.O. Pierson and A.W. Mullendore, Thin Solid Films, 95, 99 (1982).
10. H.O. Pierson and A.W. Mullendore, Thin Solid Films, 72, 511 (1980).
11. T.M. Besmann and K.E. Spear, J. Electrochem. Soc., 124 No.5, 786-797 (1977).
12. T.M. Besmann and K.E. Spear, J. Crystal Growth 31, 60 (1975).
13. T. Sato, M. Kudo, and T. Tachikawa, Denki Kagaku 55 No.7, 542 (1987).
14. J.D. Casey and J.S. Haggerty, J. Materials Science, 22, 737 (1987).
15. C. Feldman, F.G. Satkiewicz and G. Jones, J. Less-Common Met., 79, 221 (1981).
16. P. Peshev and T. Niemyski, J. Less-Common Met., 10, 133 (1965).

INSTABILITIES IN THE MECHANICAL STRESS IN DEPOSITED SiO$_2$ FILMS CAUSED BY THERMAL TREATMENTS

BHARAT BHUSHAN AND S.P. MURARKA
Rensselaer Polytechnic Institute, Center for Integrated Electronics, Troy, NY 12180

ABSTRACT

Using an in-situ stress measurement technique that measures stress as a function of annealing temperature, instabilities in mechanical stress induced by heat treatment in a variety of doped/undoped SiO$_2$ films deposited by APCVD, LPCVD and PECVD techniques have been investigated. A large hysteresis in mechanical stress, caused by first heat treatment to which the as-deposited films are subjected, has been observed in films deposited by APCVD /LPCVD techniques. No such hysteresis is obsesrved in films deposited by PECVD technique. Hysteresis in APCVD/LPCVD films is found to vanish once the films are heat treated at or above 800°C. The results are discussed in terms of oxide densification, the presence of hydrogenous species, and phosphorous.

INTRODUCION

Deposited glasses of silicon dioxide (SiO$_2$) are used as isolation materials between various levels of interconnections in a metal oxide semiconductor integrated circuit (MOS IC) [1]. The commonly used commercial systems for deposition of these glasses are based on three techniques such as Atmospheric Pressure Chemical Vapor Deposition (APCVD), Low Pressure Chemical Vapor Deposition (LPCVD) and Plasma Enhanced Chemical Vapor Deposition (PECVD). In a standard double metal MOSIC process, it has been a common practice to employ APCVD oxides or LPCVD oxides as isolation material between the interconnection levels of polysilicon/diffusion and Metal I and PECVD oxides as isolation between Metal I and Metal II (top metal). Metal I looks much cleaner (cosmetically) with PECVD oxide as intermetal dielectric rather than with APCVD oxide or LPCVD oxide. Significantly less number of hillocks are seen to develop in Metal I when it is used in conjunction with PECVD oxide in a double metal process. The reason for this phenomena is not very clear. One of the reasons could have been the deposition temperature of the glass [2,3]. The deposition temperatures for these three types of oxides are not much different. The other reason could have been the difference in the mechanical stress in these oxides. However, this also does not seem to be the reason. The stress data reported in the literature for these deposited oxides indicates a wide spread in its values and not much can be concluded from this data [2,4-8].

We have studied the stress in SiO$_2$ films deposited by various techniques (APCVD, LPCVD and PECVD) and the results are reported in the present paper.

EXPERIMENTAL PROCEDURE

All the study in the present work was done on 4" and 5" silicon <100> p type wafers. Films of APCVD and LPCVD oxides were deposited by reacting SiH$_4$, O$_2$ and PH$_3$ at a temperature of 400°C. APCVD oxides were deposited in our laboratory in Applied Materials Silox reactor. LPCVD oxide films were supplied by Gould Semiconductors and were deposited in a Anicon reactor. PECVD oxide films were also supplied by Gould Semiconductors and were deposited in a ASM PECVD reactor. After deposition of films, insitu measurement of stress as a function of annealing temperature (up to 900°C) was done in a Flexus stress measurement equipment which employs a laser scanning system for measuring the radius of curvature of a wafer. From the radius of curvature R, the stress is calculated using the procedure as given Sinha [9]. Phosphorous concentration in the films was estimated from the etch rates of films in a P-etch solution as outlined by Kern [10]. Hydrogen concentration depth profiles were measured at the 4.5 MeV Dynamitron accelerator facility at SUNY Albany. The ^{15}N + H = ^{12}C + ^{4}He + 4.4 MeV gamma-ray nuclear reaction was used for monitoring the hydrogen concentration in the present study [11].

Mat. Res. Soc. Symp. Proc. Vol. 181. ©1990 Materials Research Society

RESULTS

APCVD Oxides

Stress versus annealing temperature plots of APCVD oxide films having different concentrations of phosphorous are shown in Figure 1. Figure 2 shows the data for the same set of samples when subjected to second heat cycle. The most significant observations are: (1) The stress versus temperature plots exhibit a hysteresis during the first heat cycle to which the films are subjected to after their deposition.(2) During the first heat cycle the stress increases with temperature goes through a maximum and then decreases. On cooling stress decreases with decrease in temperature leading to a compressive room temperature stress as one would expect from the difference in thermal expansion behaviors of the oxide film and the silicon substrate.(3) The maximum stress which is developed in the film during first heat cycle and the temperature at which this maximum occurs (hereafter called relaxation temperature T_R) depend on the phosphorous concentration. Higher than phosphorous concentration, lower is the value of maximum and the room temperature stress as well as the relaxation temperature. (4) The hysteresis vanishes during the second heat cycle. (5) During second heat cycle the stress various linearly with temperature and increases towards more tensile side with increase in temperature.

To get a better insight into the nature of instability in mechanical stress which is induced by temperature, an undoped oxide film was subjected to a complex heat cycle and the data is presented in Figure 3a. While monitoring the stress insitu, the film was annealed in a isothermal mode at various temperatures. From this figure it is noteworthy that at any temperature (up to T_R) the stress increases with time. The room temperature stress following this heat cycle is found to be an order of magnitude higher than the as-deposited stress. The final room temperature stress of the film after cooling depends on the annealing temperature/time cycle to which film was subjected to during annealing. In case the annealing temperature is held below T_R, the room temperature stress in the film would tend to be more tensile after annealing. And when the annealing is carried out at temperatures higher than T_R, the stress would tend to go towards compressive side after annealing. Also the relaxation temperature which appeared to be around 650°C in Figure 1a is found to decrease to 550°C after this long annealing cycle. The maximum stress developed in the film is found to be about 6×10^9 dynes/cm².

LPCVD Oxides

Figure 4 shows the stress temperature plot of a 8 wt% phosphorus doped LPCVD oxide film. The behavior is similar to that of APCVD oxide films. In this case also hysteresis is observed in the first annealing cycle. The as-deposited film has a low tensile stress (5×10^8 dynes/cm²) which becomes compressive (1.1×10^9 dynes/cm²) after annealing. Like doped APCVD film, the relaxation temperature is found to be around 420°C. The maximum stress at T_R is observed to be around 1.8×10^9 dynes/cm² (tensile).

PECVD Oxides

Figure 5 shows the stress temperature plot a 4 wt% phosphorous doped PECVD oxide films. A remarkable difference is observed between these films and the films discussed earlier. The as-deposited films are under a compressive stress. The stress becomes less compressive with increase in annealing temperature. The final stress after annealing is less compressive than the as-deposited stress. A 4 wt% phosphorous doped oxide film does not exhibit any hysteresis during annealing. The films show a very low compressive stress (9×10^8 dynes/Cm²) at room temperature.

Densification/Hydrogen Concentration

APCVD oxide film undergoes a large densification of 5.6% as compared to 1.5% of PECVD oxide film when annealed at 850°C for 30 minutes. Also, in case of APCVD oxide films the densification is observed at temperatures as low as 500°C. Hydrogen con-

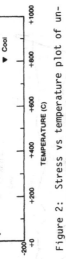

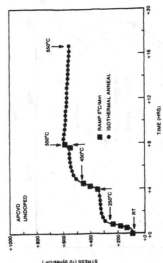

Figure 2: Stress vs temperature plot of undoped APCVD film cycle 2.

Figure 3: Stress vs time plot of APCVD undoped film during isothermal annealing.

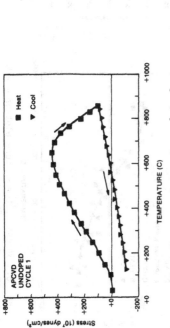

Figure 1a: Stress vs temperature plot of
undoped APCVD film

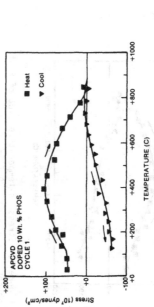

Figure 1b: Stress vs temperature plot of 10
wt% phosphorous doped APCVD film

466

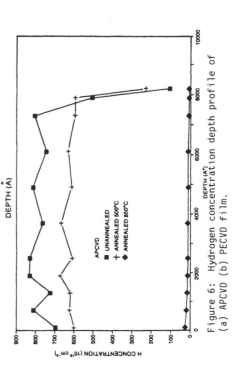

Figure 6: Hydrogen concentration depth profile of (a) APCVD (b) PECVD film.

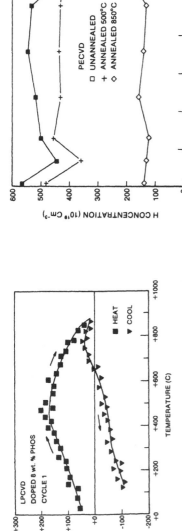

Figure 4: Stress vs temperature plot of 8 wt% phosphorous doped LPCVD film.

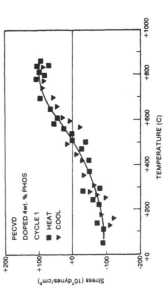

Figure 5: Stress vs temperature plot of 4 wt% phosphorous doped PECVD film.

centration profiles for an undoped APCVD oxide and 4 wt% phos doped PECVD oxide film are shown in Figure 6. The hydrogen concentration in as deposited APCVD oxide films is found to be higher ($8 \times 10^{21}/cm^3$) than PECVD films ($5 \times 10^{21}/cm^3$). The reduction in hydrogen concentration due to heat treatment is found to be more in case of APCVD films. A 30 minute heat treatment at $850°C$ reduce the concentration of hydrogen from $8 \times 10^{21}/cm^3$ to $5 \times 10^{19}/cm^3$ in APCVD films and from $5 \times 10^{21}/cm^3$ to $4 \times 10^{20}/cm^3$ in PECVD films.

DISCUSSION

Mechanical stress in a thin film deposited on a substrate comprises of an intrinsic component and a thermal component. The intrinsic stress is a consequence of the growth process of the film. The thermal component is caused by the difference in the thermal expansion coefficients of the film and the substrate. In elastic region the changes in the thermal stress with temperature can be described by the relation

$$\sigma_{th} = \left(\frac{E}{1-\nu} \right)_{film} (\alpha_{film} - \alpha_{sub})(T_2 - T_1) \tag{1}$$

where α_{film} and α_{sub} are the thermal expansion coefficients of the film and substrate respectively and T_2 the annealing temperature and T_1 the room temperature.

Among various oxide, the mechanical properties of thermally grown oxides are better understood. Therefore it will be most appropriate to compare the observed behavior of deposited oxides with thermally grown oxides. The stress at room temperature was found to be compressive and has a value of 3×10^9 dyne/cm² which agrees well with the values reported in the literature [12-14]. Slope of stress versus temperature plot ($\frac{d\sigma}{dT}$) can be calculated from equation 2. Using the value of $\alpha_{SiO_2} = 5.0 \times 10^{-7}$ C^{0-1}, $\alpha_{si} = 3.1 \times 10^{-6}$ C^{0-1} and $(\frac{E}{1-\nu})_{SiO_2} = 8.6 \times 10^{11}$ dyne/cm² [12,15], $\frac{d\sigma}{dT}$ is calculated to be 2.24×10^6 dynes/Cm²-deg. This agrees well with the measured value of 2.6×10^6 dynes/Cm²-deg for thermally grown oxide.

Now considering the case of an undoped APCVD oxide film (Figure 1a), we find that the as deposited film is under a low tensile stress. This stress is intrinsic in nature because of the fact that the thermal expansion coefficient of SiO_2 is small compared to that of silicon. Therefore, the cooling down of the film from deposition temperature should have resulted in a compressive stress rather than a tensile stress. The slope of the figure 1a during first heat cycle (from room temperature to relaxation temperature) is found to be 8.9×10^6 dynes/cm² deg. This is about 4 times higher than that a thermally grown SiO_2. Such a large slope can be explained on the basis of incorporation of hydrogeneous species in the film during deposition and their subsequent removal during heat treatment. The hydrogeneous species can be bonded with silicon oxygen networks as Silanol (SiOH), hydride SiH, or water H_2O [16]. Such hydrogeneous species can be produced as byproducts when SiH_4 and O_2 react to form SiO_2. Any post deposition heat treatment would lead to the desorption of these species and would make film to contract. If the film is not able to do so, a tensile stress would build up.

The increase in stress with temperature (up to T_R) can be explained on the basis of desorption model as described above. However, the decrease in stress beyond T_R can not be explained. The decrease in stress beyond T_R could take place if the thermal expansion coefficient of SiO_2 exceeded that of silicon. In glasses a steep rise in thermal expansion coefficient in observed around glass transition temperature Tg [17]. The reported values of Tg for undoped SiO_2 and 7 wt.% phosphorus doped SiO_2 are $1100°C$ and $750°C$ respectively. These temperatures are higher by at least $400°C$ than the observed relaxation temperature in the present case. Therefore, it is unlikely that the observed T_R is similar to Tg of glasses. Also Tg is a property of the glass and changes in thermal expansion coefficient should be observed every time glass is heated to Tg. Contrary to this, T_R in our case is observed only during first heat cycle to which the film is subjected to after its deposition. We believe that the decrease in stress beyond T_R is a consequence of stress built up in the film. With the desorption of hydrogeneous species from the film, induced by thermal treatment, the stress keeps on building up in film till it reaches a limit beyond which the film cannot withstand any more stress and starts undergoing a plastic (creep) deformation. The fact that T_R is found to be dependent on the nature of annealing cycle further supports this hypothesis. Both the desorption and the plastic deformation beyond T_R concurrently lead to the densification of the film. With the addition of phosphorous, SiO_2 matrix becomes week and as such cannot withstand as much stress as undoped SiO_2. Therefore both the T_R as well as maximum stress at T_R are found to have smaller values in doped SiO_2 films.

Out of various hydrogeneous species, water seems to be the main species responsible for observed stress behavior in APCVD/LPCVD oxide films. Absence of these water molecules in PECVD films makes these films more stable. There are two mechanisms which can render PECVD films free of water molecules: (1) The water molecules are decomposed in plasma (2) Even if they are incorporated in the film during deposition, the continuous ion bombardment keeps on desorbing them continuously from the film. This is supported by the fact that as deposited PECVD films are more denser than APCVD or LPCVD films. Our densification data also indicates the same. Whereas no or negligable densification is observed in PECVD films when annealed at 500°C for very long time (18 hours), as high as 2.7% densification is observed in APCVD films. At 850°C densification of 5.6% is observed in APCVD films against 1.5% in PECVD films. Hydrogen concentration in as deposited APCVD films have been found to be more than PECVD films. Also the change in hydrogen concentration after a heat treatment at 850°C for 30 minutes decreases the concentration in APCVD films by more than two orders of magnitude as compared to only one order change in PECVD films. Or in other words, PECVD films retain more hydrogen even after heat treatment.

As evidenced by mechanical stress versus temperature plots of annealing cycle 2 (figure 2), both APCVD and LPCVD films become stable after the heat treatment. No hysteresis or drift in mechanical stress is observed. Now the stress behavior can be explained by sample elastic laws (equation 2). Stress behavior of most of these annealed (deposited) silicon dioxide films resemble that of a thermally grown silicon dioxide film. The room temperature stress is compressive which becomes more tensile with increase in temperature. The undoped oxide films show a linear increase in stress with temperature and has a slope of 2.8×10^6 dynes/cm^2 deg which compares well with a value of 2.6×10^6 dynes/cm^2-deg obtained for thermal oxide. In case of doped oxides, a flattening in stress is observed at high temperatures. For a 8 wt % phosphorous doped SiO_2 film this flattening occurs around 700°C which corresponds to the glass transition temperature of the film as reported in literature. The flattening is due to the change in the thermal expansion coefficient of SiO_2 film at Tg.

REFERENCES

[1] A.C. Adams in "VLSI Technology" Edited by S.M. Sze, Chapter 6 McGraw-Hill Book Company (1988).
[2] W. Kern, G.L. Schnable and A.W. Fischer, RCA Rev. 37, 3 (1976).
[3] A.C. Adams, F.B. Alexander, C.D. Capio, and T.E. Smith, J. Electrochem. Soc. 128, 1545 (1981).
[4] Brad Mattson, Solid State Technology 60, Jan. (1980).
[5] G. Smolinsky and T.P.H.F. Wendling, J. Electrochem. Soc. 132, 950 (1985).
[6] H. Sunami, Y. Itoh and K. Sato, J. Appl. Phys. 41, 5115 (1970).
[7] A.J. Learn, J. Electrochem Soc. 132, 405 (1985).
[8] R. Lathlaen and D.A. Diehl, J. Electrochem. Soc. 116, 620 (1969).
[9] A.K. Sinha, H.J. Levinstein and T.E. Smith, J. Appl. Phys. 49, 2423 (1978).
[10] W. Kern, RCA Rev. 37, 55 (1976).
[11] W.A. Lanford, H.P. Trautvetter, J. Ziegler and J. Keller, Appl. Phys. Lett. 28, 566 (1976).
[12] R.J. Jaccodine and W.A. Schlegel, J. Appl. Phys. 37, 2429 (1966).
[13] I. Blech and U. Cohen, J. Appl. Phys. 53, 4202 (1982).
[14] S.P. Murarka and T.F. Retajczk, Jr. J. Appl. Phys. 54, 2069 (1983).
[15] "Thermophysical properties of Matter" Edited by U.S. Toulokian, Plenum New York (1977).
[16] W.A. Pliskin, J. Vac. Sci. Technol. 14, 1064 (1977).
[17] K. Nassua, R.A. Levy and D.L. Chadwick, J. Electrochem. Soc. 132, 409 (1985).

ELECTRICAL AND METALLURGICAL CHARACTERISTICS OF

PtSi$_x$ AND TiSi$_x$ / GaAs SCHOTTKY CONTACTS

Qian He and Luo Jinsheng
Xi'an Jiaotong University, Microelectronics Technology Division,
Xi'an, China

ABSTRACT

Electrical properties, metallurgical properties and their relationship of PtSi$_x$ / GaAs formed by sputter deposition and TiSi$_x$ / GaAs formed by e—gun multilayer evaporation have been studied. It has been found that TiSi$_x$ / GaAs contacts exhibit excellent Schottky electrical properties and good metallurgical stability after RTA (975℃ , 12 sec) but some interface unstability and degradation of electrical properties after conventional FA (800℃ ,20 min). The electrical and metallurgical properties of PtSi$_x$ / GaAs contacts begin to degrate after annealed at 500℃.

Introduction

Silicides, particularly WSi$_x$, have been widely applied to gate metallization in self—aligned GaAs MESFETs for their high temperature stability [1]. One of the shortages of WSi$_x$ as gate materials is it's high resistance [2]. The resistances of PtSi$_x$ and TiSi$_x$ are only one—half and one—third of WSi$_x$'s respectively [3]. It is therefore important to characterize the thermal stability of PtSi$_x$ and TiSi$_x$ / GaAs interface. In this paper, we report the study on PtSi$_x$ and TiSi$_x$ / GaAs contacts by AES, XPS and current—voltage measurements under different annealing conditions. PtSi$_x$ and TiSi$_x$ films were also characterized by x—ray diffraction.

Experiments

The substrates used for the experiments were SI—GaAs (100) wafers with Si$^+$ implantation. After etching the GaAs substrates in HCl:H$_2$O = 1:1 for 30 sec, PtSi$_x$ films were then deposited on them by using magnetron sputtering. Si component x varied from 0.65 to 1.6 by changing the effective sputter area of Pt. Thin layers of Ti and Si were deposited alternately on the GaAs substrates by using e—gun evaporation to form TiSi$_x$. The thickness of each Ti and Si layer (typically 100 Å) was varied to obtain TiSi$_x$ with x ranging from 1.6 to 2.9. The total thickness of silicides was about 2000 Å. The wafers were then annealed in Ar gas flow at the different conditions: rapid thermal annealing (RTA, 975℃ , 12 sec) and conventional furnace annealing (FA, 800℃ , 20 min).

X—ray diffractions were used to characterize the crystallizing properties of silicide films. AES, XPS and current—voltage measurements were used to analyze the silicides / GaAs interface. In order to obtain silicides / GaAs interface chemistry, thin films (about 100 Å) were deposited on GaAs substrates. Before XPS measurements, the thickness of the films were decreased to about 20 Å by low energy Ar$^+$ in the vacuum chamber.

Mat. Res. Soc. Symp. Proc. Vol. 181. ©1990 Materials Research Society

Results and discussions

A. Crystalization

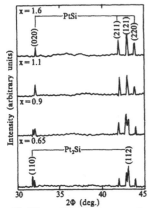

Fig.1.X—ray diffraction patterns of PtSi$_x$ annealed at 500 ℃,30 min.

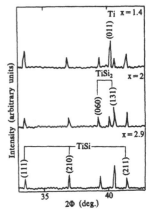

Fig.2.X—ray diffraction patterns of TiSi$_x$ after RTA.

Figs. 1 and 2 show the x—ray diffraction patterns of PtSi$_x$ and TiSi$_x$ films respectively. P-tSi$_x$ films at x=0.65 and 0.9 exhibit diffraction peaks of Pt$_2$Si and PtSi phases. Films at x= 1.1 and 1.6 exhibit PtSi phase peaks only. After annealed, TiSi$_x$ films involve Ti, TiSi and TiSi$_2$ phases. The intensity of diffraction peaks due to Ti decreases and TiSi$_2$ peaks increases with the increasing of x. Films at x=1.6 contain TiSi and TiSi$_2$ only. These results indicate that rich Ti silicides formed at the Ti / Si interface first and then, if there is enough Si, rich Si silicides formed in the multilayer Ti–Si structure.

B. Interface stability

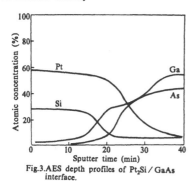

Fig.3.AES depth profiles of Pt$_2$Si / GaAs interface.

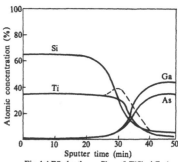

Fig.4.AES depth profiles of TiSi$_2$ / GaAs interface.

Fig. 3 shows the AES depth profiles of Pt$_2$Si / GaAs interface annealed at 500℃ for 30 min. Considerable migration of Ga and Pt was observed. This result indicates that PtSi$_x$ / Ga

As interface can only remain thermal stable till about 500℃ . Fig. 4 shows that no considerable interdiffusion of component at $TiSi_2$/GaAs interface was observed after RTA but some accumulation of Ti was observed after FA at the interface (as the dot line shown in Fig. 4) The diffusions of metals into GaAs are decided by the electronegativities of metals [4]. Refractory metals (such as Ti) have relatively smaller electronegativities than those of near-noble metals (such as Pt). Metals form intermetallic compounds with Si which can decrease the electronegativities of metals. These experiments show that the metallic bonds formed by Ti and Si are stable under RTA but the metallic bonds formed by Pt and Si can only remain stable till 500℃ .

Table I gives the results of XPS measurements. Compared with the binding energies of elemental Si, Ti, Pt and As, Ga in GaAs, no chemical shift after RTA but 0.6 eV shift of As 3d core level and 1.93 eV shift of Ti $2p_{3/2}$ core level after FA were observed at $TiSi_2$/ GaAs interface. These results show that $TiSi_2$/GaAs interface remains chemical stable under RTA but under FA some chemical reaction between Ti and As occured. At Pt_2Si/GaAs interface, 1.0 eV, 1.43 eV and 4.87 eV shifts of Ga 3d, As 3d and Si 2p core levels were observed respectively after annealed at 500℃ for 30 min. These binding energies correspond to Ga 3d peak from Ga_2O_3, As 3d peak from As_2O_3 and Si 2p peak from SiO_2 respectively. The oxidation of the interface may be induced during the $PtSi_x$ sputter deposition.

Table I. Binding energies of the elements at the interface

	Ga (3d) (eV)	As (3d) (eV)	Si (2p) (eV)	Ti $(2p_{3/2})$ (eV)	$Pt(4f_{7/2})$ (eV)
Si,Ti,Pt and GaAs	19.5	40.7	99.15	453.8	70.9
$TiSi_2$/GaAs, RTA	19.61	40.90	99.17	454.0	
$TiSi_2$/GaAs, FA	19.65	41.30	99.58	455.7	
Pt_2Si/GaAs, 500℃	20.50	42.13	104.02		71.31

C. Electrical properties

Table II and table III show the Schottky barrier height (Φ_B) and ideality factor (n) of $PtSi_x$/GaAs and $TiSi_x$/GaAs Schottky contacts respectively. It can be seen that the Schottky contact properties of $PtSi_x$/GaAs begin to degrate after annealed at 500℃ . It is suggested that the electrical degradation is directly related to the interface unstability. $TiSi_x$/GaAs contacts show excellent electrical properties after RTA but some few abnormal increases in Φ_B and n after FA. These increases can be explained by the accumulation of Ti and the chemical reaction between Ti and As at the interface which induce an interface layer. This interface layer can increase Φ_B

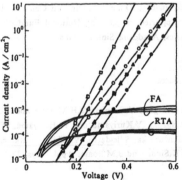

Fig.5.I–V properties of $TiSi_x$/GaAs after RTA and FA. □:x=2.9,RTA;■:x=2.9,FA; ▲:x=2,RTA;△:x=2,FA;○:x=1.4,RTA;●:x=1.4,FA.

and n. Fig. 5 shows that the reverse current of $TiSi_x$/GaAs Schottky contacts is greater after FA than after RTA. This leakage in the current–voltage characteristics after FA is also due to the interface layer. The chemical reaction at the interface induce a high concentration of deep levels which would provide recombination paths for conductive electrons.

Table II. Φ_B and n of PtSi$_x$ / GaAs Schottky contacts
annealed at 500℃ for 30 min

x	0.65	0.9	1.1	1.6
Φ_B (eV)	0.67	0.69	0.69	0.74
n	1.28	1.25	1.21	1.30

Table III. Φ_B and n of TiSi$_x$ / GaAs Schottky contacts

x	1.4		2		2.9	
	RTA	FA	RTA	FA	RTA	FA
Φ_B (eV)	0.80	0.82	0.76	0.79	0.75	0.79
n	1.14	1.21	1.09	1.19	1.12	1.19

Conclusion

We found that TiSi$_x$ / GaAs Schottky contacts are stable in terms of metallurgical and electrical properties after RTA but some degradation occured after FA. PtSi$_x$ / GaAs Schottky contacts remains stable until annealed at 500℃ . Considered low resistance, TiSi$_x$ film will be very good gate material for self–aligned GaAs MESFETs if RTA is used.

Acknowledgements

We would like to thank Ye Pin for conducting e–gun evaporation. We would also like to acknowledge Duan Lihuong and Chuei Yude for their XPS and AES analyses. This work was supported, in part, by National Fund of Natural Science and by Laboratory for Surface Physics, Academia Sinica, China.

References

[1].A.Cailegari, G.D.Spiers, J.H.Muthrie and H.C.Guthrie, J.Appl. Phys. 61, 2054 (1987).
[2].M.Suzuki, Y.Kuriyama and M.Hirayama, IEEE Electron Device Lett. 6(10), 542 (1985).
[3].S.P.Murarka, Silicide for VLSI Applications. (Academic Press, New York, 1983) Chapt. 2.
[4].A.K.Sinha and J.M.Poate, Appl. Phys. Lett. 23, 666 (1973).

SUSCEPTOR AND PROXIMITY RAPID THERMAL ANNEALING OF InP

A. Katz*, S. J. Pearton* and M Geva**
*AT&T Bell Laboratories, Murray Hill, NJ 07974
**AT&T Bell Laboratories, Solid State Technology Center, Breinigsville,PA 18301

ABSTRACT

An intensive comparison between the efficiency of InP rapid thermal annealing within two types of SiC-coated graphite susceptors and by using the more conventional proximity approach, in providing degradation-free substrate surface morphology, was carried out. The superiority of annealing within a susceptor was clearly demonstrated through the evaluation of AuGe contact performance to carbon-implanted InP substrates, which were annealed to activate the implants prior to the metallization. The susceptor annealing provided better protection against edge degradation, slip formation and better surface morphology, due to the elimination of P outdiffusion and pit formation. The two SiC-coated susceptors that were evaluated differ from each other in their geometry. The first type must be charged with the group V species prior to any annealing cycle. Under the optimum charging conditions, effective surface protection was provided only to one anneal (750°C, 10s) of InP before charging was necessary. The second contained reservoirs for provision of the group V element partial pressure, enabled high temperature annealing at the InP without the need for continual recharging of the susceptor. Thus, one has the ability to subsequentially anneal a lot of InP wafers at high temperatures without inducing any surface deterioration.

I. INTRODUCTION

Rapid thermal processing (RTP) has attracted considerable recent interest for InP device technology, and in particular for contact sintering [1,2], ion implant damage annealing and implant activation [3]. The fundamentals of the heat transfer process for the RTP allows for superior performance over the conventional heating processes, particularly the much reduced dopant redistribution [4], more reproducible contact alloying processes [5], and reduced degradation of heterostructures during high temperature processes. These advantages, however, cannot be exploited in conjunction with InP technology, unless an effective technique to restrict the incongruent evaporation of P from the surface, is provided. A similar issue was addressed while rapid thermally annealing and activating implants in GaAs [6-9], in which case action were taken to minimize the As loss. Some methods of minimizing the group V volatile elements through RTP, and thus reducing the amount of the semiconductor surface damage, were suggested. Each of the possible solutions have problems associated with them. For example, the commonly used dielectric encapsulants (Si_3N_4, SiO_2, AIN) either have adhesion problems at the high annealing temperatures involved, or else induce considerable stress into the underlying wafer. The use of an AsH_3 (or equivalent) ambient presents considerable safety concerns, although in principle this is the ideal way to minimize surface degradation [3,11]. A less effective, but more common method of reducing loss of the group V element is the so-called proximity annealing method, in which the wafer of interest is placed face-to-face with another wafer of the same type [12]. As the wafer are heated up, each begins to lose a small amount of the group V species, but an overpressure is created between the wafers which prevents further dissociation. This is ultimately an unsatisfying solution, because it relies on a loss of As, P or Sb (depending on the material being annealed). Moreover the edges of the wafer often show pitting of the surface, and movement of the wafers relative to each other at any point in the handling or annealing stages will lead to microscratches which reduce device yield. A variation of this scheme is to coat the proximity wafer with Sn to increase the group V element partial pressure [13], but such an approach is not easily amenable to the sequential annealing of large numbers of wafers.

There have been a number of reports of using an enclosed graphite cavity for annealing of GaAs [14] and InP [15]. This provides a very uniform heating environment for the wafer because of the high emissivity of the graphite and reduces problems with slip formation. Because of the larger thermal mass of the graphite cavity, the heating and cooling rates are considerably slower than for conventional proximity rapid annealing [4].

In this paper we describe the use of two different types of graphite susceptor for InP implant activation annealing, retaining all of the advantages of rapid thermal annealing (RTP) [16], while

Mat. Res. Soc. Symp. Proc. Vol. 181. ©**1990 Materials Research Society**

TABLE I: Implant conditions for the samples studied. All annealing was performed at 700°C for 10 sec.

Sample	C			P			Al			Ga		
	Dose (x 10^{14} cm^2)	Energy (KeV)	Temp (°C)	Dose (x 10^{14} cm^2)	Energy (KeV)	Temp (°C)	Dose (x 10^{14} cm^2)	Energy (KeV)	Temp (°C)	Dose (x 10^{14} cm^2)	Energy (KeV)	Temp (°C)
1	1	40	25	-	-	-	-	-	-	-	-	-
2	1	40	25	1	100	200	-	-	-	-	-	-
3	1	40	25	1	100	25	-	-	-	-	-	-
4	1	40	25	-	-	-	1	120	200	-	-	-
5	1	40	25	-	-	-	1	120	25	-	-	-
6	1	40	25	-	-	-	-	-	-	1	180	200
7	1	40	25	-	-	-	-	-	-	1	180	25
8	5	40	25	5	100	200	-	-	-	-	-	-
9	5	40	25	5	100	25	-	-	-	-	-	-
10	5	40	25	-	-	-	5	120	200	-	-	-
11	5	40	25	-	-	-	5	120	200	-	-	-
12	5	40	25	-	-	-	-	-	-	-	-	-
13	5	40	25	-	-	-	-	-	-	5	180	200
14	5	40	25	-	-	-	-	-	-	5	180	25

providing an excellent protection of the surface integrity. The superiority of the surface protection by means of the graphite susceptors, over the proximity approach is demonstrated through comparison of the electrical properties of AuGe contacts to variety of C co-implanted SI-InP substrates which were annealed prior to the metallization. In particular, the inclusion of reservoir P-containing materials within the susceptor enables the subsequential annealing of large number of InP wafers without the need for charging the graphite cavity with P prior to every run.

II. EXPERIMENTAL PROCEDURE

Implantation of carbon was performed in a non channeling direction (7° tilt of the wafers relative to the beam direction, and 15° rotation with respect to the <100> axis) at room temperature (RT). The implant energy was fixed at 40KeV and the C ion dose varied from $5x10^{12}$ to $5x10^{14} cm^{-2}$ although we concentrated on the higher dose range. The effect of co-implant of P, Al, and Ga at 25 or 200°C was also explored, since these co-implants led to either greater (in the case of P) or lesser (in the case of Ga and Al) n-type doping levels in the InP after annealing. Table I summarizes the various implant conditions, studied in this work. Each sample was halved after the ion implantation and an A-B comparison study was conducted to evaluate the annealing technique effect on the damage recovery and thus on the electrical properties of subsequently deposited AuGe contacts. A conventional A. G. Associates Heatpulse™ 410T system was used for all the annealing treatments at temperatures of 700°C or higher, prior to the deposition of the metal contact. One half of each implanted wafer was placed, face down, on a clean InP substrate in an arrangement so-called "proximity anneal". The other half of each sample was placed within a SiC coated graphite susceptor into the annealing chamber. This susceptor was charged with phosphorus prior to each anneal by heating an InP substrate at 900°C for 120 sec in order to coat with P the inside of the container[17]. The temperature of the susceptor was measured by monitoring the emissivity of the graphite with a pyrometer.

In addition to the above mentioned SiC-coated graphite susceptor, described in detail elsewhere[4,19], another type of graphite susceptor was evaluated. This different susceptor was designed to overcome the one major disadvantage of the previous susceptor, namely its need to be re-charged frequently in order to ensure an adequate partial pressure of the P for each new wafer annealed. The improvement was achieved by introducing four small reservoirs around the circumference of the lower plate. These are connected to the central cavity where the wafer is held by channels milled into the plate. The reservoirs can be filled with InP or other material that provides P partial pressure, and

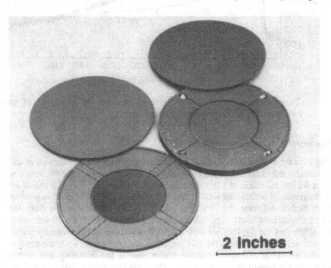

2 Inches

Figure 1   Photograph of the conventional graphite susceptor containing a 2 inch diameter wafer (left) and the susceptor with four InP-filled reservoirs.

therefore eliminates the need for continued recharging the susceptor. After the implantation damage annealing and activation, part of the samples were metallized by means of AuGe electron-gun deposition to create contact mesas, defined by means of standard photolitographs technique, and electrically measured through the transmission line method (TLM) pattern, elsewhere described[18]. These contacts were sintered by means of RTP using the same Heatpulse™ apparatus, this time under forming gas ambient (15%H$_2$) at the temperature range between 300 and 450°C for duration of 30 sec (Fig. 1).

III. RESULTS AND DISCUSSION

a. Modification of the thermal cycle

Since the thermal mass of both types of susceptor is considerably greater than that of a conventional proximity annealing arrangement, their heating and cooling rates are much slower. Figure 2 shows the temperature-time profiles for 2 inch diameter wafers annealed at 900°C for 10 sec in either type of susceptor, and in the proximity arrangement. In the latter case both the heated wafer and the protective substrate are supported on a 4 inch diameter Si substrate in which a Cr-Al thermocouple is embedded. Under most conditions, the slower heating and cooling rates for susceptor annealing are an advantage because they help to eliminate slip generation when combined with the much smaller thermal gradients compared to the proximity arrangement. It has to be noted that the susceptor with the reservoir has such a slow heating rate that there is not a well defined time spent at the designated annealing temperature. As before [4,22], we calculated that the temperature differential between center and edge of a 2 inch wafer in the susceptor is less than 7°C.

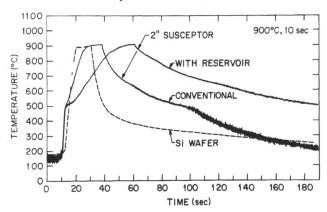

Figure 2   Temperature response as a function of time for the two types of susceptor and a conventional proximity annealing arrangement.

b. Surface morphology of annealed InP wafers

The increase in the overall thermal mass, associated with the introduction of the susceptor into the RTP chamber, led to the decreasing of the heating and cooling rates for the treated wafers, and thus can account for the reduction of slip generation in the InP wafers. Figure 3 shows an array of three optical micrographs of wafers that were etched in a molten KOH solution subsequent to the annealing, in order to further delineate the slip lines. All those wafers were taken from the same liquid-encapsulated Czochralski boule. The left one was annealed at 850°C for 10 sec within the susceptor while the middle and the right wafers were annealed for 10 sec in the proximity geometry at 800 and 850°C, respectively. The former is the only one to be almost slip free, while the small amount of weak slip lines observed in it were present in the wafers prior to the annealing, due to the thermal stress induced through the growth of the InP crystal. The latter two wafers annealed through the proximity arrangement, not only present obvious slip lines oriented in the {111} - <$\bar{1}$10> direction, but also suffered severe surface degradation. Surface pitting of the wafer annealed at 800°C is particularly observed at the periphery. The wafer that was heated at 850°C in the proximity approach suffered vastly more damage. In this one not only pits and large In droplets are observed, but also the excess In balled-up around the wafer edges. This set of results provides clear evidence that the proximity annealing method can not adequately eliminate the loss of the volatile P element from the wafer surface and edges. The almost perfect surface of the wafer annealed within the graphite susceptor, on the other hand, emphasizes once more the potential of this technique.

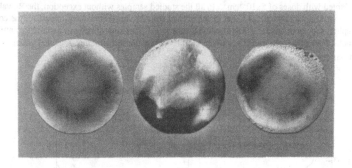

Figure 3   Optical micrographs of 2 inch-diameter InP wafers annealed within the graphite susceptor at 850°C for 10 sec. (at left), or by the proximity method at 800°C (at center), and 850°C (at right both for 10 sec. The wafers were etched in to delineate the annealing-induced defects.

c. Efficiency of the two types of the SiC-coated graphite susceptor

It is important to evaluate the number of subsequent InP wafers that can be efficiently annealed after charging the standard susceptor, before it has to be re-charged. Figure 4 shows optical micrographs of InP wafers annealed sequentially in the susceptor following an initial charging procedure. For InP this consisted of heating a sacrificial InP substrate at 850°C for 5 min to coat the inside of the susceptor with phosphorus. Separate 2 inch-diameter InP wafers were then annealed sequentially at 750°C for 10 sec without re-charging the susceptor. As shown in Fig. 4(a), the first wafer had a featureless surface morphology. The second anneal, however, led to an obvious degradation of the surface as is clear from Fig. 4(b). This deterioration becomes worse on subsequent wafers, and Fig. 4(c) shows that the third wafer has a high density of pits.

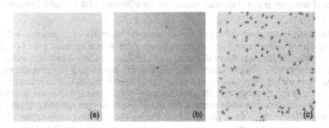

Figure 4   Optical micrographs (200 x magnification) of InP samples heated sequentially at 750°C for 10 sec after an initial charge of the conventional susceptor at 850°C for 5 min using the sacrificial InP wafer. The first wafer annealed was (a), the second (b), and the third (c).

d. Efficiency studies by means of specific contact resistance measurements

Electrical measurements of the AuGe contacted InP wafers were taken in order to evaluate in a very precise manner the efficiency of the ion implant damage removal by both the proximity approach and while annealed within the simple SiC-coated graphite susceptor. The dependence of the Rc on the alloying RTP temperature is shown in Figure 5. The superiority of the contacts to the graphite susceptor annealed InP was demonstrated for all the studied samples and is shown in this Fig for the samples that were implanted with doses of $5 \times 10^{14} cm^2$. In all the studied samples without exception, the Rc values of the as-deposited contact and after RTP at different temperatures were found to be lowest for the contacts to the graphite susceptor annealed InP implanted substrates. This superiority can certainly be attributed to the much better preservation of the surface morphology and microstructure achieved while annealed within the susceptor.

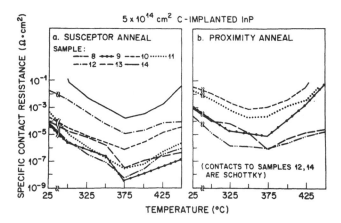

Figure 5   Specific contact resistance as a function of RTP alloying temperature (durations of 30 sec) of AuGe contacts onto InP implanted with C($5 \times 10^{14} cm^{-2}$, 40 KeV) by itself or with co-implanted species (see Table I) and activated by (a) graphite susceptor or (b) proximity method annealing.

e. Effect of different group V element sources on the efficiency of the graphite susceptor annealing

We have investigated the use of various reservoir materials in conjunction with that type of susceptor. The first experiment involved filling the reservoir with many small pieces (each of order of 1mm^3) of either InP, InAs or GaAs in the susceptor and then loading 2 inch InP wafers which were annealed at progressively higher temperatures without any precharging of the susceptor, in order to determine how well the InP surface was preserved. Figure 6 shows optical micrographs from InP wafers, annealed for 10 sec at 750°C(a), 800°C(b), or 850°C(c) using InP in the reservoir, and from similar wafers annealed at the same temperatures using InAs in the reservoir (micrograph (d), (e) and (f) respectively). The surfaces show pitting only at the highest annealing temperatures. Similar results were obtained using GaAs in the reservoir, indicating that in principle either P or As partial pressure can be used to protect InP. In practice however, the InAs melts below 570°C and forms balls within the reservoir. These InAs balls have a smaller effective surface area than the original slivers, and therefore provide a lower As partial pressure. Similarly, if GaAs is used as the reservoir material, there is always the possibility of actually incorporating As into the near-surface region. This is not desirable in some device applications.

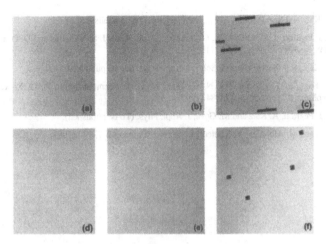

Figure 6   Optical micrographs (200 x magnification) of InP annealed for 10 sec at 750°C (a), 800°C (b) or 850°C (c) using InP in the reservoir, and InP annealed at 750°C (d), 800°C (e) or 850°C (f) for 10 sec using InAs in the reservoir.

IV. SUMMARY AND CONCLUSIONS

The use of a SiC-coated graphite susceptor containing reservoir for provision of a group V element partial pressure, enables high temperature annealing of InP with no discernible surface degradation or slip formation. The ability to provide this partial pressure without the need for continual recharging of the susceptor means that one has the ability to sequentially anneal dozens of wafers at high temperatures without inducing surface deterioration. These results suggest that annealing InP substrates within any type of graphite susceptor is superior to the proximity RTP approach.

References

[1]   A. Katz, W. C. Dautremont-Smith, P. M. Thomas, L. A. Koszi, J. W. Lee, V. G. Riggs, R. L. Brown, J. L. Zilko and A. Lahav, J. Appl. Phys. 65, 4319 (1989).

[2]   A. Katz, W. C. Dautremont-Smith, S. N. G. Chu, P. M. Thomas, L. A. Koszi, J. W. Lee, V. G. Riggs, R. L. Brown, S. G. Napholtz, J. L. Zilko, and A. Lahav, Appl. Phys. Lett, 54, 2306 (1989).

[3]   A. Lahav, R. L. Lapinsky and T. C. Henry, J. Electrochem. Soc. 136, 1096 (1989).

[4]   S. J. Pearton and R. Caruso, J. Appl. Phys. 66, 663 (1989).

[5]   A. Katz, P. M. Thomas, S. N. G. Chu, J. W. Lee and W. C. Dautremont-Smith, J. Appl. Phys. 66, 2056 (1989).

[6]   M. Nishitsuji and F. Hasegawa, Jpn. J. Appl. Phys. 28, L895 (1985).

[7]   H. Kanber, R. J. Cipolli, W. B. Henderson And J. M. Whelan, J. Appl. Phys. 57, 4732 (1985).

[8]   C. H. Kank, K. Kondo, J. Lagowski and H. C. Gatos, J. Electrochem. Soc. 134, 1261 (1987).

[9]   C. W. Farley and B. G. Streetman, J. Electron. Mater. 13, 401 (1984).

[10]   G. J. Valco and V. J. Kapoor, J. Electrochem. Soc. 134, 569 (1987).

[11]   T. N. Jackson, J. F. DeGelormo and G. Pepper, Proc. Mat. Res. Soc. Symp. 144, 403 (1989).

[12]   B. Molnar, Appl. Phys. Lett. 36, 927 (1980).

[13]   C. A. Armiento and F. C. Prince, Appl. Phys. Lett. 48, 1623 (1986).

[14]   A. Katz, C. R. Abernathy and S. J. Pearton, Appl. Phys. Lett. 56, 1028 (1990).

[15] A. Katz and S. J. Pearton, J. Vac. Sci. Technol. (To be published).

[16] S. J. Pearton, J. M. Gibson, D. C. Jacobson, J. M. Poate, J. S. Williams and D. O. Boerma, Proc. Mat. Res. Soc. Symp. 52, 198 (1986).

[17] A. Katz, S. J. Pearton and M. Soler, J. Appl. Phys. (To be published).

[18] A. Katz, P. M. Thomas, S. N. G. Chu, W. C. Dautremont-Smith, R. G. Sobers and S. G. Napholtz, J. Appl. Phys. 67, 884 (1990).

[19] S. J. Pearton, A. Katz and M. Geva, J. Appl. Phys. (To be published).

[20] A. Katz, M. Albin and Y. Komem, J. Vac. Sci. Technol. B 7, 130 (1989).

In-BASED OHMIC CONTACTS TO THE BASE LAYER OF GaAs-AlGaAs HETEROJUNCTION BIPOLAR TRANSISTORS

F. REN*, S. J. PEARTON*, W. S. HOBSON*, T. R. FULLOWAN*, A. B. EMERSON*, A. W. YANOF* AND D. M. SCHLEICH**
*AT&T Bell Laboratories, Murray Hill, NJ 07974
**Polytechnic University, Brooklyn, NY

ABSTRACT

The use of AuBe-In/Ag/Au p-ohmic contacts for the base layer of GaAs-AlGaAs heterojunction bipolar transistors (HBTs) is described. Annealing at 420°C for 20 sec produces a contact resistivity of $0.095\,\Omega\,mm$ and a specific contact resistance of $1.5 \times 10^{-7}\,\Omega\,cm^2$, and the surface morphology of the contact is excellent. The role of the silver is as a diffusion barrier to prevent Au spiking into the base layer which could degrade the HBT performance. The presence of the In layer is highly desirable in order to reduce the contact resistance, probably by forming an InGaAs phase at the metal-GaAs interface. Beryllium acts as the p-type dopant, and the top Au layer is used to lower the contact sheet resistance. Current transport through the structure is dominated by tunneling through the barrier due to field emission in the highly doped base layer at p-type doping levels above $\sim 10^{19}\,cm^{-3}$.

INTRODUCTION

There has been considerable recent interest in the use of Heterojunction Bipolar Transistors (HBTs) for high speed integrated circuits. The requisite high performance of individual HBTs is determined both by the quality of and control of the epitaxial growth and by the processing of the devices. The achievement of high quality ohmic metal contacts is essential to reduce the parasitic resistances within the device. Contact resistance dominates the total emitter resistance. This can be reduced by using a narrow bandgap, graded epitaxial InGaAs layer underneath the metal contacts [1]. For the base resistance, both the extrinsic and intrinsic base resistances have been minimized by using emitter-base self-aligned processing [2] and heavily doped base layers [3], respectively. However, the base contact resistance is still a limiting parasitic in HBT structures, and should be reduced. In this paper, we report reliable and low contact resistance In-based p-ohmic contacts. This appears to have a number of advantages over the more conventional AuBe/Au metallization in common use.

EXPERIMENTAL

In order to investigate the characteristics of our contacts, Zn-doped GaAs layers with different p-type doping levels ($p = 2 \times 10^{18} \sim 4 \times 10^{19}\,cm^{-3}$) were grown by atmospheric pressure Metal Organic Chemical Vapor Deposition (MOCVD) on semi-insulating GaAs substrates. A 3000 Å GaAs buffer layer was deposited first, followed by the 1500 Å Zn-doped GaAs layers. The electrically active Zn concentrations were obtained from Hall measurements and electrochemical capacitance-voltage depth profiling.

The contact test structures were mesa isolated using Argon ion milling to a depth of 2000 Å, followed by wet chemical etching to remove the residual milling damage. Prior to metal deposition, the wafers were chemically cleaned in ammonium hydroxide solution for 1 min and spun dry. The In-AuBe layer (400 Å) was deposited by electron-beam co-evaporation and followed by Ag (1000 Å) and Au (1000 Å) evaporation. For comparison to the standard AuBe/Au metallization scheme, we also deposited 400 Å AuBe and 1000 Å AuBe and 1000 Å Au, respectively, on companion sections from the same wafers. The contact patterns were formed by standard resist lift-off technique, and annealed at different temperatures (380-480°C) for 20 sec in a conventional AG Associates 410 rapid annealing system.

The contact resistivity and specific contact resistance measurements were performed using the Transfer Length Method (TLM) [4] and end resistance measurement (ERM) [5] to take into account the different sheet resistivity beneath the alloyed metal pads, corresponding to metal deposition on different Zn-doped epitaxial layers. The measurements were carried out as a function of temperature in the range of 25 to 150°C to give information on the relative importance of the different current transport mechanisms at each doping level in the p-type GaAs.

The contact morphology before and after annealing was examined by Scanning Electron Microscopy (SEM) and intermixing due to reactions at the metal-GaAs interface was studied by Auger Electron Spectroscopy (AES) depth profiling.

RESULTS AND DISCUSSION

Figure 1 shows AES depth profiles of the Au/Ag/AuBe-In/GaAs and Au/AuBe/GaAs specimens before and after alloying at 420°C for 20 sec. The presence of the diffusion barrier layer, Ag, minimizes the interaction of the top Au layer with the underlying GaAs. In turn, this will reduce the formation of Au spiking which results in a reduction of the base-collector breakdown voltage in an HBT. The out-diffusion of Ga and As was absent in the Ag-In based ohmic contact as illustrated from the AES survey spectrum in Fig. 2. A small amount of Be did however segregate to the surface.

SEM examination of the AuBe-In/Ag/Au metal film on GaAs was consistent with the AES results, with no apparent surface degradation following a 420°C for 20 sec alloy cycle. The contact also retained excellent edge definition as shown in Fig. 3. For self-aligned HBT processing, the separation between emitter and base metals is only a few tenths of a micron so that flowing of the contact during alloying must be minimized.

The results of the TLM and ERM measurements for the AuBe-In/Ag/Au contacts are summarized in Fig. 4. The contact resistivity and the specific contact resistance are demonstrated to be as low as 0.09 ohm-mm and 1.5 ohm-cm^2, respectively. The results are very uniform over the whole sample area (half of the two-inch wafer). The important role of In is also obviously shown as compared to the AuBe/Au contact. The AuBe/Au metallization under the same conditions gives 5% higher contact resistivities and specific contact resistances, for Zn doping level above 10^{19} cm^{-3}. For p-type doping levels below 10^{-19} cm^{-3}, the In-based metallization gave consistently lower (25-30%) contact resistance. After alloying, it appears that an In-Ga-As phase was formed at the metal-semiconductor interface, which lowers the contact barrier and enhances the current transport. Especially for low-doped material (p < 10^{19} cm^{-3}), thermionic emission dominates the current flow. Reducing the contact barrier on such material will significantly lower the contact resistance. On highly doped GaAs (p > 10^{19} cm^{-3}), tunneling is already the dominant current transport mechanism, and the presence of In has only a minor effect on the contact resistance.

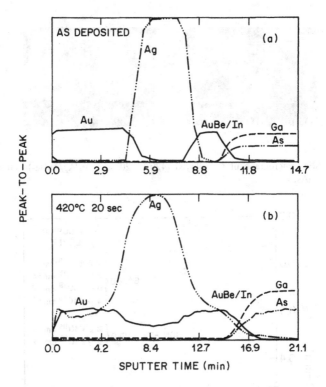

Fig. 1.   AES depth profiles of alloyed (420°C, 20 sec) Au/Ag/AuBe-In metal contacts on p-GaAs

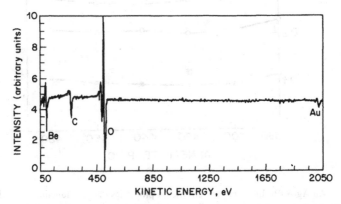

Fig. 2.   AES survey spectra from the surfaces of the alloyed Au/Ag/AuBe-In metal contacts

Fig. 3. SEM micrograph of the alloyed (420°C, 20 sec) Au/Ag/AuBe-In contacts on GaAs.

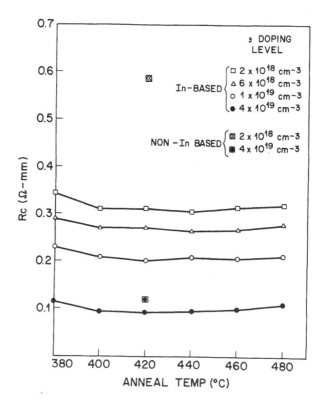

Fig. 4. Au/Ag/AuBe-In contacts electrical properties as a function of alloying temperature for different Zn doping concentration in the underlying GaAs.

It is well known [6,7] the contact resistivity is proportional to exp (ϕ_B / kT) in the pure thermionic emission case, and proportional to exp (ϕ_B / $\sqrt{p}$) in the pure field emission case, where ϕ_B is the Schottky barrier height, k is the Boltzmann constant, T is the absolute temperature, and p is the dopant concentration. The temperature dependence of contact resistivity for the In-based contacts for different doping levels in the GaAs materials are shown in Fig. 5. The predominant mode of the current flow switches from thermionic emission to field emission at a Zn concentration of 1×10^{19} cm^{-3}. Increasing the p-doping concentration definitely reduces the contact resistivity, but device performance degradation under these conditions due to p-dopant diffusion into the emitter at the higher acceptor doping levels might be expected when using Zn or Be as the base dopant. The utilization of our In-based metallization allows a reduction in the contact resistance to moderately doped ($5 \times 10^{18} \sim 1 \times 10^{19}$ cm^{-3}) base layers which are used for most conventional HBTs through the apparent formation of a low barrier In-Ga-As phase, and in addition the presence of a Ag diffusion barrier layer minimizes interfacial reactions which occur with the commonly used AuBe/Au contacts.

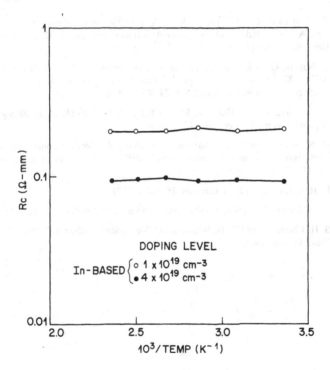

Fig. 5.   Au/Ag/AuBe-In contact resistivity as a function of (a) Doping concentration (b) Reciprocal of the measurement temperature

SUMMARY

In summary, we have studied the current transport mechanisms in the novel AuBe-In/Ag/Au p-ohmic contact for different p-doping levels in GaAs material. This metallization displays extremely low contact resistivity, is thermal stable up to 420°C and provides a reliable shallow p-ohmic base contact for GaAs based HBTs. It also has a number of advantages over the more common Au-Be contact to HBTs.

ACKNOWLEDGEMENTS

The authors acknowledge the assistance of L.J. Oster in MOCVD growth and J. Lothian in processing. The support and encouragement of A. S. Jordan and S. S. Pei is appreciated.

References

1. T. Ishibashi and Y. Yamauchi, "A possible near-ballistic conduction in an AlGaAs/GaAs HBT with a modified collector structure," IEEE Trans. Electron Devices, 35, 405 (1989).

2. K. Nagata, O. Nakajima, Y. Yamauchi, and Ishibashi, "A new self-aligned structure AlGaAs/GaAs HBT for high speed digital circuits," GaAs and Related Compounds 1985, Institute of Physics Conf. Ser. **79** 589 (1985).

3. P. M. Enquist, J. A. Hutchby, M. F. Chang, P. M. Asbeck, N. H. Sheng, and J. A. Higgins, Electronics Lett. **25**, 1124 (1989).

4. G. K. Reeves and H. B. Harrison, "Obtaining the specific contact resistance from transmission line model measurement," IEEE Electron Device Lett., vol. EDL-2, 111 (1982).

5. H. H. Berger, Solid-State Electron. **15**, 145 (1972).

6. S. M. Sze in "Physics of Semiconductor Devices," (J. Wiley, NY, 1981), pp. 255-311.

7. S. H. Rhoderick and R. H. Williams in "Metal-Semiconductor Contacts," (Clarendon Press, Oxford, 1988).

MULTILAYER METALLIZATION STRUCTURES IN A GATE ARRAY DEVICE SHOWN BY CROSS-SECTIONAL TRANSMISSION MICROSCOPY

S. F. Gong, H. T. G. Hentzell and A. Robertsson
Department of Physics and Measurement Technology, University of Linköping
S-581 83 Linköping, Sweden

ABSTRACT

A multilayer metallization structure in a gate array circuit has been investigated by cross-sectional transmission electron microscopy. Microstructures of thin films, interfaces, contacts, dislocations and step coverage are revealed. Good step coverage was observed when polyimide was used as an insulator between two metal layers.

I. INTRODUCTION

With the progressive increase of integration in integrated circuits, multilayer metallization has recently been developed in very large scale integrated circuits (VLSI), especially in gate arrays. Conventionally, silicon dioxide deposited by chemical vapor deposition (CVD) is used as insulators between different metal levels. In this work, however, we have investigated, using cross-sectional transmission electron microscopy (XTEM), a structure of multilayer metallization which uses polyimide as an insulator between two metal layers. Since polyimide can be spun onto a wafer from liquid, the topography of a polyimide film is expected to be smoother than that of a SiO_2 film deposited by CVD. Moreover, during plasma etching of a polyimide film, the slope of via hole edges can be adjusted by etching parameters in a chamber [1]. Thus, good step coverage may be expected when polyimide is used as insulating layers in multilayer metallizations.

II. EXPERIMENTAL DETAILS
A. Process of the multilayer metallization

Before metallization, wafers were fabricated with a standard processing technique for 2-μm complementary-metal-oxide-semiconductor (CMOS) integrated circuits. The overall layout of the gate array chip is a repeated arrangement of transistor cells and routings. Transistor cells are a group of identical sites which consist of p- and n-channel MOS transistors. Fig. 1 shows a cell in the circuit, in which polycrystalline silicon (poly-Si) and second metal layer lie perpendicular to the first metal layer. Via holes are round and

488

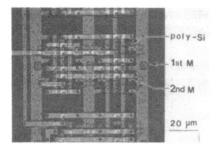

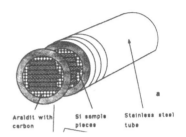

Fig. 1 *Cell in the gate array in which the lateral view of different metal levels and via holes can be seen*

Fig. 2 *Tube holding Si block with Araldite and carbon*

have a diameter of 3 μm. The first metal layer was deposited by sputtering from a target consisting of an Al and Cu alloy (4 % Cu). After patterning , liquid polyimide (PIQ-13) was spun onto the wafer. Then the layer was baked for polymerization. Reactive ion etching in a O_2^+ plasma was employed to open via holes in the polyimide layer. The second metal layer was deposited by sputtering from a target consisting of an Al and Si alloy (1 % Si). after patterning, the wafer was heat-treated in an atmosphere of H_2 + Ar (1% H_2).

B. Preparation of XTEM samples

Detailed description of preparing XTEM samples has been reported previously [2, 3]. Here we only summarize the main procedures.

Designated areas in a chip were cut under an optical microscope using a scriber. These pieces were thinned from the backside and then glued together by araldite (Ciba-Geigy AT1). The sample block was mounted into a stainless steel tube of 2 mm in inner diameter and 3 mm in outer diameter, as shown in Fig. 2. The sample block and the tube were glued together by the same araldite, but some carbon powder was added to the araldite in order to obtain good electric and heat conduction between the sample block and the tube, and to slow down the sputtering rate of the araldite during ion milling. The tube was then sliced into thin discs using a thin diamond saw. A disc at a designated location was selected and mechanically thinned down to less than 50 μm and then ion-milled by two Ar+ guns in a chamber for about 15 hours, until it is thin enough for TEM examination.

III. EXPERIMENTAL RESULTS

Fig. 3 shows a cross-sectional overview of the multilayer metallization, in which one sees, from the bottom to the top, the Si substrate, field SiO_2 (480 nm), poly-Si (410 nm), CVD SiO_2 (750 nm), first metal layer (Al+4%Cu, 570

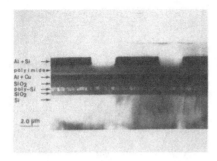

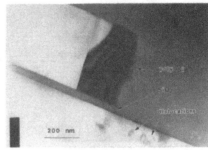

Fig. 3 *Cross-sectional overview of multilayer metallisation*   **Fig. 4** *Transistor structure*

nm), polyimide (1030 nm) and second metal layer (Al+1%Si, 960 nm). Fig. 4
shows a transistor structure, where one sees a poly-Si gate (410 nm), a gate
oxide layer (41 nm) and a source (or a drain) area. Dislocations caused by ion
implantation in the source (or drain) area can be seen. This indicates that a
high-dose ion implantation causes lattice damage, which is difficult to anneal
out.

Fig. 5 shows a contact between the first metal layer and the Si wafer.
Since the insulator between the first metal layer and the Si wafer is CVD SiO$_2$
(750 nm), the edge of the window is sharp. As a result, the thickness (180
nm) on the wall is only one-third of that (570 nm) on the SiO$_2$ layer. This is an
example of inadequate step coverage at sharp window edges.

A stacked contact between the Si substrate and the first and the second
metal layers is shown in Fig. 6. Although the via hole is very deep (1.8 µm), the
step coverage of the second metal layer is still rather good. This is due to the
planarized edge of via holes in the polyimide film. The planarization of
polyimide is also shown in Fig. 7, where the second metal layer lies rather
smoothly on the polyimide film.

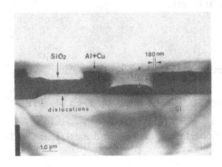

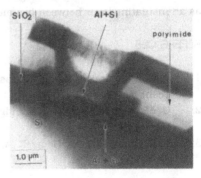

Fig. 5 *Contact between first metal layer and Si substrate*   **Fig. 6** *Stacked contact between Si substrate and first and second metal layers*

Fig. 7 *Polyimide film spun on from liquid phase showing uniform coverage of second metal layer*

IV. DISCUSSION

Good step coverage is shown by XTEM, when polyimide is used as an insulator between two metal layers. This is due to the facts that, first, spinning liquid polyimide on a wafer results naturally in a planarized film, which covers the rough topography of an already-processed wafer. Second, a sloped edge of via holes in polyimide can be achieved when proper etching parameters are chosen during reactive ion etching. This technique has successfully been used in a multilayer metallization process which was developed for Swedish Gate Array Inc. by the Center of Technology Transfer at Linköping Institute of Technology.

V. SUMMARY

XTEM has been employed to show a multilayer metallization structure in a gate array circuit. Inadequate step coverage at CVD SiO_2 windows without planarization has been shown. Dislocations in the source and drain areas have been observed. Good step coverage has been shown when polyimide was used as an insulating layer between two metal layers.

REFERENCES

1. J. E. Heidenreich III, J. R. Paraszczak, M. Moisan, and G. Sauve, J. Vac. Technol. B5, 347 (1987).
2. U. Helmersson and J.-E. Sundgren, J. Electr. Microsc. Tech. 4, 361 (1986).
3. S. F. Gong, H. T. G. Hentzell, A. Robertsson, and G. Radnoczi, IEE-G 137, 53 (1990).

ACCEPTOR DELTA-DOPING FOR SCHOTTKY BARRIER ENHANCEMENT ON n-TYPE GaAs

S. J. PEARTON, F. REN, C. R. ABERNATHY, A. KATZ, W. S. HOBSON, S. N. G. CHU AND J. KOVALCHICK
AT&T Bell Laboratories, Murray Hill, NJ 07974

ABSTRACT

The incorporation of thin C- or Zn-doped layers under metal Schottky contacts on n-type GaAs can lead to significant enhancements in the effective barrier height. A single C δ-doped layer (p = 1.3×10^{20} cm^{-3}) within 100 Å of the surface leads to a barrier height of ~0.9 eV, a significant increase over the value for a control sample (~0.75 eV). The use of two sequential δ-doped layers can lead either to a further enhancement in barrier height, or a decrease depending on whether these layers are fully depleted at zero applied bias. The temperature dependence of current conduction in barrier-enhanced diodes was measured. Both the ideality factor and breakdown voltage degrade with increasing temperature. Zinc δ-doping in a similar fashion produces barrier heights of 0.81 eV for one spike and 0.95 eV for two spikes.

INTRODUCTION

Metal contacts to n-type GaAs generally have Schottky barrier heights (Φ_B) of 0.7 - 0.8 eV due to Fermi level pinning by extrinsic surface states.[1] This is a disadvantage in many device applications where larger barrier heights would allow fabrication of digital circuits with increased noise margins and less stringent requirement on the threshold voltage uniformity of individual devices.[2,3] Crowell[4] and Shannon[5,6] first proposed increasing the barrier height on Si by tailoring the impurity profile in the depletion region near the metal-semiconductor interface. This was achieved using a thin, counter-doped, fully depleted layer under the metal contact. Significant barrier height enhancements have been demonstrated on n-type GaAs using both epitaxially grown p$^+$ layers[7] and low-energy Be ion implantation.[2] The increase in barrier height has been modeled by a number of authors using an electrostatic approach in which the introduction of acceptors under the Schottky contact leads to an increase in the potential barrier to thermionic emission of the conduction electrons in the GaAs.[2,5,7-9] The negative space charge in the fully depleted p$^+$ region is identified as the cause of the increase in effective barrier height (Figure 1 of Ref. 7 shows a band diagram for such a structure.) From these electrostatic calculations it is clear that the doping level in the p$^+$ layer, its thickness, and its distance from the metal-semiconductor interface all play important roles in determining the degree of barrier height enhancement.

Two significant features related to these issues have recently become apparent. The first is the discovery of the extremely high p-type doping levels (p > 10^{20} cm^{-3}) obtained using carbon derived from the trimethylgallium (TMG) growth precursor.[10,11] The key point about C in GaAs is its extremely low diffusion coefficient (<10^{-16} cm^2 s^{-1} at 950°C). The second important feature is the realization of atomic planar doping, which is also known as delta (δ) doping.[12] This represents the ultimate spatial control over the placement of dopants within a semiconductor structure. Based on these advances in crystal growth technology it is expected that one can examine the limits of Schottky barrier height modifications using doped interfacial layers.

Structures for these experiments were grown by two different methods utilizing two different dopants, C and Zn. All of the epitaxial layers were deposited on n$^+$-GaAs substrates (n = 2×10^{18} cm^{-3}). In the first growth method, approximately 0.5 µm of n-type GaAs (n = 1×10^{17} cm^{-3}) was deposited in a Varian gas source Gen II metalorganic molecular beam epitaxy (MOMBE) system using triethylgallium (TEG) and As$_4$. Planar, C-doped layers with full width at half maximum of ~50 Å were grown by switching from TEG to TMG and then back again. Single δ-doped layers were placed 100 Å from the surface, while with double δ-doped spikes the layers were placed at 50 and 250 Å, respectively, from the surface. In the second growth method similar structures were grown using metalorganic chemical vapor deposition (MOCVD) in an atmospheric pressure, vertical geometry reactor. The source chemicals were TMG and AsH$_3$, while diethylzinc (DEZn) was used for δ doping using the conventional growth interruption method to achieve planar doping.

Figure 1 shows a cross-sectional transmission electron micrograph from a double C δ-doped structure grown by MOMBE. The C-doped layers are visible because of the strain introduced by the very high C concentrations ($\sim 1.5 \times 10^{20}$ cm^{-3}) within the layers. The width of the spike nearest the surface is 46 Å, while the deeper one is 56 Å wide. The atomic profile of C within this sample was obtained using Cs ion sputtering in a PHI 6300 secondary-ion mass spectrometry (SIMS) system.

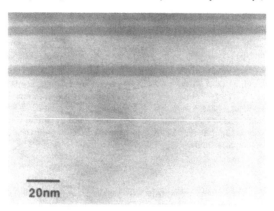

Figure 1.   Cross-sectional TEM micrograph from GaAs containing two C δ-doped layers (visible as the darker layers) near the surface.

The Cs$^+$ ion beam energy was varied from 1 - 3 keV in order to investigate possible collisional mixing effects on broadening of the δ-doping layers. The best results were obtained using 2 keV Cs$^+$ ions. Atmospheric carbon contamination is always present on GaAs samples - this was identified by monitoring the Ga-0 signal in the near-surface region, and subtracting the thickness over which this was present from the total carbon profile. The net effect of this procedure is to remove the carbon signal associated with contamination on the native oxide, while leaving the carbon signal due to the δ-doping itself. A SIMS concentration profile from which the surface carbon has been removed is also shown in Figure 2.

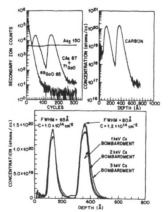

Figure 2.   SIMS atomic profiles from a double C δ-doped GaAs sample plotted as a function of Cs$^+$ ion beam sputter energy (bottom) and after removal of the C signal due to surface atmospheric contamination (top right) from the raw data (top left).

After completion of the epitaxial growth, ohmic contacts were fabricated to the rear face of the samples by electron beam evaporation of AuGeNi layer with eutectic composition, followed by alloying at 420°C for 30 sec under a flowing 90% N_2:10% H_2 ambient. The front face of the samples was deposited with circular (200 μm φ) Ti/Pt/Au Schottky contacts. The current voltage characteristics were measured as a function of temperature in the range of 20–180°C, using a temperature controlled probe station and a Hewlett-Packard parameter analyzer. The Schottky barrier height (ϕ_B) values were determined by the I-V measurements.

The current-voltage (I-V) characteristics in both reverse and forward direction are shown for one set of the C- δ-doped diodes with two C- doped spikes, measured at 40°C, in Figure 3. These measured data were fitted to the ideal diode equation:

$$J = A^{**} T^2 \exp(\phi_B/kT) \exp[(qV/nkT) - 1]$$ (1)

where J is the current density, A^{**} is the effective Richardson constant (4.4 Å $cm^{-2}k^{-2}$), T is the absolute temperature, k is the Boltzman's contact, n is the diode ideality factor, q is the electron charge and ϕ_B is the barrier height.

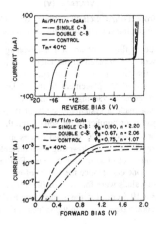

Figure 3.   I-V reverse and forward characteristics at 40°C of Au/Pt/Ti Schottky diodes on *n*-GaAs, with no C δ-doped spikes (control), one C δ-doped spike (single) and two C δ-doped spikes (double).

In these samples there is a significant increase in reverse breakdown voltage with the addition of the C- δ-doped spikes. With the very high *p*-type doping levels, the δ-doped layers are not fully depleted when two are present, and a transition is seen from barrier enhancement (reduced thermionic emission for one spike) to barrier lowering (increased tunneling with two spikes). In some samples utilizing C- δ-doping we observed very large ϕ_b values (> 1.0 eV) when two spikes were incorporated. In these cases the spikes were apparently fully depleted.

The use of Zn δ-doped layers also led to large barrier enhancements (ϕ_b = 0.81 eV, n = 1.03 for one spike and ϕ_B = 0.95 eV, n = 1.49 for two spikes) These values increased to ϕ_b = 1.03 eV, n = 1.48 and ϕ_b = 1.04 eV, n = 1.46 respectively for 900°C, 10 sec RTA. This was most likely due to diffusion of the Zn to form thicker p$^+$ layers. The carbon δ-doping was more thermally stable showing no change in ϕ_b with RTA at 900°C. The increase in reverse breakdown voltage in Zn δ-doped diodes is shown in the I-V data of Figure 4.

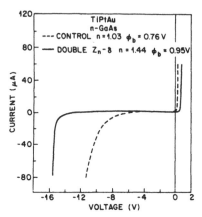

Figure 4.   I-V characteristics from TiPtAu Schottky diodes on n-GaAs. One of the samples also incorporates two Zn δ-doping spikes within 400 Å of the surface.

A comment might be made at this point about the advantage of the δ doping approach to Schottky barrier height enhancement as compared to the alternative methods of direct[2] or recoil[12] implantation of p-type dopants or the growth of a p-type epitaxial layer on n-type material.[7] Delta doping represents the ultimate control of the placement of the p+ layer for barrier enhancement and affords the maximum flexibility in terms of varying the number of doping spikes and their positions relative to each other. We have not made a systematic study of the effects of varying the δ-doping distance from the surface or the spacing or number of the spikes, but these will all have an influence on the resultant ϕ_B enhancement. For some of the C δ-doped diodes in which we observed both barrier enhancement (one fully depleted spike) or barrier lowering (two undepleted spikes) the barrier heights were also obtained from an Arrhenius plot of saturation current. Figure 5 shows the plot of I_s/T^2 as a function of $1/T$ on a semilog scale. One can deduce the values of ϕ_B from the slope of these plots and A_{eff} from the intercept. The former values were found to be 0.73, 0.91 and 0.65 eV and the latter were 6.15×10^{-3}, 3.19×10^{-4} and 5.68×10^{-5} cm^2 for the samples with zero, single and two C δ-doped spikes. These values of ϕ_b are in good agreement with the values derived from the slope of the $I_F - V_F$ curves.

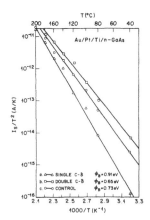

Figure 5. I_s/T^2 vs $1/T$ of Au/Pt/Ti diodes on n-GaAs containing no C δ-doped spikes (control), one C δ-doped spike (single) and two C δ-doped spikes (double).

SUMMARY

In summary we have shown that δ-doping with carbon or zinc can lead to increased Schottky barrier heights on n-GaAs. If the δ-doped spikes are not fully depleted, a p-n junction exists and the barrier can actually be lowered through an increase in tunneling.

REFERENCES

[1]   W. E. Spicer, I. Lindau, P.R. Skeath, C.Y. Su, and P. Thye, Phys. Rev. Lett. **44**, 420 (1980).

[2]   W. E. Stanchina, M. D. Clark, K. V. Vaidyanathan, R. A. Jullens, and C. R. Crowell, J. Electrochem. Soc. **134**, 967 (1987).

[3]   K. L. Priddy, D. R. Kitchen, J. A. Grzyb, C. W. Litton, T. S. Henderson, C.-K. Peng, W. F. Kopp, and H. Morkoc, IEEE Electron Devices **ED-34**, 175 (1987).

[4]   C. R. Crowell, J. Vac. Sci. Technol. **11**, 951 (1974).

[5]   J. M. Shannon, Solid-State Electron. **19**, 537 (1976).

[6]   J. M. Shannon, Appl. Phys. Lett. **25**, 75 (1974).

[7]   S. J. Eglash, S. Pan, D. Mo, W. E. Spicer, and D. M. Collins, Jpn. J. Appl. Phys. **22**, 431 (1983).

[8]   A. Van der Ziel, Solid-State Electron. **20**, 269 (1977).

[9]   S. J. Eglash, N. Newman, S. Pan, K. Shenai, W. E. Spicer, and D. M. Collins, J. Appl. Phys. **61**, 5159 (1987).

[10] K. Saito, E. Tokumitsu, T. Akasuka, M. Miyauchi, T. Yamada, M. K. Konagai, and K. Takahashi, J. Appl. Phys. **64**, 3975 (1988).

[11] N. Kobayashi, T. Makimoto, and Y. Horikoshi, Appl. Phys. Lett. **50** 1435 (1987).

[12] M. Eizenberg, A. C. Callegari, D. K. Sadana, H. J. Hovel, and T. N. Jackson, Appl. Phys. Lett. **54**, 1696 (1989).

METAL / SILICIDE INTERACTIONS IN THE Ti-Co-Si SYSTEM.

M.SETTON *, J. VAN DER SPIEGEL ** , R. MADAR #, O. THOMAS #
* University of Pennsylvania, Department of Materials Science, Laboratory for Research on the Strucutre of Matter, 3231 Walnut Street, Philadelphia PA 19104.
** Department of Electrical Engineering, 200 S 33 rd Street, Philadelphia PA 19104.
# Laboratoire des Matériaux et du Génie Physique, ENSPG, B. P. 46, Domaine Universitaire, Saint Martin d'Hères, France.

ABSTRACT

Confirming the results obtained for Ti-Co bilayers on Si and in accordance with the phase diagram, the high temperature formation and stability of three ternary silicides $Ti_{0.75}Co_{0.25}Si_2$ (T phase), TiCoSi (E) and $Ti_4Co_4Si_7$ (V) is reported. The Si rich T phase grows for Ti / $CoSi_2$ and Co / $TiSi_2$ structures. The tetragonal V compound is obtained by annealing Ti (400 Å) / Co (250 Å) / Si (900 Å) / SiO_2 whereas orthorhombic TiCoSi is prepared using Ti / CoSi / Si_3N_4 samples.

INTRODUCTION

As new stable low resistivity metallization materials for self aligned technology, $CoSi_2$ and $TiSi_2$ appear to be the most promising ones. For both systems, an impressive amount of information is available regarding resistivity, microstructure, phase sequence, formation temperature and stability with respect to dopants.

Processing schemes using two steps annealing for $TiSi_2$ and direct silicidation for Co have been demonstrated for submicron use [1,2]. There are however fundamental differences between the two systems. The nucleation of the metastable C 49 compound $TiSi_2$, isomorphous to $ZrSi_2$ results in a heavily faulted microstructure on (010) planes with a displacement of 1/2(a+c) [3]. Depending on the impurity content, film thickness and grain size, the transformation from C 49 to C 54 face centered structure occurs between 700 and 800°C [4]. During the formation of Ti silicide, Si is the main diffusing species.

The growth of cobalt silicide starts at about 325°C by the diffusion of Co atoms into the substrate to form Co_2Si. In the temperature range 325-530°C cubic monosilicide CoSi also grows but Si atomic flux is dominant. Above 550°C, $CoSi_2$ is detected and by 625°C, the reaction $CoSi+Si \longrightarrow CoSi_2$ reaches completion. This occurs mostly through the motion of metal atoms.

Since the two systems are quite different, it might be interesting to examine the characteristics and properties of the ternary Ti-Co-Si system. We have previously examined the reactions taking place for metal bilayers on Si [5].The goal of this study is to investigate the evolution of metal overlayers in contact with a silicide and select routes to form several ternary compounds. By controlling the amount of the various species we shall report the preparation of three phases and analyze the data refering to the equilibrium phase diagram.

SAMPLE PREPARATION AND ANALYSIS

All samples are annealed by Rapid Thermal Processing in vacuum bettter than 2×10^{-6} Torr. Substrate preparation, sputter deposition parameters and analytical means have been detailed in an earlier study [5]. After formation of the silicide layer, samples are sputter etched prior to metal deposition. No interface contamination is detected by AES. The bulk sample of $Ti_{0.75}Co_{0.25}Si_2$ is prepared by arc melting under Ar in a water cooled Cu crucible.

RESULTS

a-Co (150 Å) / C 54 $TiSi_2$ (1700 Å).

The as deposited film consists of polycrystalline $TiSi_2$ with an α-Co layer on top having an 002 preferential orientation. In the X-ray spectrum,the spacing d=2.046 Å indicates some strain in the metallic overlayer. As the samples are annealed. Co atoms diffuse into the Ti silicide. For

30 seconds at 750°C the front appears 700 Å deep into TiSi$_2$ (Figure 1a). Interestingly, AES shows a thin layer of pure CoSi$_2$ on top.

Raising the temperature to 925°C results in complete homogeneization of the layer and an AES profile reveals a uniform profile (Figure 1b). Table I indicates that the major phase is a ternary phase Ti$_{0.75}$Co$_{0.25}$Si$_2$ (T phase). The main peaks match those of a bulk sample of the exact same stoichiometry and correspond to what had been found for bilayers.Two minor peaks for CoSi$_2$ also detected in the thin film sample probably correspond to a surface layer.

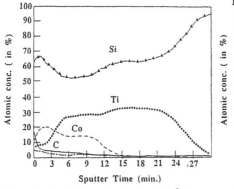 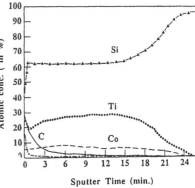

Figure 1:   1a: AES profile for Co (150 Å) /   1b: Ternary obtained for samples annealed
                TiSi$_2$ (1700 Å) / Si annealed at 750 °C.   at 925°C.

Table I

d thin film (Å)	d bulk	I bulk (%)	‖	d thin film	d bulk	I bulk (%)
3.47	3.466	40	‖	2.034	2.031	20
3.097	3.093	13	‖	1.89	CoSi$_2$ 1.88	
2.976	2.98	20	‖	1.784	1.784	25
2.249	2.248	100	‖	1.607	1.612	10
2.166	2.161	26	‖	1.197	1.198	9
2.094	2.09	17	‖	1.091	1.091	8

Table I: X-ray diffraction results obtained for a thin film of Co /TiSi$_2$ annealed at 1050°C and comparison with lattice spacings of a reference arc melted bulk sample.

b-Ti (1200 Å) / CoSi$_2$ (875 Å).

In this case also, the Ti layer is strongly 002 oriented. For a sample annealed at 480°C there is very little change apparent in the X-ray diffraction analysis. However at 600°C, we see the main peak associated with the ternary phase and by 900°C the Ti has been totally consumed. The corresponding Auger profile is shown in Figure 2 and the X-ray spectrum again indicates a single phase .

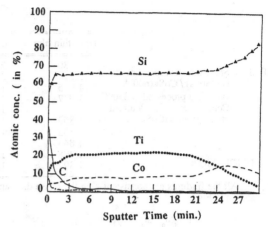

Figure 2: Ti / CoSi$_2$ processed for 30 seconds at 900°C.

c-Ti (400 Å) / Co (250 Å) / Si (900 Å) / SiO$_2$

An annealing at 850°C transforms the structure into Ti$_4$Co$_4$Si$_7$ (V phase) (see Figure 3). We are mainly concerned about the equilibrium high temperature state and did not determine the exact sequence of formation, but Co atoms probably diffuse both towards Si and Ti as previously observed for metallic bilayers. The structure of this phase is tetragonal, space group I 4/mmm, similar to Ti$_4$Ni$_4$Si$_7$ and with cell parameters a=12.513 Å and c=4.934 Å [6].

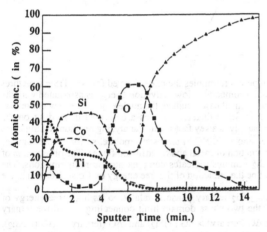

Figure 3: Ti (400Å) / Co (250 Å)/ Si (900 Å) Si on SiO$_2$ annealed at 850°C.

d-Ti (480 Å) / CoSi (600 Å) / Si$_3$N$_4$.

Since the Co/Si ratio is fixed, the equilibrium moves along a line connecting CoSi and pureTi until we reach the equiatomic compound TiCoSi (E phase). The X-ray pattern (Table II) contains most of the main peaks determined by a computer calculation based on the structure established by Spiegel et al . This phase is orthorhombic, as most of the M1M2Si compounds with parameters a=6.107 Å, b=3.72 Å, c=6.936 Å [6]. The symmetry is that of Co$_2$Si .

Table II

<table>
<tr><td></td><td>d obs. thin film</td><td>d calc.</td><td>I calc. (%)</td><td>h k l</td></tr>
<tr><td>Table II: Spectrum corresponding to
Ti (480 Å) / CoSi (600 Å)
on Si3N4 processed at 850°C.
Comparison with calculated
spectrum of TiCoSi.</td><td>2.330
2.21
1.965
1.947</td><td>2.342
2.23
1.963
1.953
1.951</td><td>100
94
50
35
14</td><td>1 1 2
2 1 1
0 1 3
3 0 1
2 1 2</td></tr>
<tr><td></td><td>1.859
1.852
1.843</td><td>1.869
1.860
1.843</td><td>23
52
13</td><td>1 1 3
0 2 0
2 0 3</td></tr>
<tr><td>DISCUSSION</td><td>1.408
1.397</td><td>1.41
1.397</td><td>15
8</td><td>1 2 3
4 0 2</td></tr>
</table>

All the findings are consistent with the phase diagram obtained by Markiv et al. [7] and shown in Figure 4.

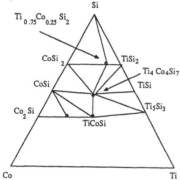

Figure 4: Partial phase diagram at 800°C according to Y. Markiv [7]

The formation of a ternary compound resembles the one observed for Ni / TiSi2 structures [8] and therefore one may think that a comparable low activation energy is associated with the grain boundary diffusion of Co atoms. In all cases studied, the nucleation of the new phase may take place at triple grain boundaries.Since different phases are formed depending on the availability of Si, we see that the Si supply is a key factor when analyzing reactions taking place during processing.At temperatures above 500°C Si atoms are mobile, so for all the samples studied one can not rule out a combination of metal and Si diffusion fluxes. The observation of $CoSi_2$ on the surface for Co / $TiSi_2$ samples indicates complex atomic redistributions. The driving force for Si diffusion could be the reduction of the free energy by Co silicidation at the interface.

The stability of the $Ti_{0.75}Co_{0.25}Si_2$ ternary phase is related to its high free energy of formation. In the phase diagram, the two phase domain limit connecting Si and the ternary silicide signify that we must have ΔG (ternary) > ΔG ($TiSi_2$) and ΔG (ternary) > ΔG ($CoSi_2$). Therefore one may place a lower boundary at 32 kcal/ metal atom. Such a high stability is rather uncommon for M1-M2-Si systems and is related to electronic factors as well as large differences in electronegativities and metallic radii. Ni and Co are chemically alike regarding silicide growth, therefore the isomorphism and comparable formation temperature of the 111 and 447 compounds are not surprizing. However an (M1,M2)Si2 compound is absent in the Ti-Ni-Si system where $Ti_4Ni_4Si_7$ is the Si richest ternary phase.

Different reaction temperatures for the various metal/silicide structures may be related to different bonding energies. For Ti-Si it has been estimated to be 1.68 eV [9] compared to a Si-Si value of 1.83 eV.

We can compare the results with those obtained for the most studied ternary systems containing Al. In Al / $CoSi_2$ [10] and Al / $TiSi_2$ [11] reactions, Al atoms first react to form Co or Ti aluminides.Ternary silicides grow at high temperatures for Al-Ti-Si and Al-Ni-Si. Even though intermetallic compound formation was observed for metallic bilayers, it is apparently unfavorable to free the silicided metal at low temperature, unlike for the Al-metal-Si systems .

Undoubtedly all the phases prepared in this study could be prepared also by reacting alloys on Si.

Ternary compounds were formed via two routes: the diffusion of Si into an intermetallic compound and through the diffusion of metal into a silicide.To our knowledge, this is the first report of the formation of $Ti_4Co_4Si_7$ and $TiCoSi$ for thin films.

CONCLUSION

We had previously studied the reactions of Ti-Co bilayers with Si and observed the formation of a ternary phase $Ti_{0.75}Co_{0.25}Si_2$. The excess metal with respect to this stoichiometry is consumed as disilicide which either segregates at the ternary / Si interface for Co or remains mixed for the low mobility Ti atoms.

This same ternary phase is obtained through the annealing of Ti / $CoSi_2$ and Co / $TiSi_2$ structures and diffusion of the metal at grain boundaries. In contrast to the Al / silicide systems, no intermetallic phases are detected when Ni or Co react with an underlying silicide.

Other compounds of the phase diagram can also be prepared via thin film reactions. When limiting the Si supply by a nitride layer in Ti / CoSi / Si_3N_4 samples, the equilibrium is reached when an equimolar orthorhombic $TiCoSi$ phase has grown.Finally, controlling the amount of the three components in Ti / Co / Si / SiO_2 structures allows the preparation of tetragonal $Ti_4Co_4Si_7$ silicide.

The Ti-Co-Si system is particularly interesting since the stoichiometries of the ternary phases can be compared to binary ones: $TiCoSi$ is similar to M_2Si, $Ti_4Co_4Si_7$ is close to monosilicide MSi and finally $Ti_{0.75}Co_{0.25}Si_2$ has the metal / Si ratio of an MSi_2 compound.

Acknowledgements: This work was funded by a PYI -NSF award ECS-8352021 and through the Laboratory for Research on the Structure of Matter by an NSF-MRL grant No 88198895.

References.

1-L. Van den hove, R. Wolters, K. Maex, R. de Keersmaecker and G. Declerck, J. Vac. Sci. Technol., B 4 (6), 1358, (1986).

2-R. D. Thompson, H. Takai, P. A. Psaras and K. N. Tu, J. Appl. Phys., 61 (2), 540, (1987).

3- A. Bourret, F. M. d'Heurle, F. K. Le Goues, A. Charai, J. Appl. Phys., 67 (1), 241, (1990).

4-I. J. M. M. Raaijmakers, A. H. van Ommen, A. H. Reader, J. Appl. Phys., 65 (10), 3896, (1989).

5- M. Setton and J. Van der Spiegel, Appl. Surf. Sci., 38 , 62, (1989).

6-W. Jeitschko, A. G. Jordan and P. A. Beck, Trans. AIME, 245, 335, (1969)

7-V. Ya Markiv, E. K. Gladyshevskii and T. I. Fedoruk, Russian Met., 3, 118, (1966).

8-M. Setton, J. Van der Spiegel and B. Rothman, J. Mater. Res., 4 (5), 1218, (1989).

9-K. Suguro, Y. Nakasaki, T. Yoshii, Appl. Surf. Sci., 41, 277, (1989).

10-G. J. Van Gurp, J. L. C. Daams, A. van Oostrom, L. M. Augustus and Y. Tamminga, J. Appl. Phys., 50 (11), 6915, (1979).

11-C. Y. Ting and M. Wittmer, J. Appl. Phys., 54 (2), 937, (1983).

FORMATION OF CoSi$_2$-SHALLOW JUNCTIONS BY ION BEAM MIXING AND RAPID THERMAL ANNEALING

L. NIEWÖHNER AND D. DEPTA
Institut für Halbleitertechnologie, Universität Hannover, Appelstr. 11A, D-3000 Hannover 1, Federal Republic of Germany

ABSTRACT

Formation of CoSi$_2$ using the technique of ion implantation through metal (ITM) and subsequent appropriate rapid thermal annealing is described. Silicide morphology is investigated by SEM and TEM. SIMS and RBS are used to determine dopant distribution and junction depth. Self-aligned CoSi$_2$/n$^+$p diodes produced in this technique are presented.

INTRODUCTION

The formation of shallow, low resistance junctions is an important requirement in VLSI in order to minimize short channel effects in MOSFETs. Low resistivity materials such as refractory metal silicides have been investigated to be used for contacts and interconnects. Two very promising metal silicides are TiSi$_2$ and CoSi$_2$ because of their low resistivities and their ability to form self-aligned (salicide) structures /1/.

Compared to TiSi$_2$, CoSi$_2$ has some important advantages. CoSi$_2$ can be formed in a self-aligned process using a single RTA-step at a relatively low temperature ($\approx$ 700°C) instead of self-aligned TiSi$_2$, which requires a formation temperature of more than 850°C in a two step RTA /2,3/. The avoidance of bridging is easier with Co, because Co is not as reactive as Ti and there is no reaction with SiO$_2$ observed at these temperatures. However, substrate cleaning prior to Co deposition is very important, because Co does not reduce any interfacial oxide as Ti does.

Several methods to form self-aligned CoSi$_2$ contacts have been reported. This paper discusses a method using the effect of ion beam mixing for silicidation which promotes a better metal-silicon reaction starting at lower temperatures compared to conventional silicidation methods /2/. Using the arsenic dopant ions as the mixing species and subsequent rapid thermal annealing (RTA) to form the silicide allows to perform both the n$^+$p junction and the low resistance CoSi$_2$ contact simultaneously.

EXPERIMENTAL

In order to investigate the fundamental characteristics of CoSi$_2$ formation by dopant implantation through metal the process was first applied on nonstructured Si -wafers. Thin films of 28.4 nm Co were deposited on p-type Si<100> substrates (resistivity 1 - 10 Ωcm) by electron beam evaporation to form 100 nm thick CoSi$_2$-layers. Prior to metal deposition the Si-wafers were not cleaned in dilute HF, i.e. the native oxide was not removed in order to study the influence of a residual oxide on the formation of CoSi$_2$ by ITM. Then As$^+$ was implanted through the Co metal (200 keV, 1.E15 cm^{-2}). Previous calculations with the simulation program TRIM87 /4/ were used to determine a suitable implantation energy which causes a maximum of ion induced damage at the interface Co/Si and a dopant maximum within the later formed CoSi$_2$-layer. After implantation the samples were RTA-processed for silicide formation and dopant activation over a range of temperatures and times. A diagram showing the CoSi$_2$-shallow junction formation process, using the ITM technique, is presented in fig. 1.

In a second run $CoSi_2/n^+p$ diodes were processed using this technique. SiO_2 patterned p-type Si<100> wafers were used to define the diode areas. The implantation and silicidation steps were carried out under the same conditions as mentioned before. Then unreacted Co was removed by selective etching for 2 min in HCl:30% H_2O_2 1:3 at room temperature. Finally a sputter deposited W layer was used for metallization.

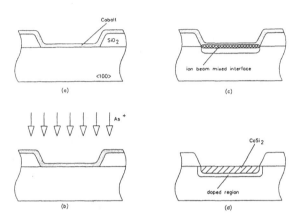

Fig. 1: Schematic diagram of the cobalt-silicidation process by ITM; (a) deposition of Co; (b) dopant implantation through the metal layer; (c) contact region after implantation, showing the effect of ion beam mixing; (d) RTA for silicide formation and dopant activation and selective etch of unreacted Co

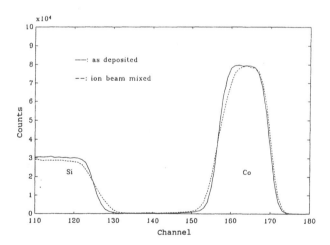

Fig. 2: RBS spectra of the 28 nm Cobalt/nat.oxide/silicon system; (–) as deposited; (--) after arsenic implantation (implantation energy 200 keV; dose 1E15 cm^{-2})

Fig. 3: Sheet resistance of the cobaltsilicide film as a function of RTA-temperature. The annealing time was 15 sec. The observed silicide phases are indicated.

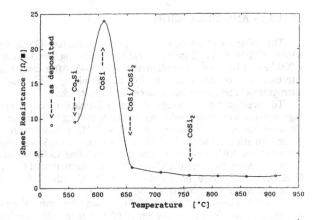

Fig. 4: SEM micrograph of a CoSi$_2$-surface annealed at 760°C for 15 sec.

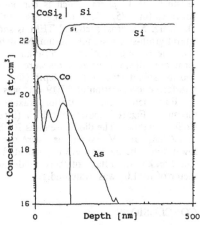

Fig. 6: SIMS depth profile of a CoSi$_2$-junction, showing the distribution of As after RTA at 760°C for 15 sec.

Fig. 5: Cross-sectional TEM micrograph of a CoSi$_2$-film annealed at 760 °C for 15 sec.

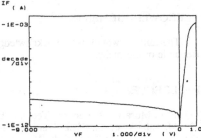

Fig. 7: I-V characteristic of a CoSi$_2$/n$^+$p diode formed by ITM; RTA at 760°C for 15 sec.

RESULTS AND DISCUSSION

The influence of ion beam mixing on the interface Co/Si was studied by means of RBS and TEM. Figure 2 shows typical backscattering spectra of an unimplanted (—) and a 200 keV, $1.E15$ cm^2 As-implanted sample (---). After implantation a smeared Co/Si interface was observed due to ion beam mixing. This was also visible in TEM micrographs. The native oxide layer has been cracked.

To investigate the kinetics of silicide formation, unpatterned wafers were used. Figure 3 shows the sheet resistance versus annealing temperature. As deposited 28.4 nm Co films had a sheet resistance of 9 $\Omega/\blacksquare$. XRD analyses were used to determine the phase formation. At about 550°C a small amount of Co_2Si was observed. With increasing temperatures XRD showed a change to crystalline CoSi, which corresponds to a high sheet resistance. Between 650°C and 750°C the formation of $CoSi_2$ starts and both phases CoSi and $CoSi_2$ were observed. Above 750°C only $CoSi_2$ was detected and sheet resistance remained constant at about 1.7 $\Omega/\blacksquare$.

The thickness of the $CoSi_2$ films was determined by cross-sectional TEM micrographs to 104 nm, so the obtained minimum sheet resistance corresponds to a resistivity of 17 $\mu\Omega cm$, which is comparable to the 18 $\mu\Omega cm$ reported for furnace-annealed films /5/. SEM micrographs of these films show a smooth silicide surface (fig. 4). In fig. 5 a cross-sectional TEM is shown. The silicide interface is sharp and the lateral grain-size is about 100 nm. No defects are visible in the Si substrate remaining from the As implantation, because all of the damaged Si was consumed during silicide formation. SIMS measurements were carried out to determine the dopant distribution and to evaluate the junction depth (fig. 6). A junction depth of about 150 nm and an interface-concentration of $5.E19$ cm^3 was achieved.

For junction characterization, leakage currents and breakdown characteristics were measured. Figure 7 shows typical I-V (foreward and reverse) curves of an ITM-formed $CoSi_2/n^+p$ diode. The diode area was $1000 \times 1000 \mu m^2$. Leakage current densities of about 10 nA/cm^2 at 5 V reverse bias were obtained. The actual leakage current should be lower since the diodes were unpassivated and edge leakage current could be possible. The breakdown voltage of these diodes was above 60 V. A foreward-current ideality factor of n = 1.06 was evaluated.

CONCLUSIONS

These results show the possibility of forming shallow $CoSi_2$-junctions without the need of a carefully executed surface cleaning process prior to Co deposition. $CoSi_2/n^+p$ diodes produced by ITM technology show good electrical properties.

ACKNOWLEDGEMENTS

The authors would like to acknowledge the support of the BMFT of the Federal Republic of Germany.

REFERENCES

/1/ S. P. Murarka; J. Vac. sci. Technol. B 4(6); 1986; p. 1325
/2/ M. Tabasky et. al.; IEEE Trans. Electron Devices; Vol. ED-34, no. 3; 1987; p. 548
/3/ N. F. Stogdale; Proc. of the 18th ESSDERC; Montpellier; 1988; p.C4-195
/4/ J. F. Ziegler, J. P. Biersack, U. Littmark; "The Stopping and Range of Ions in Solids"; Pergamon Press; New York; 1984
/5/ S. P. Murarka; "Silicides for VLSI Application"; Academic Press; New York; 1983

A NOVEL TECHNIQUE FOR DETECTING DEFECTS IN ULSI METALLIZED SYSTEMS

C.A. Pico, T. Aton, R.J. Gale, M. Bennett-Lilley, M. Harward, S. Mahant-Shetti*, T. Weaver
Texas Instruments
Semiconductor Design and Processing Center
* MOS Memory Semiconductor Group
P.O. Box 655012, MS 944, Dallas, TX 75265

Defect location and identification in the metallization systems of ultra-large-scale integrated (ULSI) devices is becoming increasingly important because of the demands of high device density. An understanding of the sources of defects is crucial to the fabrication of submicron devices. Typically, defect identification is accomplished by electrically testing large metal combs and serpents followed by scanning electron microscopy (SEM) investigation. In order to identify metallization defects quickly, we have fabricated a novel device that bypasses the need to electrically probe. This technique utilizes voltage contrast imaging in-situ SEM to locate defects typically found during ULSI device fabrication. While voltage contrast imaging has been used to locate defects in conjuction with externally applied voltages [1], our technique takes advantage of the SEM's own beam as a charging source and makes close resolution (<0.1μm) inspection unnecessary until appropriate. In this way, defects can be located and identified using ~1/20th the time presently required.

Voltage contrast imaging in the SEM takes advantage of the energy differences that exist between secondary electrons emmitted from differently biased regions. The kinetic energies of electrons emmitted from identical material but at different potentials will be shifted correspondingly   . Using an electron detector with a fixed energy window, the CRT image will show contrasting regions of light and dark representing identical material but biased differently. In general, the sample voltage bias is induced to the sample externally, such as a 5V battery, to the SEM. In our work we use the current injected from the SEM's own electron beam to charge the regions. The actual amount of voltage bias to the injected region depends on the net e- flux in, secondary e- emmitted out, and current drain paths (e.g. metal conduction). If the inspected structure is electrically insolated, then the voltage bias can be controlled by the beam voltage and current during SEM inspect. In this way, a structure can be tested as to whether it is electrically isolated or shorted to ground.

Thin metal films were deposited onto oxidized Si substrates. The films were then patterned and etched to produce metal defect-sensitive structures (Fig.1). The metal coatings used in this study are Al, Al/chemical vapor deposited (CVD)-W/Ti:W, Ti:W, and CVD-W/Ti:W. The metal defect-sensitive structures contain, primarily, three structures: 1) a 2mm long finger structure (equivalent to a continuity test), 2) a simlar finger structure with isolated metal islands (2μmX140μm) inlaid between them

Mat. Res. Soc. Symp. Proc. Vol. 181. ©1990 Materials Research Society

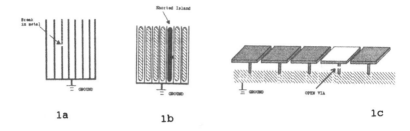

1a   1b   1c

Figure 1- Schematic drawings of scanning electron test structures
to detect a) high resistance circuits, b) metal-to-metal
shorts, and c) open vias as seen using SEM.

(equivalent to a shorting structure), and 3) metal dots (2x2 μm)
connected to ground through a 0.8μm metal via. The metal fingers
are connected to ground and do not charge during SEM inspection.
The islands, on the other hand, are electrically isolated and
charge during SEM inspection. If a metal finger has a
discontinuity (i.e. missing metal), the end of the finger will
charge during SEM inspection and appear contrasted to those
without a discontinuity. A short between an insulated metal
island and finger will cause the electrically conncected island to
appear contrasted to other islands. Lastly, a poor via will
result in a contrasted 2μmx2μm dot relative to the others.

SEM inspection of the tip of a metal continuity structure
(Fig.2a) indicates several metal lines are electrically defective.
The voltage contrasted image was used to trace down the line to
rapidly locate the origins of line corrosion (Fig.2b). In this
case, the metal was still conductive but had an increased
electrical resistance because of the corrosion. Metal shorts were
located at low magnification (X33) using the island structures
(Fig.3a). Once an island is located at low magnification, the
island was inspected closely. It took approximately 30 seconds to
find this tenth micron W filament (Fig.3b) that was causing an
island to lead short. High electrical resistance vias are clearly
identified (Fig.4) by this technique.

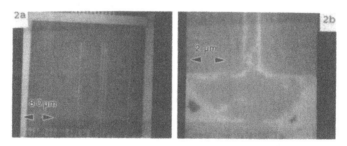

Figure 2- Scanning electron micrographs of a) the tips of an Al
finger structure indicating that five lines have a high
resistance path and b) the source (corroded Al) of the
high resistance.

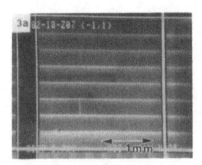
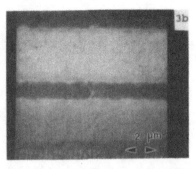

Figure 3- Scanning electron micrograph of Al/CVD-W/Ti:W patterned
structures equivalent to >20 cm length of parallel combs
showing a) the short to be located within a 200 μm length
and b) the 0.1 μm tungsten filament causing the short.

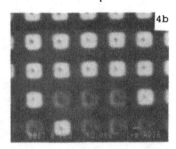

Figure 4- Scanning electron micrographs of a) a via test structure
and b) individual failing vias.

We find this method to locate and identify metal defects is
ideal. With this technique submicron defects can be located while
viewing an area greater than 3mmX3mm in the SEM. In addition,
this technique has the capability of being included as an
automated defect monitor during routine ULSI device fabrication
while reducing the area needed for parametric electrical testing.

BIBLIOGRAPHY

1) Phillip E. Russell in Materials Characterization, edited by
Nathan Cheung and Marc-A. Nicolet (Materials Research Soc.,
Pittsburgh, PA 1986), V. 69, pp. 15-22

REACTIVE ION BEAM ASSISTED EVAPORATION OF TiN FILMS FOR USE AS DIFFUSION
BARRIERS IN GaAs MMICs

T. E. KAZIOR and R. C. BROOKS*
Raytheon Company, Research Division, 131 Spring St., Lexington, Ma. 02173
*Present Address: Westinghouse Electric Corporation, Electronic Systems Group, P.O.Box
1521, Baltimore, MD 21203

ABSTRACT

We report on the development of an evaporated diffusion barrier for incorporation
into the metallizations currently used in GaAs MMICs. Results indicated that TiN films
formed by reactive ion beam assisted evaporation - the simultaneous evaporation of metal
atoms and bombardment with reactive ions - were as stable as TiN films formed by reactive
sputter deposition and far superior to films formed by reactive evaporation. In addition the
formation of these films is compatible with photoresist lift-off techniques. Chemical and
electrical analysis has shown that these films are effective barriers against Au, Ga and As
diffusion under extensive (up to 1500 hrs) elevated temperature storage (280°C). TiN-Au
gate electrodes whose electrical characteristics are comparable to conventional Ti-Pt-Au
gates have also been fabricated.

INTRODUCTION

TiN have been used extensively in Si IC technology as diffusion barriers, low barrier
Schottky diodes, and MOS gate electrodes [1,2]. In recent years, TiN films have shown
promise in GaAs technology as temperature-stable Schottky barriers for use in self aligned
gate processes [3-6] and as diffusion barriers in ohmic contacts [7,8]. However, TiN films
are traditionally either reactively sputter deposited [1,4,5,7,8], reactively evaporated at
elevated temperatures (300-600°C) [1,3] or reactively evaporated at room temperature
followed by a high temperature anneal (300-1100°C) [2,9]. Therefore, in general, the
use of these films is not compatible with conventional GaAs wafer process sequences.

The goal of this work has been to develop a reactive evaporation technique for
depositing TiN films that is compatible with photoresist lift-off techniques and is readily
inserted into existing GaAs process sequences. Conditions were established for depositing
TiN films and evaluating their use as diffusion barriers in the standard metallizations used
in GaAs MMICs. TiN-Au structures were evaluated for use as temperature-stable Schottky
barriers for GaAs MESFETs.

EXPERIMENT

All evaporations were performed in a Veeco evaporator equipped with a
Commonwealth Scientific Kaufman type ion source. The base pressure just prior to
evaporation was 5×10^{-7} torr. Two techniques for reactively evaporating TiN were
evaluated. One technique - Reactive Evaporation (RE) - entailed evaporation of Ti in a
nitrogen background (bleeding nitrogen into the vacuum chamber during deposition - the
nitrogen partial pressure was 2×10^{-4} Torr). The other technique was Reactive Ion Beam
Assisted Evaporation (RIBAE) which entailed simultaneous evaporation of Ti and
bombardment of the sample with N ions from the ion source (nitrogen partial pressure:
2×10^{-4} Torr; beam voltage: 250V; beam current: 25ma). The first technique is based on
the expectation that during deposition N becomes incorporated in the film. In the second case
ionized, energetic N reacts with the Ti on the surface of the wafer and becomes incorporated
in the film. In addition the extra energy imparted to the deposited atoms can cause bulk
atomic displacements and surface atom migration, resulting in improved film properties,
such as higher density and reduced stress.

Mat. Res. Soc. Symp. Proc. Vol. 181. ©1990 Materials Research Society

The TiN barrier layers were deposited using each technique at various thicknesses (50nm, 100nm, and 150nm) and evaporation rates (1.5 and 0.5nm/s) followed by the deposition of 10nm Ti and 400nm of Au. The purpose of the Ti layer was to ensure good adhesion of the Au layer to the barrier layer. We found that without this Ti layer the Au would delaminate. The evaporation cycles were monitored with a Residual Gas Analyzer (RGA). To ensure good adhesion the GaAs surface was subjected to an insitu Argon ion clean prior to evaporation of the barrier layer film. The TiN - Au structures were deposited on patterned resist and lifted off.

The effectiveness of the TiN films as barriers against the diffusion of Ga, As, and Au was determined electrically by monitoring the change in resistance of test structures subjected to temperature cycling. The test structures consisted of a van der Pauw pattern to measure the sheet resistance of the structure and a meander pattern (a pattern approx 11.5mm long and 5μm wide) to measure the resistance of a long narrow line. To provide good statistics, the patterns were replicated ≈ 20 times on each sample. Changes in the resistance of these structures sould be an accurate indication of the failure of the TiN diffusion barrier and the intermixing of Au, Ga, and As. To support the electrical data, Auger sputter profiling was performed on selected samples.

The test structures were measured as-deposited and after temperature storage at 280°C for extended time. Selected structures were also subjected to the ohmic contact alloy or sinter cycle (≈440°C, 30s). For comparison, evaporated Ti-Au and Ti-Pt-Au structures and reactively sputtered TiN-evaporated Au structures were fabricated. The reactively sputtered TiN films were used as the control samples for these experiments.

FETs were prepared on ion implanted active layers (135keV, $N_{peak} \approx 3 \times 10^{17}/cm^3$) formed on semi-insulating substrates. Sintered Pd-Ge-Au was used for ohmic contacts. The gate electrodes were nominally 0.5μm and defined by e-beam lithography. Following gate lithography and recess, the wafers were split in half; one half received TiN-Au gates, the other half received standard Ti-Pt-Au gates. Following post gate dc electrical test, the wafers were passivated with 200nm SiN. The DC electrical tests included measurement of the FET I-V characteristics and the gate I-V characteristics of $50 \times 150 \mu m^2$ FatFETs. The gate diode ideality factor and barrier height were calculated from the FATFET forward I-V characteristics.

RESULTS and DISCUSSION

Results of the experiments to determine optimal deposition conditions are summarized in Figure 1. Films deposited by RE at both deposition rates and films deposited by RIBAE at the high (1.5nm/s) deposition rate led to failure of the diffusion barrier at 280°C. While the data in Figure 1 is for 50nm thick TiN films, similar results were obtained for 100 and 150nm thick TiN films. In all cases, the test patterns showed a factor of four increase in resistance after ≈100 hours at 280°C, probablt due to significant intermixing of the Au and GaAs. We believe that these films failed as diffusion barriers due to insufficient N incorporation. (In these cases the films might be either Nitrogen doped Ti, Ti_2N or non-stoichiometric (or N poor) TiN. Further analysis is required to determine the exact composition of these films.) For comparison, the resistance of the Ti-Au control test structure which contains no N increased by a factor of two in only two hours.

IFigures 1 & 2 indicate that samples with 50nm thick TiN films deposited by RIBAE and a 0.5nm/s deposition rate have undergone >1500 hours of 280°C temperature storage with no change in the resistance of the meander pattern suggesting no intermixing of the Au and GaAs. We believe that these films proved to be successful diffusion barriers between Au and GaAs due to greater N incorporation. For comparison are data for evaporated Ti-Pt-Au (100nm-100nm-400nm) as well as reactively sputtered deposited TiN-evaporated Ti-Au (50nm-10nm-400nm) is included. It should be noted that the resistance of the samples with the Ti-Pt-Au structures (a standard GaAs MESFET gate metallization) increased by ≈30% in 1000 hours. This data suggests that the components of the Ti-Pt-Au structures

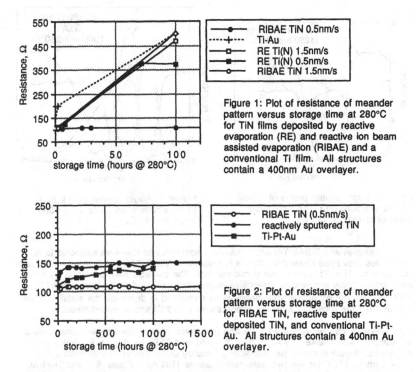

Figure 1: Plot of resistance of meander pattern versus storage time at 280°C for TiN films deposited by reactive evaporation (RE) and reactive ion beam assisted evaporation (RIBAE) and a conventional Ti film. All structures contain a 400nm Au overlayer.

Figure 2: Plot of resistance of meander pattern versus storage time at 280°C for RIBAE TiN, reactive sputter deposited TiN, and conventional Ti-Pt-Au. All structures contain a 400nm Au overlayer.

are slowly interdiffusing with time whereas the TiN-Au structures show no intermixing. For the times and temperatures used in this study the data in Figures 1 & 2 indiacte that the TiN films deposited by RIBAE are as stable barriers against Au, Ga and As diffusion as TiN deposited by reactive sputter deposition.

Auger analysis was performed on RIBAE samples as-deposited and after 1250 hours of temperature storage (see Figure 3). The Auger analysis also shows no intermixing of the Au and GaAs. This consistent with the electrical data and indicates a successful diffusion barrier has been formed. (The GaAs-TiN interface remains sharp and there is very little Au in the TiN film.) Several features should be noted: 1) The N line (384eV) overlaps one of the lines of the Ti doublet (387eV). Therefore, what is presented is the signal from the 418eV Ti line and the signal from the combined N and Ti line. 2) There appears to be a large amount of Oxygen in the TiN layer (exact % Oxygen is not known). The Oxygen is probably being gettered from the vacuum chamber chamber during deposition. However, there does not appear to be Oxygen present in Ti films evaporated in this system. Therefore, at present, neither the origin of the Oxygen nor its contribution to the effectiveness of the diffusion barrier are known. It is possible that the Oxygen is contributing by further stuffing grain boundaries or by the formation of TiO_x (in which case the films are actually of the form $Ti(O,N)$ or TiO_xN_y).

In contrast, Ti-Pt-Au structures showed significant intermixing after 1000 hours at 280°C (see Figure 4). Ga has diffused into the Pt (possibly forming a PtGa compound), Au has diffused to the Ti-GaAs interface, and Ti is diffusing into the GaAs (possibly forming a TiAs compound). (A more thorough chemical analysis is necessary to identify the presence of these compounds. Also, note that there is no detectable Oxygen in these films.) Comparison with Figure 3 shows that 50nm of TiN is a more effective barrier against Au and Ga diffusion than 100nm of Ti and 100nm of Pt.

514

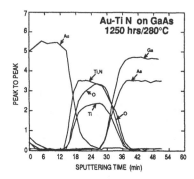

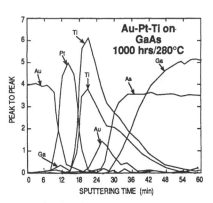

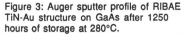

Figure 3: Auger sputter profile of RIBAE TiN-Au structure on GaAs after 1250 hours of storage at 280°C.

Figure 4: Auger sputter profile of Ti-Pt-Au structure on GaAs after 1000 hours at 280°C.

Samples with RIBAE TiN and a 0.5nm/s deposition rate that were subjected to the 440°C, 30s alloy cycle showed no change in resistance. The RIBAE TiN films were used as a substitute for Ti in Pt-Ti-Au p-type ohmic contacts. The contacts have proven to be more temperature stable and process insensitive, especially when used as the base contact for HBTs[10]. Further experimentation needs to be performed to determine the stability of the RIBAE TiN films subjected to higher temperatures (>440°C) and extended times.

Figure 5 shows typical FET I-V and gte-source reverse breakdown characteristics for samples with RIBAE TiN-Au (Figure 5 - left) and Ti-Pt-Au (Figure 5 - right) gate electrodes. Figure 6 shows the gate-source Schottky diode forward characteristics of a 50x150μm FATFET for the two gate metallizations (TiN-Au - Figure 6 - left; Ti-Pt-Au - Figure 6 - right). Figure 5 indicates that there is no significant difference in the FET DC electrical characteristics of the two metals. On the other hand, the gate diode forward I-V characteristics yield a slightly lower barrier height (Ø=0.61 for TiN-Au versus 0.66 for Ti-Pt-Au) and ideality factor (n=1.12 for TiN-Au versus 1.17 for Ti-Pt-Au) for the TiN-Au gates. These structures have been passivated with 200nm SiN and are presently undergoing 300°C temperature storage to begin to determine the stability of the TiN Schottky barriers. Preliminary data indicates no significant change in the FET or gate diode I-V characteristics of devices with TiN Schottky barriers after 500 hours at 300°C. In contrast, the FET I-V characteristics of devices with Ti-Pt-Au Schottky barriers are beginning to degrade. We believe that this is due to intermixing of the Ti-Pt-Au and GaAs, leading to intermediate compound formation. This result is consistent with the electrical and Auger measurements described above, in which the resistance increased by ≈20% in 500 hours at 280°C, and significant diffusion of Ti, Ga, and Au (and the possible formation of TiAs and PtGa compounds) was observed. These test are continuing; upon completion a more thorough chemical analysis will be performed. Companion samples are being completed to chips for bias-temperature stress reliability measurements.

Summary

The results of this work can be summarized as follows:
1)   Electrical measurements, as well as Auger analysis, have shown that the RIBAE TiN films are effective barriers against diffusion of Au, Ga, and As for the times and temperatures used in this study.
2)   TiN films deposited by RIBAE appear to have properties comparable to TiN films deposited by reactive sputter deposition.
3)   TiN films deposited by RIBAE are compatible with photoresist lift-off techniques.

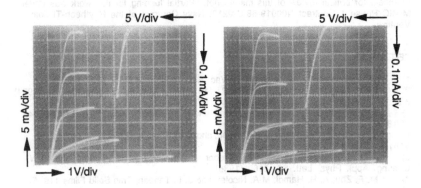

Figure 5: FET I-V and gate-source reverse breakdown characteristics of 100µm wide GaAs MESFETs with TiN-Au (left) and conventional Ti-Pt-Au (right) gate electrode.

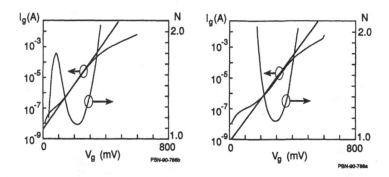

Figure 6: Gate diode forward I-V characteristics for 50x150µm² FATFETs with TiN-Au (left) and conventional Ti-Pt-Au (right) gate electrode.

4) TiN films deposited by RIBAE have been successfully used as diffusion barriers in GaAs ohmic contacts.

5) FETs formed with TiN gates have electrical characteristics comparable to conventional Ti-Pt-Au gates.

6) Preliminary data suggests that TiN-Au gate electrodes are more temperature stable than Ti-Pt-Au gate electrodes.

In conclusion, TiN films deposited by RIBAE show promise for use as diffusion barriers and Schottky barriers in GaAs MMICs. Additional electrical, chemical, and structural analysis needs to be performed to more thoroughly and extensively evaluate the temperature stability of these films and to determine their applicability as high temperature Schottky barriers for GaAs MESFET applications.

516

Acknowledgements: The authors would like to thank A. Bertrand and J. Pagliuca for wafer processing, M. Bush for dc electrical measurements, S. Hein for Auger analysis, and S. Shanfield for critical review of this manuscript. Partial funding for this work was under MIMIC Phase 1 Contract N00019-88-C-0218, Navy Contract of the Raytheon-TI Joint Venture.

REFERENCES

1)   see, for example, M. Wittmer, J. Vac. Sci. Technol. **A3(4)**, 1797 (1985).
2)   C. Y. Ting, J. Vac. Sci. Technol. **21**, 14 (1982); Thin Solid Films **119**, 11 (1984).
3)   J. R. Waldrop, Appl. Phys. Lett. **43**, 87 (1983).
4)   L. C. Zhang, S. K. Cheung, C. L. Liang, and N. W. Cheung, Appl. Phys. Lett. **50**, 445 (1987).
5)   L. C. Zhang, C. L. Liang, S. K. Cheung, and N. W. Cheung, J. Vac. Sci. Technol.. **B 5(6)**, 1716 (1987).
6)   J. Ding, Z. Lillental-Weber, E. R. Weber, J. Washburn, R. M. Fourkas, and N. W. Cheung, Appl. Phys. Lett. **52**, 2160 (1988).
7)   M. F. Zhu, A. H. Hamdi, M-A. Nicolet, and J. L. Tandon, Thin Solid Films **119**, 5 (1984).
8)   R. D. Remba, I. Suni, and M-A. Nicolet, IEEE Electron Device Lett. **6**, 437 (1985).
9)   M. Repeta, L. Dignard-Bailey, J. F. Currie, J. L. Brebner, and K. Barla, J. Appl. Phys. **63**, 2769 (1988).
10)   G. S. Jackson, E. Tong, P. Saledas, T. E. Kazior and R. C. Brooks, presented at the 1990 MRS Spring Meeting, San Francisco, CA.

DEPOSITION OF TUNGSTEN SILICIDE BARRIER LAYERS AND TUNGSTEN IN RECTANGULAR VIAS

Timothy S. Cale, Gregory B. Raupp and Manoj K. Jain
Department of Chemical, Bio & Materials Engineering and Center for Solid
State Electronics Research, Arizona State University, Tempe, AZ 85287-6006

ABSTRACT

Diffusion-reaction analysis of a two step process in which a tungsten
silicide barrier layer is deposited in a rectangular trench by low pressure
dichlorosilane reduction of tungsten hexafluoride followed by a complete
tungsten fill by low pressure hydrogen reduction of tungsten hexafluoride
reveals that high step coverage and high deposition rate can be readily
achieved with logical selection of process parameters. This fact, coupled
with the potential for accomplishing these deposition steps in the same
single wafer reactor, suggests that this two step process may offer a high
throughput alternative to blanket tungsten deposition by silane reduction.

INTRODUCTION

There is great industrial interest in adopting tungsten for plug and via
fill applications in VLSI/ULSI multilevel metallization schemes. Indeed, a
SEMATECH Phase I goal is to utilize tungsten as metal one (WAMO) in the
SEMATECH test vehicles [1]. Commercial processes have been developed to
deposit blanket tungsten films through low pressure reduction of tungsten
hexafluoride by hydrogen or silane. For WAMO applications, silane reduction
has been preferred over hydrogen reduction because the silane process
minimizes formation of wormholes or lateral encroachment in the underlying
silicon. On the other hand, step coverage of tungsten deposited by silane
reduction tend to be lower than tungsten deposited by hydrogen reduction [2].
Thus, there is renewed interest in tungsten deposition processes based on
hydrogen reduction, particularly for depositions in features of high aspect
ratios [2,3]. To prevent deleterious interactions between the reaction gases
and/or the growing film and the underlying substrate during tungsten
deposition, it is common practice to predeposit a barrier layer. Barrier
layers such as titanium nitride formed using physical vapor deposition (PVD)
have inherently poor step coverage, which effectively limits the final step
coverage by the tungsten film.

In this paper we suggest a process in which a high step coverage
tungsten silicide barrier layer is deposited through dichlorosilane (DCS)
reduction of WF_6, followed by tungsten deposition via hydrogen reduction of
WF_6. A particular advantage for low pressure chemical vapor deposition
(LPCVD) of the barrier layer is that both the barrier and the tungsten can be
sequentially deposited in the same single wafer reactor without interim
exposure to air. Step coverage and film compositional uniformity are
critical production aspects of the silicide deposition. The silicide must
also provide sufficient protection for the silicon substrate while
maintaining low contact resistance. The primary production concern regarding
the subsequent tungsten deposition is maximization of throughput subject to a
given step coverage constraint. Previous theoretical and experimental work
suggests that good step coverage can be obtained using the hydrogen reduction
system even at very high deposition rates. The presence of the silicide
barrier layer allows higher wafer temperatures during the hydrogen reduction
step, which increases deposition rate and enhances wafer throughput.

The continuum diffusion-reaction model framework [4-8] provides a
unified understanding of step coverage phenomena in low pressure chemical
vapor deposition (LPCVD) for a variety of deposition chemistries, and is used
in this paper to simulate the two step barrier/fill process. We have
presented a theoretical and experimental investigation of the effects of key
process variables on step coverage in tungsten silicide films deposited

Mat. Res. Soc. Symp. Proc. Vol. 181. ©1990 Materials Research Society

through DCS reduction of WF_6 [8]. We have also presented a theoretical study [9], supported by experimental results from other laboratories [2,3], which predicts that the step coverage of tungsten films deposited by hydrogen reduction of WF_6 can be very good even at high deposition rates.

To focus the discussion, we limit our attention to deposition of 100 nm thick tungsten silicide followed by tungsten deposition to fill 1 μm wide by 1 μm deep and 1 μm wide by 3 μm deep rectangular trenches. We show that with proper selection of process conditions, high step coverage and high throughput are not necessarily incompatible production goals.

DIFFUSION-REACTION MODEL

The assumptions and development of the diffusion-reaction model framework are described in detail elsewhere [4-8]. For multicomponent reaction systems a time-dependent one-dimensional material balance incorporating simultaneous Knudsen diffusion and r heterogeneous chemical reactions is written for each of c components over a differential volume element of the feature. If these material balance equations are made dimensionless by referencing variables to conditions at the feature mouth at time zero, the resulting equations reveal that step coverage and film uniformity are controlled by r+c *model* parameters. A step coverage modulus Φ_j arises from each of the r deposition reactions, and for rectangular trenches takes the form

$$\Phi_j = \frac{2 \, H_o^2 \, R_{jo}}{W_o D_{Ko} C_{ro}} \qquad (1)$$

where H is the trench depth, R_j is the specific molar reaction rate for reaction j, W is the trench width, D_K is the Knudsen diffusivity defined by feature geometry, C_r is the concentration of the reactant arbitrarily chosen as the reference species, and the subscript o refers to conditions at the feature mouth at time zero. These moduli represent ratios of characteristic reaction to characteristic Knudsen diffusion rates. Hence the larger the numerical values of Φ_j, the greater the effective reactant concentration gradients and hence deposition nonuniformity within the feature, and the poorer the ultimate step coverage. The other model parameters are feature aspect ratio and the c-1 reactant concentration ratios at the feature mouth.

Given established reaction stoichiometries, reaction rate expressions and feature geometry, the diffusion-reaction formulation contains no adjustable *model* parameters. With fixed initial geometry and deposition chemistry, the independent *process* parameters are wafer temperature and reactant concentrations at the feature mouth. The model parameters reveal the complex interrelationship between process variables and can be used to develop guidelines for process parameter adjustment to improve step coverage and/or throughput.

TUNGSTEN SILICIDE BARRIER DEPOSITION

For WSi_x deposition in the dichlorosilane process it appears that at least three reactions depositing W_5Si_3, WSi_2 and Si contribute to the apparent deposition rate and film composition [8]. Total observed deposition rate (thickness/time) and film composition x are then a function of the relative contributions of the three deposition reactions.

Our diffusion reaction model [8] of the tungsten silicide system has three step coverage moduli. The other *model* parameters are the DCS to WF_6 partial pressure ratio λ and the initial trench aspect ratio. The model predicts that at the high λ values typical of commercial operation, the WF_6 concentration profiles within the feature control step coverage. Increasing wafer temperature or aspect ratio and decreasing WF_6 concentration at the feature mouth will lead to more severe WF_6 gradients and will therefore degrade step coverage; this predicted behavior is in qualitative agreement with experimental measurements [8].

TUNGSTEN DEPOSITION

The stoichiometry and tungsten deposition rate equation for hydrogen reduction of WF_6 have been determined by McConica and Krishnamani [10]. For this chemistry, two *model* parameters control tungsten step coverage: (i) the step coverage modulus, corresponding to the single heterogeneous reaction, and (ii) the H_2 to WF_6 partial pressure ratio λ. For high λ values typical of commercial operation, the WF_6 concentration gradient in the feature controls the tungsten step coverage. We have shown by "availability analysis" that the optimum partial pressure ratio λ_{opt} is [9]:

$$\lambda_{opt} = (\nu_{H_2}/\nu_{WF_6}) \, (M_{H_2}/M_{WF_6})^{1/2} \approx 0.25 \quad , \qquad (2)$$

where ν_i is the generalized stoichiometric coefficient of i in the hydrogen reduction reaction and M_i is the molecular weight. If the reactants are present in this ratio at the feature mouth, they will be available in the same ratio throughout the feature. For values of λ greater than λ_{opt}, WF_6 will become more severely depleted down the depth of the feature than the co-reactant H_2, and the ultimate step coverage will be less than optimum. Use of the optimum ratio should allow significantly higher deposition rates without degrading step coverage.

RESULTS, DISCUSSION & CONCLUSIONS

Table 1 summarizes the process conditions chosen for illustration of the concepts presented in this paper. The conditions for silicide barrier layer deposition were chosen to balance deposition rate and step coverage within the film composition constraint. Tungsten deposition is carried out at conditions which yield high deposition rates, at both optimal and near commercial partial pressure ratios. In all cases feature fill is accomplished in less than 80 s total deposition time. In a single wafer reactor, additional time for purging between deposition steps, wafer loading and unloading, heating and cooling, and reactor cleaning and maintenance must be taken into account to fully evaluate throughput capabilities.

Table 2 summarizes the predicted step coverages for the two step process. For the silicide film and unity aspect ratio, step coverage is greater than 96%; for aspect ratio equal to three, step coverage is 89%, reflecting the more difficult nature of the deposition geometry. For the unity aspect ratio trench, step coverage at feature fill is high even for suboptimal $H_2:WF_6$ partial pressure ratios. However, for higher aspect ratio

Table 1
WSi_x Barrier / W Fill Process Parameters

	Silicide Barrier	Tungsten Fill
Temperature	763 K	763 K
DCS or H_2 pressure	0.17 torr	1.16 torr
WF_6 pressure	0.003 torr	4.72 or 0.116 torr
Deposition rate	190 nm/min	500 nm/min
Deposition time	31.5 s	48 s
Si:W ratio	2.02	-

Table 2
Step Coverages from WSi$_x$ Barrier / W Fill Process Simulation

Aspect Ratio	1 (1 μm x 1 μm)		3 (1 μm x 3 μm)	
Barrier Layer Step Coverage	96.6 %		89.2 %	
Tungsten Fill H$_2$:WF$_6$ Ratio	0.25	10	0.25	10
Overall Step Coverage at Feature Fill	98.1 %	95.5 %	95.8 %	84.3 %

features, adoption of the optimal ratio leads to a significantly greater final step coverage. At suboptimal ratios, high step coverages could be achieved but only at a significantly lower deposition rate. Our simulations suggest that such a penalty in SWR throughput is unnecessary.

ACKNOWLEDGEMENTS

Professors Raupp and Cale gratefully acknowledge the support of this work by the Semiconductor Research Corporation and the National Science Foundation.

REFERENCES

1. J. B. Stimmell, in Tungsten and Other Refractory Metals for VLSI Applications IV, C. M. McConica and R. A. Blewer, eds., MRS, Pittsburgh, PA, 1989, p. 1.
2. J. E. J. Schmitz, W. L. N. van der Sluis and A. H. Montree, in Tungsten and Other Advanced Metals for VLSI/ULSI Applications V, S. S. Wong and S. Furukawa, eds., MRS, Pittsburgh, PA, 1990, p. 117.
3. R. V. Joshi, E. Mehter, M. Chow, M. Ishaq, S. Kang, P. Geraghty and J. McInerney, in Tungsten and Other Advanced Metals for VLSI/ULSI Applications V, S. S. Wong and S. Furukawa, eds., MRS, Pittsburgh, PA, 1990, p. 157.
4. C. M. McConica and S. Churchill, in Tungsten and Other Refractory Metals for VLSI Applications III, V. A. Wells, ed., MRS, Pittsburgh, PA, 1988, p. 257.
5. G. B. Raupp and T. S. Cale, Chemistry of Materials 1, 207 (1989).
6. F. A. Shemansky, M. K. Jain, T. S. Cale and G. B. Raupp, MRS Symp. Ser. 146, D. Hodul, J. C. Gelpey, M. L. Green, and T. E. Seidel, eds., MRS, Pittsburgh, PA, 1989, p. 173.
7. S. Chatterjee and C. M. McConica, J. Electrochem. Soc. 137, 328 (1990).
8. G. B. Raupp, T. S. Cale, M. K. Jain, B. Rogers, and D. Srinivas, presented at the 8th Intl. Conf. on Thin Films, San Diego, CA, April 1990, and submitted to Thin Solid Films.
9. T. S. Cale, M. K. Jain, and G. B. Raupp, presented at the 8th Intl. Conf. on Thin Films, San Diego, CA, April 1990, and submitted to Thin Solid Films.
10. C. M. McConica and K. Krishnamani, J. Electrochem. Soc. 133, 2542 (1986).

PHASE STABILITY OF MOLYBDENUM- SILICON NITRIDE- SILICON MIS SCHOTTKY DIODE AT HIGH TEMPERATURES

Heungsoo Park, and C.R. Helms
Stanford Electronics Laboratory, Stanford University
Stanford, CA 94305

ABSTRACT

Previously our group[1,2] has demonstrated metal-thin insulator- silicon Schottky diode structures which allow the Si Schottky barrier height to be adjusted over nearly the full range of the silicon band gap by appropriate choice of insulator thickness and metal. However, previous attempts to achieve a structure with a high barrier height to p-type that is stable above 400C(using primarily Titanium) have failed. In this paper we report on results for a metal, Molybdenum, which has a stable tie line to SiO_2 and Si_3N_4 in metal-silicon-oxygen(nitrogen) ternary phase diagram which leads to a more stable system.

I. INTRODUCTION

The presence of a sufficiently thin(< 30Å) insulating layer at a metal-semiconductor interface has been known to result in a device that behaves electrically as a Schottky barrier.[1-6] This insulating layer is thin enough to allow current conduction via tunneling, which is necessary for the resulting device to act electrically as a Schottky barrier. However, the barrier height of such a metal-insulator-semiconductor(MIS) diode will generally differ from that obtained in the absence of an insulating layer. If an ultrathin insulating layer is present at the metal-semiconductor interface, the resulting barrier height shows a greater dependence on metal than the barrier height of an intimate contact diode.[7] Also in some cases, the barrier height has been shown to be linearly dependent on insulating layer thickness.[8] Therefore, it suggests that the barrier height of metal-thin insulator-silicon system would be more controllerble than that of metal-silicon intimate contact. Furthermore, high barrier height would be achievable in metal-thin insulator-silicon system, especially in p-type silicon Schottky diodes.

As the adjustment of its barrier height over relatively wide range is possible, a success of Si MIS Schottky diode as a device practically depends upon its thermal stability at high temperature. Thermodynamically, metal tends to react silicon and form a more stable metal silicides phase. Also, some metals like Ti have more strong tendency to form metal silicide than other metals, which results in the instability of metal-semiconductor intimate contact even at relatively low temperature.[9] In metal-thin insulator-semiconductor structure, insulating layers basically can act as a diffusion barrier, which generally improves the thermal stability of Si Schottky barrier diode. It can prevent possible interaction between metal and semiconductor which is likely to occur when they are in intimate contacts.

Another problem in metal-silicon intimate contact is the relatively small range of barrier heights that can be obtained using technologically acceptable metals. It has been demonstrated[1,2] that metal-thin insulator-silicon Schottky diode structures allow the barrier height to be adjusted over nearly the full range of the silicon band gap by appropriate choice of insulator thickness and metal. However, previous attempts to achieve a structure with a high barrier to p-type silicon that is stable above 400°C(using primarily Titanium) have failed. We suspect that this is due to the tendency of the metal to react through the insulator(SiO_2 or Si_3N_4) to form more stable oxide/metal silicide phases. Finding a stable system with high barrier height to p-type at high temperatures was the main focus of this work. In order to achieve this goal, it was necessary to find a metal which satisfies two requirements. One is the metallurgical requirement of phase stability with respect to silicon nitride, and the other is the electrical requirement of having a work function low enough to provide the high p type barrier. In this report, the phase stability of molybdenum-silicon nitride-silicon MIS Schottky diodes at high temperature(up to 600°C) will be discussed.

II. EXPERIMENTAL PROCEDURE

The samples for this study were prepared previously.[7] Nitride layers were grown by rapid thermal nitridation of silicon in pure NH_3 at temperatures ranging from 1000 C to 1250 C for times of 20 seconds. Nitridations were performed in an A.G.Associates Heat Pulse 2146[TM] rapid thermal annealing system heated by quartz lamps. Details of the sample processing and the materials properties of the nitride layers are described in detail elsewhere.[7] Diodes were formed on p-type substrates with Mo. To produce a range of nitride thicknesses, films were grown at varying temperatures, and in addition selected samples were subsequently etched in 10:1 solutions of NH_4F:HF.

Barrier height were measured at 10kHz to 1 kHz with a HP4277A[TM]LCZ bridge at room temperature. The dc electrical characteristics were measured using a HP4140B[TM] picoammeter at room temperature. Room-temperature measurements were performed in a shielded probe station. Annealing treatment of the samples was done in a tube furnace in an H_2 atmosphere for 30 minutes at 300 and 600C. Barrier heights were measured before and after annealing treatment mainly by C-V technique.

III. RESULTS AND DISCUSSION

In Figure 1, barrier height data is plotted versus nitride thickness before and after anneal treatment. These barrier heights were determined by C-V technique. The variation of C-V barrier heights across each sample was ± 0.03 V : the values reported are the averages of at least four diodes from each sample.   As expected from our early study [1,2], an almost

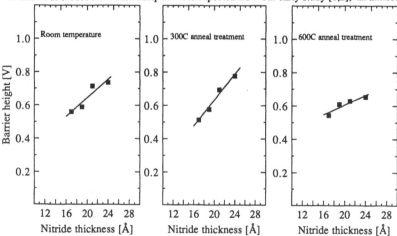

Figure 1 Barrier heights of molybdenum-silicon nitride-silicon MIS Schottky diodes versus nitride thickness (a) before anneal treatment and,(b) when 300C anneal treatment.was done, and (c) when 600C anneal treatment was done.

linear increase in barrier height with nitride thickness was found in p-type devices regardless of annealing treatment. This linear relationship between barrier height and nitride thickness suggests the presence of fixed charge near the silicon-silicon nitride interface.

To build a phase diagram from bulk thermodynamics has been known to be one of few ways to explain the reactivity of metal with other phase, especially silicon dioxide or silicon nitride, in thin-film reactions.[10] It is also useful to predict a possible system which is stable at high temperatures. In previous work, Ti-SiO$_2$-Si and Ti-Si$_3$N$_4$.-Si interfaces were shown to be relatively unstable with respect to thermal cycling.[1,2] In this study, although some reduction in barrier height was observed, much better thermal stability was obtained. In fact, since it is likely that the mechanism for the high barrier heights is at least in part associated with positive nitride fixed charge, it is possible that only minimal chemical reaction (if any) is occurring between the Mo and Si. This ,we believe, can be understood mainly with reference to the Metal-Insulator-Si phase diagrams.

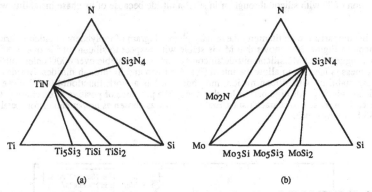

(a)   (b)

Figure 2. Calculated ternary phase diagrams for (a) Ti-Si-N, and (b) Mo-Si-N. T= 300-850°C

Some of the ternary phase diagrams for metal- silicon- nitrogen(or oxygen) system at certain temperature range are listed in Figure 2.

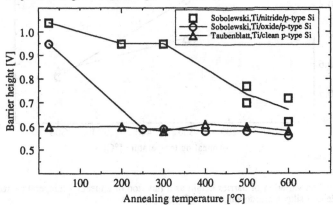

Figure 3. Barrier height as a function of annealing temperature for titanium contacts to p-type silicon for variety of surface pretreatments: (a) nitridized at 1150°C for 15 sec yielding a nitride layer 22Å thick[7]; (b) chemically oxidized, resulting in a 9Å oxide layer[9]; and (c) sputtering(followed by annealing) in ultra-vacuum to obtain an atomically clean surface.[9]

In the case of titanium-silicon-nitrogen(or oxygen) ternary system, it is obvious that titanium is unstable with respect to silicon nitride and thus it strongly tends to react with silicon nitride and make titanium silicides and titanium nitride. Previously our group has shown that Ti goes unstable with respect to silicon oxide and nitride as temperature increases up to 400°C by using Auger spectroscopy.[7] Figure 3 shows the variation of barrier height of Ti intimate contact diode and titanium-silicon oxide (or nitride)- silicon MIS diode with annealing temperature.

Titanium-silicon nitride-silicon system shows a large drop in its barrier height when annealing treatment was done at 300C, which was expected from the above thermodynamic consideration. One thing to be mentioned about here is that data for titanium-silicon nitride-silicon MIS diode having more thicker insulating layers were not available, so the decrease in barrier heights of titanium MIS diodes having less thin insulating layer might be smaller than those of titanium MIS diode having a nitride layer 22Å thick. But, it can be guessed from the ternary phase diagram of Ti-Si-N that silicon nitride can not basically prevent the reaction of Ti with silicon through or in silicon nitride because of its phase instability with Ti.

In comparison with titanium, the ternary phase diagram of molybdenum-silicon nitride-silicon in Figure 2(b) shows that Mo is stable with respect to silicon nitride over 600°C. This suggests that Mo/silicon nitride/silicon system may be stable over 600C unless nitride thickness is too thin to allow the interdiffusion of Mo and Si through nitrides. Besides its phase stability with silicon nitride, molybdenum has a work function(4.6 eV) close to silicon, which yields high barrier heights due to the presence of positive fixed charge. On the basis of these two characteristics, molybdenum was chosen as proper choice for metal of MIS Schottky diode.

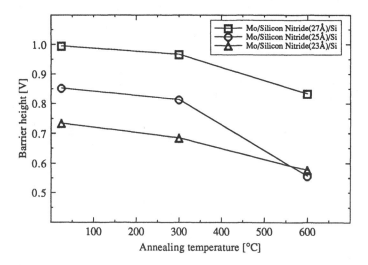

Figure 4 The variation of barrier height as a function of annealing temperature for some molybdenum-silicon nitride-silicon MIS Schottky diodes.

The thermal stability of the barrier height at higher temperature is shown in Figure 4 where the variation of C-V barrier height for some molybdenum-silicon nitride-silicon MIS diodes is plotted as a function of annealing temperature. For the case of molybdenum-silicon nitride-silicon MIS diodes, the 300C anneal treatment did not change barrier height of each sample much. This may imply that thin nitride less than 30Å can act as a diffusion barrier at 300C which does not allow interdiffusion of Mo and Si to each other to make silicides. At least, it seems that there was no significant change in the MIS structure by the 300C anneal treatment. But, there is a relatively large drop in barrier heights of diodes with thinner insulating layer when 600C anneal treatment was done. Even though overall decrease in barrier height of each sample was found when anneal treatment was done, diode having less thin insulating layer showed an interesting result. For a diode having nitride thickness of 27Å, its barrier height did not decrease so much as other diodes did even when 600°C annealing treatment was done. This means that an little increase in nitride thickness, still thin enough to allow current conduction via tunneling, may result in a diode with relatively high phase stability.

IV. CONCLUSION

Phase diagram calculated from bulk thermodynamics can provide basic information about phase stability of metal-silicon nitride-silicon MIS diode at high temperature. From ternary phase diagram of Mo-Si-N, Mo is stable with respect to silicon nitride and the barrier height of molybdenum-silicon nitride-silicon MIS diode with less thin insulating layers shows relative stability up to 600°C. The results presented here show that phase stability of metal-silicon nitride-silicon MIS diode can be achieved by proper choosing of metal contact from phase diagram calculated from bulk thermodynamics. Besides, it can be suggested by these results that chromium and tungsten may be more possible choice as metal contacts because they are stable with respect to silicon nitride and have lower work function than molybdenum.

ACKNOWLEDGEMENT

This work was supported by the U.S. Air Force RADC contract # F19628-86-K-0023.

REFERENCES

[1] Mark Sobolewski, and C.R.Helms, Appl.Phys.Lett. **54**, 638(1989).

[2] Mark Sobolewski, and C.R.Helms, J.Vac.Sci.Technol. **B7**, 971(1989).

[3] M.T.Schmidt, D.V.Podlesnik, C.F.Yu, X.Wu, R.M.Osgood,Jr., and E.S.Yang, J.Vac.Sci.Technol. **B6(4)**, 1436(1988).

[4] H.Tseng, and C.Wu, J.Appl.Phys. **61**, 299(1986).

[5] H.C.Card, and E.H.Rhoderick, J.Phys. **D4**, 1589(1971).

[6] S.Ashok, J.M.Borrego, and R.J.Gutmann, Solid-State Electron. **22**, 621(1979).

[7] Mark Sobolewski, "Structure and Electron Properties of Ultra-Thin Silicon Nitride Layers at Metal-Silicon Interfaces," Ph.D Thesis,Electrical Engineering, Stanford University, 1988.

[8] B.R.Pruniaux and A.C.Adams, J.Appl.Phys. **43**, 1980(1972).

[9]M.A.Taubenblatt,"Interface Structure and Electronic Properties of the Titanium-Silicon System," Ph.D Thesis,Electrical Engineering, Stanford University, 1985.

[10] R.Beyers, J.Appl.Phy. **56**, 146(1984).

CHARACTERIZATION OF A RAPID THERMAL
ANNEALED $TiN_xO_y/TiSi_2$ BARRIER LAYER

Sailesh Chittipeddi, Michael J. Kelly, Charles M. Dziuba, Anthony S. Oates, and William T. Cochran

AT&T Bell Laboratories, 555 Union Boulevard, Allentown, PA 18103

ABSTRACT

In this paper we characterize the thin film formed by rapid thermal anneal of a magnetron sputtered titanium film in a nitrogen atmosphere. The barrier properties of this material have been characterized for both n- and p-type junctions in our CMOS technology. We have characterized the physical properties of the film using Auger, RBS and TEM analysis in the same range of temperatures, and find that as the annealing temperature is increased a better quality TiN_xO_y film is formed. The electromigration characteristics for $A\ell/TiN_xO_y/TiSi_2$ runners, as well as the role that this system plays in minimizing failures due to stress induced voiding are examined in this study.

INTRODUCTION

The role of titanium nitride films as a contact barrier in VLSI technology has been well studied[1]. Titanium nitride films have been used as diffusion barriers between silicon and aluminum[2], as interconnect material between the drain and poly gate of the adjacent poly cell[3], as gate material for MOS transistors[4], and as an adhesion layer for tungsten deposition[5]. Additionally, TiN has also been used in thin film resistors which are fabricated by reactive evaporation[6], as well as for surface hardening and wear reduction[7]. Less attention, however, has been paid to the role that the TiN system plays in improving electromigration[8] and minimizing failures due to stress induced voiding for metal runners in the $A\ell/TiN_xO_y/TiSi_2$ system.

EXPERIMENTAL

Titanium films were deposited on Si, SiO_2 and $TiSi_2$ substrates by magnetron sputtering with the Varian 3180 system.The Ti films on the various substrates were then annealed in nitrogen using the A. G. Associates Heatpulse Model 2101 annealer at temperatures ranging from 600°C to 1000°C for durations ranging from 20 s to 45 s. The temperature was monitored using a calibrated thermocouple. After the $TiN/TiSi_2$ film was formed, it was exposed to the atmosphere resulting in the formation of $TiN_xO_y/TiSi_2$ films.

The $A\ell$ was sputter deposited from a composite target of $A\ell$-Si-Cu. TEM analysis was performed using a Philips TEM system at 200 keV. To evaluate the contact resistances, source drain resistances and device performance, we fabricated both n- and p-type MOSFET's using our twin tub CMOS technologies. The contact windows were opened to salicided junctions, then Ti was deposited and subsequently annealed in N_2. The contact resistance tester measured the resistance between the salicide and silicon interface, whereas the salicide contact resistance tester measured the resistance between the metal and the salicided junction. Electromigration results were obtained on metal I meander patterns, with 0.5 micron thick aluminum and with a nominal width of 0.9 μ. The metal was stressed at 270°C with a current density of $3x10^6$ A/cm². Stress voiding studies were carried out by by subjecting the wafers to a temperature of 180°C and 210°C for 40 hours and then studying the electromigration characteristics.

RESULTS AND DISCUSSION

The thicknesses of $TiN_xO_y/TiSi_2$ formed at different nitridation temperatures was studied using Auger analysis and Rutherford BackScattering. The Auger analysis was made on a Perkin-Elmer PHI-595 system with a 5 keV electron beam. RBS studies were performed using 2.25 MeV ^4He ($\Theta = 164°$) ions. The RBS data is summarized in Table I. All the samples were found to have a layer of nitride on the surface. The ratios shown in the table were obtained at the center of the film and change by a ratio of ~0.25 from the surface to the interface. Auger analysis also reveals similar $TiSi_x$ thicknesses with approximately a 150 Å TiN_xO_y layer on the surface. Cross-section TEM analysis shows that as the temperature of the nitridation is increased the TiN_xO_y and the $TiSi_2$ films tend to show more of a distinction indicating that one forms a better quality TiN_xO_y film. XPS analysis did not reveal much information even at grazing incidence, since the 150 Å of TiN_xO_y was not sufficient to stop the the the penetration of the incident $CuK\alpha$ radiation.

Table I: RBS data for various times and temperatures of anneal for 600 Å Ti on Si.

RBS Data RTA Temperature / Time Split for 600 Å Ti on Si		
Split	Si/Ti Ratio ± 0.04	$TiSi_x$ Thickness ±100 Å
800°C / 20 sec.	1.84	1100
800°C / 30 sec.	1.95	975
800°C / 45 sec	1.99	900
850°C / 20 sec.	1.95	1000
850°C / 30 sec.	1.95	940
850°C / 45 sec.	1.88	975
900°C / 20 sec.	1.90	975
900°C / 30 sec.	1.92	875
900°C / 45 sec.	1.92	940

The sheet resistances obtained at various temperatures on Si and SiO_2 substrates in the presence of nitrogen are shown in Figures 1a and 1b for times ranging from 10 s to 30 s. It can be seen from the figure that at temperatures greater than 800°C, the sheet resistance drops below 2 ohms/sq. and does not change significantly at higher temperatures for the Ti annealed on the Si substrate. The difference in sheet resistances at temperatures above and below 700°C can be ascribed to the fact that at temperatures below 640°C the C_{49} phase of $TiSi_2$ is formed, whereas above that temperature, the C_{54} phase results[11]. On the other hand for the Ti on SiO_2 substrate, the sheet resistance decreases monotonically with increasing temperature. This behavior can be explained by the fact that at temperatures greater than 700°C the SiO_2 reacts with Ti resulting in the formation of a disordered phase $TiSi_x$ film[12].

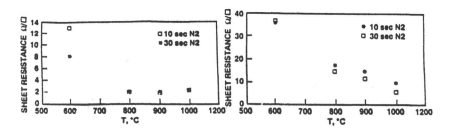

Figure 1a: Sheet resistance vs. RTA temperature for 600Å Ti on Si.   **Figure 1b:** Sheet resistance vs. RTA temperature for 600Å Ti on SiO_2

The source/drain resistance tester which was used had window openings of 1 μ to p+ and n+ salicided contacts. The results of the measurements to p+ junctions are summarized in Table II, whereas, the measurements for the n+ system are summarized in Table III. From Table II we find that as the RTA temperature is increased, the salicide contact resistances as well as the source drain resistance decrease. As is seen from Table III, the n− salicide contacts also decrease as the RTA temperature increases, however the S/D resistance does not change significantly. Further, as the RTA temperature is increased the gains for the n-channel as well as p-channel nominal devices increase. One possible explanation for the difference between the p+ and the n+ salicide contacts could lie in the fact that boron has more affinity for $TiSi_2$ as opposed to arsenic which tends to move away from the $TiSi_2$, the so called "snowplow effect"[9].

Table II: P channel device parameters for the $TiN_xO_y/TiSi_2/TiSi_2$ system.

RTA Temperature Split for ($TiN_xO_y/TiSi_2/TiSi_2$) − 600 Å Ti				
Split	P Sal. Cont.	P+ Cont.	P Ch. S/D Res.	P Ch. Nom. Beta
800 ° C, 20 sec.	.578	125.65	138.75	825.0
850 ° C, 30 sec.	.543	88.75	115.0	834.0
900 ° C, 30 sec.	.539	62.5	106.41	835.0

Table III: N-channel device parameters for the $TiN_xO_y/TiSi_2/TiSi_2$ system.

RTA Temperature Split for the ($TiN_xO_y/TiSi_2/TiSi_2$) − 600 Å Ti				
Split	N Sal. Cont.	N+ Cont.	N Ch. S/D Res.	N Ch. Nom. Beta
800 ° C, 20 sec.	.555	1.913	71.5	2895.8
850 ° C, 30 sec.	.517	1.464	67.6	2935.7
900 ° C, 30 sec.	.526	.5430	70.2	3041.7

In Table IV we summarize the electromigration results obtained on metal runners 0.9 microns in width. The results obtained for Ti under $A\ell$ with and without RTA are summarized therein. We find that the MTF (mean time to failure) for the $A\ell/TiN_xO_y$ system is vastly improved over that obtained for the conventional $A\ell$-Si-Cu. The electromigration results obtained for contact structures will be the subject of a more detailed study later[10].

Table IV: Comparison of electromigration results for the TiN_xO_y / Al and standard Al-Si-Cu systems.

Electromigration for Ti Process Options		
Process Option	t_{50} (95% Limits) Hr.	Std. Dev. (95% Limits)
Standard Al-Si-Cu	64.8 (30, 140)	1.09 (0.64, 1.87)
TiN_xO_y / Al	286 (127, 646)	0.93 (0.44, 1.92)

In Table V we summarize some preliminary stress induced voiding results. Prior to the unbiased bake the two lots examined showed comparable void densities (< 25% voids for all wafers in the patterns examined). After an unbiased bake at 180°C and 210°C for a period of 40 hours, electromigration studies were performed. It is seen from the table that the mean time to failure for the $A\ell/TiN_xO_y$ system improves by a factor of 5 to 6 over the $A\ell$-Si-Cu system, thus showing that the $A\ell/TiN_xO_y$ system plays a rather effective role in minimizing the failures due to stress induced voiding.

Table V: Electromigration data obtained after an unbiased bake at 180°C and 210°C.

Bake	Split	t_{50} (95% Limits) Hr.
210°C	TiN_xO_y / Al	105 (89, 124)
	Al	13.6 (9.2, 20)
180°C	TiN_xO_y / Al	109 (85, 140)
	Al	20 (14, 18)

SUMMARY

The $A\ell/TiN_xO_y/TiSi_2/TiSi_2$ structure has been studied. We have found that as the RTA temperature is increased one can lower the contact resistance and improve the device performance. Electromigration and stress induced voiding results show significant improvement over conventional $A\ell$-Si-Cu systems. Hence, the long term reliability of CMOS devices is significantly improved. At higher temperatures we have found that the quality of the $TiN_xO_y/TiSi_2$ film formed is enhanced.

ACKNOWLEDGEMENTS

We wish to thank the personnel in the Device Development Line for technical support and the text processing facility for help in preparing this manuscript.

REFERENCES

[1] C. W. Nelson, Proc. of Int. Symposium on Hybrid Microelectronics, Dallas Texas, pp. 413 (1969). C. Y. Ting, S. S. Iyer, C. M. Osburn, G. J. Hu, and A. M. Schweigart, in Proc. of the First Int. Symp. on VLSI Sci. and Technol. (Electrochemical Society, New York 1982), pp. 224 and references therein.

[2] S. Kannmori, Thin Solid Films 136, pp. 195-214,(1986).

[3] T. Tang, C.C. Wei, R. Haken, T. Holloway, C. F. Wan, and M. Douglas, IEDM Tech. Digest, pp. 590 (1985).

[4] M. Wittmer, J. Vac. Sci. Tech., A3(4), pp. 1797 (1985).

[5] V. V. S. Rana, J. A. Taylor, L. H. Holschwander, N. S. Tsai, Proc. of Workshop on Tungsten and other Refractory Materials for VLSI applications, pp. 187 (1986).

[6] G. Beensh-Marchwika and T. Berlicki, Thin Solid Films 62, pp. 267 (1980).

[7] A. Matthews and D. G. Teer, Thin Solid Films 52, pp. 415 (1980).

[8] T. Okamoto, K. Tsukamoto, M. Shimuzu, Y. Mashito and T. Matsukawa Dig. Symp. VLSI Tech., pp. 51 (1986), K. Y. Fu, E. Travis, S.W. Sun, C. L. Grove, R. L. Pyle, F. Pintchovski, P. Schani Proc. VMIC Conf. pp. 439 (1989). and P. A. Gargini, C. Tseng, and M. H. Wood, Proc. 20th Ann. Rel. Phy. Symp. pp. 66 (1982).

[9] M. Wittmer and T. E. Seidel, J. Appl. Phys., 49, 5287 (1978).

[10] A. S. Oates, (unpublished).

[11] R. Beyers and R. Sinclair, J. Appl. Phys, 57, 5240 (1984).

[12] A. E. Morgan, E. K. Broadbent and A. H. Reader, Mat. Res. Soc. Symp., 52, 279 (1986).

CHARACTERIZATION OF THE Al/RuO$_2$ INTERFACE UPON THERMAL ANNEALING

QUAT T. VU, E. KOLAWA AND M-A. NICOLET
California Institute of Technology, Pasadena CA 91125

ABSTRACT

We have characterized the Al/RuO$_2$ interface after annealing at temperatures in the range 450°C-550°C for durations up to several hours by backscattering spectrometry, cross-sectional transmission electron microscopy, and electrical four point probe measurement of specially designed structures. The electrical measurement yields the specific contact resistance of the interface by applying a transmission line type model developped for this purpose. An interfacial aluminum-oxygen polycrystalline compound is shown to grow with annealing temperature and duration, with a concurrent reduction of a thin layer of RuO$_2$. However, the specific contact resistance between Al and RuO$_2$ is found to decrease with annealing duration at 500°C. This last result indicates that the interfacial reaction does not lead to an insulating interface as could have been expected if the growth were pure and dense Al$_2$O$_3$.

1. INTRODUCTION

Reactively sputtered RuO$_2$ has been shown to be an effective diffusion barrier between Al and Si up to annealing temperatures of 600°C for 30 minutes [1,2]. However, it was shown that after heat treatment a polycrystalline layer of a new phase is formed at the interface which has tentatively been identified by one group [3] as some form of polycrystalline Al$_2$O$_3$ or possibly some ternary phase of Ru-O-Al. Another group found this growth to be amorphous Al$_2$O$_3$ [4,5]. Since Al$_2$O$_3$ is insulating, this layer growth can have deleterious effects on the electrical conduction properties of the metallization system.

In this report, we monitor the behavior of the Al/RuO$_2$ interface by backscattering spectrometry, cross-sectional transmission electron microscopy and electrical four point probe measurement of specially designed structures. Values of the specific contact resistance of the interface are derived from the electrical measurements using a transmission line type model developed for four point probe measurements of multilayered structures.

2. EXPERIMENTAL PROCEDURES

Substrates of <111> oriented n-type silicon with resistivity around 5 mOhmcm were used to prepare the <Si>/RuO$_2$/Al structures for 2.1 MeV ^{4}He$^+$ backscattering spectrometry studies and the <Si>/SiO$_2$/Al/RuO$_2$/Al structures for the specific contact resistance studies.

The RuO$_2$ was deposited by reactive rf magnetron sputtering using a planar Ru cathode of 7.5 cm diameter. The substrate holder was placed about 7 cm below the target and was neither cooled nor heated externally. The background pressure was 5x10^{-7} Torr or better prior to deposition. The sputtering gas is a mixture of argon and oxygen. The flow ratios of Ar to O$_2$ and total gas pressure were adjusted by mass flow controllers and monitored with a capacitive manometer in a feedback loop. The total gas pressure is kept at 10 mTorr and the total gas flow is around 60cc/min at a forward power of 300 W. The ratio of O$_2$ to total gas flow is 20%, 30% and 50%. To cross check with former results [2,5], a deposition was made with RuO$_2$ deposited at a total gas pressure of 10 mTorr with an Ar-O$_2$ gas mixture having an initial partial pressure of O$_2$ in Ar of 50%.

The Al layers were sputter-deposited in the same chamber as RuO_2 without breaking the vacuum, at a pressure of 5 mTorr of Ar and a forward power of 300W with the substrate biased at -50V. Since an accurate control of the deposited Al thickness is required for the specific contact resistance measurements, a separate study of the deposition rate of Al under these conditions was made. With the substrate holder rotating at constant speed under the target, the deposition rate was determined to be 6.6 nm/min. This knowledge allowed us to tailor the Al layer thickness to the desired values.

The annealings were performed in vacuum at a pressure in the range of 5 to 9×10^{-7} Torr.

For the specific contact resistance measurements, specially designed structures of the sandwich type $<Si>/SiO_2/Al/RuO_2/Al$ were made. The RuO_2 layers were deposited at 50% O_2 flow ratio to a thickness of about 300 nm. The Al layers were deposited with well-controlled thicknesses in the range of 100 to 600 nm. To characterize independently the annealing behavior of the Al layers, these same Al layers were deposited on separate SiO_2 substrates and annealed at the same time as the sandwich structure. Each set of samples composed of one sandwich structure and of the two corresponding Al on SiO_2 structures was annealed at $500°C$ repeatedly, in steps of 30 min each. After each annealing step, the samples were measured at room temperature using the four point probe setups. Details on the electrical measurements are given in ref. 6 and are summarized in the Electrical Results section below.

3. RESULTS AND DISCUSSIONS

3.1 Backscattering Spectrometry and Cross-Sectional Transmission Electron Microscopy

In Fig. 1, we show the 2.1 MeV $^4He^{++}$ backscattering spectra obtained on the structure $<Si>/RuO_2/Al$. The RuO_2 film of this sample was deposited at a 20% O_2 flow ratio to a thickness of about 270 nm. The Al layer was sputtered without breaking the vacuum and is about 170 nm thick. The annealing temperature was kept at $500°C$ and the total annealing duration increased in steps of 30 min. To increase the apparent thickness of the interfacial growth, the spectra were taken with the sample tilted at an angle of 45 degrees with respect to the sample normal.

For short annealing durations up to 30 min, the backscattering spectrum is practically identical to the as-deposited spectrum. As we increase the annealing duration, we start to observe changes corresponding to a reaction at the $Al-RuO_2$ interface. The leading edge of the Ru signal develops a peak that grows with annealing time. This indicates a reduction of RuO_2. At the same time, the leading edge of the oxygen signal which rides on top of the Si signal shifts toward higher energies while the top of the oxygen signal develops a small dip corresponding to the reduced RuO_2 layer. These signal changes are indicative of a movement of the oxygen from RuO_2 toward Al leaving behind a reduced layer of RuO_2 at the interface. A close look at the Al signal also reveals a smeared-out low-energy edge consistent with an interfacial growth of an Al-oxygen compound. We note also a slight oxidation of the aluminum surface after annealing reflected in a small oxygen peak at the surface energy edge. The same behavior is observed on structures made with RuO_2 layers deposited under the other conditions mentioned in the preceding section.

A study of the interfacial growth as a function of annealing temperature clearly shows that the higher the temperature, the thicker the growth. Also, a backscattering spectrum taken on a sandwich structure $<Si>/SiO_2/Al/RuO_2/Al$ after anneal shows very clearly a splitting of the O_2 signal into two small peaks corresponding to the two interfacial reactions.

A cross sectional TEM study of a $<Si>/SiO_2/Al/RuO_2/Al$ structure annealed at $550°C$ for 30 min was made. The results clearly show an interfacial layer which is polycrystalline, very non-uniform in thickness with thickness values in the range of tens of nm. The non-uniformity seems to be related to the Al grain boundaries. We have been unable to clearly identify the phase of this layer [3].

Our observations by backscattering spectrometry and cross-sectional transmission electron microscopy are in agreement with previous studies made by our group [1,3]. Similar studies of the Al-RuO$_2$ interface have been carried out by another group [2,4,5] with the conclusion that the interfacial layer is amorphous Al$_2$O$_3$.

3.2 Electrical Studies

To study the electrical behavior of the Al-RuO$_2$ interface before and after heat treatment, we focused on the measurement of the specific contact resistance of the interface by applying a model developped for four point probe measurement of bilayered structures [6]. For this purpose, we consider structures of the sandwich type SiO$_2$/Al/RuO$_2$/Al. Fig. 2(a) shows a schematic of it. Fig. 2(b) represents the electrical circuit model for a small element of the structure. p_1 and p_2 are the sheet resistances of the top and bottom Al layers respectively. p_3 is the sheet resistance of the RuO$_2$ layer. In the direction perpendicular to the film, the resistance to current flow per cm^2 of the film is composed of the specific contact resistance p'_c of the two Al-RuO$_2$ interfaces assumed to be identical, and of the resistances of the film themselves p_{A1}, p_{A2} and p_R.

The resistivity of Al and of RuO$_2$ before (and after) heat treatment is about 3.5 microOhmcm and 150 microOhmcm respectively. Thus with the thicknesses in the range of values specified for our structures, we have p_3 at least 15 times p_1 or p_2. Since they are in parallel, p_3 can be neglected. In the perpendicular direction p_{A1} and p_{A2} are negligible compared to p_{R2}. p_R is calculated to be around 10^{-8} Ohmcm^2. Thus if $p_c = 2p'_c$ is 10^{-7} Ohmcm^2 or larger, as is the case for our samples, then p_R can be neglected also. The final circuit model can thus be simplified to two sheet resistances in parallel separated by an interface characterized by the specific contact resistance p_c, equal to twice the value p'_c of the specific contact resistance of the interface between Al and RuO$_2$ (Fig. 2(c)). After annealing, the growth of the interfacial layers affects our model only in a change of the value of p'_c which then includes the resistance in the perpendicular direction to the structure of the newly grown interfacial layers. Whatever the real nature of these interfacial layers, the value of p'_c as measured is also that which will be relevant in a real contact.

A four point probe measurement calculation based on that model, which is basically a transmission line type model, has been made. Details are published elsewhere [6]. We summarize here some relevant results which allow us to derive the value of the specific contact resistance p_c out of such a measurement.

We call p_o the parallel combination of p_1 and p_2 ($p_o = p_1 p_2 / (p_1 + p_2)$). The potential drop between the two inner probes of a four point probe setup having a probe spacing of s will be given as follows with I being the current sourced in and out of the outer probes

$$V = I p_o \ln 4 / 2\pi + I p_o F p_1 / 2\pi p_2 , \qquad (1)$$

where

$$F = 2[K_o(s/b) - K_o(2s/b)] ,$$

with

$$b = [p_c / (p_1 + p_2)]^{1/2} \qquad (2)$$

the characteristic length of the problem and a measure of p_c. K_o is the modified Bessel function of the 2nd kind of zeroth order.

We can check that in the limits of p_c equal 0 or infinity, equation 1

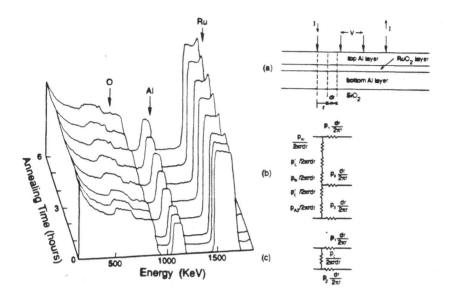

Fig.1 2.1 MeV ^{4}He^{++} backscattering spectra of a Si /RuO$_2$/Al structure annealed repeatedly at 500°C in steps of 30 min showing a reaction at the Al/RuO$_2$ interface. The tilt angle is 45°, the scattering angle 170°

Fig.2(a) Schematic of sandwich structure.(b) Electrical model of an element of the structure. (c) Simplified circuit model applicable to our structures.

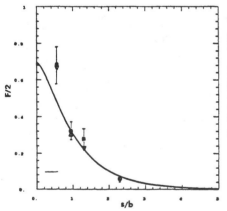

Fig.3 Plot of the factor F/2 as a function of s/b (solid line). The various data points corresponding to structures with different Al thicknesses, are values obtained on as-deposited structures using for p$_c$ the common value of 1.2x10^{-4} ohmcm^2.

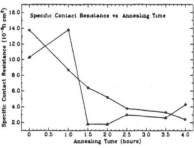

Fig.4 Evolution of the specific contact resistance p$_c$ as a function of annealing duration at 500°C showing a decrease from an initial value to below the sensitivity limit. The sensitivity limit of the structure represented by the squares (the triangles) is around 2x10^{-5} ohmcm^2 (8x10^{-5}). Below the limit, the data are not significant.

gives the correct results.

We write

$$R=V/I \ , \ R_o=V_o/I, \ V_o=Ip_o \ln 4/2\pi$$

and

$$n=p_1/p_2$$

then we have

$$(R-R_o)/R_o=nF/\ln 4 \qquad (3)$$

This last expression represents a percentage increase of R over R_o. Thus, when p_c is non-zero, R will be larger than R_o by a term which is mainly determined by F. Fig. 3 shows the evolution of $F/2$ as a function of s/b. We note that F is $\sim$ 0 for s/b $\geq$ 5. With experimental uncertainties estimated at $\pm 15\%$ [6], we can say that R is significantly larger than R_o if the percentage increase is $\geq 20\%$. This occurs if b > s/3 for a structure having n=2. This sets a limit on the sensitivity of the method. For n=1/2, the sensitivity is much worse. This is reflected in fig. 4 for the two structures used in the annealing study below. In principle, the sensitivity can be increased by optimal structure design. This point is discussed in ref.6.

When R is larger than R_o, then the value of F is determined. p_1 and p_2 were measured independently of the sandwich structure, on the SiO_2/Al samples from which R_o and n were calculated after each annealing step. These repeated measurements were necessary to account for the changing resistivity of the Al films. For a known F, the curve in Fig. 3 will then yield a value for s/b. With s known from the measurement setup (s=250 microm for our smallest probe spacing setup and s=1240 microm for our largest probe spacing setup), b can be calculated. Using equation (2), p_c can be derived.

Measurements were made on as-deposited structures designed with various Al thicknesses. The different symbols in fig. 3 represent different structures. The F factor was derived from the measured values. The experimental results fit the theoretical curve well when a common value of 1.2×10^{-4} ohmcm$_2$ is used for p_c for all structures (fig. 3). This confirms the validity of the model. Since p_c is twice the specific contact resistance p'_c between Al and RuO_2, we obtain then the specific contact resistance value between as-deposited Al and RuO_2 as equal to 6×10^{-5} ohmcm^2.

A series of measurements upon annealing were performed on two types of structures. Structure 1 (square symbol in fig. 4) has $p_1=137$ mohm/square and $p_2=65$ mohm/square, structure 2 (triangle symbol) has $p_1=65$ mohm/square and $p_2=137$ mohm/square so that n=2 for structure 1 and n=1/2 for structure 2. Fig. 4 shows the evolution of the specific contact resistance as a function of annealing duration at 500°C. It decreases from its initial value around 10^{-4} Ohmcm^2 to below the sensitivity limits for these structures. This result shows that the interfacial layer that grows at the Al-RuO_2 interface upon annealing does not impede the current flow as would have been expected if this layer were insulating Al_2O_3. It shows rather that this layer is conducting. The sensitivity of backscattering spectrometry (about 5 at.%) does not allow us to clearly pinpoint the presence of Ru in the interfacial layer. We surmise that the interfacial layer may be some form of conducting compound related to aluminum oxide or a ternary of Ru-Al-O since Ru is a well known component of thick film hybrids [7].

4. CONCLUSIONS

It is shown that upon annealing an interfacial layer grows bwtween Al and RuO_2. The experimental results found are not consistent with a layer composed of pure insulating Al_2O_3. Since Ru is a well known component used in thick film hybrids, we surmise that there is Ru in this interfacial layer.

The electrical results are also inconsistent with post-annealing I-V characteristics of shallow junctions incorporating RuO_2 as a diffusion barrier showing an increase in the forward voltage drop. Further work on the

Si or silicide interface with RuO_2 is needed to clarify this point.

Acknowledgements

The financial support for this work was provided principally by Intel Corporation through a grant, and partly by the U.S. Army Research Office under Contract DAAL03-89-K-0049. In addition, the first author is deeply indebted to Intel Corporation for a personal fellowship.

REFERENCES

1- E. Kolawa, F.C.T. So, E.T-S. Pan and M-A. Nicolet, Appl.Phys.Lett. 50, 854 (1987).
2- L. Krusin-Elbaum, M. Wittmer and D.S. Yee. Appl.Phys.Lett.50,1879 (1987).
3- C.W. Nieh, E. Kolawa, F.C.T. So and M-A. Nicolet, Mat.Lett 6, 177 (1988).
4- A. Charai, S.E. Hornstrom, O. Thomas, P.M. Fryer and J.M.E. Harper, J.Vac.Sci.Technol. A7, 784 (1989).
5- J.M.E. Harper, S.E. Hornstrom, O. Thomas, A. Charai and L. Krusin-Elbaum, J.Vac.Sci.Technol.A7, 875 (1989).
6- Quat T. Vu, E. Kolawa, L. Halperin and M-A. Nicolet, submitted to Solid State Electronics.
7- C.R.S. Needes, in Proc.1982 Int. Microelectronics Conf.,1982, Tokyo, Japan pp. 94-101.

INTERACTION OF COPPER FILM WITH SILICIDES

Yow-Tzong Shy*, Shyam P. Murarka*, Carlton L. Shepard† and William A. Lanford†
* Materials Engineering Department, Rensselaer Polytechnic Institute, Troy, NY 12180
† Physics Department, State University of New York at Albany, Albany, NY 12222

ABSTRACT

Bilayers of Cu with $TiSi_2$ and $TaSi_2$ were tested by furnace annealing at temperatures from 200 to 500°C. Rutherford Back Scattering (RBS) technique was used to investigate the interaction between various films and determine the stability of Cu on silicide structures. The sheet resistance was also monitored. The results show that Cu on $TiSi_2$ and $TaSi_2$ structures are extremely stable structures at annealing temperatures in the range of room temperature to 500 °C. In such structures, therefore, there will not be a need of any diffusion barrier between Cu and the silicide films.

INTRODUCTION

The advent of submicron technology has intensified the demand for a high conductivity metal, for applications as the interconnect both on and off the chip. This has brought attention to copper for use as interconnect and wiring metallization. $TaSi_2$ and $TiSi_2$, with low resistivity and latter with the advantage of the self-aligned processing, have found applications in silicon integrated circuit as gate ($TaSi_2$ and $TiSi_2$) and contact ($TiSi_2$) metallization.[1,2] Thus in future developments may lead to the use of copper as upper level interconnection, directly on $TaSi_2$ or $TiSi_2$ used as gate and contact metallization. On the other hand there is considerable concern about the use of copper which may diffuse through the contacting silicides into active regions of the devices where it is known to become a recombination-generation center. Therefore, the study of the stability (interactions and diffusion) of copper on $TaSi_2$ and $TiSi_2$ is important. In view of the above, we have investigated the interactions between copper film on $TaSi_2$ and $TiSi_2$ using sheet resistance measurements and Rutherford backscattering (RBS) techniques. It is found that copper on $TaSi_2$ and $TiSi_2$ is stable at least up to 500°C for anneals carried out in argon/3% hydrogen ambient and no detectable interaction is observed.

EXPERIMENTAL

$TaSi_2$ (2000Å) films were prepared by co-sputtering Ta and Si in the desired Si/Ta ratio onto either the polycrystalline silicon film on oxidized silicon or on single-crystal silicon wafers. $TiSi_2$ (2500Å) films were also prepared by co-sputtering Ti and Si in the desired Si/Ti ratio only onto polycrystalline silicon film on oxidized silicon wafers. All samples were then subjected to a 2 min rapid thermal anneal (RTA) at 900°C in an AG Associates 210 annealer in argon/3% hydrogen gas mixture ambient. The $TaSi_2$ films after the RTA treatment were subjected to an exposure to 30:1 buffer HF solution for 30 sec followed by de-ionized (DI) water rinse and spin drying. $TiSi_2$ surfaces were exposed only to the DI water rinse. Approximately 1300Å of copper were then deposited on the top of $TaSi_2$ and $TiSi_2$ films by DC magnetron sputtering, with a base pressure of 5×10^{-7} Torr and the substrates at room temperature. It is estimated that the layer thickness varied across a sample or from sample to sample by ±5%. All wafers were then cut into smaller (~6-7 cm²) pieces. The smaller pieces were annealed in the temperature range from 200 to 500°C at 50°C intervals, in a furnace in the ambient of argon/3% hydrogen gas mixture. The annealing time was 30 min at each temperature. The sheet resistance of all the samples were measured by a 4-point probe at room temperature before and after each anneal. The RBS technique was used to analyze the samples that were (a) unannealed (b) annealed, and (c) after removal of copper from surface by etching in dilute nitric acid. The latter group of samples were made from both group (a) and (b) to enhance the sensitivity of RBS method.

Mat. Res. Soc. Symp. Proc. Vol. 181. ©1990 Materials Research Society

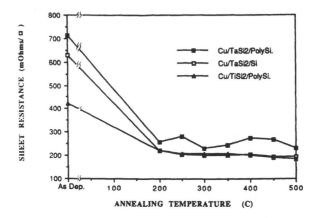

Figure 1. Sheet resistance vs. annealing temperature - sheet resistance of the samples remains constant vs. annealing temperature indicating that copper is not consumed by diffusion.

RESULTS AND DISCUSSION

Unreacted copper films have very low resistance compared to underlying silicide films ($\sim 1.8\mu\Omega$-cm vs. 50 and 20-25$\mu\Omega$-cm for co-sputtered and annealed $TaSi_2$ and $TiSi_2$ respectively[1,3]). Any significant interdiffusion or reaction between the copper and the silicide will lead to an increase in the measured sheet resistance of copper. Thus the monitoring of the sheet resistance of the copper-on-silicide samples provides a very simple and sensitive way to follow the interaction.[4] Table I identifies the samples used in these experiments. Figure 1 shows the plot of the sheet resistance of different samples as a function of annealing temperatures. Note the initial, as deposited, variations in the measured sheet resistance which reflects contribution from copper, silicide, and underlying substrate and small ($\pm 5\%$) thickness variation on wafers in a run. Also note the initial value is very high compared to the annealed sheet resistance. The initial drop in the resistance, at temperature $\leq 200°$C, is associated with the annealing of the defects and release of trapped gases in the sputtered metallic films.[5] For anneals at temperatures higher than 200°C the sheet resistance, within experimental error, is independent of the annealing temperature, indicating no significant consumption of copper and effect of interdiffusion if any. From these results it would appear that $TaSi_2$ and $TiSi_2$ films are suitable diffusion barriers for copper, when annealed in argon-hydrogen ambient. However RBS data shown below show slightly different behavior for $TaSi_2$ films on polysilicon.

Table I. Samples used in these experiments.

Sample number	Descriptions
1	$Cu/TaSi_2/Polysilicon/SiO_2$
2	$Cu/TaSi_2/Si$
3	$Cu/TiSi_2/Polysilicon/SiO_2$

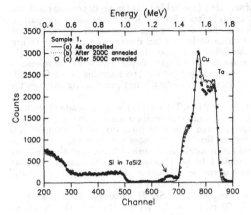

Figure 2. RBS profiles of sample 1 (see Table I.) for (a) as- deposited film (b) after a 200°C/ 30 min anneal and (c) after a 500°C/ 30 min anneal.

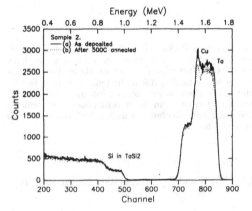

Figure 3. RBS profiles of sample 2 (see Table I.) for (a) as- deposited film (b) after a 500°C/ 30 min anneal.

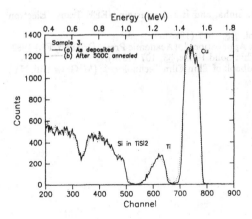

Figure 4. RBS profiles of 1300Å Cu on 2500Å TiSi₂ on polysilicon for (a) as-deposited film (b) after a 500°C/ 30 min anneal. There is no apparent indication of Cu diffuse into TiSi₂.

Figure 2 shows the RBS profiles of sample 1 (see Table I.) (a) in as- deposited condition and (b) after a 200°C/ 30 min anneal and (c) after a 500°C/ 30 min anneal. A comparison of the three profiles show that annealing at 200 - 500°C range led to small copper diffusion in $TaSi_2$. The diffused amount i.e. the area under the tail increasing with temperature. Interestingly, however, the depth of the diffusion appears to be the same in Figure 2(b) and 2(c) indicating no detectable increase in the diffusion depth which is approximately one half the thickness of $TaSi_2$ layer. This would suggest either a roughing of the Cu-$TaSi_2$ interface or a defect aided diffusion. In latter case defect is not penetrating through the thickness of the silicide.

Figure 3 shows the RBS profiles of sample 2 (see Table I.) (a) in as- deposited condition and (b) after a 500°C/ 30 min anneal. Lower temperature data are not shown since there are no differences in the profiles of as-deposited or annealed (up to 500°C) samples.It is clear then that copper, within the detection limit of RBS, does not show any penetration in this sample of $TaSi_2$ on silicon. This result would then appear to support the observed penetration of copper in $TaSi_2$ on sample 1 to be associated with the sample artifact (roughness or defect). Note that amount of copper diffused in sample 1 (Figure 2.) is still small, not affecting the sheet resistance of the sample significantly.

Figure 4 shows the RBS profiles of copper on $TiSi_2$ film (sample 3) (a) in an as-deposited film and (b) after a 500°C/ 30 min anneal. Within the detection limit, no interdiffusion is discernible.

CONCLUSION

The interaction of copper films deposited on $TaSi_2$ and $TiSi_2$ films on polysilicon or silicon substrates has been investigated using sheet resistance measurements and RBS analysis. Within the detection limits of these techniques copper on these silicide films seems to be stable when annealed in argon containing 3% hydrogen. Electrical testing of the devices, utilizing these silicides and copper as interconnection metal, will verify if small amounts of copper, undetectable by changes in sheet resistance measurements and RBS technique, diffused in active areas and affected the device performance. Such experiments are in progress at the present time.

ACKNOWLEDGMENTS

Authors would like to thank SEMATECH and SRC corporation for the support of this work.

REFERENCES

1. S.P. Murarka, D.B. Fraser, A.K. Sinha, and H.J. Levinstein, IEEE Trans. Electron Devices, ED- 27, 1409 (1980).
2. R.A. Haken, J. Vac. Sci. Technol. B3, 1657 (1985).
3. S.P. Murarka, "Silicides for VLSI Applications", (Academic Press, Orlando, FL) 1983.
4. P.M. Hall and J.M. Morabito, Thin Solid Films, 33, 107 (1976).
5. L.I. Maissel and R. Glang, "Handbook of Thin Film Technology", (McGraw Hill, NY) 1970.

METAL SURFACE MORPHOLOGY CHARACTERIZATION USING LASER SCATTEROMETRY*

S. M. GASPAR, K. C. HICKMAN, J. R. MCNEIL,
University of New Mexico, Albuquerque, NM 87131;
R. D. JACOBSON, Sandia Systems, Inc., Albuquerque, NM;
G. P. LINDAUER, Department of Materials Science, University of Florida, Gainesville, FL;
Y. E. STRAUSSER, Instruments, S. A., Inc., Santa Barbara, CA:
and E. R. KROSCHE, Intel Corporation, Albuquerque, NM.

ABSTRACT

We have applied angle-resolved laser scatterometry to characterize the morphology of metals deposited under various conditions. Scatterometry is a rapid, noncontact and nondestructive diagnostic which yields surface statistics including rms roughness and power spectral density of the microstructure.

* Work supported in part by SEMATECH.

INTRODUCTION:

In many areas of microelectronics fabrication it is becoming increasingly important to characterize the microstructure of smooth surfaces. The surface morphology of metal interconnect material must be controlled for several reasons, including the effect morphology has during lithography, the influence of morphology in creating interlevel shorts, and its relation to electromigration and device failure. Use of laser scatterometry to characterize the angular distribution of light scattered from a surface is a promising surface diagnostic technique which is rapid and noncontact. The sample requires no special preparation (e.g., etching or overcoating with a conducting film) and therefore is not destroyed as when using other techniques such as scanning electron microscopy or transmission electron microscopy. The technique has been used extensively to examine reflecting optical samples.[1-5]

RELATION BETWEEN SCATTERED LIGHT AND SURFACE MICROSTRUCTURE

The relation between scattering of electromagnetic radiation and surface topography has been studied for many years, originally in the radar field.[18-20] In general the connection is complicated and depends critically upon the actual surface topography. However, for a perfectly reflecting surface, and in the smooth-surface limit when the heights of the surface irregularities are much smaller than the wavelength of the scattered light, the relation becomes simple. Here we review E. L. Church's treatment[2] in relating scattered light to the surface roughness, which is described in more detail in Ref. 5.

Vector scattering theories describe the differential light scatter dI_s as[2]

$$\frac{1}{I_o} \frac{dI_s}{d\omega_s} = \frac{C}{\lambda^4} Q\left(\theta_o, \phi_o, \theta_s, \phi_s, \hat{n}\, \chi_o, \chi_s\right) W(p,q) , \tag{1}$$

where C is a constant, I_o is the intensity of the incident light, and $d\omega_s$ is the solid angle of the detection system. The scattered intensity is proportional to λ^{-4} as in Rayleigh and Mie scattering. The quantity Q in Eq. (1), called the "optical factor," is independent of the surface condition and is a function of the polar and azimuthal angles of incidence, θ_o and ϕ_o, and scatter angles, θ_s and ϕ_s, complex index of refraction n of the surface, and polarization states of the incident and scattered light, χ_o and χ_s, respectively. The "surface factor," $W(p, q)$, is the power spectral density (PSD) of the surface roughness.

If the surface (i.e., the best fit plane) is in the x-y plane, and $Z(x,y)$ is the surface height variation (surface roughness) relative to that plane, the PSD is given by

$$W(p,q) = \frac{1}{A} \left[\frac{1}{2\pi} \int \int dx \, dy \, e^{i(px+qy)} \, Z(x,y) \right]^2$$

(2)

where A is the area of the scatterer, and p and q are the surface spatial frequencies in the x and y directions. In other words, W is the average square magnitude of the two-dimensional Fourier transform of the surface roughness.

Although the PSD contains information about the surface roughness, it is often convenient to describe the surface roughness in terms of specific surface-finish parameters to easily compare measurement results. The most commonly used surface-finish parameter is the rms roughness σ. The rms roughness which is measured is given in terms of the instrument bandwidth and MTF, $G(p, q)$, as

$$\sigma^2 = \int_{pmin}^{pmax} dp \int_{qmin}^{qmax} dq \, G(p,q) W(p,q) ,$$

(3)

where $W(p, q)$ is given in Eq. (2). Different values of σ^2 will result if the integral limits (i.e., bandwidths) or MTF's of instruments differ. The MTF characteristic of the optical scatterometer is approximately unity. Because σ depends on the range of surface spatial frequencies, f, included in the measurement, the measurement is thus said to be band limited.[5] Alternately, the surface spatial wavelength, $d=1/f$, can be used to describe lateral surface structure.

EXPERIMENTAL ARRANGEMENT OF THE LASER SCATTEROMETER

The instrument used for the measurements reported here is shown in Fig. 1 and has been described previously.[5] The light source is a linearly polarized He-Ne laser (wavelength 633 nm) with the plane of polarization perpendicular to the plane of incidence (s-polarized light); other laser sources are also used in the instrument. The spatially filtered laser output is focused ~ 26 cm beyond the sample onto an apertured photomultiplier (PMT) detector which rotates about the illuminated spot on the sample surface. The spot size on the sample is typically 0.5 - 2 mm. However, smaller spot diameters may be used if only high frequency structure is to be characterized. Scattered light intensity typically ranges over ten orders of magnitude. Because of the large change in scattered light intensity which occurs over a small angular range (e.g., three orders of magnitude within the first 5°), it is especially important to perform measurements at many values of θ_s to fully characterize a sample. In the system used in this investigation, scattered light is typically measured at increments in θ_s of 0.1° to 0.4°, and the bandwidth of the instrument is approximately 0.64 μm to 70 μm in spatial wavelength.

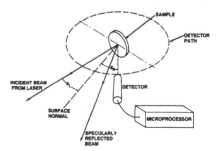

Fig. 1 Experimental arrangement of the scatterometer

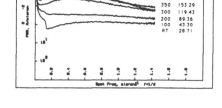

Fig. 2 Scatterometer analysis of Al-2% Si material

MEASUREMENT RESULTS

We examined the surface morphology of Al-2% Si material which was deposited at different substrate temperatures. We were easily able to distinguish samples deposited at substrate temperatures which differed by 50° C. The samples became rougher, and average grain size increased as the deposition temperature increased. This is illustrated in Figure 2 by observing the PSD characteristics of the samples. As the substrate temperature increased, the area under the curves increased. In addition the secondary peak of the curves shifted to lower spatial frequencies, and this indicates an increasing lateral dimension of the predominant structure on the surfaces of the samples which is characterized by the scatterometer. These results are consistent with those of scanning tunnelling microscope (STM) and SEM examination. Figure 3 illustrates how the grain size determined by SEM analysis and the rms roughness determined by the scatterometer are very closely correlated. However, the peaks in the PSD characteristics in Figure 2 indicate larger grain sizes than the SEM analysis. The reason for this, we believe, is that the scatterometer is not sensitive to the actual grain boundaries because of their small lateral extent. They are much smaller than the wavelength of light and therefore outside the spatial frequency bandwidth of the technique. The technique detects height differences between crystals. If two or more adjacent crystals are of nearly the same height, the scatterometer characterizes these as a single, larger crystal.

We applied scatterometry to characterize Al-2% Si samples at a number of points to determine uniformity of the material across the wafer. The rms roughness calculated from the PSD characteristics of each measurement is plotted in Figure 4, and it can be seen that the samples are relatively uniform.

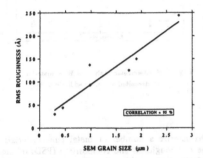
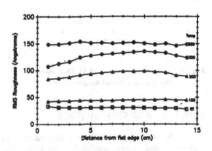

Fig. 3 Grain size analysis of Al-2% Si material determined by scatterometry and SEM techniques

Fig. 4 rms roughness profiles of Al-2% Si samples

Samples of Al-2% Si coated with a TiN AR layer were characterized using the scatterometer. The technique detected similar PSD and rms roughness characteristics for portions of the same sample which had the TiN coating and portions which were stripped of the TiN. Other samples of the same material were profiled across their diameter and had widely varying microstructure characteristics. The ability to accurately characterize the morphology of samples which are AR coated is very desirable in the case of a deposition process in which several layers are deposited without breaking vacuum.

We also applied the technique to examine Al-2% Si material deposited in two different sputter systems: Perkin-Elmer and Anelva. This was performed in an attempt to explain differences in processing the materials during lithography. Figure 5 illustrates the difference in rms roughness characteristics of the two materials. For the film deposition conditions used for these samples, the Perkin-Elmer material was rougher and had larger grain structure. This is consistent with STM analysis of the material and might be the result of a higher substrate temperature in the Perkin-Elmer system during film deposition. The Perkin-Elmer system employs no direct substrate heating and temperature control, whereas the Anelva system does. Samples from both systems were approximately 80% uniform across their diameters.

Figure 6 illustrates the characteristics of WSi$_2$ material. Two samples were characterized: as-deposited and WSi$_2$ after 500 Å of oxide growth. The as-deposited sample is smooth and has a featureless PSD characteristic, indicating a very fine microstructure. The sample with SiO$_2$ is significantly rougher. In addition, the PSD characteristics of this sample indicates significant structure at large (2 micron) lateral dimension. In addition, the PSD characteristic begins to increase at the right end, indicating significant short-range structure is present. We are in the process of characterizing these same samples using shorter optical wavelengths, which will extend the sensitivity of the technique to better characterize this structure.

We also applied the technique to characterize CVD W. The material was rough with an rms of 160-180 Å, and PSD characteristic was featureless. The latter is indicative of surface microstructure composed of all lateral dimensions within the bandwidth of the scatterometer.

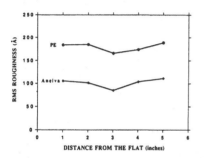

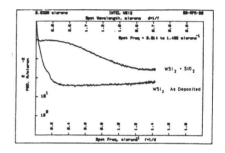

Fig. 5 Scatterometer analysis of Al-2% Si material deposited in different sputter systems

Fig. 6 Scatterometer analysis of WSi$_2$ material as-deposited and after addition of SiO$_2$

SUMMARY AND CONCLUSIONS

We have applied laser scatterometry to characterize a number of metal film materials. The technique is very useful for characterizing the rms roughness and composition (PSD) of the surface microstructure. Scatterometry is a nondestructive, rapid technique which might be implemented for *in-situ* analysis in some applications.

REFERENCES:

1. G. A. Al-Jumaily, S. R. Wilson, J. J. McNally, J. R. McNeil, J. M. Bennett, and H. H. Hurt, "Influence of Metal Films on the Optical Scatter and Related Microstructure of Coated Surfaces," Appl. Opt. 25, 3631-3634 (1986).
2. E. L. Church, H. A. Jenkinson, and J. M. Zavada, "Measurement of the Finish of Diamond-turned Metal Surfaces by Differential Light Scattering," Opt. Eng. 16, 360-374 (1977).
3. L. D. Brooks and W. L. Wolfe, "Microprocessor-based Instruments for Bidirectional Reflectance Distribution Function (BRDF) Measurements from Visible to Far Infrared (FIR)," in *Radiation Scattering in Optical Systems I*, W. H. Hunt, ed., Proc. Soc. Photo-Opt. Instrum. Eng. 257, 177-183 (1980).
4. P. Roche and E. Pelletier, "Characterizations of Optical Surfaces by Measurement of Scattering Distribution," Appl. Opt. 23, 3561-3566 (1984).
5. R. D. Jacobson, S. R. Wilson, G. A. Al-Jumaily, J. R. McNeil, J. M. Bennett, and Lars Mattsson, "Microstructure Characterization by Angle-resolved Scatter and Comparison to Measurements Made by Other Techniques," submitted to Applied Optics.

THE EFFECTS OF Si ADDITION ON THE PROPERTIES OF AlCu FILMS
USED IN MULTILEVEL METAL SYSTEMS

S.R. Wilson, D. Weston and M. Kottke
Advanced Technology Center, Motorola Semiconductor Products Sector, 2200 W.Broadway Rd., Mesa, AZ 85202

ABSTRACT

This paper reports a systematic study of the properties of Al(Cu 0.5-1.5%) and Al(Cu 0.5-1.5%, Si 0.5-1.5%)alloys. The effects of deposition temperatures ranging from room temperature to 475°C were studied as well as whether the films were deposited on Ti-W or SiO_2. The Cu and Si profiles were measured as well as the presence of Al_2Cu and Si precipitates. The effects of post deposition bake treatments are also discussed.

INTRODUCTION

As integrated circuits (ICs) have become larger and more complex, the complexity of the metal interconnect system has also increased. In addition, many of the short range interconnections that were previously performed with polycrystalline Si or silicides, now require metal interconnects to reduce the sheet resistance and increase speed. Because of their many desirable properties, Al and Al alloys have been used in the great majority of metal interconnect applications.

In most IC applications Al(Si), Al(Cu) or Al(Si,Cu) have been used as the interconnect alloy for a variety of reasons. Si was added to Al films to reduce junction spiking in the Al-Si contacts.[1] However, Si precipitation in the contacts caused high contact resistance[2] and subsequently barrier layers were developed to prevent the interactions between Al and Si. This led many IC manufacturers to remove the Si from the Al. Cu was added to Al to reduce hillocks and electromigration. However, Cu in Al can form Al_2Cu precipitates, which can increase the corrosion of the metal after it has been patterned by reactive ion etching.[3] Pitting corrosion of Al(Cu) films occurs due to the galvanic action between the precipitate and surrounding Al when the film is exposed to an electrolyte such as water. We have shown that the residue of this pitting corrosion can lead to a residual metal after RIE that can result in shorts in tight pitched lines.[4] This paper reports a systematic study of the addition of Si to Al(Cu) films. The effects of deposition conditions, percent of Si and Cu in the target and subsequent thermal treatments are reported. In addition, the effects on the film properties of a TiW barrier layer versus a dielectric layer under the film will be presented.

PROCESSING CONDITIONS

The films studied were: (1) Al(1.5%Cu), (2) Al(1.5%Cu, 1.5%Si), (3) Al(0.5%Cu) and (4) Al(0.5%Cu, 0.5%Si). The films were deposited on oxidized Si wafers or on oxidized wafers coated with a 0.15μm Ti-W layer. The Al films were sputter deposited in a single wafer deposition system with the wafers heated to temperatures ranging from no external heating (room temperature) to 475°C. The deposition rates were ~1.1μm/min. and the wafers were not biased during the deposition. To simulate post deposition wafer processing, wafers were selected from each group and subjected to bakes at 250°C, 350°C, 400°C, 430°C and 450°C.

The Cu and Si profiles at the film interfaces and throughout the bulk of the film were determined using Auger electron spectroscopy (AES). The Cu and Si precipitate densities and sizes were determined by delineation etches and inspection in an optical microscope or an SEM.

Deposited films were also subjected to standard photoresist and RIE processes to study the effects of these processes on the different alloys. Snake and comb test patterns were etched into the films and tested electrically for shorts and opens. These patterns were also inspected for the effects of corrosion.

RESULTS

Effects of Deposition Temperature and Sublayer Type

As expected, increasing the deposition temperature, produced an increase in the average grain size and range of grain size as seen in Table I. At a deposition temperature of 305°C the addition of Si appears to reduce the average grain size slightly, but it is impossible to say if the result is statistically significant. However, at 435°C it is clear that Si reduces both the average grain size and the range of grain sizes. We observed the same trend comparing Al(1.5%Cu) with Al(1.5%Cu, 1.5%Si) and these results are consistent with those reported by Kobayashi et al.[5] The films deposited above 400°C on SiO_2 have an average grain size that is ~30% greater than that of the films deposited on Ti-W. This is probably due to the incorporation of Ti into the Al films.[6]

Table I: Grain size and grain size distribution for Al(0.5%Cu) and Al(0.5%Cu, 0.5%Si).

Deposition Temperature °C	Alloy	Average Grain Size μm	Range μm
305	Al(0.5%Cu)	1.3	0.0-2.8
305	Al(0.5%Cu, 0.5%Si)	1.2	0.0-2.8
435	Al(0.5%Cu)	4.5	0.0-14.0
435	Al(0.5%Cu, 0.5%Si)	2.8	0.0-8.0

The Cu and Si distributions of the various alloys deposited at temperatures ranging from room temperature to 465°C were determined by AES. Due to the large number of samples not all films were measured at each temperature, but enough films were profiled to determine trends. Cu profiles were essentially independent of the sublayer (Ti-W or SiO$_2$) that the films were deposited upon. The Cu profiles in the Al(0.5%Cu) and Al(0.5%Cu, 0.5%Si)were uniform throughout the bulk of the film for deposition temperatures ranging from room temperature to 435°C. However, the films containing 1.5%Cu show a pileup at the sublayer interface for deposition temperatures ranging from room temperature through 325°C. The pileup reaches a maximum at ~200°C as shown in Fig.1. At ~350°C and above, the Cu distribution was uniform throughout the film.

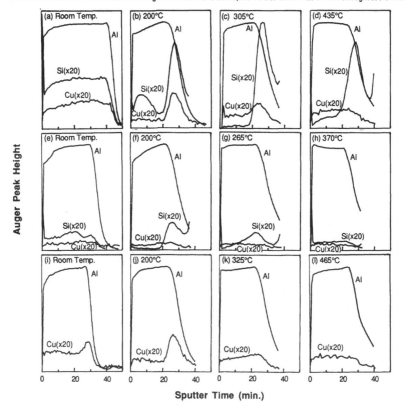

Sputter Time (min.)

Figure 1. AES profiles for Al(1.5%Cu, 1.5%Si) (a-d), Al(0.5%Cu, 0.5%Si)(e-h) and Al(1.5%Cu) (i-l) deposited at different temperatures.

The Si profiles showed a strong dependence on deposition temperature and sublayer type. At room temperature the Si profile was uniform throughout the film for both concentrations of Si and both substrates. Fig. 1 shows the Cu and Si profile for the Al(0.5%Cu, 0.5%Si) and Al(1.5%Cu, 1.5%Si) films deposited on Ti-W. The film deposited at 200°C shows a pileup of Si at the Ti-W/Al interface and very little Si throughout the bulk of the film. As the deposition temperature increases, the amount of Si at the Ti-W interface decreases and the concentration of Si in the bulk of the film increases. For depositions at 370°C) the Si is uniformly distributed. However, in the case of films deposited on SiO$_2$, the Si in the Al(0.5%Cu, 0.5%Si) film remains piled up at the SiO$_2$ interface for temperatures up to 475°C. The Al(1.5%Cu, 1.5%Si) films show far less dependence on the sublayer type. Both sublayers have large pileups of Si at the Al sublayer interface for all temperatures above room temperature (which shows a flat profile). To verify that the Si pileup was associated with Si precipitates, the films were etched in solution B, which removes Al but not Ti-W or Si, and examined with an SEM. As shown in Fig. 2 the density of precipitates decreases and the size increases from ~0.1-1.0μm. The same trend is seen on the SiO$_2$ substrates, but the density of Si precipitates is ~100 times greater. The size of the precipitates is essentially equal at 200°C regardless of the sublayer, however at 285°C the Si precipitates on SiO$_2$ substrates are only one third the size of those on Ti-W.

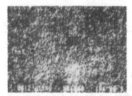  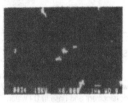

200°C **245°C** **285°C**

Figure 2. SEM micrograph showing Si precipitates at the Ti-W interface after etching Al(0.5%Cu, 0.5%Si) films in a nitric, phosphorous and acetic acid etch to remove the Al and Cu.

After deposition at the different temperatures the surfaces of the various alloys were subjected to a delineation etch and examined in an SEM for the presence of Al$_2$Cu precipitates near the surface. (As we discussed in another paper[4], these precipitates near the surface can cause pitting corrosion of the film during wet processing.) A typical Al$_2$Cu precipitate is shown in Fig. 3 and the results of this examination are presented in Table II. Note for all alloys that the Al$_2$Cu precipitate s are present at 200°C. No Al$_2$Cu was observed for the room temperature deposition. At the higher temperatures, the lower Cu alloys are less likely to have precipitates at the surface and the presence of Si seems to reduce the occurrence of Al$_2$Cu precipitates at the surface.

Table II: Pressence of Al$_2$Cu precipitates for the different alloys and deposition temperatures.

Alloy	Temperature °C	Precipitate
Al(0.5%Cu)	R.T.	No
	200-305	Yes
	435	No
Al(0.5%Cu, 0.5%Si)	200	Yes
	285-435	No
Al(1.5%Cu)	R.T.-465	Yes
Al(1.5%Cu, 1.5%Si)	R.T.	No
	200	Yes
	305-435	No

Figure 3. A typical Al$_2$Cu precipitate.

Effects of Post Deposition Bakes

To examine the effects of post deposition thermal treatments, the films were subjected to a series of bakes ranging from 250 to 450°C for 1hr. The films that were baked at higher temperatures than they were deposited hillocked. No increase in the native oxide thickness was observed by AES after any of the bakes. The Al(1.5%Cu) film deposited at 325°C showed no change in the as-deposited profile (see Fig. 1) after a 250°C bake. A 350°C bake caused the Cu profile to be uniform throughout the bulk of the film and a 450°C bake caused the Cu to be uniform throughout the bulk, and the beginning of a pileup peak at the surface was observed. Films deposited in the intermediate temperature range (300-350°C) and the high temperature range(>425°C) were examined for Al$_2$Cu precipitates after the bakes. The data is presented in Table III and can be compared to the data in Tables II. The 250°C bake caused precipitates on both of the alloys with 0.5% Cu even though precipitates were not detected after deposition on the higher temperature films. The 350°C or 450°C bake reduces the presence of Al_2Cu precipitates in the films with 0.5%Cu regardless of the as-deposited condition. However, both the 350°C and 450°C bakes cause precipitates on the Al(1.5%Cu, 1.5%Si) films which did not show precipitates after deposition.

Table III: Al_2Cu precipitates detected after 1 hr bakes.

Alloy	Deposition Temp.	Bake Temps. 250°C	350° C	450° C
Al(0.5%Cu)	305°C	Yes	No	No
	435°C	Yes	No	No
Al(0.5%Cu, 0.5%Si)	305°C	Yes	No	No
	435°C	Yes	No	No
Al(1.5%Cu)	325°C	Yes	Yes	Yes
	465°C	Yes	Yes	Yes
Al(1.5%Cu, 1.5%Si)	305°C	Yes	Yes	Yes
	435°C	Yes	Yes	Yes

SUMMARY

In another paper[4], we have shown that the presence of Al_2Cu precipitates correlates well with the pitting corrosion which occurs on the surface of an Al film film that is subjected to standard wet processing steps including photoresist development and rinse in D.I water. This pitting corrosion could lead to metal residues and shorts between metal patterns after RIE. In that paper we also reported that higher deposition temperatures reduced the occurrence of pitting and that the addition of Si also reduced pitting.

In this paper we have discussed the effects of deposition temperature, concentration of Cu or Si and sublayer type on properties of several Al alloys. A Ti-W sublayer or the addition of Si reduce the grain size. For the 0.5% Cu films the Cu profile is essentially uniform throughout the film independent of deposition temperature. The 1.5%Cu films show a pileup of Cu at the bottom interface for temperatures above R.T. and below ~400°C. The 0.5%Si profiles depended upon both the deposition temperature and sublayer type, with substantial pileup at the SiO_2 /Al interface. This pileup correlated with Si precipitates. The 1.5%Si profiles were independent of the sublayer and showed more Si throughout the bulk of the film. Many of these effects, plus the effects of the bakes can be explained by referring to the solid state phase diagram for Al-Cu. [7] As the temperature is increased more of the Cu should be in solution and in addition the larger grain size reduces the number of grain boundaries as a possible site for precipitation. The presence of Al_2Cu precipitates on the 0.5% Cu films deposited at the higher temperatures may be due to the fact that the deposition process occurs quickly and the film does not reach equilibrium. The post deposition bakes however allow the films to reach equilibrium. The maximum in precipitate density after the 250°C bake is due to the Cu concentration exceeding the solid solubility by the maximum amount. At the higher temperature bakes of 350 and 450°C the Cu in the 0.5%Cu films is all soluble, but for the 1.5%Cu films the copper content exceeds solubility even at 350°C. For the 1.5%Cu films baked at 450°C, where all the Cu is expected to go into solution, precipitation is still observed because the films cool slowly enough to allow diffusion and formation of precipitates.

ACKNOWLEDGEMENTS

We would like to thank D. Johnston and J. Adams for assistance in developing the deposition conditions. S. Bell for help in processing many of the samples. J. Helbert for help with and discussions on photoresist processing and its interactions with metal layers. We would also like to express our appreciation of C. Tracy and S. Thomas for their support of this program , for proof reading the manuscript and helpful discussions on the various topics. Finally we would like to thank M. Park and S. Krause of Arizona State University for supplying some of their initial data on the films being studied here.

REFERENCES

1. P.A. Totta, and R.P. Sopher, IBM J. Res. & Dev. 13, 226 (1969).
2. S.R. Wilson, Simon Thomas and S.J. Krause, Inst. Phys. Conf. Ser. No. 87, p375 (1987).
3. Wen-Yaung Lee, J.M. Eldridge and G.C. Schwartz, J. Appl. Phys., 52, 2994 (1981).
4. D. Weston, S.R. Wilson, and M. Kottke, J. Vac. Sci. Technol. A, in press.
5. T. Kobayashi, H. Kitahara and N. Hosokawa, J. Vac. Sci. Technol., A5, 2088 (1987).
6. S.R. Wilson, R.L. Duffin, D. Weston, and R.J. Mattox, to be presented Spring 1990 Meeting of the Electrochemical Society, Montreal.
7. D.T. Hawkins and R. Hultgren Metals Handbook, 8th edition, T. Lyman, ed., (American Society for Metals, Metals Park,Ohio, 1973) p. 259.

THE EFFECT OF CHEMICAL TREATMENT OF THE GaAs SURFACE ON THE OHMIC CONTACT PROPERTIES

BALÁZS KOVÁCS, MARGIT NÉMETH–SALLAY AND IMRE MOJZES
Research Institute for Technical Physics of the Hungarian Academy of Sciences,
H–1325 Budapest P.O.Box 76, Hungary

ABSTRACT

Electrical properties of metal/GaAs system are known to be influenced very much by GaAs surface condition before the metal deposition. The surface condition of GaAs epitaxial layer can be affected by the wet chemical treatment of the surface. The influence of different processes were compared to choose the best technique for a high quality ohmic contact technology. Applying a chemical treatment which contains degreasing process, a light surface etching and a rinsing process, the carbon contaminations, 20–25 nm GaAs surface layer and the native oxide layer are removed from the GaAs surface. The benefit of the rinsing step is to produce a reproducible, stable surface condition before the metallization. Since the importance of the finishing step was assumed, the investigated processes differed in this step. The compared methods finished with either alkaline, like NH_4OH, or acidic, like HCl, etchants resulted higher specific contact resistance than the process finished with neutral, high purity (18 MΩcm) water. In the latest case the obtained specific contact resistance was $(6.4 \pm 2.7) \times 10^{-6}$ Ωcm^2 on a GaAs epitaxial layer with the doping concentration 1.5×10^{17} cm^{-3}.

INTRODUCTION

Reliable low resistance ohmic and high quality Schottky contacts on compound semiconductor epitaxial layers are required especially for high frequency devices. Several authors [1–6] have investigated the surface preparation for GaAs but the surface processing of GaAs has not yet been fully established.

The present paper concentrates on the effect of the last technological steps before the metallization. As a method for measuring the effect of the different etching and rinsing processes the specific contact resistance was applied using Process Control Monitor (PCM) patterns formed on MESFET like epitaxial structures.

SAMPLE PREPARATION AND MEASUREMENTS

The GaAs epitaxial layer structures for MESFET purposes were prepared by Effer method using a Cr doped, <100> oriented semiinsulator GaAs substrate. The layer structure consists of a 2μm thick buffer layer ($N_D < 10^{15}$ cm^{-3}) and a 0.3μm thick, S–doped active layer ($N_D \approx 10^{17}$ cm^{-3}).

After having opened the window for ohmic metallization the surface of the GaAs should be very carefully cleaned. For these purposes an etching in NH_4OH :H_2O_2 : H_2O = 1:1:100 was used at room temperatures for 10 sec. (All components were of electronic grade, the water was of 18 MΩcm.) The chemical surface preparation was finished in five different ways which are listed in Table I.

Method	Rinsing	Etching	Rinsing
1	3x H_2O t = 10 s	1 : 10 = HCl : H_2O T = 25°C, t = 15 s	————

Method	Rinsing	Etching	Rinsing
2	3x H_2O t = 10 s	1 : 10 = HCl : H_2O T = 25°C, t = 15 sec	3x H_2O t = 10 s
3	———	1 : 10 = NH_4OH : H_2O T = 25°C, t = 2x 30 s	———
4	———	1 : 10 = NH_4OH : H_2O T = 25°C, t = 2x 30 s	3x H_2O t = 10 s
5	3x H_2O t = 10 s	———	———

Table I: The finishing steps of the surface preparations

The ohmic contact metallization was prepared by vacuum evaporation at a pressure lower than 10^{-4} Pa. The metallization contains AuGe (eutectic)/Ni/Au layers with thickness 75 nm, 12.4 nm and 20 nm respectively. After the vacuum evaporation the final contact's layout was formed by 'lift—off' process. The finishing step of ohmic contact formation was a 400°C heat treatment.

Previous investigations [7], concerning of the homogeneity of the GaAs epitaxial layer, showed that the used growing process results layers, in which the local values of the free carrier concentration and the layer thickness could alternate up to 100 percentage. Since the specific contact resistivity is a function of the carrier concentration [8], beside of the specific contact resistivity the value of the carrier concentration was obtained in each PCM chips applying a FATFET pattern [9]. The specific contact resistivity values as the function of the carrier concentration are shown in Figure 1.

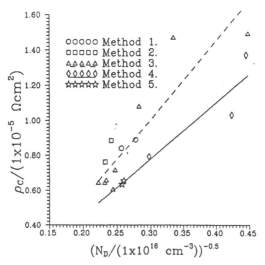

Figure 1 : The obtained specific contact resistivities versus the carrier concentration of the epitaxial layers

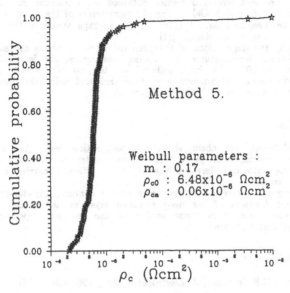

Figure 2 : The cumulative probability of the specific contact resistivity along a half of a 2" GaAs wafer

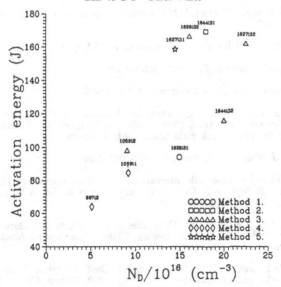

Figure 3 : The activation energies of the ohmic contact degradation.

Each points represent averaged values obtained on minimum 25 Cross Bridge Kelvin Resistor pattern [10]. Calculating the parameters of the Weibull distribution function of the measured data (Figure 2), the average values of the investigated electrical parameters were obtained [11].

To study the degradation of the prepared ohmic contacts the samples were heated at different temperatures. The ageing was carried out in flowing forming gas (30% H_2 + 70% N_2) from 240 °C up to 450 °C. Following the heat treatments the contact parameters were measured and using the Arrhenius plot the activation energies were obtained. (Figure 3)

CONCLUSION

Five different wet chemical surface preparation were studied. Methods, finished with neutral, high purity (18 MΩcm) water resulted the best ohmic contacts taking into consideration the effect of different carrier concentration values of the samples.

Studying the activation energy value of the contact degradation, it proved to be a linear function of the used epitaxial layer carrier concentration. This dependence was much more pronounced than the assumed dependence on the chemical treatment methods.

REFERENCES

1. A.C. Adams, B.R.Pruniaux, J. Electrochem. Soc., 120, 408 (1973)

2. T.Oda, T.Sugano, Jap. J. Appl. Phys., 15, 1317 (1976)

3. I.Shiota, K.Motoya, T.Ohmi, N.Miyamoto, J.Nishizawa, J. Electrochem. Soc., 124, 155 (1977)

4. I.Mojzes, Proc. 23rd Int. Sci. Coll. Ilmenau (GDR), 5, 103 (1978)

5. E.Kohn, J. Electrochem. Soc., 127, 505 (1980)

6. D.A.Figueredo, Solid State Electron., 29, 959 (1986)

7. B.Kovács, Sz.Varga, K.Somogyi, Proc. 1st. International Conference on Epitaxial Crystal Growth, 1–7 April, 1990., Budapest, Hungary, Ed. E.Lendvay, to be published by Trans Tech Publications Ltd, Switzerland

8. N.Braslau, J. Vac. Sci. Technol., 19, 803–807, (1981)

9. I.Mojzes, B.Kovács, Proc. 5th International School on Microwave Physics and Technique, 29 Sept. – 3. Oct. 1987. Varna, Bulgaria. Ed. A.Y.Spasov, World Scientific, Singapore, pp. 269–304.

10. B.Kovács, I.Mojzes, Proc. Int. Conf. Electrical Contacts and Electromechanical Components, May 9–12, 1989, Beijing, China., International Academic Publishers, pp. 513–518. (1989)

11. K.Sawa, M.Hasegawa, K.Miyachi, Proc. Int. Conf. Electrical Contacts and Electromechanical Components, May 9–12, 1989, Beijing, China., International Academic Publishers, pp. 454–460. (1989)

PARAMETERS AFFECTING THE PROCESS WINDOW FOR LASER PLANARIZATION OF ALUMINUM.

Ivo J. Raaijmakers, Harren Chu†, Edith Ong†, Shi-Qing Wang, and Ken Ritz
Philips R&D center, Philips Components - Signetics, Sunnyvale, CA94088-3409;
(†) XMR Inc., 5403 Betsy Ross Dr., Santa Clara, CA95054.

ABSTRACT

Irradiation with excimer laser pulses is demonstrated as a technique to fill contact/via holes for integrated circuits with Al-(1 wt.%)Cu. We discussed the influence of the size, shape and aspect ratio of the contact hole, and the metal thickness on the process lattitude for laser planarization. The most dominant parameters found to influence the planarization process are the shape of the contact, and the shape of the as-deposited metal surface ("step coverage"). The application of a Cu anti-reflection coating increased the process window substantially. Although a planar Al-Cu surface was easily formed with laser irradiation, void free filling of contact holes was a more demanding task. Conditions for which completely filled contacts can be obtained were discussed.

INTRODUCTION

Pulsed laser irradiation has been demonstrated to be a potential technique to planarize thin films of Au [1], Cu [2] and Al alloys [3-5] by momentarily melting them. Planarization can be achieved within a few microseconds due to the high surface tension and low viscosity of liquid metals. Both excimer lasers (pulse duration of 10 - 100 ns) and dye-lasers (pulse duration of 0.1 - 1 μs) have been used. During laser melting and flow, the dielectric acts as a thermal barrier, preventing exposure of underlying devices to the high temperatures prevailing in the liquid metal.

With increasing laser energy, the metal film has been found to proceed through the following stages: melting, planarization, contact fill and ablation [4]. Planarization of the metal film does not necessarily imply that the contact is completely filled with metal: the range of energies (process windows) for which contact fill occurs can be substantially smaller than that for planarization [4].

Much of the efforts have concentrated in technology feasibility, materials characterization, and device characterization [5-9]. Now, focus is towards expanding the process windows for planarization and contact fill [3,4,10,11]. In this paper we will report the influence of contact hole size, shape, and aspect ratio, and metal thickness on the laser planarization process. The impact of a Cu anti-reflection coating, and substrate heating will be discussed. First, we will describe planarization of the film. Then, we will turn our attention to complete filling of contacts (or vias).

EXPERIMENTAL

Contact holes to Si with diameters ranging from 0.6 to 1.5 μm were etched in layers of boro-phospho-silicate glass. Glass thicknesses of 0.5, 1.0 and 1.5 μm were used, resulting in contact holes with aspect ratios (the ratio of the depth of the contact hole to its nominal diameter at the bottom of the hole) ranging from 0.33 to 2.5. Contact holes were either etched with an anisotropic plasma etch, yielding vertical walls (dry-etched contacts), or with an isotropic wet etch for half of the oxide depth, followed by an anisotropic plasma etch (wet/dry-etched contacts). A 0.1 μm $Ti_{0.22}W_{0.37}N_{0.41}$ layer was deposited onto the wafers, followed by a 1.0 μm Al-Cu(1 wt.%) layer without breaking vacuum. Details of the deposition processes are described elsewhere [12]. Some samples received a 7 nm thick pure Cu anti-reflection coating in another magnetron sputter deposition system. Prior to deposition of the Cu layer, samples were precleaned by an *in situ* Ar$^+$ sputter etch, which removed 10 nm SiO_2 from an oxidized Si wafer.

The XMR model 7100 aluminum planarization system was used for reflow of Al-Cu thin films. The system comprises of a pulsed XeCl (308 nm) excimer laser source, capable of delivering up to 500 mJ per pulse at a repetition frequency of 300 Hz. The pulse duration is about 45 ns (full width at half maximum). The laser beam was focussed on

Mat. Res. Soc. Symp. Proc. Vol. 181. ©1990 Materials Research Society

to the wafer through a beam homogenizer, which converts the quasi-Gaussian laser beam profile into a nearly square profile having a uniformity of ±5% across the beam. To process an entire wafer, a grid of partially overlapping spots was produced by stepping the laser beam over the wafer. The process chamber was evacuated by a turbomolecular pump to a base pressure of about 1×10^{-4} Pa. The substrate was heated by a W filament radiant heater. Wafer temperature was measured by a pyrometer at a wavelength of 5 μm. The pyrometer was calibrated with a thermocouple mounted on an Al-Cu coated Si-wafer.

RESULTS AND DISCUSSIONS

Planarization

Planarization of the metal film and filling of the contact holes was concluded from plan view scanning electron microscopy. The minimum fluence needed to planarize the Al-Cu layer at a substrate temperature of 250 °C is shown in Fig. 1 as a function of aspect ratio. It can be concluded that aspect ratio does not have a significant influence on the minimum fluence for planarization, at least not in the range of oxide thicknesses and hole sizes considered here. The wet/dry-etched contacts need about 25% less fluence to planarize than dry-etched contacts. This suggests that the initial shape of the metal surface ("step coverage") is important. The same conclusions were reached for planarization at room temperature, 250 °C, and 350 °C substrate temperature.

Baseman et al. [3] report a slight increase in the minimum fluence for planarization with increasing contact size (from 2.7 ± 0.4 Jcm^{-2} for 2 μm vias to 3.3 ± 0.6 Jcm^{-2} for 6 μm vias, at a substrate temperature of 30 °C). We were unable to detect a significant increase in minimum fluence for planariza- weak dependencies of planarization on oxide tion with size for our smaller (0.6 to 1.5 μm) contact holes. Tuckerman et al. [1] estimate that, for typical planarization conditions, features of lateral size less than a characteristic flow length (L) of about 20 μm may planarize in about 1 μs. The flow length is large compared to typical dimensions of the contact holes, so it is understandable that the size of our small contact holes does not affect planarization behaviour to a great extent. Since cooling of the metal film occurs by thermal conduction through the oxide and metal layer, the thickness of the oxide may influence the time (t) during which flow of liquid metal is possible. The flow length is only weakly dependent on time $(L \propto t^{\frac{1}{4}})$, however, which indicates that the oxide thickness can not exert a large influence on the planarization behaviour. The

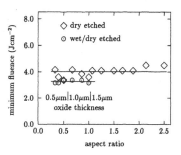

Fig. 1. Minimum fluence needed to planarize a 1μm Al-Cu film at 250 °C as a function of aspect ratio.

thickness and hole size explains our observation that the minimum fluence for planarization is practically independent of aspect ratios.

Since we did not detect effects of aspect ratio or oxide thickness on planarization, we combined results for different oxide thickness in a single substrate temperature - fluence plot (Fig. 2). Data for 0.5, 1.0 and 1.5 μm oxide for 1 μm dry-etched contact holes are shown. The solid lines give the transitions between partial and complete planarization (planarization limit, PL), and between planarization and ablation (ablation limit, AL). Assuming that the liquid metal needs to reach the same maximum temperature in order to observe similar effects in the filmafter cooling, the slope of these lines should be equal to $-C_p/\alpha$, where C_p is the specific heat capacity of the Al-Cu thin film, and α the fraction of incident laser energy absorbed in the metal. Approximating these quantities by constants $(C_p \approx 2.7 \times 10^{-4}$ Jcm^{-2}K^{-1} for a 1 μm Al-Cu film; $\alpha \approx 0.1$) over the temperature range of interest (20 - 400 °C), one calculates for the slope of the lines a value of 2.7×10^{-3} Jcm^{-2}K^{-1}. From Fig. 2 we determine a value of about 3.6×10^{-3} Jcm^{-2}K^{-1}. The agreement is reasonable, considering the crude model used.

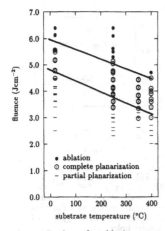

Fig. 2. Regimes for ablation, complete planarization and partial or no planarization for a 1 μm Al-Cu film over 1 μm, dry-etched contact holes as a function of substrate temperature.

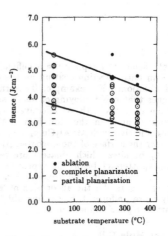

Fig. 3. Regimes for ablation, complete planarization and partial or no planarization for a 1 μm Al-Cu film on 1 μm wet/dry-etched contact holes as a function of substrate temperature.

The process window is defined as the quotient $(AL - PL)/(AL + PL)$. Since the slopes of the lines in Fig. 2 are equal and negative, the process window is expected to increase as the substrate temperature increases. In fact, we find the process window to increase from about 12% at room temperature, to about 18% at a substrate temperature of 350 °C. These process windows are consistent with those reported by other investigators [3,4].

Fig. 3 shows a graph similar to the one presented in Fig. 2, but now for planarization of wet/dry-etched contacts. One would not expect the ablation limit to be affected very much by this change in profile, as indeed can be observed from a comparison of Figs. 2 and 3. The minimum fluence for planarization, however, is lower. This yields process windows of 21% and 24% for planarization at room temperature and 350 °C, respectively. The fact that these values are significantly larger than those obtained for the dry-etched contacts suggests the importance of initial shape of the metal surface.

It has been reported that the application of an anti-reflection coating (ARC) on the Al film enlarges the process window [1,4,6]. Cu or Ti are commonly incorporated in Al layers up to concentrations of a few atomic percent to improve electromigration resistance. We tried the application of a thin (about 7 nm) pure Cu anti-reflection coating atop the Al-Cu film. From typical diffusivities in molten metals (10^{-4} cm^2/s) and the melt duration (of the order of 2 μs) we calculate a characteristic diffusion length of about 0.2 μm. Diffusion of Cu in the underlying metal will thus certainly occur. Alloying of the Cu ARC with the Al-Cu film was confirmed with SIMS measurements. Complete alloying of a 7 nm Cu ARC with a 1 μm Al-(1 wt.%)Cu film, yields a total of about 3 wt.% of Cu in the Al-Cu film.

Fig. 4 shows the minimum fluence needed to planarize such a Cu coated Al-Cu layer as a function of aspect ratio. In this case, too, the aspect ratio does not influence the minimum fluence to planarize the layer. Application of the ARC decreased the fluence needed for planarization by about a factor of 2. If the only effect of the ARC is to increase absorption of incident laser energy, one would expect the fluence for ablation to decrease by the same factor as that for planarization, and there would be no net effect on the process window. Fortunately, application of the Cu anti-reflection coating decreased the fluence for ablation only by a factor of about 1.4 (at 250 °C, it decreased from about 5 Jcm^{-2} to about 3.6 Jcm^{-2}). The smaller shift of the ablation limit as compared to that of the planarization limit increased the process window to about 30 - 35%.

Several reasons may explain the beneficial effect of an anti-reflection coating on the onset of ablation. For example, Tuckerman [1] argues that rapid alloying of his Si antireflection coatings with Al effectively caps the energy input in the thin film. Also, patches of native oxide, or Al_2Cu precipitates at or near the surface of the uncoated Al-Cu film may be sites which ablate more readily. The sputteretch before Cu deposition removes the native aluminum oxide, and the Cu layer may render the absorption of laser energy more uniform, thus causing ablation to start at higher temperatures.

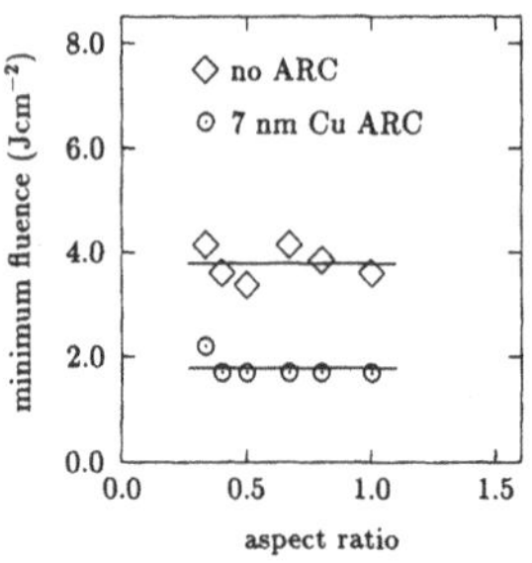

Contact Hole Fill

Fig. 4. Minimum fluence needed to planarize a 1 μm Al-Cu film at 250 °C with and without a 7 nm Cu anti-reflection coating as a function of aspect ratio over dry-etched contact holes.

In the previous sections it was shown that a process window of appreciable size can be obtained for planarization of an Al-Cu film over 0.6 μm to 1.5 μm diameter contacts. In this section we will show that the process window for completely filling contacts can be substantially smaller.

Figs. 5a and b show cross-section SEM micrographs from a 1 μm Al-Cu film, covering 1 μm contacts with an aspect ratio of 1, as-deposited and irradiated with a laser pulse of 4.3 Jcm^{-2} at 250 °C, respectively. The energy of this pulse is more than sufficient to planarize the surface of the sample (cf. Fig. 2). However, the contact is not filled by the Al alloy. Instead it appears that the Al-Cu film closes at the top of the contact hole, thus creating a planar Al-Cu layer but leaving a void in the contact hole. Although improvements were obtained by raising the substrate temperature or by application of an ARC it was difficult to fill these contacts completely, even at fluences close to the ablation limit. The problem gets worse if the diameter of the contact hole decreases, keeping metal and oxide thickness constant. This is shown in Fig. 6. With decreasing contact hole size, the Al-Cu film retracts further from the bottom of the contact hole. In addition, we observed the problem to be more severe for thicker metal films.

In contrast, it is possible to fill the shallow contacts (aspect ratios smaller than 1), but the process window for contact fill is narrower than that for planarization. In addition, wet/dry-etched contacts (Fig. 5c and d) could be filled at the same or even lower fluences which only planarize the metal over the dry-etched contacts. In Figs. 5e and f contact fill with a thinner (0.5 μm) metal film is shown. The laser fluence has now been decreased to 2.9 Jcm^{-2}, to account for the smaller heat capacity of the thinner metal film. From Fig. 5f it can be observed that the contact with the thinner metal film is filled after the laser pulse, although occasionally a small void was observed near the bottom of the contact.

The above described results for contact fill are in qualitative agreement with those presented by other investigators. Liu et al. [4], and Pramanik et al. [8] show complete filling of dry-etched, 0.8 μm diameter contact holes with aspect ratio's of about 1, using thin metal films (about 0.5 μm). Mukai et al. [6] show filling of a submicron contact hole with tapered walls.

Apparently, the shape of the as-deposited metal in the contact hole (or "step coverage") is important. Marella et al. [14] show that the above described observations can be explained from calculations of the mass flow of the metal on top of the contact hole. Although the configuration of a completely filled hole should be more favourable because of minimum surface energy, the formation of a void in the contact hole and a planar surface is favoured kinetically. Once a void is formed it is difficult to fill it, even by increasing the incident energy (which should increase the flow time). Good results can be obtained for

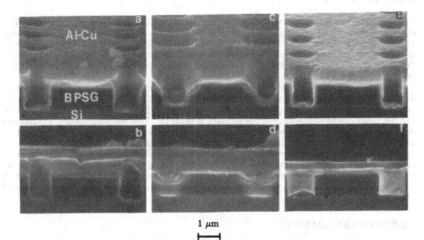

Fig. 5. Cross-section SEM micrographs of Al-Cu contact fill. (a) and (b) dry-etched contacts with 1 μm metal, as-deposited and irradiated with 4.3 Jcm^{-2} at 250 °C, respectively; (c) and (d) wet/dry-etched contacts with 1 μm metal, as-deposited and irradiated with 4.3 Jcm^{-2} at 250 °C, respectively; (e) and (f) dry-etched contacts with 0.5 μm metal, as-deposited and irradiated with 2.9 Jcm^{-2} at 250 °C, respectively.

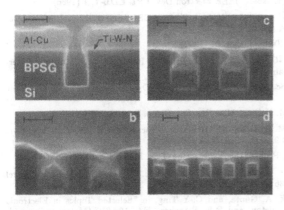

Fig. 6. (a) As-deposited 1μm Al-Cu film on a 0.8 μm diameter 1.5 μm deep dry-etched contact hole; (b) same as (a) but after irradiation with a laser pulse of 4.1 Jcm^{-2} at 400 °C; (c) same as (b) for a 0.7 μm contact hole; (d) same as (b) for a 0.6 μm contact hole. The bar in each micrograph represents a length of 1 μm.

tapered or wet/dry-etched contacts, and metal films with thickness smaller than the diameter of the contact. These conditions correspond to a relatively large distance of the metal film surfaces on opposite sides of the contacts. Such a configuration of the as-deposited metal film in the contact shows less tendency to close on top of the contact hole, and to bury a void beneath its surface. One may also be able to change the shape of the metal surface on top of the contact hole in a beneficial way by improving the step coverage of the metal films, by e.g. bias sputtering.

558

CONCLUSIONS

Process windows for excimer laser planarization of Al-Cu films over contacts for VLSI structures have been investigated. The process lattitude for planarization can be made quite large with the application of a Cu anti-reflection coating. The most remarkable result is, that over nearly one order of magnitude variations in aspect ratio (0.3 - 2.5), and a factor of three in oxide thickness (0.5 - 1.5 μm), the minimum fluence for planarization does not change appreciably.

Complete filling of a high aspect ratio contact with vertical walls with Al-Cu is a demanding task. Cross-section SEM micrographs showed that contact holes with aspect ratio of about 1 could be completely filled only with metal films which are thin compared to the diameter of the contact, or if the contact is tapered. The most important parameters determining the flow behaviour of the metal films appear to be the profile of the contact and the shape of the as-deposited metal surface ("step coverage").

For dry-etched, high aspect ratio contacts and metal thicknesses comparable to or exceeding the contact diameter, the formation of a planar metal surface and a void in the contact is consistently observed. It is anticipated that one may be able to avoid this effect by e.g. bias-sputtering, increasing the deposition temperature, or tapering the walls of the contact.

ACKNOWLEDGMENTS

The contributions of Pamla Bittencurt, Shonna Close, Susan McArthur, and Shirley Mc-Neel for wafer processing, and of Sandy Ooka for scanning electron microscopy are greatly appreciated. We would like to thank Sheau Chen for discussions.

REFERENCES

[1] D.B. Tuckerman and A.H. Weisberg, IEEE Electron Dev. Lett. **EDL-7**, 1 (1986).

[2] S.Q. Wang, and E. Ong, to be published in Proc. 7th Int. IEEE VLSI Multilevel Interconnection Conf., Santa Clara, CA (1990).

[3] R.J. Baseman, J. Vac. Sci. Technol. **B-8**, 84 (1990).

[4] R. Liu, K.P. Cheung, W.Y.-C. Lai, and R. Heim, Proc. 6th Int. IEEE VLSI Multilevel Interconnection Conf., p. 329, Santa Clara, CA (1989).

[5] S. Chen, and E. Ong, Proc. of the SPIE **1190**, "Laser/Optical Proccessing of Electronic Materials", p. 207, (1989).

[6] R. Mukai, N. Sasaki, and M. Nakano, Mat. Res. Soc. Symp. Proc. **74**, 229 (1987).

[7] B. Woratschek, P. Carey, M. Stolz, and F. Bachmann, Proc. 6th Int. IEEE VLSI Multilevel Interconnection Conf., p. 309, Santa Clara, CA (1989).

[8] D. Pramanik, and S. Chen, Proc. of the Int. Electron. Dev. Meeting, p.673 (1989).

[9] E.K. Broadbent, K.N. Ritz, P. Maillot, and E. Ong, Proc. 6th Int. IEEE VLSI Multilevel Interconnection Conf., p. 336, Santa Clara, CA (1989).

[10] R. Baseman, J. Andreshak, A. Gupta, and C.-Y Ting, in "Selected Topics in Electronic Materials", ed. by B.R. Appleton, and D.K. Biegelsen, **EA 18**, 259 (Materials Research Society, Pittsburgh 1989).

[11] R. Baseman, and J. Andreshak, to be published in Mat. Res. Soc. Symp. Proc. **158** (1990).

[12] I.J.M.M. Raaijmakers, T. Setalvad, A.S. Bhansali, B.J. Burrow, L. Gutai, and K.-B. Kim, submitted for publication in J. Electron. Mater. (1989).

[13] R. Mukai, K. Kobayashi, and M. Nakano, Proc. 5th Int. IEEE VLSI Multilevel Interconnection Conf., p. 101, Santa Clara, CA (1988).

[14] P.F. Marella, D.B. Tuckerman, and R.F.W. Pease, submitted for publication in Appl. Phys. Lett. (1990).

INTERFACE MORPHOLOGY, NUCLEATION AND ISLAND FORMATION OF TiSi$_2$ ON Si(111).

Hyeongtag Jeon, C. A. Sukow, J. W. Honeycutt, T. P. Humphreys,
R. J. Nemanich, and G. A. Rozgonyi
Department of Physics and Department of Materials Science and Engineering,
North Carolina State University, Raleigh, NC 27695-8202

Abstract

In this study we investigate the formation mechanisms and morphology of TiSi$_2$ formed by deposition of Ti on atomically clean silicon substrates. Ti films of 50-400 Å thickness were deposited in ultra-high vacuum on Si (111) wafers and annealed to temperatures between 500-900°C. Films were monitored in situ with AES and LEED, and post preparation characterization was accomplished with SEM, TEM and Raman scattering. The results show that for films of thickness ≤ 100 Å the C49 TiSi$_2$ phase is stable over the entire 600-800°C temperature range. However, for films of 200-400Å thickness, the C49 to C54 phase transition occurs at temperatures varying from 700 to 800°C dependent upon film thickness. The high temperature annealing results in flat interface structures, and island formation is observed for all films with the C54 structure. The interface morphology and the mechanisms of TiSi$_2$ island and phase formation are discussed in terms of surface and bulk free energies considerations based on nucleation theory.

Introduction

In recent years titanium silicides have received considerable attention as low resistivity contacts in very large scale integration device applications. TiSi$_2$ has several advantageous material properties among the transition metal silicides, including its low sheet resistivity and high temperature stability [1,2]. However, a significant problem in the TiSi$_2$ material system is the nucleation of islands, resulting in a rough surface morphology [3,4,5].

Two different crystal structures of TiSi$_2$ have been identified after thin film reaction of Ti on Si [6]. These are TiSi$_2$ of the C49 and C54 crystal structures. The C49 structure is metastable (ie. it does not occur in the binary phase diagram), and it is formed during low temperature annealing at a temperature of ~ 450-550°C. In comparison, the transition to the stable C54 phase occurs at a higher temperature, typically > 650°C for Ti films of > 400 Å thickness. We have previously proposed that the nucleation of the C49 phase is due to a lower surface and interface energy than the C54 phase.

In this paper we describe the formation mechanisms and morphology of TiSi$_2$ films formed from thin film reaction of Ti on atomically clean silicon substrates. Moreover, the role of film thickness and substrate deposition temperature in determining the formation of the C49 and C54 TiSi$_2$ phases is investigated. The surface structure and surface chemical contaminants were examined by in-situ low energy electron diffraction (LEED) and Auger electron spectroscopy (AES). Phase identification and the surface and interface morphologies were examined by the ex-situ Raman spectroscopy, scanning electron microscopy (SEM) and transmission electron microscopy (TEM), respectively.

Experimental

Silicon (111) oriented substrates (25mm dia.) with resistivities of 0.8 ~ 1.2 Ω-cm (n-type, P doped) were used. The wafers were cleaned by UV-ozone exposure and spin etched with HF+H_2O+ethanol, 1:1:10 [7]. The wafers were then transferred into the UHV deposition chamber and heated to a temperature of 800°C for 10 minutes to desorb the residual oxide and hydrogen. The base pressure of the UHV system was typically <1x10^{-10} Torr. Following in-situ cleaning, the LEED showed the 7x7 Si(111) surface reconstruction. Oxygen and carbon contaminants were below the detection limits of AES. Ti films between 50Å and 400Å thickness were deposited by evaporation from a Ti filament on the atomically clean Si substrates at ambient temperature and annealed or the films were deposited directly onto substrates at temperatures in the range of ~500-900°C. The thickness was monitored using a quartz oscillator. After each Ti film deposition or annealing, the sample was cooled to room temperature and in situ LEED and AES measurements performed. Post preparation characterization of the grown layers included investigation by Raman spectroscopy, SEM, and TEM.

Results

Phase identification for each TiSi$_2$ film deposited at different substrate deposition temperatures were carried out using Raman spectroscopy. Illustrated in Table 1 is a summary of the results pertaining to the crystal structure of the TiSi$_2$ films in the temperature range of 500-900°C. For Ti film thickness ≤ 100Å the metastable C49 phase over the entire 600-800°C temperature range, and the transition to the C54 structure is not observed. However, for films of 200Å Ti deposition, the transition temperature to C54 occurs at 800°C, and for the thicker Ti films, such as 400Å, the phase transition to the C54 phase occurs after 700°C. It is apparent for Ti films between 200-400Å there is a trend correlating the TiSi$_2$ phase transition temperature with film thickness. In particular, the transition temperature from the metastable C49 to the C54 phase increases with decreasing film thickness.

Table 1 Summary of the TiSi$_2$ structure (C49 or C54) characterized by Raman spectroscopy for Ti films of 100 to 400Å and deposition or annealing temperatures in the 500-900°C range.

Thickness	500°C	600°C	700°C	800°C	900°C
50Å	-	C49	C49	C49	-
100Å	-	C49	C49	C49	-
200Å	C49	C49	C49	C54	C54
400Å	C49	C49	C54	C54	C54

Consider now the interface morphology during the phase formation process. Immediately after all thicknesses of Ti deposition on room temperature substrates, the AES shows no evidence of the 92eV Si(LVV) peak [3]. This indicates that the films uniformly cover the surface. The morphology of the interface for those samples deposited at room temperature shows an abrupt flat profile with a relatively uniform Ti film. The changes of interface morphology are displayed in the cross-sectional TEM micrographs shown in Fig. 1. The initial of 400Å Ti films deposited on Si(111) substrates at temperatures of (a) 500°C and (b) 700°C, respectively. A rough interface morphology is observed for Ti films deposited or annealed at 500°C to 600°C. These temperatures correspond to the formation of the C49 phase of TiSi₂. In contrast, for films deposited or annealed at >700°C, the morphology changes, and a sharp, flat interface forms. The C49 films deposited at low temperature show a high density of planar defects, but for the C49 TiSi₂ phase formed at high temperatures (≥800°C) the defects are significantly reduced. In fact high resolution TEM shows that the films are epitaxial and pseudomorphic.[8,9]

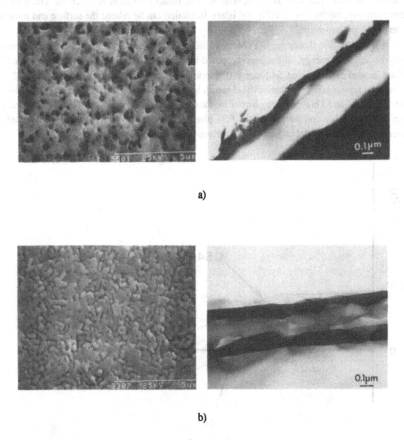

a)

b)

Fig.1 SEM and TEM micrographs of the 400Å Ti deposited on a 500°C (a) and a 700°C Si(111)
(b) substrates.

The surface morphology of the films also changes during the formation process. This was displayed both in the SEM and LEED. Immediately after deposition, no LEED pattern was observed.   However, after deposition or annealing at 700°C, the LEED showed a weak pattern characteristic of the 7x7 reconstructed Si(111) surface.   Indeed, the reappearance of the reconstructed substrate LEED pattern is a clear indication of the onset of island formation. Further evidence pertaining to the morphological properties of the islands can be obtained from an inspection of the corresponding SEM micrographs shown in Fig. 1.

Discussion

The results of this study address the nucleation of a metastable phase and transformation to stable phase. The basic results are that the transition temperature is dependent on thickness, the interface becomes flat at higher temperatures, and island formation is observed. The interface morphologies, nucleation of TiSi$_2$ and island formation can be related the surface and interface energies of the structures.

In nucleation theory the free energy of the nucleus is the summation of the surface energy and the bulk free energy. It has been proposed that the initial nucleation of the metastable C49 phase instead of the stable C54 phase is due to a lower surface or interface energy [3,10,11]. A schematic representation of this model is shown in Fig. 2. There is a lower free energy barrier for the C49 phase and a higher energy barrier for the C54 phase, and this is attributed to an increase in the surface energy component for the C54 phase. It should be noted that the usual "surface energy" term of the model includes both a surface and interface contribution.

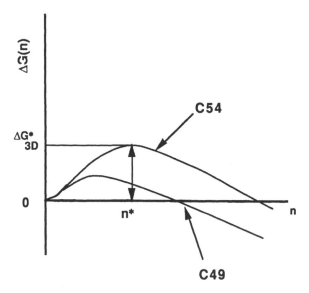

Fig.2   A schematic of free energy vs nucleus size.

The higher surface or interface energy of the C54 phase is also consistent with the island formation aspects. The results indicate island formation for the C54 phase. It is interesting to note that the observed 7 x 7 LEED indicates that the Si regions between the islands are atomically clean. The contact angles of the island structures have previously been used to estimate the surface and interface energies of the structures [5].

The character of the rough interface associated with the low temperature C49 phase is probably due to kinetic limitations. It has been established that at ~200°C, interdiffusion occurs which is followed by the nucleation of the C49 phase. It is likely that the rough interface associated with the C49 phase is due to different diffusion character for the random Ti grains which results after the deposition. At high temperatures, the interface of the TiSi$_2$ is much smoother with larger grain sizes (see Fig.1). The smooth interface and the island coalescence results in a reduction in interface area. Indeed, the reduction of the surface-to-volume ratio in structures is due to diffusion which allows minimizing the surface energy.

A thickness dependence for the TiSi$_2$ phase transition temperature is observed for films of thickness $\leq$ 400Å. In order to account for the C49 phase obtained for Ti films depositions of $\leq$100Å, we propose that there is insufficient free energy to overcome the free energy barrier (ΔG) to form the C54 phase. The interface energy and surface energy of TiSi$_2$ play an important role in the TiSi$_2$ formation. Two aspects are important: the surface to volume ratio and the pseudomorphic character of the C49/Si(111) interface. In the thin and thick film cases, the surface area of thin film is much larger than that of the thick film. Therefore, the nuclei formed after thin film deposition ($\leq$100Å) exhibit a higher surface energy component. In contrast, the nuclei formed after thick film deposition will exhibit a lower surface to volume ratio, and the interface energy will not play as large a role. The pseudomorphic interface of the C49/Si(111) is likely to have a lower energy than that of the C54 on Si. This will also contribute to stabilize the C49 thin film structures. The increase in the free energy barrier height of the C54 phase results in a higher transition temperature. In the case of the Ti depositions of $\leq$100Å, the phase transition from the C49 phase to the C54 phase is not observed.

Summary

The surface and interface morphologies, phase transition temperature and island formation of TiSi$_2$ films have been examined. The results clearly show that for films of $\leq$ 100 Å thickness the metastable C49 TiSi$_2$ phase is stable up to 800°C. For films between 200Å and 400 Å, the transition temperature to the C54 TiSi$_2$ is dependent upon film thickness. In particular, it has been demonstrated for films of 200 Å and 400 Å thickness the transition temperature decreases from 800°C to 700°C. Island formation is observed for the C54 films and the epitaxial C49 thin films formed at >800°C. It is proposed that a higher surface and/or interface energy of the C54 phase plays a role in all three of the effects.

Acknowledgment

This work was supported by the National Science Foundation through grants DMR 8717816 and CDR 8721505.

References

1.   S. P. Murarka, J. Vac. Sci. Technol. **17**, 775 (1980).
2.   R. J. Nemanich, R. Fiordalice, and Hyeongtag Jeon, IEEE. J. Quant. Elect. **25**, 997 (1989).

3.   Hyeongtag Jeon, R. J. Nemanich, Thin Solid Films. **184**, 357 (1990).

4.   P. Revesz, L. S. Zheng, L. S. Hung, and J. W. Mayer, Appl. Phys. Lett. **48**, 1591 (1986).

5.   Hyeongtag Jeon, R. J. Nemanich, J. W. Honeycutt, and G. A. Rozgonyi, Mat. Res. Soc. Symp. Proc. **160** (1989) (in press).

6.   R. Beyers, and R. Sinclair, J. Appl. Phys. **57**, 5240 (1985).

7.   D. B. Fenner, D. K. Biegelsen, and R. D. Bringans, J. Appl. Phys. **66**, 419 (1989).

8.   Hyeongtag Jeon, J. W. Honeycutt, C.A. Sukow, T.P. Humphreys, R.J. Nemanich, and G.A. Rozgonyi, Mat. Res. Soc. Symp. Proc.,(Symposium **V**, 1990 Spring MRS Meeting).

9.   M. S. Fung, H. C. Cheng, and L. J. Chen, Appl. Phys. Lett. **47**, 1312 (1985)

10.   R. J. Nemanich, C. M. Doland, R. T. Fulks, and F. A. Ponce, Mat. Res. Soc. Symp. Proc. **54**, 255 (1986).

11.   R. J. Nemanich, C. M. Doland, and F. A. Ponce, J. Vac. Sci. Technol. **B5**,1039 (1987).

THE EFFECT OF AQUEOUS CHEMICAL CLEANING PROCEDURES ON SCHOTTKY CONTACTS TO N-TYPE GAAS

M.L. Kniffin and C.R. Helms
Department of Electrical Engineering, Stanford University, Stanford, CA 94305

ABSTRACT

The effect of interfacial oxides and impurities left by aqueous chemical cleaning procedures on the n-type Schottky barrier heights of various metal-GaAs contacts have been examined as a function of annealing temperature. The as-deposited barrier heights of metals which are expected to be reactive with respect to the native oxides of GaAs (Mn, Cr, Al and Ti) were the most sensitive to variations in the residual oxide thickness. Omitting the final oxide etch from the cleaning sequence resulted in a 50 to 60 meV reduction in the as-deposited barrier height of these metals. However annealing these contacts at temperatures as low as 175 °C results in a barrier height that is independent of the initial surface clean. In contrast, for metals which are expected to be unreactive with respect to the native oxides of GaAs (Ni, Au, Cu and Ag) the pre-deposition cleaning procedure had little effect on either the as-deposited or the annealed Schottky barrier heights.

INTRODUCTION

Conventional GaAs MESFET processes typically include a wet chemical cleaning sequence, aimed at minimizing oxides and contaminants on the wafer surface, prior to depositing the contact metallization. Though this step results in a substantial reduction of chemical and oxide residues, wafers which have been immersed in an aqueous chemical solution and subsequently exposed to air, will invariably show some oxygen and carbon contamination [1-4]. This residual oxide layer is ultimately incorporated into the metal-semiconductor interface, and in some instances can have a significant influence on contact properties and interfacial chemistry [4-7]. As substrate preparation is one of the least controlled aspects of diode fabrication, the sensitivity of contact properties to differences in interfacial layer thickness and composition has important implications in terms of device reproducibility and uniformity. Despite the importance of understanding such phenomenon, a clear and consistent picture of how GaAs surface chemistry can influence Schottky contacts has yet to emerge. If fabrication processes are to be appropriately modeled and controlled, a better understanding of the role of oxides and impurities in junction formation is needed. This is the motivation behind the research presented in this paper.

In this study, we have systematically investigated the effects of three different aqueous chemical cleans on the n-type Schottky barrier properties of metal-GaAs contacts. Eight different metal-GaAs systems were examined. The metals (Ti, Cr, Mn, Al, Ni, Au, Cu, and Ag) were chosen to represent a wide range of reactivity and metal work function.

EXPERIMENT

The samples were cut from (100) n-type (Si 5×10^{16} cm^{-3}) GaAs wafers. Prior to any other processing steps, Ni:Au:Ge thin films were deposited on the backsides of the wafers and annealed to provide an ohmic contact for subsequent electrical measurements.

All of the samples were degreased with a standard organic solvent clean, then etched in a $NH_4OH:H_2O_2:H_2O$ (1:1:10) solution. After etching in the ammonium hydroxide-peroxide solution for one minute, the wafers were rinsed in static DI water for 30 seconds and blown dry in flowing nitrogen This chemical cleaning procedure removes a considerable amount of material from the wafer surface, resulting in a nominally clean but oxidized GaAs surface.

After the ammonium hydroxide-peroxide clean, the wafers were divided into three groups, each receiving a different final clean. One set received no additional chemical treatment. The others were cleaned in one of the following aqueous solutions aimed at reducing the residual oxide thickness: $HCl:H_2O$ (1:1) followed by a 15 second DI water rinse or $NH_4OH:H_2O$ (1:10). The samples were then dried under flowing nitrogen and inserted into the either the electron-beam evaporator or the surface analysis chamber.

Mat. Res. Soc. Symp. Proc. Vol. 181. ©1990 Materials Research Society

The residual oxides and impurities left on the GaAs surface by these different substrate preparations were examined using a combination of Auger electron spectroscopy and X-ray photoelectron spectroscopy. The results will be reported in greater detail elsewhere .

For electrical measurements, circular Ti, Mn, Cr, Al, Au, Ag, Cu and Ni diodes were deposited onto the chemically etched GaAs substrates. The diodes were defined by placing a stainless steel shadow mask in front of the wafer during the evaporation. The barrier heights of the contacts were measured both before and after each of a series of sequential low temperature anneals, using conventional current-voltage and capacitance-voltage measurement techniques.

RESULTS AND DISCUSSION

Auger scans of the GaAs surface following each of the aforementioned cleans are shown in figures 1 and 2. One can see that carbon and oxygen are the major contaminants, and that the extent to which the surfaces is oxidized varies depending upon the surface clean.

The surfaces etched in the ammonium hydroxide-peroxide solution were fairly heavily oxidzed, as evidenced by the relatively large oxide component visible in the As_{LMM} signal. Quantitative analysis of the Auger and XPS spectra, indicates that the ammonium hydroxide-peroxide etch leaves a relatively thick residual oxide (8 to 10 Å) layer containing both gallium and arsenic oxides. Typically less than half of a monolayer of carbon was present on the surface of the peroxide-etched samples.

As one can see from figure 2, when GaAs is cleaned in a dilute acid or base the residual oxide layer is much thinner (< 4-5 Å). These surfaces tended to getter impurities upon exposure to air, with the carbon contamination typically exceeding half a monolayer. Trace quantities of sulfur, chlorine, and nitrogen were also sometimes observed on the HCl-cleaned surface. Both of the oxide-stripped surfaces were somewhat arsenic-rich, although we found that the HCl-based clean tended to leave a much larger quantity of excess arsenic. These findings agree with the literature [1-3].

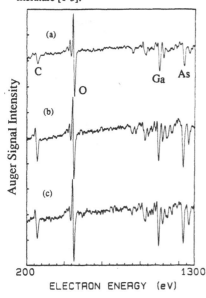

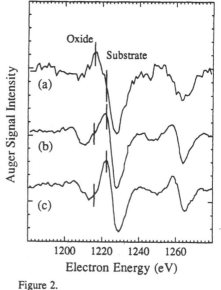

Figure 1.
Auger spectra of chemically etched GaAs surfaces:
(a) NH$_4$OH:H$_2$O$_2$:H$_2$O (1:1:10), (b) NH$_4$OH:H$_2$O (1:10) and (c) HCl:H$_2$O (1:1).

Figure 2.
As_{LMM} Auger spectra of clean and chemically etched GaAs surfaces:
(a) NH$_4$OH:H$_2$O$_2$:H$_2$O (1:1:10), (b) NH$_4$OH:H$_2$O (1:10) and (c) sputter cleaned. Vertical lines mark the positions of features corresponding to the substrate and chemically shifted oxide components.

Table I. The effect of chemical cleaning procedure on the Schottky barrier heights of as-deposited and annealed (350 °C) contacts to n-type GaAs.

		$NH_4OH:H_2O:H_2O_2$		$HCl:H_2O$		$NH_4OH:H_2O$	
		n	$\Phi_b{}^{IV}$	n	$\Phi_b{}^{IV}$	n	$\Phi_b{}^{IV}$
Ti	As-deposited	1.07	0.72	1.06	0.77	1.09	0.76
	Annealed	1.09	0.82	1.08	0.82	1.07	0.82
Cr	As-deposited	1.06	0.68	1.07	0.73	1.06	0.73
	Annealed	1.08	0.77	1.06	0.76	1.07	0.77
Mn	As-deposited	1.07	0.71	1.07	0.77	1.06	0.77
	Annealed	1.06	0.81	1.08	0.81	1.06	0.80
Al	As-deposited	1.08	0.73	1.09	0.78	1.08	0.78
	Annealed	1.05	0.88	1.06	0.88	1.05	0.87
Ni	As-deposited	1.10	0.77	1.09	0.77	1.09	0.77
	Annealed	1.08	0.83	1.08	0.82	1.09	0.83
Cu	As-deposited	1.10	0.91	1.09	0.90	1.08	0.91
	Annealed	1.08	0.86	1.07	0.86	1.08	0.86
Au	As-deposited	1.10	0.81	1.09	0.82	1.09	0.82
	Annealed	1.06	0.76	1.07	0.77	1.06	0.77
Ag	As-deposited	1.10	0.90	1.09	0.88	1.09	0.89
	Annealed	1.06	0.82	1.06	0.82	1.06	0.83

The results of the IV measurements are summarized above in table I, where both the as-deposited barrier heights and the barrier heights following the highest temperature anneal (350 °C) are listed. In general the barrier heights determined by current-voltage measurements were in excellent agreement with those determined from capacitance-voltage measurements after correcting for image force lowering [5].

Variations in the levels of carbon contamination and excess surface arsenic concentration that arise from the choice of oxide stripping solution appear to have little effect on the Schottky barrier properties of metal-GaAs contacts. For all of the metal-GaAs systems that we examined, contacts to NH_4OH-water and HCl-water cleaned substrates had identical or nearly identical IV and CV barrier heights.

The residual oxide thickness proved to be a significantly more important parameter, particularly for metals such as Mn, Cr, Ti and Al which are expected to react to some extent with the native oxide upon deposition [9]. Typical current-voltage characteristics for Mn contacts to n-type GaAs are shown in figure 3 (a). Analogous behavior was observed for Ti, Al and Cr.

On substrates which received an oxide strip as the final clean, the as-deposited IV and CV barrier heights for these metals were generally within 0.02 eV of the values reported by Waldrop for heat-cleaned (100) GaAs [8]. This is not surprising as these metals are expected to reduce several angstroms of native GaAs oxides upon deposition [9]. It is possible that the metal reacts through the thin oxide layer to the substrate, thus mitigating the effects of the oxide on the electrical and chemical properties of the interface.

The thicker residual oxide present after a peroxide clean results in a noticeable decrease in the n-type barrier height of these metals. A 60 meV reduction in barrier height was observed for the Mn-GaAs interface and similar decreases in Schottky barrier height were observed for manganese, aluminum and titanium. This represents a shift towards ideal Schottky behavior. A similar shift in the barrier height of low work function metals on n-type GaAs due to the presence of a thick oxide layer has been observed by other authors [6,7], and has been attributed to a reduction in interface state density.

After annealing at 175 °C for 30 minutes, the barrier heights of contacts fabricated on wafers cleaned without an oxide strip increase to equal those of the contacts to NH_4OH-water and HCl-water substrates. This most likely due to the reduction of the interfacial oxide layer by the metal. Sequential anneals at 350 °C increased the barrier heights of all of the "reactive" contacts. Again, the annealed barrier height was independent of the initial clean.

In contrast, for Ni and the noble metals (Cu, Ag and Au), which are expected to be unreactive with respect to the native oxides of GaAs [8], variations in the pre-deposition cleaning procedure had a much smaller impact on the as-deposited Schottky barrier height, although diodes fabricated on the more heavily oxidized substrates tended to have slightly higher ideality factors. Typical current-voltage characteristics for Ag contacts to n-type GaAs are shown in figure 3 (b). Generally omitting the oxide strip resulted in less than a 20 meV variation in the barrier heights of these diodes, which is well within the bounds of measurement error.

Variations in the residual oxide thickness also had no effect on the thermal stability of these contacts. Upon annealing, we observed a monotonic decrease (increase) in the Schottky barrier heights of the noble metal (Ni) contacts. As in the as-deposited case, the pre-deposition cleaning procedure did not effect the barrier height.

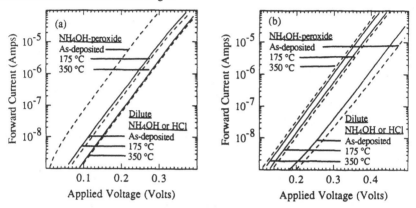

Figure 3. Forward IV characteristics of both as-deposited and annealed Schottky contacts to chemically etched n-type GaAs: (a) Mn-GaAs and (b) Ag-GaAs.
- - - - $NH_4OH:H_2O_2:H_2O$ etch, ——— $NH_4OH:H_2O$ or $HCl:H_2O$

ACKNOWLEDGEMENTS

This work was supported by the Joint Services Electronics Program at Stanford University Contract No. DAAL03-88-C-0011 and by DARPA Contract No. DAAL01-88-K-0828.

REFERENCES

1. C.C. Chang, P.H. Citrin and B. Schwartz, J. Vac. Sci. Technol. 14, 943 (1977).

2. R.P. Vasquez, B.F. Lewis and F.J. Grunthaner, J. Vac. Sci. Technol. B1, 791 (1983).

3. Z.H. Lu, C. Lagarde, E. Sacher, J.F. Currie and A. Yelon, J. Vac. Sci. Technol. A7, 646 (1989).

4. C.M. Garner, C.Y. Su, W.A. Saperstein, K.G. Jew, C.S. Lee, G.L. Pearson and W.E. Spicer, J. Appl. Phys. 50, 3376 (1979).

5. N. Newman, Z. Liliental-Weber, E.R. Weber, J. Washburn and W.E. Spicer, Appl. Phys. Lett. 53, 145 (1988)

6. M.T. Schmidt, D.V. Podelesnik, H.L. Evans, C.F. Yu, E.S. Yang and R.M. Osgood, Jr., J.Vac. Sci. Technol. A6, 144 (1988).

7. Brian Leslie Smith, PhD Thesis, University of Manchester Institute of Science and Technology, 1969.

8. J.R. Waldrop, Appl. Phys. Lett. 44, 1002 (1984).

9. Steven P. Kowalczyk, J.R. Waldrop and R.W. Grant, J. Vac. Sci. Technol. 19, 611 (1981).

STRUCTURAL AND SUBSTRUCTURAL ANALYSIS OF GLASS SURFACE

ZENON BOCHYŃSKI
Non-Crystalline Materials Division, Institute of Physics, Adam Mickiewicz University, 60-780 Poznań 2, Grunwaldzka 6, Poland.

ABSTRACT

A structure of surface and near surface layers in inorganic glasses used as substrated in metallization was studied. The analysed samples were made of melted quartz, technical and optical quartz glass and sodium-silicate glass.
The sample surface were at first treated mechanically ground and polished and then chemically etched.
In the investigation we applied the following methods:
- X-ray and electron /mean voltages 35...65 kV/ diffraction,
- electron emission microscopy,
- optical /UV-VIS/ and infrared /IR/ microspectrophotometry and microscopy.
For the studied series of quartz based /SiO_2/ glasses we determined the size of inhomogeneous regions and the distortion degree of the elementary tetrahedra /basic structural elements/ and estimated the thickness of surface and near-surface layers.
The obtained structural parameters enabled us to propose structural models of the surface of real glasses used in electrotechnology, electronics and microelectronics.

INTRODUCTION

A structural found in any quartz glass /and in group of quartz materials/ is a tetra of the SiO_4 type. Its presence is independent of the composition and chemical formula of silicate.
Considerably big oxygen ions /O^{2-}/ of a radius $r_0 = 1.40$ Å surround tetrahedrally small ion of silicate of radius $r_{Si} = 0.41$ Å. The ratio $r_{Si}/r_0 = 0.293$ enables us to assing a coordination number of 4 oxygen ions to silicon ion.
All major crystalline forms of SiO_2 quartz:
- quartz α and β /below 870 °C, hexagonal system,
  $\rho = 2.66$ g/cm^3/,
- tridymite /from 870 to 1.470 °C, hexagonal system,
  $\rho = 2.30$ g/cm^3/,
- cristobalite /from 1.470 to 1.705 °C, cubic system,
  $\rho = 2.27$ g/cm^3/,
are composed of SiO_4 tetrahedras linked by their corner oxygen ions each of which belongs to two silicon atoms; thus the stechiometric composition of SiO_2 is obeyed.
Thus each crystal of quartz is a gigantic macromolecule /$SiO_2/_n$ composed of only two kinds of atoms: of silicon and oxygen.
Crystalline silicate are divided into particular structural groups according to the type of bonding between silico-oxygen tetrahedra SiO_4 as the letter often differ considerably in the anion charge per Si iodate /0... -4/ or 0 iodate /0... -1/.
Mineralogy raports structures in which SiO_4 tetrahedrons have either island or group packing. Moreover, they form ring, chain, orlayer structures. On the other bond, in the spatial

structure of silicate all four oxygen atoms of each tetrahedrons /SiO$_4$/ belong at the same time to 2-tetrahedrons /SiO$_4$/.
All corner-adjucent tetrahedrons form a three-dimensional coordination lattice typical of all polymorphic types of quartz
/SiO$_2$/$_n$ in which the ration between silicon to oxygen Si:O =
= 1:2.

In particular types of monocrystals all oxygen-silicon
molecules SiO$_2$ under the influence of the surroundings form
a structure yielding either dextro- or laevorotary system.

On the other hand, in the fused quartz /SiO$_2$/ or in
quartz glass /SiO$_2$-R$_2$O$_3$/ all oxygen-silicon molecules SiO$_2$
form a nodal lattice of almost homogenous structural nods distribution in the space.

However, in the homogeneous lattice those may be formed
local areas of inhomogeneity which may occur in the surface
layer and in the whole volume of the sample under study.

A similar nodal lattice is formed by oxygen-silicon molecules SiO$_2$ in oxide glass /SiO$_2$-Na$_2$O-CaO-MgO-R$_2$O$_3$/ made from
quartz glass in which SiO$_2$ takes 72% wt.

SCHEME OF STRUCTURAL ANALYSIS

For the structural analysis of simple /and rendered sim-
ple/ atom systems of partial internal ordering, F. Zernike and
J.A. Prins /1927/ introduced for the first time analytical integral expression.

Zernike-Prins equation still widely used and modified
enables us to correlate the intensity of diffraction radii
with meat atom distributed in space.

Then W.H. Zachariasen /1932/, B.E. Warren /1942/, E.A.
Poray-Koshits et al. /1942/, J. Krogh-Moe /1956/ and N. Norman
/1957/ developed more and more advanced methods allowing for
analysis of more complex system including hetero- and poly-
atomic ones.

Quantitative structural analysis of relatively simple
quartz glass and a series of complex sodium silica glass were
performed by a method elaborated /1978/ and modified /1981...
...1983/ by the author.

By modification of equations obtained by the author and
by introduction of differential /by substraction/ radial distribution functions together with subsequent approximations
of effective molecule, more precise structural parameters of
nodal lattice of the studied molecular systems of surface layers were obtained.

First, at the first stage from the function of intensity
angular distribution $\bar{I}$/S/, for particular effective molecules,
summary

$$4 \pi r^2 \sum_{1}^{2 \to 5} \bar{K}_m \cdot \bar{\rho}_m /r/ =$$

and differential

$$4 \pi r^2 \left[\sum_{1}^{2 \to 5} \bar{K}_m \cdot \bar{\rho}_m /r/ - \sum_{1}^{2 \to 5} \bar{K}_m \cdot \bar{\rho}_o \right] =$$

functions of radial distribution were determined.

At the second stage, upon analyzing complex molecular systems /of different degress of local ordering/ of glass, differential functions of radial distribution in the form as follows may by successfully applied.

As a result, we obtain mean coordination number and mean distances:
- at the first stage only for particular effective molecules of the analyzed system,
- at the second stage only for particular complex atom groups /composed of complexes/.

EXPERIMENT

The samples of X-ray structural and IR spectrophotometric studies were plane-parallel plates of an optimum thickness, made of solid or powdered material.

The samples for microscopic optical and electronemission studies as well as for density measurements were plates made of a solid material.

The surfaces of the samples subjected to studies were either natural, mechanically treated by grinding or optical polishing, or chemically treated.

For X-ray structural studies - in the range of broad and mean scattering angles - automatic registration goniometers HZG 3 and HZG 4 produced by Freiberger Präzisionsmechanik - C. Zeiss /Jena, GDR/ were applied. The goniometers worked in the Bragg-Brentano system /upon transmission and at reflection/ and in the Debye-Scherrer-Hull system coupled with highly efficient monochromators constructed by us. Monochromators produced very monochromatic radiation /of the order ±0.0012 Å and ±0.0021 Å corresponding to the radiation 1.5418 Å of K_αCu and 0.7107 Å of K_αMo/.

For electron-emission mapping of surface layers an emission electron microscope EF Z2...Z6 C. Zeiss /Jena, GDR/ was used. A direct picture of the mapped surface was obtained in three methods:
- thermal emission of source by electron bombardment /resolution up to 150 Å/,
- secondary electron emission by ion bombardment /resolution up to 150 Å/,
- secondary electron emission by electron bombardment /resolution up to 300 Å/.

In microscopic optical studies 3 systems of high standard C. Zeiss microscopes /Jena, GDR/ from the AMPLIVAL series were used. Thereby for observation on transmission only PERAVAL interphako was employed, for both transmission and reflection - - AMPLIVAL pol interphako, and only for reflection - EPIVAL interphako. As for as resolution power in these systems is concerned, on transmission it attains the order of 2.000...3.000Å, whereas on reflection, 1.000...2.000 Å. The enlargement of surface representation possible to obtain when using the optical systems attained 1.250 : 1, and for a high quality contrast up $1/500\lambda$.

These apparatus enabled us to perform measurements of thickness with the accuracy of ±0.002 μm and of the changes in the coefficient of refraction of light, with the accuracy of $\pm$ 0.0002.

In spectrophotometric studies of surface and internal layers, a sound registration spectrophotometer SPECORD 75 IR - C. Zeiss /Jena, GDR/ was employed.

RESULTS AND CONCLUSION

Quartz glass many be formed as a result of melting of different forms of crystalline quartz or of silica.

In all quartz glasses and in the whole group of oxide silicate glasses, there occurs a structural unit. It is a tetrahedron of the SiO_4 type and its presence - in a more or less distorted form - is independent of the chemical composition of glass.

In the crystalline forms of quartz silicon ion $/Si^{4+}/$ may be easily assigned a coordination number of 4 oxygen ions $/O^{2-}/$, whereas on the other hand, in quartz glasses the coordination number as smaller, ranging from 2.86 to 3.58 of oxygen ions $/O^{2-}/$.

Thus, as follows from the quantitative X-ray analysis, silicon ion $/Si^{4+}/$ may be surrounded at most by 4 oxygen ions $/O^{2-}/$, whereas a medium size silicon ion $/Si^{4+}/$ by 2.86...3.58 oxygen ions $/O^{2-}/$.

The lowest oxygen limit based on structural model may equal to 2...2.5 oxygen ions $/O^{2-}/$.

Due to this, each oxygen ion may occur at best between the two closses lying silicon ions /belonging in 1/4 to each of them/. Usually, however, it is placed between less than 2 ions /between fractionel values/ of silicon belonging to the

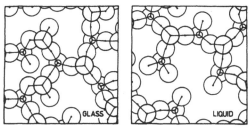

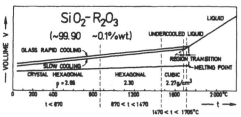

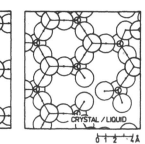

Fig. 1. Models of nodal lattice and parameters of phase transitions for quartz glass.

closest neighbourhood. Thus the bonding between the two ions of silicon and oxygen make a maximum distance in the whole tetrahedrons /SiO_4/ or their fragments /SiO, SiO_2, SiO_4/ of different degree of distortion in a linear arrangement of -Si-O-Si-ions. Macroscopic density of quartz glass SiO_2-R_2O_3/ ρ_1 = = 2.19923...2.20000 g/cm^3/ is also considerably lower than the density of monocrystalline quartz SiO_2 /ρ_2 = 2.65210 g/cm^3/.

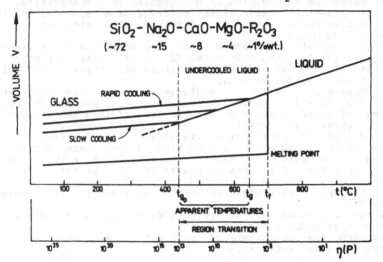

Fig. 2.   Parameters of phase transitions for sodium-silicate glass.

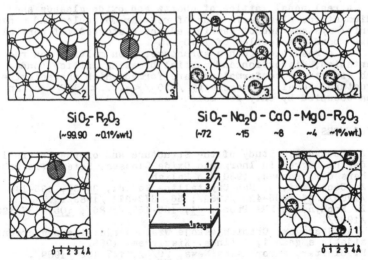

Fig. 3.   Aperiodic models /structural and substructural/ of nodal lattice of quartz and sodium-silicate glass surface.

As a result of melting of monocrystalline quartz and a subsequent formation of quartz glass, the mean volume accessible in the whole glass mass for the tetrahedron molecules of SiO_4 /or for their equivalent fragments/ is greater by about $17.0468...17.0759\%$.

It proves that in the quartz glass of $SiO_2-R_2O_3$ 5 tetrahedron molecules SiO_4 /or their equivalent fragments/ take the same volume as is occupied by 6 molecules in a monocrystal.

Sodium-silica oxide glass is formed as a result of melting crystalline forms of wuartz of forms of silica with oxides /Na_2O, CaO, MgO/.

Structural units formed in the sodium-silica glass /from quartz glass in which SiO_2 makes 72 wt. %/ are tetrahedrons of the SiO_4 type with different degrees of distortion together with bonded Na, Ca, Mg ions.

Macroscopic density of sodium-silica glass $SiO_2-Na_2O-CaO-MgO-R_2O_3$ /$\rho = 2.499_5...2.500_6$ g/cm^3/ is also smaller /despite introduction of 28 wt. % of metalic oxides/ than the density of monocrystalline quartz SiO_2 /$\rho_2 = 2.65210$ g/cm^3/.

Due to the melting of crystalline quartz of silica with appropriate oxygen and to a formation in next stage of oxide sodium-silica glass, the mean weight accessible on the whole glass mass for tetrahedron molecules of SiO_4 /or for their equivalent fragments/ is about $5.708...5.769\%$ greater.

Through Fourier analysis we have obtained for the studied samples:
- structural parameters of averaged distribution of threedimensional nodal lattice,
- substructural parameters of local extreme sequences /in the nodal lattice/.

RECAPITULATION

A real nodal lattice of quartz and oxide glasses and sodium-silica glasses is made of regular molecules and atomic /ionic/ systems and of reoriented and distorted /of partially broken bonds/ molecules of /$SiO_2/_n$.

As a result of such a structure, upon melting /on reorientation, distortion and partial breaking lattice bonds/ such structural conformations are formed which characterize structural local inhomogeneity /SiO_2/ < ... < /SiO_4/ of an area occupied by only few molecules.

REFERENCES

1. Z. Bochyński, Study of the Structure and of the Internal Ordering Degree in Inorganic Oxide Glasses, Scientific Publishers UAM, Poznań, 1980 /in Polish/.
2. Z. Bochyński, J. Non-Crystalline Solids, 38-39, 135-140 /1980/; 46, 405-425 /1982/; 56, 373-374 /1983/.
3. Z. Bochyński, MRS Proc., 61, 135-157 /1986/; 105, 187-190 /1988/.
4. W.A. Szaragow, Chimichjeskoje wzaimodjejstwije powjerchnosti stjekla s gazami, Sztinca, Kiszinjew, 1988.
5. H. Gleiter, Europhysics News, 20/9/, 117-136 /1989/.

Low Resistance α-Ta Film for Large Area Electronic Devices

Shigeru Yamamoto,Takehito Hikichi and Toshihisa Hamano

Fuji Xerox,Hongo 2274 Ebina-shi Kanagawa Japan

ABSTRACT

A New method of the low resistivity α-Ta deposition technique has been developed. The sputtered Ta film deposited on TaMo alloy has bcc structure (α-Ta) in contrast to tetragonal of the Ta mono layer (β-Ta), and shows the resistivity as low as $22 \mu\Omega \cdot cm$. The mechanism of this transformation is not explained by simple epitaxial growth. X-ray diffraction analysis and RBS analysis indicated that the TaMo alloy layer, due to Mo diffusion from the under layer, acts as a seed plane of growth. The Ta/TaMo layered film is suitable for Large Area Electronic devices for its low resistivity.

I. INTRODUCTION

Recently, the Amorphous Silicon Thin Film Transistor (a-Si TFT) technology has been applied for Large Area Electronic Devices such as a linear image sensor[1] and active matrix LCDs. In the inverted staggered TFT, the resistivity of the first metal layer as the gate line is important. Switching delay and signal distortions caused by CR constants of the gate lines, limit panel size and performance of devices. In another sights, production yield strongly depends on break down defects of the gate insulator at TFTs and crossings in the multiplex circuit. Therefore the thickness of the gate metal is required as thin as possible. Tantalum is suitable for its ability to form TaOx by anodization. The double insulator of SiNx and anodized TaOx increases yields and shows good TFT characteristics. However, it is a serious problem that sputtered Ta film is tetragonal β-Ta and shows high resistivity about $180 \mu\Omega \cdot cm$. Recently, alloy films based on Ta such as TaMo[2] and TaW[3] were reported to have bcc structure same as that of bulk Ta and show relatively low resistivity $40 \sim 50 \mu\Omega \cdot cm$. And the sputtered Ta film on the TaMo layer (Ta/TaMo)[4] is transformed to bcc structure and shows lower resistance.

We have been investigated the mechanism of this transformation and got lower resistivity comparable to that of bulk phase α-Ta. In this paper the deposition method and the growth mechanism are described.

II. EXPERIMENT

RF magnetron sputtering machines was used. It has two RF magnetron 5 inch cathodes and a rotational substrate stage. The equipment scheme is shown in Figure 1. The background pressure is less than 2×10^{-6} Torr. TaMo alloy films were deposited from an alloy target whose composition of Ta is 65%. The purity of targets are 99.99%. To avoid oxidation of surface, two layers are deposited successively in vacuum. RF power and pressure were determined for reducing stress. The resistivity of upper Ta layer is calculated from the sheet resistance of underlayer measured by four point probe technique. The structure of films are determined by X-ray diffraction

Mat. Res. Soc. Symp. Proc. Vol. 181. ©1990 Materials Research Society

analysis. The diffusion at the interface between two layers was estimated by RBS analysis.

III. RESULTS

Three configurations, Ta/Ta(N) , Ta/Mo and Ta/TaMo, were studied to intend to grow epitaxially. Figure 2 shows each X-ray diffraction pattern of mono layer deposited on the glass substrate. These X-ray spectra indicated that Ta(N), TaMo and Mo have same structure of bcc and (110) oriented. Table I shows 2θ values originated in (110) of these films. By X-ray

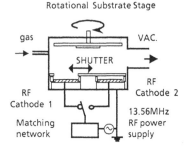

Rotational Substrate Stage

Fig. 1 Schematic configuration of the sputtering machine.

diffraction analysis, Ta films on TaMo and Mo were determined as α-Ta in contrast to β-Ta of Ta/Ta(N). Figure 3 shows typical X-ray diffraction patterns of α-Ta/TaMo and β-Ta. Figure 4 shows the changes of the resistivity of Ta upper layer with the variation of its thickness. The resistivity of Ta/Mo and of Ta/TaMo decrease and saturate to $22\mu\Omega\cdot$cm with a increase of thickness. Ta/TaMo shows lower resistivity than that of Ta/Mo especially in thiner region. Figure 5 shows shift of (110) peak with a increase of Ta thickness. Peak shifts from extremely lower value at the beginning of the deposition and rapidly increases until thickness is less than 500Å, and then saturates. Figure 6 shows RBS spectra of Ta/TaMo and of Ta/Mo. Mo diffuse to upper Ta layer about 300Å in Ta/TaMo , in contrast to about 100Å in Ta/Mo.

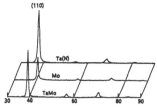

Fig.2 Xray diffraction patterns of mono layer

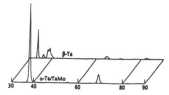

Fig.3 X-ray diffraction pattern of α-Ta/TaMo and β-Ta

Table I properties of Sputtered films

films	2θ of (110) [degree]	resistivity [$\mu\Omega\cdot$cm]
TaMo	38.20	104
Mo	40.62	31
Ta(N)	37.75	93
Ta/Mo	38.52	25
Ta/TaMo	38.20	22
Ta/Ta(N)	tetragonal	200

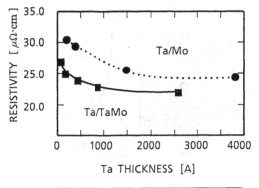

Fig.4 Changes of resistivity and depend on Ta thickness in Ta/Mo and Ta/TaMo layered structure

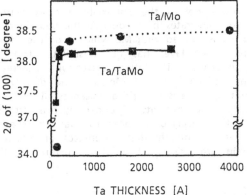

Fig.5 Changes of 2θ (110) depend on Ta thickness in Ta/Mo and Ta/TaMo layered structure.

IV. DISCUSSION

From these results, it is clear that the mechanism of α-Ta growth is not the simple epitaxial growth ordered by the lattice of underlayer in such as semiconductor case. Because Ta on Ta(N) didn't transformed to α-type, although the lattice mismatch of Ta(N) to α-Ta is smaller than that of Mo. X-ray diffraction analysis indicates that the lattice constants is not shifted continuously from that of the under layer. And there is the intermediate layer that is largely strained in both case of

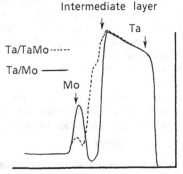

Fig.6 RBS spectra of Ta/TaMo and Ta/Mo.

Ta/TaMo and Ta/Mo. Accordingly it is suggested that α-Ta cannot grow epitaxially under the condition only small mismatch exists. The thickness of intermediate layers are comparable to Mo diffusion length shown by RBS data.

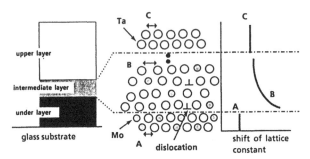

Fig.7 A growth model with intermediate layer

We propose the growth model of the epitaxial-like growth from intermediate alloy layer. Figure 7 illustrates the model. At the beginning of the deposition, TaMo alloy layer is formed with Mo which is diffused from underlayer. This intermediate TaMo layer has bcc crystalline structure naturally, but also has many dislocation. The lattice relaxes with a increase of Ta thickness, because the concentration of Mo and the density of dislocations decrease. Accordingly there is the critical thickness, at which α-Ta is able to grow epitaxially. The resistivity of upper layer is strongly depend on the grain size and the density of dislocations, that is succeeded from the intermediate layer. It is probably that differences of resistivity between Ta/Mo and Ta/TaMo are originated in the structure of the intermediate layer.

V. CONCLUSION

Low resistance α-Ta films by layered structure of sputtered Ta/TaMo and Ta/Mo were studied. We got the results as follows.

1 Sputtered Ta films on TaMo and Mo have bcc (α-type) crystalline structure in contrast to tetragonal (β-type) of conventional sputtered Ta films on glass substrate.

2 The growth mechanism of α-transformation is not simple epitaxial growth.

3 The intermediate layer, due to Mo diffusion from the underlayer, is needed to transform to α-Ta. This intermediate layer acts as a seed plane of the α-Ta epitaxial-like growth.

ACKNOWLEDGMENT

The authors gratefully acknowledge Dr.Kurosaki for RBS analysis.

REFERENCE

1) H.Ito,T.Suzuki,M.Nobue, Y.Nishihara, Y.Sakai,T.Ozawa and S.Tomiyama, Proc. MRS Symp.,95 , 437 (1987)

2) M.Dohjo, T.Aoki, K.Suzuki, M.Ikeda, T.Higuch and Y .Oana, S.I.D '88 Digest, 330 (1988).

3) T.Nomoto, Sekodo, I.Abiko, K.Kaminishi, Proc. Japan Display '89, 502 (1989)

4) M.Ikeda, M. Dohjo and Y.Oana, J.Apple.Phys. 66 (5), 2052 (1989)

PROPERTIES OF LASER PLANARIZED ALUMINUM ALLOY FILMS

Seshadri Ramaswami and Jonathan Smith
Advanced Micro Devices, Sunnyvale, CA 94086

ABSTRACT

An excimer laser (XeCl, 308 nm) has been used to locally melt and reflow aluminum alloy films into contacts of VLSI device structures resulting in contact filling and local planarization. This technique has the potential to be the answer to the 'poor' aluminum step coverage problem. The first order problems are that of ablation and voids. Voids have been categorised as 'line voids' and 'contact voids'. Various processing conditions of the barrier, Al-Si-Cu alloy and Al-Cu alloy metallizations were explored to obtain the optimum process with the desired film characteristics.

INTRODUCTION

The use of laser energy for the planarization of aluminum films has been described in earlier publications [1-3]. In this study, laser planarization conditions of various 1 μm aluminum alloy films has been investigated. The intent was to address the first order problems and determine the metal films(s) and the planarization conditions that result in the widest process window. The fluence should be high enough to provide void free filling of contacts without causing ablation. The typical spot size of the laser beam on the wafer is 3 mm x 3 mm with an incident energy of 325 mJ per pulse. The pulse width is 25 ns FWHM with a nominal 80 ns base to base width. The pressure in the process chamber is typically 5.0 x 10⁻⁶ torr and the process temperatures are in the range of 250 °C to 400 °C.

EXPERIMENTAL

For structural studies, wafers with contact sizes of 1.2 μm, 1.0 μm and 0.8 μm were prepared. The contact profiles were either sloped (using a wet/dry contact etch or reflowed borophosphosilicate glass) or vertical with aspect ratios (contact height/diameter) of 1.3 : 1.0 to 1.0 : 1.0. The process sequence is that of barrier metal deposition, aluminum alloy deposition followed by laser planarization. Optical microscopy (under white light, 633 nm and 408 nm wavelength conditions) and scanning electron microscopy were used to study the topographical characteristics of these films.

The morphology of the planarized aluminum alloy films was studied on flat test wafers and on wafers with topographical structures. The underlying films were either barrier metallization or oxide. The findings outlining the methodologies to minimize 'contact' void formation are outside the scope of this paper. The reflow induced re-distribution of the constituent elements was studied using SIMS, Auger and Rutherford backscattering spectroscopy. These films were characterized extensively for stress, hardness, etch rate, crystallographic orientation, roughness, reflectivity, change in bulk resistivity and resistance to electromigration. Data on some of these parameters are presented in this paper.

RESULTS and DISCUSSION

A. Heat sink effects :

A contact to substrate acts as a heat sink for the over-lying metal. It has been determined that a metal mass within proximity of a contact to substrate is free of 'line' voids. Consequently, metal lines in high contact density (array) areas are less prone to voids as compared to lines in the periphery. Also, enhanced void formation is observed if the planarization is done under high substrate temperature (400 °C) conditions.

Material 'segregation' has been observed in areas that are over a heat sink to the substrate (fig 1). These appear as areas of contrast in a scanning electron microscope image. Using Auger microprobe analysis, this has been determined, in part, to be due to (a) differences in the native aluminum oxide thickness and (b) the presence of sub-surface copper rich phase. In some cases, the cooling mechanism forces the metal volume in the vicinity of a contact to form one aluminum grain.

B. Sensitivitiy to metallization schemes:

The process window was found to be independent of the barrier metallization thickness (in the range of 200 A° to 1000 A°), but strongly dependent on its stoichiometry. A good wetting surface (to aluminum) is desirable. The barrier metal - aluminum alloy interaction is also dependent on whether these depositions are done sequentially in-situ (in vacuum) or after an interim exposure to the ambient.

Al-Si-Cu films may be used if steps are taken to ensure that silicon precipitate formation is minimized during the deposition process. Silicon precipitates cause increased localized absorption of energy, which in turn enhances the propensity for ablation and/or void formation [4]. In general, film thickness uniformity, composition, grain size and reflectivity need to be well controlled. So as to eliminate voiding within vertical wall contacts, it is essential to minimise the as-deposited overhang of the aluminum film around the contact.

Films deposited at lower (e.g 100 °C) temperatures are more sensitive to pattern density (on planarization), than their hotter (e.g. 400 °C) counterparts. So, if a process is optimized for a dense contact array, this pattern sensitivity causes rheological problems in the peripheral structures.

C. Re-distribution of elements:

SIMS, Auger and RBS were used in an attempt to determine the extent of inter-mixing (on laser planarization) of the barrier/AlSiCu, barrier/AlCu films. In order to minimise ion mixing effects and other interferences and increase detection sensitivity, 1 μm films were etched back to thinner layers. In the case where silicon was precipitated at the aluminum alloy/barrier metal interface during the deposition process, a re-distribution was observed after planarization. Using these spectroscopic probes, it was difficult to acertain the zone of inter-diffusion. However, using the 'bulk' techniques outlined in the section below, it could be surmised that some inter-diffusion had taken place. Below the depth of 500 A° from the top surface, there was no detectable difference in the concentration of oxygen in the aluminum layer between the planarized and control samples.

D. Comparisons of 'control' and 'planarized' films for 'aluminum alloy' and 'barrier-aluminum alloy' metallizations:

These tests were performed on metallizations on test wafers with no topographical structures. The legends 'alloy-1' and 'alloy-2' in the figures refer to Al-Si-Cu and Al-Cu alloys. The legend 'alloy' refers to a furnace heat treatment at 350°C for 30 min in a Ar-10%H$_2$ ambient. The wafers that were part of the four studies mentioned below had a modified barrier stoichiometry. This stoichiometry was chosen to provide an enhanced wetting surface for the aluminum alloy film. In all cases, the standard deviation within a data set were comparable and hence not indicated in the figures.

Etch rate in a phosphoric acid based metal etch:

After planarization, the wet etch rate reduced by 10% on an aluminum alloy film and by 25% on a barrier - aluminum alloy film. This is indicative of barrier-aluminum alloy inter-

25% on a barrier - aluminum alloy film. This is indicative of barrier-aluminum alloy inter-mixing during planarization. On the other hand, 'control-alloyed' films exhibited a reduction of approximately 20%. Using this information, it is possible to hypothesize on the nature of grain modification for the alloying and planarization processes.

Surface roughness as measured by a stylus profilometer:

Upon the absorption of laser energy, the melting and quenching processes cause the roughness of 'alloy-1' and 'alloy-2' films to be higher by a factor of 4 as compared to the as-deposited films (300 A° in comparison to 75 A°, fig 2). Scanning electron micrographs show ridge formation and grain boundary grooving. When a barrier layer underlies these metal films, the resulting planarized film remains smooth (50 A°). This supports the inter-diffusion hypothesis, and corroborates the work of Woratschek et. al. [4], who have reported a smoother topography on planarized Ti/Al-1% Si films.

Sheet resistance:

Barrier-aluminum alloy films exhibit a 5% increase in the sheet resistance after planarization followed by a similar increase after the alloy step. However, the sheet resistance of the control samples remains relatively unchanged.

Film stress:

The stress of barrier - aluminum alloy films increased by a factor of 2 on planarization (fig 3). A subsequent stress relaxation was observed after the alloy cycle. However, stress of aluminum alloy films remained unchanged. Also, on the control samples, there was no change in film stress between the alloy and the deposition steps for the aluminum alloy and barrier-aluminum alloy metallizations. A typical stress-temperature hysterisis curve is shown in Fig 4. The curves for all the metallizations tested as part of this study were fairly similar, possibly indicating the occurence of similar micro-structural modifications in these samples.

CONCLUSIONS

Voids within metal lines have been observed in peripheral structures. The density of occurence of these voids can be controlled by choosing the appropriate metallizations. Silicon precipitation needs to be minimized. Al-Cu films may be the metallization of choice. Material segregation has been observed. This could lead to pitting due to preferred etching in subsequent plasma based process sequences. Inter-diffusion between the barrier and the aluminum alloy metallization has been detected for barriers with certain stoichiometries. Stoichiometries should be chosen such that the electrical integrity of the contact is not degraded after planarization.

Fig 1: Segregation in array

Fig 2: Roughness (A°)

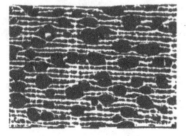

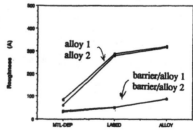

Fig 3: Tensile stress

Fig 4: Stress - temperature hysterisis

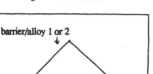

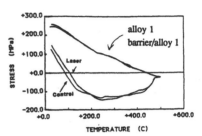

REFERENCES

[1]   D.B. Tuckerman and R.L. Schmitt, Proc. 2nd IEEE V-MIC Conf., Santa Clara, CA, p. 24, (1985).

[2]   R. Liu, K.P. Cheung, W.Y.-C. Lai, and R. Heim, Proc. 6th IEEE V-MIC Conf. , Santa Clara, CA, p. 329 (1989).

[3]   S. Chen and E. Ong, Proc. of the SPIE 1190, 'Laser/Optical processing of electronic materials', p. 207, (1989).

[4]   B. Woratschek, P. Carey, M. Stolz and F. Bachmann, Proc. 6th IEEE V-MIC Conf. , Santa Clara, CA, p. 309 (1989).

PART X

W and W-Alloys
Deposition Techniques

EFFECTS OF DWELL TIME AND CURRENT DENSITY
ON ION-INDUCED DEPOSITION OF TUNGSTEN

KHANH Q. TRAN*, YUUICHI MADOKORO†, TOHRU ISHITANI†,
AND CARY Y. YANG*

* Microelectronics Laboratory, Santa Clara University, Santa Clara, CA 95053
† Central Research Laboratory, Hitachi, Ltd., Kokubunji Tokyo 185, Japan

ABSTRACT
30-keV focused Ga^+ ion beam was used for induced deposition of small-area tungsten thin films from $W(CO)_6$ on Si and SiO_2. Deposition yield, calculated assuming pure tungsten depositions, depends on dwell time (beam diameter/scan speed) and beam current density. High current density and/or long dwell time are known to cause low deposition yield because of the depletion of adsorbed gas molecules during ion beam irradiation. Based on a model taking this effect into account, numerical fitting was carried out. The reaction cross-section was estimated to be 1.4×10^{-14} cm^2. For doses below 10^{17} ions/cm^2, film resistivity decreases with increasing dose. This was confirmed for several dwell times. However, for doses above 10^{17} ions/cm^2, film resistivity remains independent of dose. In this "high"-dose range, variation of beam current density has little effect on film resistivity. AES analyses revealed a consistency between film composition and resistivity. For a "high"-dose film with a resistivity of 190 $\mu\Omega$-cm, the approximate tungsten content was 50 at%.

INTRODUCTION
Applications of focused ion beam (FIB) in microelectronics processing technology are of great interest. Maskless implantation[1,2] and etching[1,3], as well as low-proximity-effect lithography[4], have been demonstrated. Another emerging technique, ion beam induced deposition (IBID), has applications in the area of x-ray lithography mask and microcircuit repair[5]. The deposition of tungsten is widely studied because it is suitable for this purpose. For these applications, repair of local defects is achieved by milling and subsequent deposition.

The mechanism for deposition is thought to involve ion/substrate interaction leading to the dissociation of adsorbed organometallic molecules. However, the kinetics of IBID have yet to be fully understood. In addition to determining parameters that affect growth, the need to correlate these parameters with film quality is of vital importance, especially where low-resistivity conductive film is desirable.

In this present study, beam current density and dwell time are varied to demonstrate their effects on deposition yield and film resistivity. Auger Electron Spectroscopy (AES) analyses of some samples have also been performed. From the results obtained, a simple model is adopted to estimate the reaction cross-section.

EXPERIMENTAL SETUP AND PROCEDURES
A modified scanning ion microprobe using Ga liquid metal ion source was used. The base pressure of the sample chamber was about 2×10^{-6} torr. Granular $W(CO)_6$ was heated to about 45°C and vaporized. The vapor was introduced through a small nozzle located at approximately 1mm above the sample and 100μm laterally away from the irradiated area. Thus, a localized organometallic ambient was created. The computer system controlled the scan speed, irradiation time, and scan area. The ion beam was

TABLE I. Experimental Conditions

Ion Species	Ga$^+$	
Accelerating Voltage	30 kV	
Gas Species	W(CO)$_6$	
Substrate	Si	SiO$_2$
Substrate Temperature	Room Temperature	
Dose	$6 \times 10^{16} \sim 10^{18}$ ions/cm^2	
Beam Current	$150 \sim 1000$ pA	
Beam Diameter (B.D.)	$1.5 \sim 3$ μm	
Scan Speed	$0.8 \sim 8$ cm/s	
Dwell Time	$20 \sim 260$ μs	
Beam Current Density	$4 \sim 30$ mA/cm^2	
Ion Source Chamber Pressure	3.1×10^{-6} torr	
Sample Chamber Pressure	1.7×10^{-5} torr	2.3×10^{-5} torr
Gas Source Temperature	41°C	45°C
Scan Area	(B.D. $\times$ 40) μm^2	[(B.D. + 6) $\times$ 70] μm^2

repetitively scanned to fabricate patterns with desired ion doses. Si and thermally oxidized Si substrates (100) were maintained at room temperature. The latter was covered with probe structures to measure the resistance of deposited films. Film thickness was obtained by stylus profilometry. From the film dimensions, resistivities were calculated.

Beam diameter was estimated by measuring the 10-90% amplitude of the secondary electron signal resulting from scanning over a Ni mesh. Beam current was measured by a Faraday cup. From the above, the beam current density and dwell time (beam diameter/scan speed) could be determined. Physically, the dwell time is a measure of momentary time exposure of the substrate area corresponding to the beam spot. Variations in beam current density and dwell time were achieved by careful adjustment of the beam current, beam diameter, and scan speed. All experimental parameters are summarized in Table I.

RESULTS AND DISCUSSION

Results of film growth as a function of dwell time are shown in Fig. 1 for Si and SiO$_2$ substrates. Film thickness varied from 150 to 1500 Å and was proportional to dose. As indicated, beam current densities for the substrates were kept at approximately 4 and 5 mA/cm^2, respectively. The yield, in units of deposited atoms/ion and representing a measure of ion efficiency, was calculated on the assumption that the deposit consists of pure tungsten. For both substrates, the yield decreases monotonically with increasing dwell time. Yield for Si ranges from 0.25 to a maximum of 0.83, while for SiO$_2$ the result was from 0.92 to 1.71. The yield disparity for the two substrates can be attributed to changes in W(CO)$_6$ ambient pressure (see Table 1) and difference in substrate. In comparison, yields of $1 \sim 2$ atoms/ion were reported by Stewart et al[6].

The role played by beam current density is similar to that of dwell time in affecting the local concentration of surface adsorbates. Fig. 2 shows the decrease in yield for increasing beam current density for both substrates. The dwell times were maintained at about 42 and 30 μs for Si and SiO$_2$ substrates, respectively. At maximum beam current density, deposition yield for Si decreased to 0.05 atom/ion, while for SiO$_2$, a minimum of 1.30 atoms/ion was reached. Higher beam current densities could result in a complete depletion of surface adsorbates, leading to sputtering of the substrate.

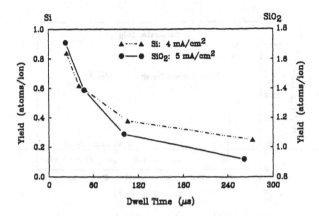

FIG. 1. Yield vs dwell time for constant beam current densities for Si and SiO₂ substrates.

Fig. 3 shows film resistivity as a function of dose. For doses below 10^{17} ions/cm², the beam current density was kept relatively constant with varying dwell time. The resistivities decreased sharply with increasing dose and ranged typically from 180 to 2500 $\mu\Omega$-cm. A similar trend was also observed by Madokoro et al[7]. Variation in dwell time was observed to have a minimal effect on the corresponding resistivity values. The minimum resistivity achieved is about 130 $\mu\Omega$-cm, roughly 25 times that of bulk tungsten.

In Fig. 3, the role of beam current density on resistivity is also depicted for doses above 10^{17} ions/cm². The film resistivity in this dose range remains constant and unaffected by increases in beam current density. For high doses, film resistivity is about 190 $\mu\Omega$-cm. This value is similar to those reported by Ref. 6. As the dose increases, a transition from disconnected tungsten islands to a more continuous film structure had been observed[8]. This transition is consistent with our observed drop in resistivity in the "low"-dose region.

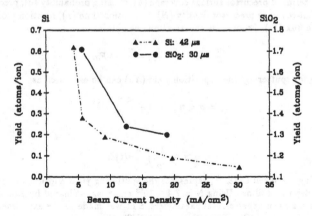

FIG. 2. Yield vs beam current density for constant
dwell times for Si and SiO₂ substrates.

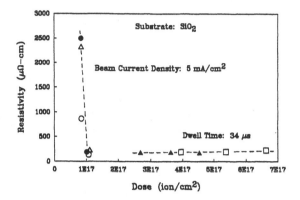

FIG. 3. Film resistivity vs dose for deposition on
SiO$_2$. Beam current density and dwell time were kept
constant for doses below and above 10^{17}ions/cm^2, respectively.

AES depth profile for a film deposited with a dose below 10^{17} ions/cm^2 reveals a
carbon-rich content. On the other hand, a film deposited with a dose of 6.6×10^{17} ions/cm^2
has tungsten and carbon atomic percentages of 50 and 35, respectively, with other constituents being oxygen and gallium. These results are consistent with the resistivity values
obtained.

In order to estimate the reaction cross-section, the model used by Scheuer et al[9] for
electron beam induced deposition was adopted. The assumptions put forth by Rüdenauer
et al[10] were also incorporated. These assumptions state that the precursor gas surface
coverage is confined to a monolayer and that only metal atoms can be sputtered from the
instantaneous top layer of the film. The model expresses the time rate of precursor surface
coverage in terms of precursor surface coverage (n), sticking probability (c), precursor gas
flux (F_g), full-coverage precursor density (N), desorption time (τ), reaction cross-section
(σ), and ion flux (F_i). From mass conservation one obtains

$$\frac{dn}{dt} = cF_g(1 - \frac{n}{N}) - \frac{n}{\tau} - \sigma F_i n \tag{1}$$

Accounting for sputtering, the deposition yield (Y_d) can be expressed as

$$Y_d = \sigma <n> -(1 - \frac{<n>}{N})Y_s \tag{2}$$

where

$$<n> \equiv \frac{1}{\tau_D} \int_0^{\tau_D} n(t)dt \tag{3}$$

with τ_D being the dwell time. Y_s represents the sputtering yield and was determined by
sputtering deposited films. Using $N = 7.4 \times 10^{13}$ cm^{-2}, as estimated by Koops et al[11],
together with known experimental parameters, a fit was made to the experimental yield
versus dwell time for SiO$_2$. From the numerical fit, the reaction cross-section was determined to be 1.4×10^{-14} cm^2. This means that on the average, adsorbates within a

6.7Å radius of the impacting ions are dissociated. In comparison, a reaction cross-section estimate of 3.6×10^{-13} cm^2 was obtained[7] based on the assumption of steady-state surface coverage.

CONCLUSION

A study of the effects of dwell time and beam current density on the ion-induced deposition of small-area tungsten thin films from $W(CO)_6$ has been performed. The results obtained indicate that the deposition yield decreases for increasing dwell time and increasing beam current density. High beam current density and/or long dwell time leads to depletion of adsorbates and possible sputtering. For doses below 10^{17} ions/cm^2, film resistivity decreases with increasing dose. Variation in dwell time did not appear to affect significantly film resistivity. For doses above 10^{17} ions/cm^2, the resistivity value is essentially independent of beam current density. AES analyses show that for a film deposited with a dose below 10^{17} ions/cm^2, the content is carbon-rich. On the other hand, for a film deposited with a dose of 6.6×10^{17} ions/cm^2, approximate tungsten and carbon atomic percentages are 50 and 35, respectively, with other constituents being oxygen and gallium. Finally, numerical fitting to a simple model yields a reaction cross-section of 1.4×10^{-14} cm^2.

ACKNOWLEDGEMENT

The authors thank Mr. Y. Kawanami for many useful discussions and K.T. is grateful to Dr. H. Obayashi and Mr. Y. Kuwahara for their support during his stay at Central Research Laboratory, Hitachi, Ltd.

REFERENCES

1. G.W. Rubloff, J. Vac. Sci. Technol., B7 1454 (1989).
2. E. Miyauchi and H. Hashimoto, Nuclear Instr. and Methods in Phys. Res., B6 851 (1985).
3. Y. Ochiai et al, J. Vac. Sci. Technol., B5 423 (1987).
4. S. Matsui et al, J. Vac. Sci. Technol., B4 845 (1986).
5. J. Melngailis, J. Vac. Sci. Technol., B5 469 (1987).
6. D.K. Stewart, L.A. Stern, and J.C. Morgan, SPIE Proc., 1089 18 (1989).
7. Y. Madokoro, T. Onishi, T. Ishitani, 20th Symp. on Ion Implantation and Submicron Fabrication, Rikagaku Kenkyusho, 133 (1989).
8. Y. Takahashi, private communication.
9. V. Scheuer, H. Koops, and T. Tschudi, Int. Conf. Microcircuit Engineering, 5 423 (1986).
10. F.G. Rüdenauer and W. Steiger, J. Vac. Sci. Technl., B6 1542 (1988).
11. H.W.P. Koops, R. Weiel, and D.P. Kern, J. Vac. Sci. Technol., B6 477 (1988).

DEPOSITION OF TUNGSTEN BORIDE BY ION BEAM SPUTTERING

F.MEYER, D.BOUCHIER, V.STAMBOULI AND G.GAUTHERIN
Institut d'Electronique Fondamentale d'Orsay. U.R.A. D0022 CNRS.
Université Paris XI - Bât.220 - 91405 ORSAY CEDEX FRANCE.

ABSTRACT

Refractory metal compounds, such as nitrides or borides, are attractive candidates for diffusion barrier between silicon and aluminium in VLSI technology. We studied tungsten boride films deposited on silicon (100) by ion beam sputter deposition (IBSD).

The tungsten boride films were prepared by sputtering a W_2B_5 target by argon ions with energy ranging from 0.5 to 2keV. The substrate temperature was varied from room temperature to 630°C. Finally, the films were patterned by selective wet etching in order to characterize the resulting Schottky diodes. We observed that a boron loss occurs during deposition, probably due to the backscattering of sputtered boron on previously deposited W atoms. By *in situ* AES analysis, we verified that a 5 nm thick layer acts as a diffusion barrier for silicon up to about 630°C, for all deposition conditions. The films properties were found to depend weakly on the primary ion energy and on the substrate temperature. All the films have resistivity at room temperature in the range of about 250 $\mu\Omega$ cm. The measured density, in the range of 12 g/cm^3, is very close to that of WB_2 bulk material, while the intrinsic stress of the films remains compressive and in the range of -1GPa. This value is notably lower than what we measured for pure tungsten prepared under similar deposition conditions.

I-INTRODUCTION.

The borides, carbides, nitrides and alloys of refractory metals are attractive candidates for diffusion barrier in contact metallizations, because of their high temperature stability and their rather good electrical conductivity [1,2].

Various diffusion barriers that are used in VLSI circuits involve tungsten. Thus Ti-W [3], W-Re [4] and W-N [4,5,6] have been widely investigated and some researchers have worked on W-C layers [7]. Reactive evaporation, reactive sputtering, cosputtering or sputtering from a compound target are mostly used to produce diffusion barrier films [4]. But to our knowledge, the study of tungsten borides is limited to the formation of these compounds during either chemical vapor deposition (CVD) of boron onto tungsten filaments for boron fiber manufacturing [8] or reaction between W and boron implanted silicon wafers [9]. In this last case, Torrès et al [9] have noted that tungsten boride formation may strongly slow down the reaction between silicon and tungsten. This result suggests that tungsten boride may be efficient to prevent interdiffusion or reaction between thin metal overlayers and silicon.

The aim of this study is to investigate the preparation of tungsten boride layers by ion beam sputter deposition (IBSD) in a UHV system. Properties such as morphology, structure, density, stress and resistivity have been studied as a function of deposition parameters. We also report preliminary results on the thermal stability of very thin tungsten films onto silicon substrates and on effects of thermal treatments on Schottky diode characteristics. The properties of IBSD W-B-films and W-films are compared.

II-EXPERIMENTAL DETAILS.

The tungsten boride films were formed by ion beam sputtering [10] using a 99.5% pure W_2B_5 target in a UHV system. The total pressure of gaseous impurities (H_2O, CO, CO_2, N_2...) remained below 5×10^{-7}Pa. The ion focusing system was optimized, to allow an ion beam energy down to 200eV with a beam current of 2mA and a working pressure (mainly noble gas) of about 2×10^{-4}Pa. In this work, the energy of argon ions were varied from 0.5keV to 2keV. Additional equipment attached to the deposition chamber includes a residual gas analyzer and an Auger electron spectrometer (CMA).

Before their introduction into the deposition system, 1-5 Ωcm p-type (100) oriented silicon wafers were degreased and chemically cleaned with the purpose of growing a very thin silicon oxide layer to protect the Si surface. The samples were kept in ethanol until their loading into the apparatus via a load lock chamber. Just prior to metallization oxide was eliminated *in situ* by flash heating at 900°C for 2min. The efficiency of these treatments was checked *in situ* by means of AES.

A Dektak profilometer was used to determine either the thickness of the films or the curvature of the substrates due to the intrinsic stress in the sputtered layers. Film density was evaluated by weighing the films by means of an ultra micro balance (Sartorius 4504 MP8). The microstructure of the films was estimated from X-ray measurements. XRD patterns were obtained using a CGR θ-2θ diffractometer CoKα radiation. The resistivity was measured at room temperature using a four point probe. Circular Schottky diodes (0.15×10^{-2} cm^2 area) were fabricated by photolithographic patterning and selective etching of the tungsten boride film. Capacitance-voltage (C-V) and current-voltage (I-V) measurements were performed to determine the barrier height.

In situ AES measurements were done to investigate the thermal stability of very thin tungsten boride films (thickness $\approx$ 5nm) on silicon during sequential annealing treatments for 30min at different temperatures.

To get additional informations, some analyses were done using a JEOL 840 scanning electron microscope (SEM) equipped with an energy dispersive spectrometer (EDS).

III- RESULTS AND DISCUSSION.

1) Incident ion energy effects.

a) Deposition rate.

The growth rate of a IBSD film depends on the rare gas (atomic mass and energy) and on the target used and is proportional to the ion current I$^+$. We can observe (Fig.1) that the deposition rate of tungsten boride is lower than that of pure tungsten for incident ion energy ranging from 0.5 to 2keV and is larger than that of pure boron for 2keV-Ar$^+$.

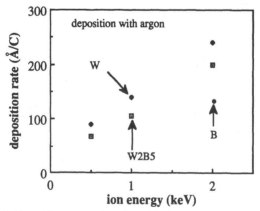

Figure.1. Deposition rate as a function of the ion energy.

b) Properties of films deposited at room temperature.

An Auger spectrum of a tungsten boride film is shown in Fig.2a. The spectrum is not modified when the primary ion energy varies from 0.5keV to 2keV. Therefore, we can speculate that the composition of the films may be only weakly dependent on the energy of the primary ions even if a severe overlapping between the main boron KLL (179eV) peak (Fig.2c) and the main tungsten NVV (169eV and 179eV) peaks (Fig.2b) prevents any simple quantitative analysis.

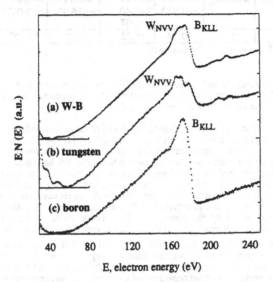

Figure.2. Auger spectra in the E N(E) mode, recorded from a) tungsten boride, b) pure tungsten, c) pure boron.

Changing the incident ion energy did not affect the film properties either. As indicated in Table-I, the density of tungsten boride films remain close to 12g/cm^3 for experimental parameters investigated. According to the fact that sputtering results in films as dense as bulk material [11], we have verified this assumption for IBSD-W-films (Table-I) in particular, it is possible to estimate the composition of the tungsten boride layers. Table-II shows that the measured density of the IBSD films is higher than that of the target material (W$_2$B$_5$) and agrees better with that of WB$_2$. This result indicates a loss of boron in the films with respect to the target . This loss of boron could evidence different angular distributions of ejected components or else a preferential resputtering or a reflection of a fraction of the arriving boron flux on the film. This last effect is very likely to occur as boron atoms are very light with regard to tungsten atoms.

Table-I. Properties of W-B and W films prepared by IBSD as a function of the incident ion energy. (All the stresses are compressive)

	tungsten boride ion energy (keV)			tungsten ion energy (keV)		
	0.5	1	2	0.5	2	20
density (g/cm^3)	12.3±0.5	12.0±0.5	11.0±0.5	19.3±0.5	19.3±0.5	19.3±0.5
stress (GPa)	1.1±0.2	1.2±0.2	1.0±0.2	2.5±0.5	2.5±0.5	2.5±0.5
resistivity (μΩcm)	250±25	250±25	260±25	33±3	60±6	40±4
argon content (at.%)	-	-	2	1.6	4.6	2.5

Table-II. Structure, lattice parameters and density of tungsten boride compounds

	α-W	γ-W$_2$B	β-WB	δ-WB	WB$_2$	ϵ-W$_2$B$_5$
structure	cubique	tetragonal	orthorombic	tetragonal	hexagonal	hexagonal
lattice parameters (Å)	a=3.16	a=5.564 c=4.740	a=3.19 b=8.40 c=3.07	a=3.115 c=16.93	a= 3.09** c= 3.04**	a=2.982 c=13.87
density (g/cm^3)	19.3	17.14	15.72	15.74	12.75*	9.27

(from ref-12, * from ref.13, ** this study).

The stress of all tungsten and tungsten boride films is compressive, and it is worth noting that stress in W-B films is twice as weak as in sputtered W and appears to be lower than that of tungsten nitride whatever the nitrogen concentration [5].

We can observe (Table-I) that the electrical resistivity of tungsten boride films is rather high and is much higher than that reported for bulk material (20-40$\mu\Omega$) [1]. This last result is often observed in thin films and is commonly related to the presence of defects, impurities or else to the texture. Argon is the only impurity we have detected in our films. We have measured by EDS the argon content in a film prepared with 2keV argon ions. We observed a lower argon incorporation in W-B than in IBSD-W films prepared under similar conditions (2% against 4.6%, respectively) (Table-I). Since rare gas incorporation in sputtered films is due to reflection of primary ions from the target, the lower incorporation of argon in W-B-films may be related to the decrease of the reflection coefficient caused by the presence of light boron atoms in the target. We have demonstrated [10] that argon atoms act as the most efficient electron scattering centers in IBSD W films which always exhibit the bcc α-structure with a strong <110> prefered orientation. The linear relation between resistivity of tungsten films and argon concentration leads to a strong dependence of resistivity with primary ion energy E_0 (Table-I). The fact that the resistivity of tungsten boride films remains constant regardless of the energy E_0 suggests that the resistivity of the present films correlates better with the amount of boron and/or the amorphous structure, as established by X-ray diffraction, than with argon contamination.

The etchability of thin films is another important property to determine their device applicability. We have verified that W-B-films can be patterned in a W etch (KH$_2$PO$_4$, KOH, K$_3$Fe(CN)$_6$, H$_2$O) at rate similar to that of W.

2) Interactions between W-B and silicon.

AES measurements taken immediately after the deposition did not detect any silicon on 100nm thick tungsten boride films grown at temperature up to 630°C. However SEM reveals that some defects developped in the films prepared at the highest temperature. These defects are few in number and do not alter the resistivity of the films that remains in the range of 250$\mu\Omega$cm over the whole temperature range. These results suggest that the films are rather stable, do not undergo important chemical and structural changes, and that interfacial reaction may only have occured.

X-Ray diffraction measurements support this assumption. It was found that the films consist of amorphous material with fine-grain (below 100Å) polycristalline material even for films prepared at temperature above 400°C. The two diffraction peaks (2θ =39° and 2θ= 53°) cannot be identified by comparison with published crystallographic data [12]. They correspond to an hexagonal phase with the lattice parameters a=3.09Å and c=3.04Å, and we speculate that they can be related to the hexagonal WB$_2$ compound. This assumption is supported by the fact that this compound exhibit a density close to 12g/cm^3 [13] and lattice parameters close to those of other hexagonal diborides [12].

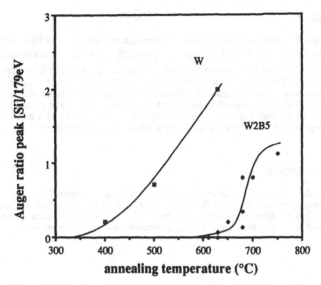

Figure.3. Auger peak ratio [90eV]/[179eV] as a function of annealing temperature of tungsten and tungsten boride thin films (thickness ≈ 5nm) on silicon.

In order to study the interfacial reaction of tungsten boride with silicon Auger measurements taken after sequential annealing of very thin films (t≈5nm) have been carried out. Fig.3 depicts the variation of the peak height ratio between the 179eV(W+B) and 92eV(Si) Auger lines with annealing temperature for W-films and W-B films. Two important aspects are revealed. First, the use of tungsten boride instead of tungsten results in an improvement of the barrier performance of the film. Second, our results further demonstrate the importance of oxygen on the thermal stability of metal barrier [4,14]. Indeed the low temperature for reaction between silicon and tungsten we obtained may be explained by the fact that the metallization is done on a clean silicon surface under ultrahigh vacuum conditions [14]. And the relative scattering of experimental results we have reported on Fig.3 accounts, even in a UHV system, for the oxygen contamination of the top of the film that occurs if the time between the deposition and the annealing is too much important. We have also observed such a behavior during the study of isothermal annealing of W on silicon [15], a fraction of a monolayer of oxygen on the top of a thin W-film is even able to inhibit the silicide formation up to 550°C.

Some electrical measurements have been performed on Schottky diodes fabricated by photolithographic patterning of the W-B film. First, increasing the deposition temperature seems to anneal out the radiation effects induced in the substrate during the metallization. The diodes prepared at 500°C exhibit rather good electrical characteristics. Barrier height of 0.64eV and ideality factor of 1.5 were determined by I-V measurements. The curve $1/C^2$-V is linear and its slope agrees with the doping of the substrate ($N_D=10^{15}cm^{-3}$); nevertheless, the rather high value of the barrier height determined by C-V (1.1eV) suggests that defects remain at the interface. Beyond 580°C the leakage currents increase and prevent any characterization of the diode. This effect may be attributed to an interfacial reaction between W-B film and silicon leading to the formation of a compound which is not patterned in the W etch. This problem can be avoided by patterning mesa structures.

596

IV-CONCLUSION.

Tungsten borides films were prepared by ion beam sputter deposition in a UHV system. The composition, the density ($d \approx 12 g/cm^3$), the resistivity ($\rho \approx 250 \mu \Omega cm$) and the compressive stress ($\sigma \approx -1 GPa$) of the films do not depend on the energy of the primary ion. The films deposited on silicon are able to withstand higher temperature than IBSD-W films (630°C against 450°C, respectively). The properties of tungsten boride films are comparable to those of other borides, carbides or nitrides [3-7]. For a complete picture, additional experiments will be needed to study the effectiveness of this material as a diffusion barrier between silicon and a metal overlayer (aluminum or copper) and the behavior and the role of boron at high temperature.

ACKNOWLEDGEMENTS

M.Eizenberg is gratefully acknowledged for suggesting this study. The authors wish to thank the Laboratory of Metallurgy of l'Ecole Nationale des Arts et Métiers de Paris for XRD measurements. They are also grateful to A.Chabrier and F.Fort for their technical assistance and the cooperation of R.Laval in performing EDS measurements was greatly appreciated.

REFERENCES.

1. M.A.Nicolet, Thin Solid Films, 52, 415 (1978).
2. M.Wittmer, J.Vac.Sci.Technol., A2, 273 (1984).
3. J.O.Olowafe, C.J.Palmstrøm, E.G.Colgan and J.W.Mayer, J.Appl.Phys. 58, 3440 (1985).
4. J.M.E.Harper, S.E.Hörnström, O.Thomas, A.Charai and L.Kruzin Elbaum, J.Vac.Sci.Technol., A7, 875 (1989).
5. F.C.T.So, E.Kolawa, X.I.Zhao and M.A.Nicolet, Thin Solid Films, 153, 507 (1987).
6. H.P.Kattelus, E.Kolawa, K.Affolter and M.A.Nicolet, J.Vac.Sci.Technol. A3, 2246 (1985).
7. H.Y.Yang, X.A.Zhao and M.A.Nicolet, Thin Solid Films, 158, 45 (1985).
8. N.Ohmae, A.N.Akamura, S.Koike and M.Umino, J.Vac.Sci.Technol., A5, 1367 (1987).
9. J.Torrès, J.C.Oberlin, R.Stack, N.Bourhila, J.Palleau, G.Goltz and G.Bomchil, Appl.Surf.Sci., 38, 186 (1989).
10. F.Meyer, D.Bouchier, V.Stambouli, C.Pellet, C.Schwebel and G.Gautherin, Appl. Surf.Sci. 38, 286 (1989).
11. S.M.Rossnagel and J.J.Cuomo, Thin Solid Films, 171, 143 (1989).
12. A. Taylor and B.J.Kagle, in "Crystallographic Data on Metal and Alloy Structures", Dover publication , New York (1963).
13. Handbook of Chemistry and Physic , 43rd edition , the Chemical Rubber Publishing Co., (1961)
14. G.Bomchil, G.Goeltz and J.Torrès, Thin solid films, 140 , 59 (1986).
15. G.Leusink and F.Meyer (private communication).

LOW TEMPERATURE TUNGSTEN DEPOSITION
BY ArF-LASER INDUCED PHOTO-CVD

Rutger L. Krans, Arjan Berntsen, and Wim C. Sinke
FOM-Institute for Atomic & Molecular Physics, Kruislaan 407, 1098 SJ Amsterdam,
The Netherlands

ABSTRACT

Laser-induced Chemical Vapor Deposition of tungsten on Si(100) using WF_6 and H_2 has been investigated using a high-vacuum system comprising a cold-wall reactor. The activation source is a pulsed ArF-excimer laser. The deposition rate depends linearly on the repetition rate, when H_2 is used as a reducing agent. When no H_2 is used the laser radiation suppresses deposition.

At deposition temperatures down to 200 °C laser deposited layers have resistivities better than 20 $\mu\Omega$·cm. Thick layers have resistivities down to 8 $\mu\Omega$·cm. There is a direct relation between layer thickness and resistivity. X-ray diffraction revealed the layers to consist of α-tungsten. β-tungsten was only obtained for those thermally deposited layers where growth was slower than expected.

Nuclear reaction analysis of fluorine showed that most fluorine is present near the W-Si interface, and that the amount of fluorine relative to the amount of tungsten in the layer decreases markedly with deposition temperature.

1. INTRODUCTION

The rationale for depositing tungsten lies mainly in its excellent suitability for VLSI purposes: being a refractory metal, tungsten is not sensitive to electromigration at the high current densities encountered in submicron devices; furthermore its contact resistance to silicon is an order of magnitude lower than that of aluminum, and its resistivity is 5.35 $\mu\Omega$·cm, which is almost as good as that of aluminum. Finally tungsten can be used as a diffusion barrier between silicon and aluminum, which becomes increasingly important with decreasing device dimensions.

An advantage of chemical vapor deposition is that uniform step coverage can be obtained. $W(CO)_6$ has been used as a source gas for chemical vapor deposition, but gives rise to high carbon and oxygen concentrations in the deposited tungsten. This in turn deteriorates the resistivity of the layers [1]. Layers grown from WF_6 do not have such high levels of impurities.

In our experiments WF_6 has been used in conjunction with H_2. When depositing on a silicon substrate two reactions can take place:

$$2\,WF_6\,(g) + 3\,Si\,(s) \longrightarrow 2\,W\,(s) + 3\,SiF_4\,(g) \tag{1}$$

and

$$WF_6\,(g) + 3\,H_2\,(g) \longrightarrow W\,(s) + 6\,HF\,(g) \tag{2}$$

The first (silicon reduction) reaction proceeds quickly, but stops when the deposited tungsten layer prevents silicon to come into contact with the WF_x. The second (hydrogen reduction) reaction is much slower, but can proceed independent of the thickness of the deposited layer [2].

One difference between thermal and laser-induced chemical vapor deposition (CVD) is the way in which atomic hydrogen, necessary for reaction (2) to proceed, is formed. In thermal CVD H_2 dissociates on the tungsten surface and consecutively reacts with WF_6. In case of ArF-laser induced deposition the 6.4 eV photons are thought to dissociate gaseous WF_6 into WF_5 or perhaps WF_4; this produces fluorine radicals that react with H_2 in the gas phase to form HF and atomic hydrogen. The hydrogen radical further reacts with the WF_x [3-5]. Because H_2 does not dissociate on Si or on SiO_2 an initial tungsten layer must be formed by reaction (1) in the case of thermal CVD, whereas for laser-induced deposition the second reaction can take place straight away. A second difference is that the WF_6 is partially decomposed by the photons, so less silicon is needed to further strip this WF_x of fluorine atoms in reaction (1).

These characteristics of laser-induced CVD may have two important consequences: in the first place tungsten can be deposited on surfaces by laser deposition where deposition is impossible thermally [6]; in the second place the silicon reduction reaction may be suppressed, resulting in less silicon consumption, i.e. less damage to the silicon substrate [5].

2. EXPERIMENTAL

The depositions are performed in a cold wall reaction chamber. A quadrupole mass spectrometer is mounted in a turbo pumped chamber. This chamber is connected to the reaction chamber by a 50 μm diameter orifice. Both chambers can be baked out to obtain a base pressure of 10^{-8} Torr.

Samples are sized 11 by 11 mm^2, and are mounted on sample holders consisting of a resistively heated molybdenum block, which is attached to a housing via sapphire rods in order to provide thermal isolation. This results in the block and the sample being the only warm parts in the reaction chamber during deposition. The temperature is monitored by a Pt-100 resistance mounted in the block. The thermal contact to the sample has been calibrated with a special silicon sample having a thermocouple fitted in the silicon. In order to obtain good thermal reproducibility the molybdenum block was polished and doubly polished samples were used. The sample to sample error in the temperature determination is 10 °C. In order to prevent floating potentials the sample is earthed. The sample holder enters the reaction chamber through a sample loading chamber, in which the sample can be baked out beforehand.

During depositions the pressure is 1 Torr, as measured with a pressure transducer. The process gasses are pumped with a rotary pump. A partial WF$_6$ pressure of 0.105 Torr is used for all depositions. H$_2$ and WF$_6$ flows are regulated by mass flow controllers. The windows are mounted on 10 cm long tubes, and a laminar flow of nitrogen from the windows to the reaction chamber keeps the windows clear from deposits. The nitrogen flow is of the same order as the flow of the reaction gasses. The WF$_6$ contains less than 20 ppm SiF$_4$ impurities; the other impurities in the process gasses amount to less than 1 ppm.

In the experiments described below an ArF excimer laser is employed as a source of 6.4 eV photons (193 nm). The yield is typically 140 mJ per pulse at repetition rates up to 50 Hz. The ArF-laser beam is slightly focussed to a beam 9 mm wide and 1.5 mm high, and passes over the sample surface at a height of 5 mm.

Samples were cut from doubly polished three-inch Si(100) wafers. The samples were ultrasonically cleaned in demineralized water. Consequently they were dipped in a 1% HF-solution for one minute, cleaned in demineralized water and spin dried. The time elapsed between etching and mounting in the loading chamber was kept below six minutes.

Elsewhere the set up has been described in more detail [7,8].

3. RESULTS AND DISCUSSION

3.1. Rutherford backscattering analysis

We investigated the influence of the repetition rate of the laser on the tungsten deposition rate, for various partial H$_2$ pressures. The deposited layers were examined with Rutherford backscattering spectrometry (RBS) [9]. We used a primary beam of 2 MeV He$^+$ ions and detected the ions at a scattering angle of 120 degrees. At these conditions a good depth resolution is obtained. The number of deposited tungsten atoms per cm^2 can be determined from the height and width of the tungsten peak, and the magnitude of film thickness variations can be deduced from the slope of the trailing edge of the peak. An example of an RBS spectrum is shown in figure 1, together with a simulation made with the RUMP program [10].

As can be seen in figure 2 the deposition rate increases linearly with the laser repetition rate. If the H$_2$ partial pressure is increased the deposition rate also increases. If no H$_2$ is used the reaction is known to be self-limiting within 200 seconds for thermal CVD, with most of the tungsten being deposited in the first ten to twenty seconds [11]. In this case the laser has a markedly different influence: the total amount of deposited tungsten decreases with the repetition rate. Thus the WF$_5$ or WF$_4$, which is formed in the gas phase by the 6.4 eV photons, presumably plays an important role in suppressing the silicon reduction reaction. The largest effect is visible going from 0 to 5 Hz laser repetition rate.

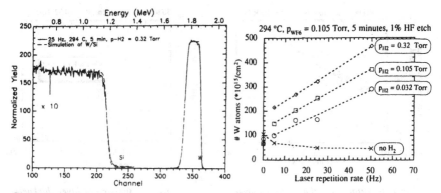

Fig. 1
RBS spectrum of a tungsten layer on silicon.
The dashed line is calculated by simulating a
layer tungsten with $322 \cdot 10^{15}$ at/cm^2, with
thickness variations having a FWHM value
of $65 \cdot 10^{15}$ at/cm^2.

Fig. 2
Number of deposited tungsten atoms per
cm^2, as a function of partial H$_2$ pressure and
laser repetition rate.

The time dependance of the deposition is shown in figure 3; The laser-induced growth rate is approximately 70% larger than the thermal growth rate. The deposition rates for laser-induced deposition increase strongly as a function of temperature (figure 4). This has also been observed by Shintani [6] and by Deutsch [4].

For HF-dipped samples grown at various partial H$_2$ pressures, laser repetition rates and temperatures the thickness variations have been determined from RBS spectra. The amount of variations only depends on layer thickness (figure 5). The total thickness variations could be fitted as the quadratic sum of two contributions, the first having a FWHM value 21 % of the total thickness, the second being constant at a FWHM value of 69 Å. From SEM analysis we were able to identify both contributions. The proportional part of the thickness variations arises from the columnar structure that is deposited; these variations are present at the surface. The part of the variations that is independent of layer thickness is present at the interface.

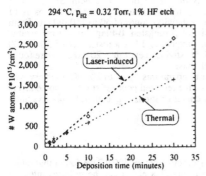

Fig. 3
Number of deposited tungsten atoms per
cm^2, as a function of total deposition time.
The laser enhances the growth rate by 70 %.

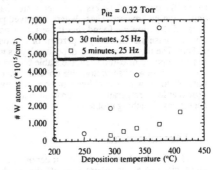

Fig. 4
Number of deposited tungsten atoms per cm^2
as a function of substrate temperature, for
laser-induced deposition.

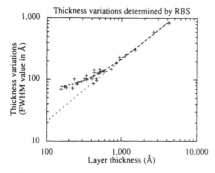

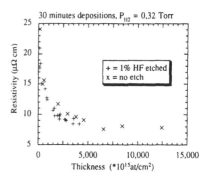

Fig. 5
Thickness variations in the deposited layers, as a function of tungsten layer thickness. The bulk density of tungsten was used to convert from atoms/cm² to Å. The curve is the geometrical sum of constant variations of 69 Å (at the interface) and bulk variations 21 % of the layer thickness.

Fig. 6
Resistivity as a function of layer thickness. The + denote laser-induced layers on etched samples. The x denote layers grown by laser-induced and thermal CVD, on unetched silicon.

3.2. Resistivity

The resistivity of the deposited tungsten has been determined by measuring the sheet resistance with a four point probe, and by comparing this with the amount of tungsten atoms determined from RBS. The bulk density of tungsten was used to convert the number of atoms per cm² to thickness. Density variations are thus not taken into account.

There is a direct relation between layer thickness and resistivity, for all layers deposited at temperatures between 200 and 420 °C (figure 6). Layers that are two micron thick have a resistivity of 8 $\mu\Omega$·cm; at a layer thickness of 130 Å the resistivity is 20 $\mu\Omega$·cm. This increase by a factor 2.5 can at least partially be attributed to thin film effects; as shown above this layer has a FWHM value for the thickness variations 57% of the layer thickness.

Apart from this thickness dependance we observe no difference in resistivity between layers grown at various partial gas pressures, temperatures or repetition rates. At temperatures from 400 °C good resistivities are generally reported, but at lower temperatures observations diverge [4,6,12]. Our laser deposited layers all had resistivities better than 20 $\mu\Omega$·cm. Some of our thermally deposited layers however were markedly thinner than expected on the basis of deposition parameters. These layers displayed poor resistivity values of 50 to 150 $\mu\Omega$·cm. From X-ray diffraction we found that these layers consist of β-tungsten. β-tungsten is a metastable crystal phase stabilized by a percent of carbon or oxygen, and has a resistivity of 50 to 300 $\mu\Omega$·cm. The thermal and laser-deposited layers with resistivities better than 20 $\mu\Omega$·cm were found to be composed of α-tungsten, the stable bcc crystal phase.

From these observations we conclude that for temperatures down to 200 °C it is very well possible to grow tungsten layers of good resistivities; the reported poor resistivity of some layers grown at low temperatures are probably caused by contaminations, e.g. caused by poor vacuum conditions or etching problems.

3.3. Laser pulse energy

Performing laser deposition at deposition temperatures below 300 °C we found that under some circumstances layers peel off during deposition. This effect can be circumvented by reducing the energy per pulse. At 200 °C and a repetition rate of 50 Hz the layers are well adherent at energies up to 6 mJ per pulse (laser energy above the sample). If the energy is slightly increased the layer starts to peel off. If the energy is then decreased the peeling stops. This process can be repeated during one deposition. If we go to a repetition rate of 25 Hz the pulse energy can be roughly doubled before peeling begins. At higher temperatures the same effects occur for higher pulse energies. Sometimes a layer which is deposited at pulse energies near this critical value starts to peel off after deposition when it is cooling down to room temperature.

3.4. Fluorine measurements

The amount of fluorine in thermally and laser-deposited layers has been determined as a function of deposition time and temperature, using the $^{19}F(p,\alpha)^{16}O$ reaction [13]. In one set of experiments an H_2^+ beam of 1.7 MeV was used, resulting in 850 keV protons entering the sample. In the second set of experiments the set-up was changed to use 1270 keV protons. At this energy tungsten layers up to $8 \cdot 10^{18}$ at/cm^2 can be probed rather uniformly, because between 1100 and 1270 keV the cross section for the nuclear reaction is constant within 20% [14]. We obtained depth resolution from the fact that the 6.9 MeV α-particles undergo stopping in the tungsten layers.

Because the yield of the reaction is rather low a large detector was positioned close to the sample, at a scattering angle of 150°. Drawback is a relatively large opening angle (half angle 11.5°). In order to stop the backscattered protons a 35 μm thin mylar foil was mounted in front of the detector. The same accelerator was used as the one used for the RBS measurements. The determined yields were compared with the yield from a calibration sample, consisting of a thin layer of MgF$_2$ deposited on Si. Using RBS with 2 MeV helium ions the F content of this sample was determined with an accuracy of 3%.

The absolute fluorine content of the layers deposited thermally and by laser deposition increases with deposition time (figure 7); however the average F concentration, defined as the number of fluorine atoms divided by the number of tungsten atoms, decreases. Thus most fluorine is incorporated in an early stage of the deposition.

For deposition times of 5 and 30 minutes the dependance on deposition temperature is shown in figure 8. The average F concentration rapidly decreases as a function of temperature, from 2.4 % at 200 °C to 0.1 % at 375 °C. The absolute amount of fluorine however increases with temperature at temperatures below 290 °C, because at these temperatures the deposition rates are very low. There is no marked difference between thermally and laser deposited layers.

The layer grown at 294 °C for 5 minutes contains $7 \cdot 10^{15}$ at/cm^2. This is comparable to the $4.8 \cdot 10^{15}$ at/cm^2 [15] finds in a layer grown thermally at 285 °C. Kuiper [11] finds fluorine amounts an order of magnitude lower ($3 \cdot 10^{14}$ at/cm^2) for a sample deposited by silicon reduction at 300 °C.

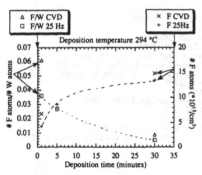

Fig. 7
Fluorine measurements as a function of deposition time. The x and + denote the total number of fluorine atoms per cm^2 in the layers (right hand axis). The Δ and $\square$ denote the number of fluorine atoms divided by the number of tungsten atoms (left hand axis).

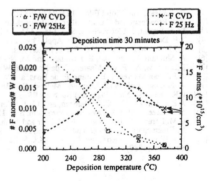

Fig. 8
Fluorine measurements as a function of deposition temperature. The x and + denote the total number of fluorine atoms per cm^2 in the layers (right hand axis). The Δ and $\square$ denote the number of fluorine atoms divided by the number of tungsten atoms (left hand axis).

Fig. 9
Depth resolved fluorine concentration from nuclear reaction analysis. The α-particles originating from reactions at the surface arrive at the detector with an energy of 2.5 MeV, after having lost 4.7 MeV in a thin mylar foil in front of the detector. The α-particles created near the interface lose extra energy while traveling through the tungsten layer, and arrive at the detector with 1.2 MeV less energy. This spectrum shows a laser-induced CVD sample grown at 375 °C for 30 minutes. It has a higher fluorine concentration near the tungsten-silicon interface than in the rest of the tungsten layer.

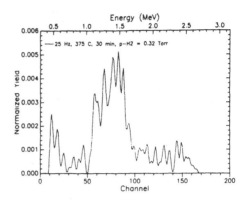

We have obtained depth resolved spectra for thick layers (figure 9). Here we see that the fluorine concentration is highest near the silicon-tungsten interface, confirming the conclusion from the series as a function of deposition time that most fluorine is incorporated in an early stage of the deposition.

ACKNOWLEDGEMENTS

We would like to thank J. Toth of the Centrum voor Submicron Technologie (CST) in Delft for performing SEM. The work described here was performed as part of the research program of the Stichting voor Fundamenteel Onderzoek der Materie (FOM), with financial support from the Nederlandse organisatie voor Wetenschappelijk Onderzoek (NWO) and the Dutch ministry of economic affairs within the framework of the IOP-IC program.

REFERENCES

[1]   Y. Pauleau, Thin Solid Films 122, 243 (1984).
[2]   E.K. Broadbent and C.L. Ramiller, J. Electrochem Soc 131, 1427 (1984).
[3]   W.A. Bryant, J. Electrochem. Soc. 125, 1534 (1978).
[4]   T.F. Deutsch and D.D. Rathman, Appl. Phys. Lett. 45, 625 (1984).
[5]   A.J.P. van Maaren and W.C. Sinke, Mater. Res. Soc. Proc. 168 (1990), in press.
[6]   A. Shintani, S. Tsuzuku, E. Nishitani, and M. Nakatani, J. Appl. Phys. 61, 2365 (1987).
[7]   R.L. Krans, A.J.P. van Maaren, E. de Haas, and W.C. Sinke, J. Dutch Vac. Soc. 1989(4), 109.
[8]   A.J.P. van Maaren, R.L. Krans, E. de Haas, and W.C. Sinke, Appl. Surf. Sc. 38, 386 (1989).
[9]   A. Polman et al., Nucl. Instr. and Methods B37/38, 935 (1989).
[10]   L.R. Doolittle, Nucl. Instr. and Methods B9, 334 (1985).
[11]   A.E.T. Kuiper, M.F.C. Willemsen, and J.E.J. Schmitz, Appl Surf. Sc. 38, 338 (1989).
[12]   M.L. Green and R.A. Levy, J. Electrochem. Soc. 129, 973 (1982).
[13]   G.M. Lerner and J.B. Marion, Nucl. Instr and Meth. 69, 115 (1969).
[14]   D. Dieumegard, B. Maurel, and G. Amsel, Nucl. Instr. and Methods 168, 93 (1980).
[15]   H.J. Whitlow, Th. Eriksson, M. Östling, C.S. Petersson, J. Keinonen, and A. Anttila, Appl. Phys. Lett. 50, 1497 (1987).

MODIFIED LINE-OF-SIGHT MODEL FOR
DEPOSITION OF TUNGSTEN SILICIDE BARRIER LAYERS

Timothy S. Cale and Gregory B. Raupp
Department of Chemical, Bio & Materials Engineering and Center for Solid State
Electronics Research, Arizona State University, Tempe, AZ 85287-6006

ABSTRACT

A modified line-of-sight model for transport and deposition during LPCVD
is used to predict step coverage and film composition uniformity of tungsten
silicide barrier layers. Predictions are compared with experimental results
for 2 μm wide by 6 μm deep trenches with barrier layers of 0.2 μm nominal
thickness. Model predictions are in quantitative agreement with those of a
diffusion-reaction model and are in qualitative agreement with experiment.

INTRODUCTION

Step coverage is a critical concern for tungsten silicide barrier layers
deposited by low pressure reduction of tungsten hexafluoride (WF_6) by
dichlorosilane (DCS). Because tungsten silicide LPCVD involves deposition of
at least three solid phases (W_5Si_3, WSi_2 and Si) [1], an additional concern is
the potential spatial variation in film composition within a feature.

In this paper we present a modified line-of-sight model to predict step
coverage and film composition uniformity of LPCVD tungsten silicide barrier
layers. We have derived the integro-differential equations which govern
transport and deposition in rectangular trenches for any fraction of feature
mouth closure [2]. In that work, we limited our attention to systems where
the heterogeneous deposition reaction depends only on temperature and the
concentration of one component; e.g., decomposition reactions. Film profiles
predicted by this modified line-of-sight model are in good agreement with the
results of Monte Carlo simulations [3]. In this paper we extend the analysis
to multicomponent, multireaction systems, using DCS reduction of WF_6 as an
example. Model predictions are compared with those of a diffusion reaction
model and to experimental results [1].

TUNGSTEN SILICIDE DEPOSITION

In this study we assume the following reaction stoichiometries:

$$5\ WF_6\ +\ 11\ SiH_2Cl_2\ \rightarrow\ W_5Si_3\ +\ 22\ HCl\ +\ 7\ SiF_4\ +\ SiF_2$$
$$WF_6\ +\ 4\ SiH_2Cl_2\ \rightarrow\ WSi_2\ +\ 8\ HCl\ +\ SiF_4\ +\ SiF_2 \tag{1}$$
$$SiH_2Cl_2\ \rightarrow\ Si\ +\ 2\ HCl$$

We have previously correlated [1] the apparent film deposition rate and film
composition dependences on wafer temperature and reactant partial pressures
using deposition rate expressions of the form (see **NOMENCLATURE**)

$$R_m\ -\ k_{om}\ \exp\ (-E_m/RT)\ P_{DCS}^{\beta_m}\ P_{WF_6}\ /\ (1\ +\ K_m\ P_{WF_6}) \tag{2}$$

where R_m is the molar deposition rate of solid m per unit area. The parameter
values are listed in Table 1. Note that these parameter values have not been
established through careful kinetic studies in gradientless reactors.

The concentration dependences in Equation 2 written in terms of reactant
partial pressures can be rewritten in terms of the flux of each species to the
surface. The flux of the i-th species to a point on a surface η_i is
conveniently expressed in terms of the volume number density n_i of the i-th

Table I
Kinetic Parameters for WSi$_\chi$ Deposition

m	Solid	k_o (mol/cm^2-s-mtorr$^{\beta+1}$)	E (kcal/mol)	β	K (mtorr^{-1})
1	W_5Si_3	3	40	0.5	0
2	WSi_2	1.8×10^{24}	120	1	1
3	Si	6.6×10^{13}	90	2	10

species with which the surface is in contact as

$$\eta_i = n_i \, \bar{c}_i/4 \quad . \tag{3}$$

The overall sticking coefficient for the i-th species σ_i is the probability that a molecule of the i-th species, after colliding with the surface, reacts to become part of the growing film. Given the heterogeneous kinetics of a CVD system and operating conditions, the sticking coefficient for species i in multireaction chemistries can be written as the following function of local fluxes of the reacting species to the surface:

$$\sigma_i(T,\underline{n}) = [\sum_m \nu_{im} R_m(T,\underline{n})]/\eta_i \tag{4}$$

where ν_{im} is the stoichiometric coefficient of reactant i in the reaction depositing solid m as given in equation (1). The probability that a particle will desorb without reacting is $1-\sigma$, thus the flux of molecules of the i-th species leaving the surface is less than the impinging flux by this factor:

$$\eta_i^\ell = (1-\sigma_i)\eta_i. \tag{5}$$

The local rate of change in film thickness at a point on the surface is

$$\frac{\partial T}{\partial t} = \sum_m V_m R_m(T,\underline{n}) \tag{6}$$

where V_m is the molar volume of the solid produced in reaction m. The local instantaneous silicon to tungsten ratio χ deposited on the surface is written in terms of the three reaction rates as

$$\chi = (3 R_{W_5Si_3} + 2 R_{WSi_2} + R_{Si}) / (5 R_{W_5Si_3} + R_{WSi_2}) \quad . \tag{7}$$

MODIFIED LINE-OF-SIGHT MODEL

Consider a long rectangular trench of width W and depth H as an idealized model of a via cut into the surface of a patterned wafer. The coordinate system origin is fixed at the top edge of the left wall. The x axis runs down the trench wall, the y axis runs across the trench and the z axis runs along the length of the trench. Molecules enter the trench from the source volume and collide with the walls perhaps many times before reacting or exiting.

The local rate of formation of each solid phase is formulated in terms of the flux of each reactant to and from the film surfaces using the following set of assumptions: 1. The frequency of molecule-molecule collisions is negligible relative to molecule-wall collisions; i.e., we are in the Knudsen regime typical of LPCVD on patterned wafers. In that case, each species in the gas phase can be treated separately. 2. The open end of the cylinder is exposed to a low pressure, ideal gas with number density n_v and a Maxwellian distribution of velocities. 3. The molecules reflect perfectly diffusely from the trench surfaces. These surfaces are at the same temperature as the source gas volume; *i.e.*, the molecules leaving the surface possess the same distribution of velocities as those in the gas. 4. Surface diffusion on the interior surface of the feature is negligible. These assumptions completely specify the transport in the trench and specify the problem to be symmetric about the trench plane of symmetry. In addition to these four assumptions for line-of-sight transport, three additional assumptions are added for deposition: 5. Sticking coefficients do not depend on the angle which the molecules strike the surface. 6. Sticking coefficients do not depend on the collision history of the molecules. 7. The film grows slowly relative to the redistribution of fluxes to the walls caused by the changes in geometry. For all but first order reactions, sticking factors depend on fluxes, which in turn are spatially dependent. Any reaction products which desorb will have their own spatially dependent fluxes. Because we assume that molecule-molecule collisions are rare relative to molecule-wall collisions, these gaseous reaction products need not be considered in this study.

We have presented the derivation of the governing equations based on these model assumptions [2]. A critical step in the derivation is the reduction of the problem to two dimensions by integrating out the z dependence. There is no z dependence in this solution, by the assumption of an infinite trench. The flux of molecules of the i-th species to a differential length of the left wall dx, independent of the number of previous impacts by these molecules, is

$$\eta_{iw} = \int_0^W \eta_{iv}^\ell q_{wv}' dy + \int_{x_1}^{x_2} \eta_{ir}^\ell q_{wr}' dx + \int_{y_1}^{y_2} \eta_{ib}^\ell q_{wb}' dy \qquad (8)$$

if the transport between points on the same wall are ignored. The first integral represents the flux contribution from the source, the second contribution from the opposing trench wall, and the third from the trench base. Similarly, the flux of molecules of species i to a differential length of the base dy is

$$\eta_{ib} = \int_0^W \eta_{iv}^\ell q_{bv}' dy + \int_{x_1}^{x_2} \eta_{iw}^\ell q_{bw}' dx + \int_{x_1}^{x_2} \eta_{ir}^\ell q_{br}' dx \qquad (9)$$

if the base surface is assumed to be concave downward. The limits of integration x_1 and x_2 in Equations 8 and 9 are the surface of the film on the exterior of the feature and the intersection of the wall and base, and y_1 and y_2 are the intersections between each wall and the base. The two dimensional "transmission probability" between points j and k on the surface is expressed as (see NOMENCLATURE and Ref. 2)

$$q_{kj} = \frac{\kappa_{kj}}{2s_{kj}^3 \cos\gamma_j} [n_j \cdot S_{kj}] [n_k \cdot S_{kj}] \qquad (10)$$

where $\cos\gamma_j$ relates the differential length of surface to the differential in x or y. The visibility factor κ_{kj} is unity if the points can "see" each other and zero if they cannot.

Numerical solutions to these governing equations are obtained by dividing the surface of the walls and base, as well as the trench mouth, into short segments. These areas are tracked and updated during the simulation. Segments are generated along the walls and removed from the base as the feature narrows and becomes deeper during the deposition. Equation 8 for the wall becomes

$$\eta_{iw} = \sum_j \int_{1_j} \eta_{ij}^\ell q_{wj}' dl_j \qquad (11)$$

where w represents a point on the wall, and the summation in j represents integrations over segments of the source and feature surfaces. If the segments are assumed short enough to approximate the integral as the value of the integrand at the segment midpoint times the segment length, Equation 11 becomes

$$\eta_{iw} \approx \sum_j \eta_{ij}^{\ell} \, q_{wj}' \, l_j \qquad (12)$$

where η_{iw} is the flux to the midpoint of the wall segment identified by subscript w and l_j is the length of the segment of source or feature surface represented by subscript j. Similarly, for the base

$$\eta_{ib} \approx \sum_j \eta_{ij}^{\ell} \, q_{bj}' \, l_j \qquad (13)$$

where the summation is over the source segments and the two wall segments.

Equations 12 and 13 are incorporated into differential expressions of the form given by Equation 6 to give the deposition rates. The change in position of segment endpoints is taken as the average of the changes in adjacent segments. The differential equations are conveniently integrated using Euler's rule, with the fluxes determined by iteration. This iterative process converges rapidly in all cases considered to date. To verify that the surface is divided into enough segments, the flux profile is determined for rectangular trenches in the absence of deposition, which Cale et al. [2,4] have shown to be uniform. Using 50 points each for the mouth, walls and base, the uniform solution is obtained with a maximum error of 1%. This level of error is better than the measurement errors involved in the microscopy which would be used to test the model predictions.

RESULTS AND DISCUSSION

Depositions were performed in a Spectrum CVD single wafer, cold wall, rapid thermal reactor [1]. In all experiments, inlet flow rates of DCS and WF_6 were fixed at 120 and 3.8 sccm, respectively. Total pressure was fixed at 180 mtorr. Deposition rates were time-averaged values estimated from weight gain on blank wafers, using measured film composition and assuming a nominal film density. Compositions were measured by Rutherford backscattering. Step coverage, defined as the ratio of film thickness at the base of the wall to the film thickness on the surface exterior to the trench, were estimated from scanning electron micrographs. Composition uniformity is defined as the fractional change in Si to W ratio between the same two points. Details of sidewall composition measurements by Auger spectroscopy are given in Ref. 1

The well mixed reactor approximation is used in combination with observed deposition rates to estimate reactant partial pressures in the reactor as a function of temperature for the given flow rates. These partial pressures were used to determine the kinetic parameter values given in Table I such that observed deposition rates and film compositions are correctly predicted.

Predictions of step coverage and film composition uniformity using the modified line-of-sight model are in quantitative agreement with predictions of the diffusion reaction model [1]. Figure 1 shows (solid line) the modified line-of-sight (and diffusion reaction) model predictions of step coverage for 0.2 μm films deposited in 2 μm wide by 6 μm deep trenches as a function of deposition temperature. The triangles are measured step coverages in features of the same nominal size and 0.2 ± 0.02 μm films. Both LPCVD models predict the correct trend in step coverage as a function of temperature. Quantitative discrepancy between the predicted and measured values is likely due largely to our use of the perfectly mixed reactor model, which gives low estimates of the WF_6 partial pressure in the reactor [1].

The dashed line in Figure 1 is the composition uniformity predicted by both models as a function of reactor temperature for 2 μm by 6 μm trenches. Similar, though less dramatic composition nonuniformity is predicted for 1 μm by 1 μm trenches. Auger sidewall composition profile measurements of a film deposited at 773 K in a 1 μm by 1 μm trench indicate that silicon to tungsten

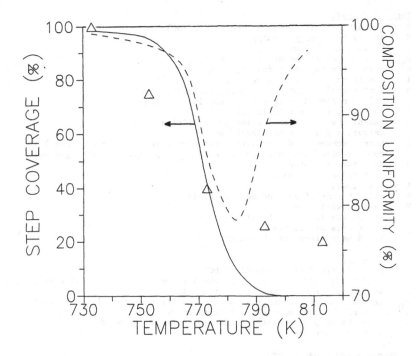

Figure 1.   Modified line-of-sight model predictions of step coverage and film composition uniformity as a function of deposition temperature with fixed feed flow rates and assuming a well mixed reactor.   The triangles are experimental step coverages.

ratio increases with depth in the trench [1].   Again, qualitative agreement is indicated in that the predicted ratio also increases.   More sidewall experiments are needed to determine whether the models' predictions of the temperature dependence of composition uniformity is quantitatively correct.

CONCLUSIONS

Modified line-of-sight model predictions of step coverage and composition uniformity of LPCVD tungsten silicide barrier layers are in excellent agreement with those of our diffusion-reaction model [1].   Model predictions are in qualitative agreement with the limited experimental information available.   Step coverage degrades as temperature is increased and the reactor becomes starved for WF_6; silicide film composition may become nonuniform.

ACKNOWLEDGEMENTS

Professors Raupp and Cale gratefully acknowledge the partial support of the National Science Foundation and the Semiconductor Research Corporation.

NOMENCLATURE

English Symbols

c	average velocity of each species (cm/sec)
E	activation energy (kcal/gmole)
H	feature depth (μm)
k_o	preexponential factor (see Table I)
K^o	adsorption parameter for WF_6 (1/mtorr)
l	length of surface segment (μm)
$\underline{n}$	unit normal vector representing surface segment
n	volume number density of each species (molecules/cm^3)
P	partial pressure (millitorr)
q'	two-dimensional transmission probabilities
R	universal gas constant (kcal/gmole/K)
R	heterogeneous molar deposition rate for reaction (gmole/μm^2/sec)
$\underline{S}$	two dimensional projection of vector between source and target points
s	magnitude of $\underline{S}$ (μm)
T	film thickness (μm)
t	time (sec)
V	molar volume of each phase (cm^3/gmole)
W	feature width (μm)
x	coordinate down trench wall
y	coordinate across trench
z	coordinate along trench length

Greek symbols

β	reaction order with respect to DCS
κ	visibility factor; 1 if two points can see each other, 0 if they cannot
σ	overall sticking factor for reactant i
ν_{im}	generalized stoichiometric coefficient for component i in reaction m
η	flux of the each species (molecules/μm^2/sec)
γ	angle between differential surface element and dx or dy
χ	silicide Si:W atomic ratio

Subscripts and Superscripts

b	subscript indicating point on base
r	subscript indicating point on opposing wall
v	subscript indicating source volume
w	subscript indicating point on wall
i	subscript indicating reactant i
m	subscript indicating solid m or reaction producing solid m
ℓ	superscript indicating flux leaving surface

REFERENCES

1. G. B. Raupp, T. S. Cale, M. K. Jain, B. Rogers and D. Srinivas, presented at 8th Intl. Conf. Thin Films, San Diego, CA, April 1990, and submitted to *Thin Solid Films*.

2. T. S. Cale and G. B. Raupp, submitted to *J. Vac. Sci. Technol. B*.

3. A. Yuuki, Y. Matsui and K. Tachibana, *Jap. J. Appl. Phys.* 1989, 28, 212.

4. T. S. Cale, G. B. Raupp and T. Gandy, accepted for publication in *J. Applied. Phys.*

EFFECT OF RESIDUAL STRESS ON SI-CONSUMPTION DURING W-DEPOSITION BY LPCVD

S. H. LEE*, J. J. LEE*, AND Dong-Wha KUM**
* Seoul National University, Dept. of Metallurgical Engineering, 56-1 San, Kwanakgu, Seoul 151-742, Korea
** Korea Institute of Science and Technology, Division of Metals, P.O.Box 131, Cheongryang, Seoul, Korea

ABSTRACT

The effect of stress on W-deposition at low pressure by Si-reduction of WF_6 has been studied on patterned wafers utilizing cross-section SEM and cross-section TEM. Si-consumption and encroachment formation at the Si/SiO_2 interface was analyzed as a function of the oxide thickness, which changes the residual stress of the Si-substrate beneath the SiO_2-layer. At the edge of the SiO_2-layer, significant enhancement of W-deposition was observed from the early stage of W-metallization. Penetration of W into the Si-substrate beneath the edge showed strong dependency upon the oxide thickness, while penetration along the SiO_2/Si interface (i.e. encroachment) did not. It has been analyzed that early stage of W-deposition was strongly affected by residual stress in the Si-substrate, and contact between WF_6 and Si will be much easier through porous W-film and the W/Si interface, where the existence of microvoids is expected.

INTRODUCTION

Research activities on W-metallization are increasing due to potential applications for interconnects, via-hole filling, diffusion barriers and contact materials in advanced VLSI [1-4]. W is often deposited by reactions between WF_6 and Si, SiH_4 or H_2. When Si reduction of WF_6 is used for selective W-deposition or W plug technology, excess Si-consumption and encroachment delay earlier application for advanced VLSI. Stresses at Si/SiO_2 interface have been suspected to enhance the formation of encroachment [5,6], but the origin of encroachment has not been clearly resolved. Even though it is known that SiO_2 on Si creates residual stresses in the SiO_2 film, at the SiO_2/Si interfaces and in the underlying Si due to lattice mismatch and different thermal coefficients, the significance of stresses in the Si-substrate is not jet recognized. Stresses in the Si-substrate were measured by X-ray diffraction topography, Raman scattering, and photoreflectance method [7-9]. Blech et al. [10] calculated the distribution of stresses in the Si-substrate based on linear elastic theory. According to the calculation, stress increases linearly with the oxide thickness and stress concentration is maximum just beneath the edge of oxide layer, which fact agrees well with experimental observations [7,8].

In this study, effects of stresses in the Si-substrate on W-deposition and defect formation were investigated when WF_6 is reduced by Si. For the purpose, SiO_2 layers with different thicknesses were deposited on Si-substrate.

EXPERIMENTAL

SiO$_2$ was deposited on phosphorous doped Si (100) wafers at 400°C by CVD, annealed at 800°C for 30 min., and then patterned by photo lithography. The oxide thicknesses were 300 nm, 600 nm and 1000 nm, respectively. Small slices (15mm x 20mm) were cleaned in DI water and freshly prepared H$_2$O/HF solution to get rid of native oxides on the surfaces. W was deposited in a cold-wall type low pressure reactor at 360°C. Flow rates for WF$_6$ and Ar-gas were 4 sccm and 100 sccm, respectively. Three slices with different oxide layers and one bare Si-wafer were processed at the same time. Three different deposition times were chosen : 1, 8 and 24.5 min. SEM and TEM were used to observe cross-sections of the specimens after deposition. W-thickness on the patterned specimen was measured by the cross-sectional SEM, and W-layer on the bare Si-specimen was measured by the β-backscattering method. Both measurements are known to give similar results within reasonable error range [11].

RESULTS AND DISCUSSION

Fig. 1 shows cross-sectional view of two specimens having W deposited for one minute ; (a) and (b) are specimens with oxide layers of 600 nm and 1000 nm, respectively. Both specimens show that significantly thick W is formed beneath the edge of the SiO$_2$ (indicated by arrows), and that it decreases to a very thin layer on the Si-surface between oxide patterns. (The specimen with 300 nm oxide is not shown, because W-layer was so minute that it could not be resolved clearly by the cross-sectional SEM.) This observation indicates that deposition of W is the fastest at the edge. Fig. 1 also shows that encroachment has been formed from the early stage of deposition. The size of encroachment (measured as distance from the edge) is about 400 nm for both specimens. When reaction time increased, the same phenomenon is observed. Fig. 2 shows specimens after 8 min. deposition ; (a), (b) and (c) are specimens with 300 nm, 600 nm and 1000 nm oxides, respectively. The thicker W-layers beneath the edge are quite apparent. The maximum thickness increases monotonically as the thickness of oxide increases ; the values are 160 nm, 310 nm and 380 nm, respectively. However, the

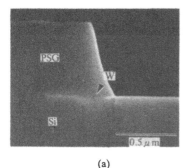

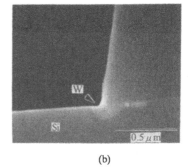

(a)   (b)

Fig. 1   SEM micrographs of the deposited tungsten on oxide-patterned silicon wafers
(deposition temp. : 360°C, deposition time : 1 min., WF$_6$: 4 sccm, Ar : 100 sccm).
(a) d (oxide thickness) = 600 nm, (b) d = 1000 nm.

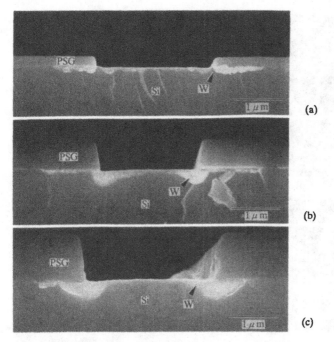

Fig. 2 SEM micrographs of the deposited tungsten (temperature : 360℃, time : 8 min).
(a) d = 300 nm, (b) d = 600 nm, (c) d = 1000 nm.

oxide thickness does not affect the length of encroachment ; they are 1010 nm, 1230 nm and 1035 nm long, respectively.

Fig. 3 shows specimens after 24.5 min. deposition. Enhanced deposition at the edge have extended laterally such a way that the thick W-layer covered whole span of the exposed Si, when the oxides are thicker than 600 nm. Thicknesses of W-layer are (b) 600 nm and (c) 700 nm. For the case of 300 nm oxide specimen (a), thick deposition of W is seen only at both edges of the SiO$_2$. Small opening at the original tips of SiO$_2$/Si interface after long deposition should be noticed. Experimental observations are summarized in Fig. 4. In Fig. 4 (a) maximum thickness of W-film at the edge is plotted as a function of the oxide thickness. Fig. 4 (b) shows that encroachment increases with deposition time and does not change much with the thickness of patterned oxide.

The enhanced deposition of W at the edge shows strong dependency on the thickness of oxide layer. And the thickness variation of W at the edge observed in this study is very similar to the "equi-$\Delta\theta$" contours in the Si-substrate calculated by Blech et al. [10]. Referring to the fact that stress is the highest at the edge, and dies out as one goes away from the edge, it can be deduced that W-deposition at the beginning is closely related with stress state in the Si-substrate.

Once W-film is formed, the initial stress state will be modified. Then a question, how the initial shape of W-layer could have maintained as the deposition progressed, is aroused. Microstructure of the W-film and the W/Si interface were observed by cross-sectional TEM, and are shown in Fig. 5. The rough and porous W/Si interface is noticed in this micrograph.

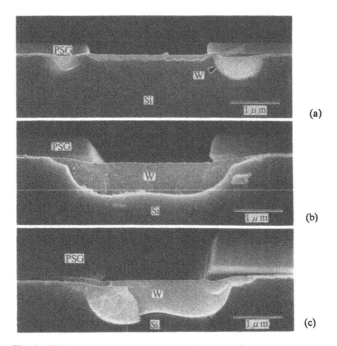

Fig. 3 SEM micrographs of the deposited tungsten (temperature : 360℃, time : 24.5 min).
(a) d = 300 nm, (b) d = 600 nm, (c) d = 1000 nm.

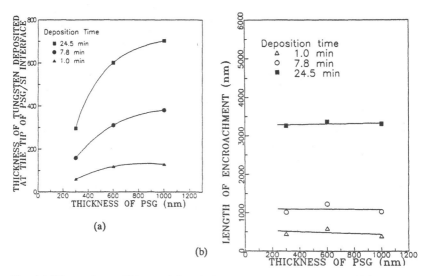

Fig. 4 (a)The maximum thickness of deposited tungsten, vs. the thickness of oxide patterns.
(b)The length of encroachments, vs. the thickeness of oxide patterns.

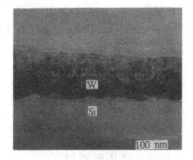

Fig. 5 Cross-sectional TEM micrograph
of the deposited tungsten.

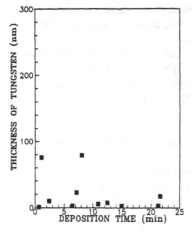

Fig. 6 The thickness of tungsten films
deposited on bare silicons, vs.
deposition time (other process
conditions are same as by oxide
-patterned silicon wafer).

This fact implies that penetration of WF_6 along the interface had controlled the growth of W-film, not the Si diffusion through the W-film. Unlimited growth of W by the WF_6 penetration through porous W-layer and boundary was also reported by Wong et al. [12]. W deposited on bare Si-substrate at the same CVD condition show the limiting thickness of about 20 nm except a few specimens (Fig. 6). In contrast, W-deposition on thick oxide-patterned specimen showed no limiting thickness.

CONCLUSION

The stress in Si-substrate, which is induced by oxide layer, plays an important role at the beginning of W-deposition by Si-reduction of WF_6. It enhances the deposition reaction significantly and causes porous and rough W/Si interface. Contact between WF_6 and Si will be much easier through porous W-film and the W/Si interface. The reducing reaction of WF_6 by Si can thus continue.

ACKNOWLEDGEMENT

This work was supported by the Korea Institute of Science and Technology (# 4-2E10590), and the Korean Science and Engineering Foundation (# 88040805).

REFERENCES

1. Tungsten and Other Refractory Metals for VLSI Applications, edited by R.S.Blewer (Mater. Res. Soc. Conf. Proc. Pittsburgh, PA 1986).
2. Tungsten and Other Refractory Metals for VLSI Applications II, edited by E.K.Broadbend (Mater. Res. Soc. Conf. Proc. Pittsburgh, PA 1987).
3. Tungsten and Other Refractory Metals for VLSI Applications III, edited by V.A.Wells (Mater. Res. Soc. Conf. Proc. Pittsburgh, PA 1988).
4. Tungsten and Other Refractory Metals for VLSI Applications IV, edited by R.S.Blewer and C.M.McConica (Mater. Res. Soc. Conf. Proc. Pittsburgh, PA 1989).
5. R.J.Mianowski, K.Y.Tsao, and H.A.Waggener, in ref. 1, p.145.
6. M.L.Green and R.A.Levy, J.Electrochem. Soc. 132, 1243 (1985).
7. E.S.Meieran and I.A.Blech, J. Appl. Phys. 36, 3162 (1966).
8. J.R.Patel and N.Kato, J. Appl. Phys. 44, 971 (1973).
9. J.T.Fitch, C.H.Bjorkman, G.Lucovsky, F.H.Pollak, and X.Yin, J.Vac. Sci. Technol. B7, 775 (1989).
10. L.A.Blech and E.S.Meieran, J. Appl. Phys. 38, 2913 (1967).
11. C.S.Ryu, J.J.Lee, D.W.Kum, and S.K.Joo, J. Korean Inst. Met. 27, 995 (1989).
12. M.Wong, N.Kobayashi, R.Browning, D.Paine, and K.C.Saraswat, J. Electrochem. Soc. 134, 2339 (1987).

THE DEPENDENCE OF ALUMINUM/TUNGSTEN REACTION
ON CRYSTALLINE PHASES OF CVD TUNGSTEN

Y.Harada, H.Onoda and S.Madokoro
VLSI R&D Center, Oki Electric Industry Co., Ltd.,
550-1 Higashi-asakawa, Hachioji, Tokyo 193 Japan

ABSTRACT

The reaction between Al and CVD-W films has been studied. Al/α-W/Si and Al/β-W/Si structures were prepared by deposition on different Si substrates by changing deposition conditions using silane reduction of tungsten hexafluoride, followed by Al-Si-Cu alloy film sputter deposition. The sheet resistance of Al/α-W/Si structure is higher than that of Al/β-W/Si structure after 500°C annealing. RBS measurements show that the W diffusion into Al occurs in both structures after annealing, and the reaction between α-W and Al takes place easily compared with that between β-W and Al. This causes the sheet resistance difference. The activation energies for the W diffusion into Al, however, are almost the same in both structures. When CVD-W films are exposed to air after removal from the reactor, the sheet resistance of β-W film increases according to the exposure time, while that of α-W film does not. AES measurements indicate that the β-W film absorbs more oxygen than the α-W because of the difference of grain structures. The resistance increase of β-W film is caused by the oxygen that is absorbed from air. Our results indicate that the oxygen in the β-W layer suppresses the W diffusion into Al. Once the reaction begins, however, the diffusion into Al does not depend on crystalline phases of CVD-W.

INTRODUCTION

As device geometries have shrunk, interest in tungsten metallization by Low Pressure Chemical Vapor Deposition (LPCVD) has increased[1,2]. The tungsten CVD method has been found to be effective for the formation of contact plugs, interconnection and diffusion barriers. The tungsten CVD method using silane reduction has a high deposition rate, a low temperature, and a negligible influence from the underlying substrate for selective contact plug formation, in comparison with that using hydrogen reduction[3,4]. Recently, α-phase tungsten (α-W) or metastable β-phase tungsten (β-W) have been observed in tungsten films produced by changing deposition conditions[5,6]. Our research was focused on the reaction between aluminum (Al) and tungsten (W) films that had different crystalline phases by changing CVD conditions using silane reduction of tungsten hexafluoride. And we have found that the reaction between Al and W depended on the crystalline phase of W film.

EXPERIMENTAL

The substrates used in this experiment were 6-inch Si(100) wafers. The wafers were dipped in a 1% hydrofluoric acid solution to remove a native oxide just prior to the W film deposition. W films were deposited using silane (SiH_4) reduction of tungsten hexafluoride (WF_6) in a cold-wall, single-wafer, LPCVD system equipped with a load lock. The crystalline phase of LPCVD W depended upon the deposition conditions. α-W and β-W films were deposited under the following deposition conditions. In the α-W sample, gas flow ratio of SiH_4/WF_6 was 0.6 for 6sccm flow of SiH_4 at temperature of 235°C. In the β-W sample, gas flow ratio of SiH_4/WF_6 was 1.0 at 275°C. Flow rates of H_2 and Ar as carrier gases were 1000sccm and 15sccm, respectively. Total pressure was constant through the experiment (250mtorr). Deposition rates of α-W and β-W were 45nm/min and 100nm/min, respectively. The sample sheet resistance was measured by a four-point probe. The crystalline phase of W were analyzed by X-ray diffraction (XRD), and Auger Electron Spectroscopy (AES) was used to analyze the depth profile of impurities in the W film. Aluminum-1%Si-0.5%Copper (Al-Si-Cu) film was then sputtered onto the samples. The Al thickness was 200-500nm. After the Al/W/Si structure was formed, the samples were annealed in a nitrogen gas ambient at temperatures of 410°C, 450°C, and 500°C. The sheet resistances of the samples were then measured. And Al and W interdiffusion was analyzed by Rutherford Backscattering Spectrometry (RBS) using 1.5MeV He ions.

RESULTS AND DISCUSSION

X-ray diffraction measurements of as-deposited α-W and β-W on the Si substrates resulted in figure 1. X-ray peaks of the α-W film show α-W(110), (200) and (211) in the b.c.c. pattern of W. In the β-W sample, which is an A_3B compound with A15 structure[7], the peaks of β-W(200), (210), (211) and (400) are observed. After Al films were deposited on the above W films, the samples were annealed in a furnace tube. The sheet

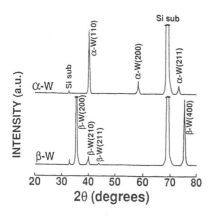

Figure 1. X-ray diffraction patterns of as-deposited α-W and β-W films.

resistance change of the samples as a function of the annealing temperature
is plotted in figure 2. At below 450°C, both samples show little change in
sheet resistance. At 500°C, however, whereas the sheet resistance of the
β-W sample has a little increase, that of the α-W sample increases sharply.
Figure 3 shows the RBS spectra of Al(200nm)/α- or β-W(150nm)/Si structure
(dotted line: as-deposited, solid line: after annealing at 500°C for 30
minutes). In both α-W and β-W samples, the insignificant difference in the
slope of the leading edge of the Si spectrum between the as-deposited and
500°C anneal cases indicates that W does not interact with Si. As can be
seen, however, the W diffusion into Al occurs in both structures after 500°C
annealing. In α-W sample, it is evident from the leading edge of the W
spectrum that W diffuses into Al and reaches the Al surface after 500°C
annealing. Moreover, X-ray spectrum of the sample exhibits lines
characteristic of WAl_{12} and the interaction has clearly begun between W and
Al. In β-W sample, however, the RBS spectrum of the sample after annealing
at 500°C, has a slight difference from the spectrum of the as-deposited

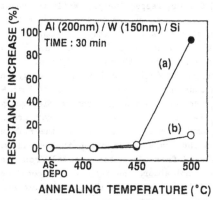

Figure 2. Sheet resistance increase
of Al/W/Si structure as a function
of annealing temperature.
(a):Al/α-W/Si, (b):Al/β-W/Si.

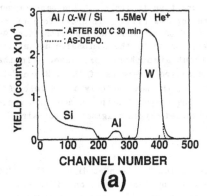

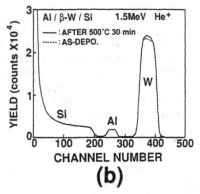

Figure 3. RBS spectra of Al/W/Si structures as-deposited and after 500°C
for 30 min. (a):Al/α-W/Si, (b):Al/β-W/Si.

film. X-ray peaks of WAl_{12} are not observed. As a result, it seems that the difference in the sheet resistance change between α-W and β-W samples, is caused by W diffusion into Al.

The thicker Al film(500nm) on α- or β-W(150nm)/Si structures was prepared in order to get the W depth profile in the Al layer. The samples were annealed in N_2 ambient at temperatures of 410°C, 450°C, and 500°C for 30-3240 minutes. The diffusion coefficient was estimated by W depth profile in Al using RBS measurements of the structure. Figure 4 shows the diffusion coefficient for the α-W and β-W in Al as a function of reciprocal temperature. The diffusion coefficient of α-W in Al is larger than that of β-W. The activation energies for the W diffusion into Al, however, are almost the same (2.2eV) in both structures. This indicates that the mechanism of W diffusion into Al is the same in both structures.

The properties of the α-W and β-W films were examined to consider these phenomena. Figure 5 shows the sheet resistance change of α- and β-W films as a function of time when the W films were exposed to air after removal from the reactor. The initial resistivities of as-deposited α-W and β-W film are approximately $14\mu\Omega\cdot cm$ and $490\mu\Omega\cdot cm$, respectively. While the sheet resistance of β-W film increases drastically according to the exposure time and begins to saturate in a few days, that of α-W film stays constant.

AES depth profiles are presented in figure 6 for as-deposited α-W and β-W films. The β-W film contains more oxygen than the α-W film. The oxygen content of the α-W film is below the detection limit. The oxygen incorporation into β-W film did not occur during W deposition because both W films were deposited with the same reactor. Scanning Electron Microscopy (SEM) observation shows that the β-W film has a fine columnar grain structure, and the α-W film has a granular one. It seems that the W film grain structure has an influence on the oxygen absorption in the films. The oxygen in the β-W film is absorbed from air after removal from the reactor. The α-W film, however, does not absorb the oxygen. We conclude that the resistance increase of β-W film is caused by the absorbed oxygen. Hashimoto et al. reported that the oxygen content in reactive sputtered TiN film increases according to the exposure time to air[8]. The same oxygen absorption seems to occur in the case of β-W film. In order to confirm that oxygen incorporation into β-W film does not occur during deposition, α-W(45nm) on β-W(300nm)/Si structure was formed by the successive deposition of W films in the same reactor without atmospheric exposure. In the stacked W structure, XRD measurements showed the peaks of the α-W and β-W. In this structure, the oxygen content was the same as only α-W film by AES measurement. This result shows that the oxygen absorption does not occur during β-W film deposition, and the absorbed oxygen originates from air.

CONCLUSION

The reaction between Al and CVD-W films has been studied. Al/α-W/Si and Al/β-W/Si structures were prepared by deposition on different Si substrates by changing deposition conditions using silane reduction of tungsten hexafluoride, followed by Al-Si-Cu alloy film sputter deposition.

Figure 4. Diffusion coefficient for
α-W and β-W in Al as a function of
resiprocal temperature.

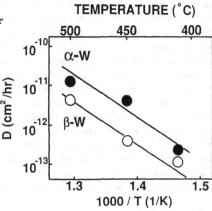

Figure 5. Sheet resistance increase
of α-W and β-W film as a function
of the exposure time to air.

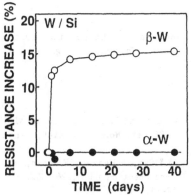

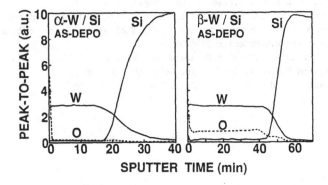

Figure 6. AES depth profile of as-deposited α-W film and β-W film.

The sheet resistance of Al/α-W/Si structure is higher than that of Al/β-W/Si structure after 500°C annealing. The W diffusion into Al occurs in both structures after annealing, and the reaction between α-W and Al takes place easily compared with that between β-W and Al. This causes the sheet resistance difference. The activation energies for the W diffusion into Al, however, are almost the same in both structures. When CVD-W films are exposed to air after removal from the reactor, the sheet resistance of β-W film increases according to the exposure time, while that of α-W film does not. The β-W film absorbs more oxygen than the α-W because of the difference in grain structure. The resistance increase of β-W film is caused by the oxygen that is absorbed from air. Our results indicate that the oxygen in the β-W layer suppresses the W diffusion into Al at the initial stage of reaction. Once the reaction begins, however, the diffusion into Al does not depend on crystalline phases of CVD-W.

ACKNOWLEDGEMENTS

The authors would like to thank S.Inomata and M.Kinoshita for the estimation of oxygen content in W films by AES. They are also grateful to S.Ushio for his encouragement.

REFERENCES

[1]   C.Kaanta,W.Cote,J.Cronin,K.Holland,P-I.Lee,and T.Wright, Proc. of VLSI Multilevel Interconnection Conf.,p.21 IEEE cat. No.88CH-2624-5, Santa Clara,CA,June 13-14,(1988).

[2]   D.Moy, M.Schadt, C-K.Hu, F.Kaufman, A.K.Ray, N.Mazzeo, E.Baran, and D.J.Pearson,Pro. of VLSI Multilevel Interconnection Conf.,p26 IEEE cat. No.89TH-0259-2, Santa Clara, CA, June 12-13,(1989).

[3]   T.Ohba, S.Inoue, and M.Maeda,Proc.IEDM Tech.Dig.,213(1987).

[4]   H.Kotani, T.Tsutsumi, J.Komori, and S.Nagao, ibid.,217.

[5]   J.E.J.Schmitz, M.J.Buiting, and R.C.Ellwanger, Proc. of the Workshop on Tungsten and Other Refractory Metals for VLSI Applications, edited by R.S.Blewer and C.M.McConica 27 (1988).

[6]   T.Ohba, T.Suzuki, T.Hara, Y.Furumura, and K.Wada,ibid,17.

[7]   C.C.Tang and D.W.Hess, Appl. Phys. Lett. 45, 633(1984).

[8]   K.Hashimoto and H.Onoda, Appl. Phys. Lett. 54, 120(1989).

DEPOSITION MECHANISM OF TUNGSTEN SILICIDE FILMS BY LOW PRESSURE CVD

JAE H. SONE AND HYEONG J. KIM
Seoul National University, Dept. of Inorganic Materials Engineering, Seoul, Korea

ABSTRACT

WSi_x thin films were deposited on SiO_2/Si substrates by Low Pressure Chemical Vapor Deposition (LPCVD) using WF_6 and SiH_4 gases. The deposition mechanism has been studied by measuring the thickness, resistivity and composition of the films by varying deposition temperature and gas flow rate at a constant total reactant gas pressure. Below 300°C, the surface chemical reaction was the rate-limiting process and the deposition rate increased exponentially with temperature having a thermal activation energy of 3.2 kcal/mol. Meanwhile, above 300°C the reaction was governed by the mass transfer step in the gas. The deposition rate in this range is insensitive to the deposition temperature but shows dependence of the flow rate of reactant gases. AES and RBS analyses were performed to determine the stoichiometry of WSi_x thin film. The Si content in film gradually increased as the deposition temperature increased. The resistivity of as-deposited WSi_x film has dependence on both deposition temperature and Si/W ratio, and exponentially increased with a moderate slope. However, temperature insensitive behavior of resistivity appeared in the mass transfer controlled region. Such resistivity changes with temperature were discussed with the Si/W ratio and the microstructure of films.

INTRODUCTION

Tungsten silicide (WSi_2) has replaced poly-silicon as gate level or multi-level interconnections in VLSI devices. Its fabrication process, properties, and applications have been extensively studied in many papers.[1-3] However, deposition mechanism of WSi_2 thin films has not been understood well yet.

The chemical reactions for WSi_x from SiH_4 and WF_6 should take place spontaneously in the thermodynamic consideration.[4] But the actual reaction seems to be governed by kinetics. Dependence of deposition rate on the deposition temperature and the flow rates of reactant gases was studied by others,[4-5] but they measured only at narrow temperature region and did not discuss mechanism of deposition. However, they reported temperature insensitivity of film resistivity for films deposited in the optimum deposition temperature region, without a proper explanation.

Since as-deposited WSi_x film by LPCVD is an amorphous nonstoichiometric compound, film properties, especially resistivity, must have a strong relationship with deposition temperature and Si/W ratio. In this work it is proposed that as-deposited WSi_x might consist of low resistance WSi_2 or W_5Si_3 microcrystallines and high resistance amorphous Si on the basis of both results of compositional analyses from AES and RBS and resistivity changing behavior from four-point probe measurement.

EXPERIMENTAL

Deposition of WSi_x film on SiO_2/Si substrates were carried out in a commercially available GENUS 8402 cold wall type LPCVD reactor. Experimental variables are substrate temperature and WF_6 flow rate. Deposition temperature changed from 150 to 550°C with an interval 50°C in the flow rates of 16 sccm WF_6, 1900 sccm SiH_4, and 500 sccm Ar, and in the total pressure of 200 mTorr in order to study the effect of substrate temperature on WSi_x thin film deposition. And WF_6 gas flow rate changed from 5 sccm to 30 sccm in the deposition condition: 1900 sccm SiH_4, 500 sccm Ar, 200 mTorr of total pressure and 360 °C of deposition temperature.

To determine the deposition rate of as-deposited WSi_x thin films we measured film thickness with tape test technique and SEM. Four-point probe technique was used to measure resistivity. AES analysis was powerful technique to identify composition of thin film with standard sample identified by RBS technique. Average value from AES depth profiling data was used to eliminate instrumental error because AES was surface-sensitive method.

RESULTS

Deposition mechanism during CVD process is governed by thermodynamic reaction of vapor phases, reactor shape and deposition condition such as deposition temperature, deposition time, gas flow rate and total pressure. Since all these parameters are closely related one another, it is so difficult to determine deposition mechanism. But it is well known that deposition rate is generally determined by two deposition mechanisms in CVD. One is the surface chemical reaction controlled mechanism and the other is the mass transfer controlled mechanism.[6-7] In the former the overall deposition rate is controlled by chemical reaction on substrate surface and has exponential dependence of temperature since chemical reaction and adsorption and desorption of vapor molecules or atoms are thermally activated processes. In this case apparent activation energy can be obtained from a slope in ln(deposition rate) vs. 1/T plot. In the latter deposition rate is sensitive not to temperature but to reactant gas flow rate because reactant gas flow rate determines the amount of reactant species involved in the reaction. Dependence of deposition rate of WSi_x films on deposition temperature is shown in Fig. 1. Up to 300°C deposition rate increases with increasing temperature, which indicates that surface reaction governs the overall reaction in this temperature region. The apparent thermal activation energy of reaction obtained from slope of plot is 3.2 kcal/mol, which is much lower value than 11.0 kcal/mol measured for the activation energy in W-film deposition.[8] However, in the temperature region from 300°C to 500°C, since deposition rate is independent of deposition temperature, this region is controlled by mass transfer mechanism. Above 500°C, deposition rate decreases as temperature increases, of which behavior must be associated with deficit of reactant gases due to homogenous gas phase nucleation at high deposition temperature.

The resistivity of as-deposited WSi_x films is shown as a function of deposition temperature in Fig. 2. The resistivity increases monotonically as temperature increases, except in mass transfer limited region, where resistivity is relatively independent of temperature. Temperature dependence of resistivity can be explained by Si/W atomic ratio and microstructure of films. Si/W ratio of as-deposited film, obtained by AES and RBS, as a function of deposition temperature is shown in Fig. 3. Si/W ratio increases exponentially with increasing temperature,

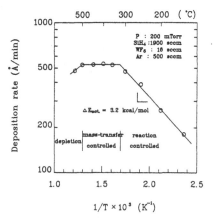

Fig. 1   Arrhenius plot of the deposition rate of WSi_x thin film as a function of deposition temperature, showing the reaction activation energy of 3.2 kcal/mol.

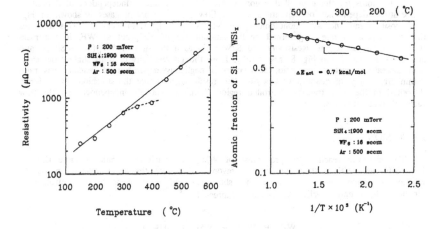

Fig. 2 The WSi$_x$ thin film resistivity as a function of the deposition temperature.

Fig. 3 Arrhenius plot of atomic fraction of Si in WSi$_x$ thin film as a function of deposition temperature.

which implies that decomposition and incorporation of SiH$_4$ is thermally activated process. Using the relation between deposition temperature and Si/W ratio, resistivity change of films was plotted again as a function of Si/W ratio, as shown in Fig. 4. Curve is divided into three parts, which are closely associated to three deposition mechanism regions. In the low Si/W ratio region, i.e., in surface reaction controlled region, resistivity change as a function of Si content in WSi$_x$ shows composited metallic behavior. This result leads that film is a mixture of metallic WSi$_2$ and W$_5$Si$_3$ microcrystallines[9]. In the mass transfer controlled region, gentle slope was shown, while steep slope appears in high Si/W ratio region. Although no crystalline phase was identified by TEM analysis as shown in Fig. 5, WSi$_x$ film is likely to consist of fine microcrystallines.

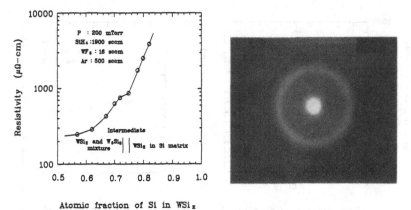

Fig. 4 The WSi$_x$ thin film resistivity as a function of atomic fraction of Si in WSi$_x$.

Fig. 5 TEM diffraction pattern of as-deposited WSi$_x$ film

In the mass transfer controlled region, where deposition rate is independent of deposition temperature, dependence of deposition rate on flow rate of reactant gases is shown in Fig. 6. Deposition rate increases monotonically as $WF_6/(WF_6+SiH_4)$ ratio increases. This indicates that the deposition rate is governed by the amount not of SiH_4 but of WF_6 in the mass transfer controlled region. Resistivity and Si/W ratio as a function of $WF_6/(WF_6+SiH_4)$ ratio are shown in Fig. 7 and Fig. 8, respectively. Even though resistivity of films slightly decreases with increasing reactant gas ratio, Si/W ratio is negligibly dependent of reactant gas ratio within error range of quantitative analysis. This result suggests that the stoichiometry of films, deposited in the mass transfer controlled region, remains relatively independent on the flow rates of reactant gases.

DISCUSSION

The chemical reaction for production of WSi_x, of which Si/W ratio is more than 2, from SiH_4 and WF_6 is thermodynamically favored. But since the actual reaction needs the reaction activation energy of 3.2 kcal/mol shown in Fig. 1, it seems to be controlled by kinetics. The overall reaction is generally expressed as[4]

$$WF_6 + 2SiH_4 = WSi_2 + 6HF + H_2$$

But it is likely that this reaction takes place not spontaneously but stepwise since resistivity and Si/W ratio of as-deposited films have deposition temperature dependence, as shown in Fig. 2 and 3. Two reactant molecules, WF_6 and SiH_4, decompose separately, as following forms:[4,10]

$$WF_6 + 3H_2 = W + 6HF \text{ (hydrogen reduction)}$$

or

$$2WF_6 + 3Si = 2W + 3SiF_4 \text{ (Si reduction)}$$

and

$$SiH_4 = Si + 2H_2$$

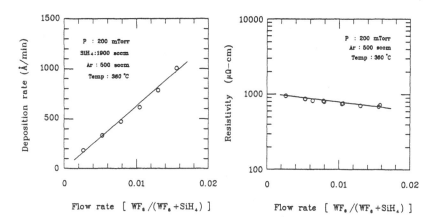

Fig. 6 The deposition rate of WSi_x thin film as a function of WF_6 flow rate.

Fig. 7 The WSi_x thin film resistivity as a function of WF_6 and SiH_4 flow rate.

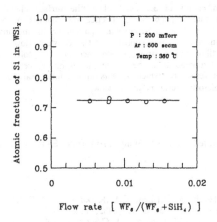

Fig. 8 Atomic fraction of Si in WSi$_x$ thin
film as a function of WF$_6$ and SiH$_4$
flow rate.

In the initial stage of reaction, no silicon deposition takes place at the deposition temperature
since decomposition reaction of SiH$_4$ is favorable above 600°C if no catalysts exist. But after
W is produced, decomposition reaction of SiH$_4$ will be favorable since W acts as catalyst.
In the next stage silicon reduction of WF$_6$ is a dominant process because the silicon reduction
is much faster than hydrogen reduction. And it was reported that the overall reaction
probability of the WF$_6$ with silicon was not sensitive to the deposition temperature from
200 to 700°C[8]. Consequently, we can say that decomposition reaction of the WF$_6$ takes place
immediately and is independent on temperature even if silicon exists. However, decomposition
of the SiH$_4$ is thermally activated process as shown in Fig. 3, where the incorporation of
Si into films increases with increasing temperature.

Since the SiH$_4$ decomposition rate is much low at the lower deposition temperature,
the Si/W ratio of film should be low. In this region film contains a large amount of W
and a small amount of Si, that is, it has W-rich composition. Film consist of amorphous
phase Si with a few of W$_5$Si$_3$ or WSi$_2$ microcrystallines. Resistivity of film is determined
by its Si/W ratio. As deposition temperature increases, both the Si/W ratio and the resistivity
increase. This behavior was shown in Fig. 3 and 4. Consequently, the overall deposition
rate is limited by the decomposition of SiH$_4$ and increases with temperature in the surface
reaction controlled region.

In the higher temperature region more Si is incorporated in WSi$_x$ film and film has
the almost stoichiometric WSi$_2$ compound with Si matrix. In this region decomposition of
WF$_6$ is rate-limiting reaction since there are enough Si-supplying source. Therefore, deposition
rate is dependent not on temperature but on WF$_6$ flow rate, as already shown in Fig. 1
and 6. The films deposited in this range might consist of a lot of WSi$_2$ microcrystallines,
of which size is less than 4 nm and existence is not able to be identified by TEM and
X-ray. A small amount of excess Si due to the increase of Si/W ratio is able to segregate
in grain boundary without changing resistivity. This result gives a possible explanation for
temperature insensitivity of resistivity in the intermediate region. As temperature increases
higher, Si incorporation will be vigorous and film changes to Si-rich composition.
Microstructurally Si may segregate in the boundary of WSi$_2$ microcrystallines and disconnect
them. This behavior results in rapidincrease of resistivity of film.

SUMMARY and CONCLUSIONS

The rate-limiting mechanism in WF$_6$/SiH$_4$ system appears to be the decomposition of

SiH_4 with an activation energy of 3.2 kcal/mol in the surface chemical reaction controlled region. The deposition rate displays a dependence on changes in deposition temperature. Meanwhile, WF_6 flow rate is a rate-controlling parameter in the mass transfer controlled region, where deposition rate remains constant at the constant flow rates of reactant gases.

The resistivity of films has a strong dependence on the stoichiometry of films, which gradually changes with temperature in the whole measuring temperature region. Temperature independent behavior of resistivity, which appears in the mass transfer controlled region, can be explained in the microstructural point of view.

ACKNOWLEDGMENT

This work is supported by 1989 Grant-in-Aid from ETRI in Korea.

REFERENCES

1. H.J.Geipel,Jr., N.Hsieh, M.H.Ishaq, C.W.Koburger, and F.R.White, IEEE J.Solid-State Circuits, vol.SC-15, no.4, p.482, (1980)
2. D.L.Bror, J.A.Fair, K.A.Monnig, and K.C.Sarawat, Solid State Technology, vol.26 p.183 (1983)
3. T.Yanai, I.Kai, T.Kobayashi, and K.Yoshioka, Proc. VLSI Multilevel Interconnection Conference, p.516 (1986)
4. D.L.Bror, J.A.Fair, K.Monnig, and K.C.Sarawat, Semiconductor International p.184, May (1984)
5. T.Hara, H.Takahashi, and Y.Ishizawa, Solid State Science and Technology, vol.134, no.5, p.1302 (1987)
6. W.Kern and V.S.Ban, Thin film processes edited by Vosen and Kern (Academic Press, London, 1978), pp.267-27
7. A.S.Groove, Physics and Technology of Semiconductor devices, (John Willey and Son's, New York, 1967), pp.11-13
8. M.L.Yu, B.N.Eldridge, and R.V.Joshi in Tungsten and Other Refractory Metals for VLSI Applications IV, edited by R.S.Blewer and C.M.McConica (Mat. Res. Soc. Proc., Pittsburgh, PA, 1989) p.221
9. S.P.Muraka, J.Vac.Sci.Technol., vol.17, no.4, p.776 (1980)
10. E.K.Broadbent, in Tungsten and Other Refractory Metals for VLSI Applications, edited by R.S.Blewer (Mat. Res. Soc., Pittsburgh, PA, 1985) p.365

Author Index

Subject Index

MATERIALS RESEARCH SOCIETY SYMPOSIUM PROCEEDINGS

ISSN 0272 - 9172

Volume 1—Laser and Electron-Beam Solid Interactions and Materials Processing, J. F. Gibbons, L. D. Hess, T. W. Sigmon, 1981, ISBN 0-444-00595-1

Volume 2—Defects in Semiconductors, J. Narayan, T. Y. Tan, 1981, ISBN 0-444-00596-X

Volume 3—Nuclear and Electron Resonance Spectroscopies Applied to Materials Science, E. N. Kaufmann, G. K. Shenoy, 1981, ISBN 0-444-00597-8

Volume 4—Laser and Electron-Beam Interactions with Solids, B. R. Appleton, G. K. Celler, 1982, ISBN 0-444-00693-1

Volume 5—Grain Boundaries in Semiconductors, H. J. Leamy, G. E. Pike, C. H. Seager, 1982, ISBN 0-444-00697-4

Volume 6—Scientific Basis for Nuclear Waste Management IV, S. V. Topp, 1982, ISBN 0-444-00699-0

Volume 7—Metastable Materials Formation by Ion Implantation, S. T. Picraux, W. J. Choyke, 1982, ISBN 0-444-00692-3

Volume 8—Rapidly Solidified Amorphous and Crystalline Alloys, B. H. Kear, B. C. Giessen, M. Cohen, 1982, ISBN 0-444-00698-2

Volume 9—Materials Processing in the Reduced Gravity Environment of Space, G. E. Rindone, 1982, ISBN 0-444-00691-5

Volume 10—Thin Films and Interfaces, P. S. Ho, K.-N. Tu, 1982, ISBN 0-444-00774-1

Volume 11—Scientific Basis for Nuclear Waste Management V, W. Lutze, 1982, ISBN 0-444-00725-3

Volume 12—In Situ Composites IV, F. D. Lemkey, H. E. Cline, M. McLean, 1982, ISBN 0-444-00726-1

Volume 13—Laser-Solid Interactions and Transient Thermal Processing of Materials, J. Narayan, W. L. Brown, R. A. Lemons, 1983, ISBN 0-444-00768-1

Volume 14—Defects in Semiconductors II, S. Mahajan, J. W. Corbett, 1983, ISBN 0-444-00812-8

Volume 15—Scientific Basis for Nuclear Waste Management VI, D. G. Brookins, 1983, ISBN 0-444-00780-6

Volume 16—Nuclear Radiation Detector Materials, E. E. Haller, H. W. Kraner, W. A. Higinbotham, 1983, ISBN 0-444-00787-3

Volume 17—Laser Diagnostics and Photochemical Processing for Semiconductor Devices, R. M. Osgood, S. R. J. Brueck, H. R. Schlossberg, 1983, ISBN 0-444-00782-2

Volume 18—Interfaces and Contacts, R. Ludeke, K. Rose, 1983, ISBN 0-444-00820-9

Volume 19—Alloy Phase Diagrams, L. H. Bennett, T. B. Massalski, B. C. Giessen, 1983, ISBN 0-444-00809-8

Volume 20—Intercalated Graphite, M. S. Dresselhaus, G. Dresselhaus, J. E. Fischer, M. J. Moran, 1983, ISBN 0-444-00781-4

Volume 21—Phase Transformations in Solids, T. Tsakalakos, 1984, ISBN 0-444-00901-9

Volume 22—High Pressure in Science and Technology, C. Homan, R. K. MacCrone, E. Whalley, 1984, ISBN 0-444-00932-9 (3 part set)

Volume 23—Energy Beam-Solid Interactions and Transient Thermal Processing, J. C. C. Fan, N. M. Johnson, 1984, ISBN 0-444-00903-5

Volume 24—Defect Properties and Processing of High-Technology Nonmetallic Materials, J. H. Crawford, Jr., Y. Chen, W. A. Sibley, 1984, ISBN 0-444-00904-3

Volume 25—Thin Films and Interfaces II, J. E. E. Baglin, D. R. Campbell, W. K. Chu, 1984, ISBN 0-444-00905-1

MATERIALS RESEARCH SOCIETY SYMPOSIUM PROCEEDINGS

MATERIALS RESEARCH SOCIETY SYMPOSIUM PROCEEDINGS

Recent Materials Research Society Proceedings listed in the front.

Tungsten and Other Refractory Metals for VLSI Applications, Robert S. Blewer, 1986; ISSN 0886-7860; ISBN 0-931837-32-4

Tungsten and Other Refractory Metals for VLSI Applications II, Eliot K. Broadbent, 1987; ISSN 0886-7860; ISBN 0-931837-66-9

Ternary and Multinary Compounds, Satyen K. Deb, Alex Zunger, 1987; ISBN 0-931837-57-X

Tungsten and Other Refractory Metals for VLSI Applications III, Victor A. Wells, 1988; ISSN 0886-7860; ISBN 0-931837-84-7

Atomic and Molecular Processing of Electronic and Ceramic Materials: Preparation, Characterization and Properties, Ilhan A. Aksay, Gary L. McVay, Thomas G. Stoebe, J.F. Wager, 1988; ISBN 0-931837-85-5

Materials Futures: Strategies and Opportunities, R. Byron Pipes, U.S. Organizing Committee, Rune Lagneborg, Swedish Organizing Committee, 1988: ISBN 1-55899-000-3

Tungsten and Other Refractory Metals for VLSI Applications IV, Robert S. Blewer, Carol M. McConica, 1989; ISSN 0886-7860; ISBN 0-931837-98-7

Tungsten and Other Advanced Metals for VLSI/ULSI Applications V, S. Simon Wong, Seijiro Furukawa, 1990; ISSN 1048-0854; ISBN 1-55899-086-2

High Energy and Heavy Ion Beams in Materials Analysis, Joseph R. Tesmer, Carl J. Maggiore, Michael Nastasi, J. Charles Barbour, James W. Mayer, 1990; ISBN 1-55899-091-7

Physical Metallurgy of Cast Iron IV, Goro Ohira, Takaji Kusakawa, Eisuke Niyama, 1990; ISBN 1-55899-090-9

Atom Probe Microanalysis: Principles and Applications to Materials Problems, M.K. Miller, G.D.W. Smith, 1989; ISBN 0-931837-99-5

MATERIALS RESEARCH SOCIETY INTERNATIONAL SYMPOSIUM PROCEEDINGS

International Meeting on Advanced Materials
Sunshine City, Ikebukuro, Tokyo, Japan
May 30-June 3, 1988

Executive Editors
Masao Doyama, Shigeyuki Sōmiya, Robert P.H. Chang

Volume 1—Ionic Polymers/Ordered Polymers for High Performance Materials/Biomaterials, Senior Editors: Norio Ise, Eishun Tsuchida/Shohei Inoue, Minoru Matsuda/Hideki Aoki, Yohji Imai, Ishi Miura, 1989, ISBN: 1-55899-030-5

Volume 2—Hydrogen Absorbing Materials/Catalytic Materials, Senior Editors: Shuitiro Ono, Yasuo Sasaki, Seijirau Suda/Yoshihiko Moro-oka, 1989, ISBN: 1-55899-031-3

Volume 3—Powder Preparation/Rapid Quenching, Senior Editors: Kazuo Akashi, Yoshiharu Ozaki, Tohoru Takeda,/Akihisa Inoue, Tsuyoshi Masumoto, Takeyuki Suzuki, 1989, ISBN: 1-55899-032-1

Volume 4—Composites/Corrosion-Coating of Advanced Materials, Senior Editors: Shiushichi Kimura, Akira Kobayashi, Sokichi Umekawa/Kazuyoshi Nii, Yasutoshi Saito, Masahiro Yoshimura, 1989, ISBN: 1-55899-033-X

Volume 5—Structural Ceramics/Fracture Mechanics, Senior Editors: Yoshiteru Hamano, Osami Kamigaito/Teruo Kishi, Mototsugu Sakai, 1989, ISBN: 1-55899-034-8

Volume 6—Superconductivity, Senior Editors: Koichi Kitazawa, Kyoji Tachikawa, 1989, ISBN: 1-55899-035-6

Volume 7—Superplasticity, Senior Editors: Masaru Kobayashi, Fumihiro Wakai, 1989, ISBN: 1-55899-036-4

Volume 8—Metal-Ceramic Joints, Senior Editors: Nobuya Iwamoto, Tadatomo Suga, 1989, ISBN: 1-55899-037-2

Volume 9—Shape Memory Materials, Senior Editors: Kazuhiro Otsuka, Ken'ichi Shimizu, 1989, ISBN: 1-55899-038-0

Volume 10—Multilayers, Senior Editors: Ryoichi Yamamoto, Tomeji Ohno, 1989, ISBN: 1-55899-039-9

Volume 11—Microstructure-Property Relationships in Magnetic Materials, Senior Editors: Motofumi Homma, Yasuo Imaoka, Masuo Okada, 1989, ISBN: 1-55899-040-2

Volume 12—Photoresponsive Materials, Senior Editor: Shigeo Tazuke, 1989, ISBN: 1-55899-041-0

Volume 13—Advanced Cements and Chemically Bonded Ceramics, Senior Editors: Masaki Daimon, Shigeyuki Sōmiya, Giichi Sudoh, Kunihiro Takemoto, 1989, ISBN: 1-55899-042-9

Volume 14—Biosensors, Senior Editor: Isao Karube, 1989, ISBN: 1-55899-043-7

JSAP-MRS INTERNATIONAL CONFERENCE ON ELECTRONIC MATERIALS

Shigaku-Kaikan, Tokyo, Japan
June 13-15, 1988

Proceedings of First International Conference on Electronic Materials, Editors: Takuo Sugano, Robert P.H. Chang, Hiroshi Kamimura, Izuo Hayashi, Takeshi Kamiya, 1989, ISBN: 1-55899-044-5

Printed in the United States
By Bookmasters